HOW TO ORDER THIS BOOK

BY PHONE: 800-233-9936 or 717-291-5609, 8AM–5PM Eastern Time

BY FAX: 717-295-4538

BY MAIL: Order Department
Technomic Publishing Company, Inc.
851 New Holland Avenue, Box 3535
Lancaster, PA 17604, U.S.A.

BY CREDIT CARD: American Express, VISA, MasterCard

BY WWW SITE: http://www.techpub.com

SCIENCE & TECHNOLOGY

Edited by

Catharina Y. W. Ang, Ph.D.

Division of Chemistry
National Center for Toxicological Research
U S. Food and Drug Administration

KeShun Liu, Ph.D.

Soyfoods Laboratory
Hartz Seed, A Unit of Monsanto

Yao-Wen Huang, Ph.D.

Department of Food Science and Technology
University of Georgia

LANCASTER · BASEL

Asian Foods
a **TECHNOMIC**® publication

Technomic Publishing Company, Inc.
851 New Holland Avenue, Box 3535
Lancaster, Pennsylvania 17604 U.S.A.

Printed in the United States of America
10 9 8 7 6 5 4 3 2 1

Main entry under title:
Asian Foods: Science and Technology

A Technomic Publishing Company book
Bibliography: p.
Includes index p. 531

Library of Congress Catalog Card No. 98-83242
ISBN No. 1-56676-736-9

Table of Contents

Preface

IN recent years, there has been a growing interest in Asian prepared foods, as well as the health and cultural aspects of Asian diets, both in the East and West. There are many reasons for this interest, including business development, food industry expansion, cultural exchange, tourism development, medical findings that link health benefits with some Asian foods and Asian dietary habits, and simple curiosity. This has created an urgent need for a book on the subject. *Asian Foods: Science and Technology* meets this need in a timely manner by providing readers with basic, yet up-to-date, information regarding many types of Asian prepared foods with respect to their origin, preparation methods, processing principles, technical innovation, quality factors, nutritional value, and market potential. It also includes information on Asian dietary habits, the health significance of Asian diets, and some traditional functional foods.

The book starts with rice products, the major staple in most Asian countries. This is followed by wheat products: noodles in Chapter 3 and breads, cakes, cookies, pastries, and dumplings in Chapter 4. Chapter 5 covers Asian foods from other grains and starchy materials. Traditional soyfoods are thoroughly treated in Chapter 6. The next three chapters deal with foods from animal sources, including meat products, poultry and egg products, and fishery products. Asian fruit and vegetable products are discussed in Chapters 10 and 11, respectively, while fats and oils are treated in Chapter 12 and Chinese alcoholic beverages in Chapter 13. Chapter 14 discusses traditional Chinese functional foods. The final two chapters, which are integral parts of the book and rather unique, examine cultural aspects of Asian dietary habits and health implications of Asian diets, respectively. Extensive lists of references are included at the end of each chapter for those who want additional reading.

This book is the first of its kind to cover various types of Asian prepared foods, more than 400 varieties, along with discussion of cultural and health aspects of Asian diets, in a single English volume with up-to-date scientific and technological information. It combines the traditional art of Asian food preparation with modern science and processing technology. Each chapter has been written by expert(s) in the particular products and subjects and has been reviewed by one or two professionals who also specialize in the field, as well as by the editors, to ensure the accuracy of the information. The book should serve as a unique reference for food professionals who pursue product development, dieticians who are interested in Asian diets and dietary habits, business developers who seek the market potentials of Asian prepared foods, college students majoring in food sciences and human nutrition who need supplemental information, and anyone else who is interested in Asian foods or Asian dietary habits, for whatever reasons. We would be thrilled and honored if this book contributes to the readers' general understanding of Asian prepared foods and dietary habits, induces or assists in business development, and promotes cultural exchanges among nations and ethnic groups around the world.

We are fully aware that it is invariably the result of teamwork to prepare and publish a professional book of this magnitude. We would like to express our sincere appreciation to all the chapter authors for their valuable contribution and to those who served as chapter reviewers for their efforts toward excellency of this volume.

Our special thanks go to Dr. Eleanor S. Riemer, Publisher, Food Sciences, Ms. Kimberly J. Martin, Copy Editor, and Ms. Susan G. Farmer, Managing Editor, Mr. Stephen C. Spangler, Production Manager, of Technomic Publishing Co., Inc., for their editorial advice and excellent assistance in bringing this project to completion. Thanks are also extended to those who have supplied useful photos, given permission for reprints, and provided encouragement and suggestions. And, lastly, we are grateful to our spouses, Mr. Peng Ang, Mrs. Yan Li (Liu), and Mrs. Ping-Yuan C. Huang, for their support.

List of Contributors

Catharina Y. W. Ang, Ph.D.
National Center for Toxicological Research
Food and Drug Administration
HFT-230, 3900 NCTR Road
Jefferson, AR 72079

Monisha Bhattacharya, Ph.D.
Cereal Science Laboratory
Department of Botany
University of Hong Kong
142 Pokfulam Road, Hong Kong

Tien Chi Chen, Ph.D.
Professor Emeritus
Computer Science and Engineering
The Chinese University of Hong Kong
Hong Kong

Tsun-Chieh Chen, Ph.D.
Poultry Science Department
Mississippi State University
P.O. Box 9665
Mississippi State, MS 39762

Guangseng Cheng
Institute of Microbiology
Chinese Academy of Sciences
Beijing, China

Harold Corke, Ph.D.
Cereal Science Laboratory
Department of Botany
University of Hong Kong
142 Pokfulam Road
Hong Kong

Chung-Yi Huang, Ph.D.
Department of Food Science
National I-Lan Institute of Agriculture and Engineering
I-Lan, Taiwan 26015

Sidi Huang
Asian Food Laboratory
Bread Research Institute of Australia
SCIRO Gate 1, Riverside Corporate Park
Delhi Road
P.O. Box 7
North Ryde NSW 2113, Australia

Yao-Wen Huang, Ph.D.
Center for Food Safety and Quality Enhancement
Department of Food Science and Technology
University of Georgia
Athens, GA 30602

KeShun Liu, Ph.D.
Hartz Seed
A Unit of Monsanto Co.
P.O. Box 946
Stuttgart, AR 72160

Xingqiu Lou
Department of Animal Sciences
University of Kentucky
Lexington, KY 40546

Bor S. Luh, Ph.D.
Professor Emeritus
Department of Food Science and Technology
University of California
Davis, CA 94516

Seiichi Nagao, Ph.D.
Flour Millers Association
Wheat Flour Institute
15-6 Kabuto-cho, Nihon-bashi
Chuo-ku, Tokyo 103, Japan

Jacqueline M. Newman, Ph.D.
Department of Family, Nutrition, and Exercise Sciences
Queens College—CUNY
Flushing, NY 11367

John X. Q. Shi, Ph.D.
Southern Crop Protection and Food Research Center
Agriculture and Agri-Food Canada
Guelph, Ontario, N1G 2W1 Canada

Michael Tao, Ph.D.
Tao, Tao and Chiu, Inc.
533 Cuesta Drive
Aptos, CA 95003

Peter J. Wan, Ph.D.
Southern Regional Research Center
United States Department of Agriculture
P.O. Box 19687
New Orleans, LA 70179

Samuel L. Wang, Ph.D.
Horticultural Research Institute of Ontario
University of Guelph
4890 Victoria Avenue, North
Vineland Station, Ontario, L0R 2E0 Canada

Youling Xiong, Ph.D.
Department of Animal Science
206 Garrigus Building
University of Kentucky
Lexington, KY 40546

Fang-Qi Yang
Department of Food Science and Engineering
Wuxi University of Light Industry
Jiangsu, China

List of Reviewers

Sam K. C. Chang, Ph.D.
Professor
Department of Food and Nutrition
North Dakota State University
Fargo, ND 58105

Tsun-Chieh Chen, Ph.D.
Professor
Department of Poultry Science
Mississippi State University
P.O. Box 9665
Mississippi State, MS 39762

Harold Corke, Ph.D.
Associate Professor
Cereal Science Laboratory
Department of Botany
University of Hong Kong
142 Pokfulam Road
Hong Kong

Keith W. Gates, M.S.
Assistant Public Service
Marine Extension Service
University of Georgia
715 Bay St.
Brunswick, GA 31522

Thomas Heinze, M.S.
Chemist
Division of Chemistry
National Center for Toxicological Research
U.S. Food and Drug Administration
Jefferson, AR 72079

Sidi Huang, M.S.
Senior Research Scientist
Asian Food Laboratory
Bread Research Institute of Australia
SCIRO Gate 1
Riverside Corporate Park, Delhi Road
P.O. Box 7
North Ryde NSW 2113, Australia

Bor S. Luh, Ph.D.
Professor Emeritus
Department of Food Science and Technology
University of California
Davis, CA 94516

Don R. McCaskill, M.S.
Director
Research and Development
Riceland Foods, Inc.
Stuttgart, AR 72160

Finlay MacRitchie, Ph.D.
Professor
Grain Science & Industry
Kansas State University
Manhattan, KS 66506

Robert Shewfelt, Ph.D.
Professor
Department of Food Science and Technology
University of Georgia
Athens, GA 30605

Chuck E. Walker, Ph.D.
Professor
Grain Science & Industry
Kansas State University
Manhattan, KS 66506

Peter J. Wan, Ph.D.
Food and Feed Processing Research Unit
Southern Regional Research Center, USDA
P.O. Box 19687
New Orleans, LA 70179

Ron-Fu Wang, Ph.D.
Research Microbiologist
Division of Microbiology
National Center for Toxicological Research
U.S. Food and Drug Administration
Jefferson, AR 72079

Wallace Yokoyama, Ph.D.
Research Scientist
U.S. Department of Agriculture
Western Regional Research Center
Albany, CA 30605

Louis L. Young, Ph.D.
Research Food Technologist
U.S. Department of Agriculture
Russell Research Center
Athens, GA 30605

Youling L. Xiong, Ph.D.
Department of Animal Science
206 Garrigus Building
University of Kentucky
Lexington, KY 40546

CHAPTER 1

Introduction

KeSHUN LIU
CATHARINA Y. W. ANG
YAO-WEN HUANG

ASIA is the largest continent in both size and population. It covers almost a third of the world's land area and has about three-fifths of its people, with a population of approximately 3.5 billion. There are dozens of ethnic groups, both large and small, who live in nearly 50 countries (Pye et al., 1997). Large Asian population groups include those who live in and are from the countries or regions of Bangladesh, Cambodia, China, Hong Kong, India, Indonesia, Japan, North and South Korea, Laos, Malaysia, Nepal, Pakistan, the Philippines, Singapore, Thailand, Taiwan, and Vietnam.

For thousands of years, Asian ethnic groups have developed their unique cultures, parts of which are the chief foods that these groups eat, as well as the way they are prepared and preserved. Both aspects differ widely not only from other parts of the world but also among the Asian ethnic groups. Even within the same ethnic group, there are regional differences, yet in spite of such differences, because of geographical closeness, there are many similarities among Asian foods and dietary habits.

First, Asian foods are rich in tradition. Many have a history of several hundred to several thousand years. Tofu, a food made of soybean, for example, was believed to be first made in China in the second century B.C. Second, they are unique in preparation. For example, *pidan,* a Chinese delicacy of preserved duck eggs, known as 1000-year-old eggs in the West, is prepared by curing the eggs in the shell for about 6 months in a mixture of ashes, mud, lime, salt, tea, and other ingredients. The curing process makes the eggs taste like cheese. In most cases, the seemingly traditional art of Asian food processing has a solid scientific basis behind it, even though this basis was unknown at the time of creating the food product. Thus, there exists a great potential for business development of many Asian foods, when the traditional art of preparation can be combined with modern processing technology. Third, many Asian foods are not only delicious but also healthy. Soyfoods, for example, have been shown to have protective effects against many chronic diseases that prevail in Western countries, including cancer, heart disease, kidney disease, and osteoporosis (Messina, 1997).

Dietary habits are an integrated part of an ethnic culture, referring to coherent eating patterns with respects to choice of food, methods of eating, preparation, number of meals per day, time of eating, size of portions eaten, and food for special occasions. As will be discussed in Chapter 15, many Asian ethnic groups have long history and the dietary pattern of each group is rather unique and deep in tradition. For business and pleasure, it is particularly important to understand the cultural aspects associated with food consumption in Asia, where social and business arrangements are often initiated or finalized at the table.

Because of the features just discussed, along with the growth of tourism, the development of modern transportation and communication systems, and the influence of Asian immigrants, Asian foods have become very popular throughout the world. At the same time, Asian dietary habits are becoming a subject of much interest, particularly within the scientific community. As will be discussed thoroughly in Chapter 16, more than ever, people are realizing that what they eat does make a difference, not only in the way they look and feel, but also in the length and quality of their lives. Therefore, foods and diets that promote health and prevent diseases are of the utmost, universal interest of the 21st century. In this regard, Asian foods, as well as Asian dietary habits, are of particular importance.

More importantly, Asia is one of the fastest growing regions in today's world economy. This is particularly true for China, the largest country in the world, where double-digit growth has been maintained for years. As the economy steadily grows and more and more people can afford to buy processed foods, many developing countries in Asia, including China, are strengthening their own food industries. By combining the traditional art of food preparation with modern science and processing technology, many traditional Oriental foods, which used to be prepared at household and village levels, are now being transformed into large-scale industrial production with high and consistent quality. At the same time, as the Asian market is opened wider, many food companies, as well as food service corporations, in industrialized countries have made inroads in the regions, trying to establish new business ventures there. Understanding Asian foods, as well as the dietary culture, would maximize the business potential of global food industries, regardless of where business entities come from.

This book provides information about Asian prepared foods, as well as health and cultural aspects of Asian diets. Major categories of Asian prepared foods include products from rice, wheat and other starchy grains, soybeans, pork, beef, poultry and eggs, fish, fruits, and vegetables. Most products under discussion have wide consumption and are processed for distribution or have marketing potential. Processing techniques may include all types of preservation methods, traditional and modern in a broad sense, such as heating, dehydration, salting, fermentation, canning, and freezing. Although fats and oils and alcoholic beverages, as well as traditional functional foods, are also included, fresh produce, raw foods, and foods prepared through culinary techniques at home or in restaurants for immediate consumption are generally excluded.

For each key prepared Asian food product, emphasis will be placed on formulation of ingredients, preparation methods, processing principles, technical innovation, quality factors, nutritional values, utilization, and market status or potential. Many chapters begin with a brief introduction to the origin and classification of the products under discussion. Within many of the product categories, there is additional discussion of the differences in preparation and varieties among Asian ethic groups, such as Chinese, Japanese, Korean, Indonesian, and so on.

Together, more than 400 varieties of Asian foods are treated in this volume, along with Asian dietary habits and health significance, yet since this book is about foods and dietary habits of about 60% of the world's population, it cannot be, nor is it intended to be, all-inclusive. Differences in the length of discussion among many ethnic groups are somewhat arbitrary but, for the most part, are based on the popularity of the foods, as well as the availability of information. Therefore, after reading this book, readers are advised to consult the references listed at the end of each chapter, as well as literature of other sources for more information.

REFERENCES

Pye L. W., Weiner, M., and Zonis, M. 1997. Asia. In: *The World Book Encyclopedia.* World Book, Inc., Chicago, IL, pp. 770–812.

Messina, M. J. 1997. Soyfoods: Their role in disease prevention and treatment. In: *Soybeans: Chemistry, Technology, and Utilization,* K. S. Liu, ed., Chapman and Hall, New York, (now acquired by Aspen Publishers, Inc., Gaithersburg, Maryland) pp. 442–447.

CHAPTER 2

Rice Products

BOR S. LUH

1. INTRODUCTION

RICE is one of the world's most important cereals for human consumption. In the densely populated countries of Asia, especially China, India, Indonesia, Japan, Korea, Pakistan, Sri Lanka, and Bangladesh, rice is an important staple food. As much as 80–90% of the daily caloric intake of people in these Asiatic countries are derived from rice. It is also consumed in the form of noodles, puffed rice, breakfast cereals, rice cakes, fermented sweet rice, and snack foods made by extrusion cooking (Mercier et al., 1989). Rice is also used in making beer, wine, and vinegar. Some oriental foods require the use of glutinous rice (sweet rice), which consists largely of amylopectin in the starch fraction, in contrast to nonglutinous rice that contains both amylopectin and amylose. When the actual extraction rates of the cereals are considered, rice is calculated to produce more food energy per hectare of land than other cereals. Total food protein production per hectare is also high for rice, second only to that for wheat. China contributes 38% of the world's rice production on 24% of the world's growing area (Herdt and Palacpac, 1983). There are many forms of rice products that are produced as alternatives for nutritive values and special sensory attraction. Literature on rice research and development has been published by Houston (1972), Juliano (1985), Luh (1991), and Marshall and Wadsworth (1994).

2. RICE SNACK FOODS

In most Asiatic countries, rice cereals are consumed as cooked rice and are served simultaneously with prepared vegetable dishes, poultry, beef, seafood,

TABLE 2.1 Physicochemical (Quality) Characteristics of Conventional Cooking and Processing of Long-, Medium-, and Short-Grain Rice Types.

Milled Rice Characteristics	Conventional Cooking and Processing Type		
	Long	Medium	Short
Apparent amylose content, %	21–23	15–20	15–20
Alkali spreading value, average	3–5	5.5–7	5.5–7
Gelatinization temperature,[a] °C	69–72	64–68	64–68
Gelatinization temperature type	Intermediate	Low	Low
Protein ($N \times 5.95$), %	6–8	6–8	6–8
Amylographic paste viscosity, BU[b]			
Peak	650–850	700–900	700–900
Hot	350–400	350–400	350–400
Cool	650–850	650–800	650–800
Breakdown	300–400	350–450	350–450
Setback	−150–+150	−250–−150	−250–−50

[a]Amylographic gelatinization temperature.
[b]Bu = Brabender Units.
Source: Based on measurements of fully developed mature grains of conventional varieties within each grain type. Results of tests conducted at the Regional Rice Quality Laboratory, Beaumont, TX; adapted in part from Webb et al. (1985, 1989).

and others. There are many kinds of rice snack foods, prepared for more attractive taste, texture, and aroma. They are served in some cases for special occasions, for some special tastes, and for convenience.

Some rice snack foods are made from either glutinous rice (sweet or waxy rice) containing largely amylopectin (98% of total starch), but very little amylose (less than 2% of total starch), while others are made from both types. Table 2.1 lists the physicochemical (quality) characteristics of long-, medium- and short-grain rice. A typical glutinous rice flour contains 11.0–13.5% moisture, 1% ash (max.), 75–80% total starch, 5.5–6.5% protein, and 0.5% total fat. Glutinous rice flour is often used in making snack foods since the sticky characteristics of high amylopectin content are necessary in many specialty rice foods. Another reason for application of glutinous rice in baked and popped snacks is that glutinous rice flour expands readily and produces a more porous texture.

2.1 RICE CRACKER PROCESSING

Rice cracker is a Japanese snack food made from rice. *Arare* and *senbei* are the traditional rice crackers in Japan. Rice crackers made from glutinous rice are generally called *arare* or *okaki.* They are softer in texture and can be easily dissolved in the mouth. Rice crackers made from nonglutinous rice are called *senbei.* They have harder and rougher texture than *arare.* Both products are consumed in Japan, as well as exported to the United States and other countries. In choosing between glutinous or nonglutinous rice, one has to pay attention to uniformity in quality, rate of water absorption, extent of refinement, and absence of objectionable odors and taste. When a starchy raw material is used, potato starch is a suitable replacement. In order to improve the flavor and appeal of snack foods, seaweed, sesame, red peppers, sugar, pigments and spices are added. For oil-fried crackers, the oil must be of good quality and well refined.

A flowchart for processing *arare* rice crackers from glutinous rice is presented in Figure 2.1. Glutinous rice is washed in a washing machine and soaked for 16 h in water at temperatures below 20°C. After draining, the rice,

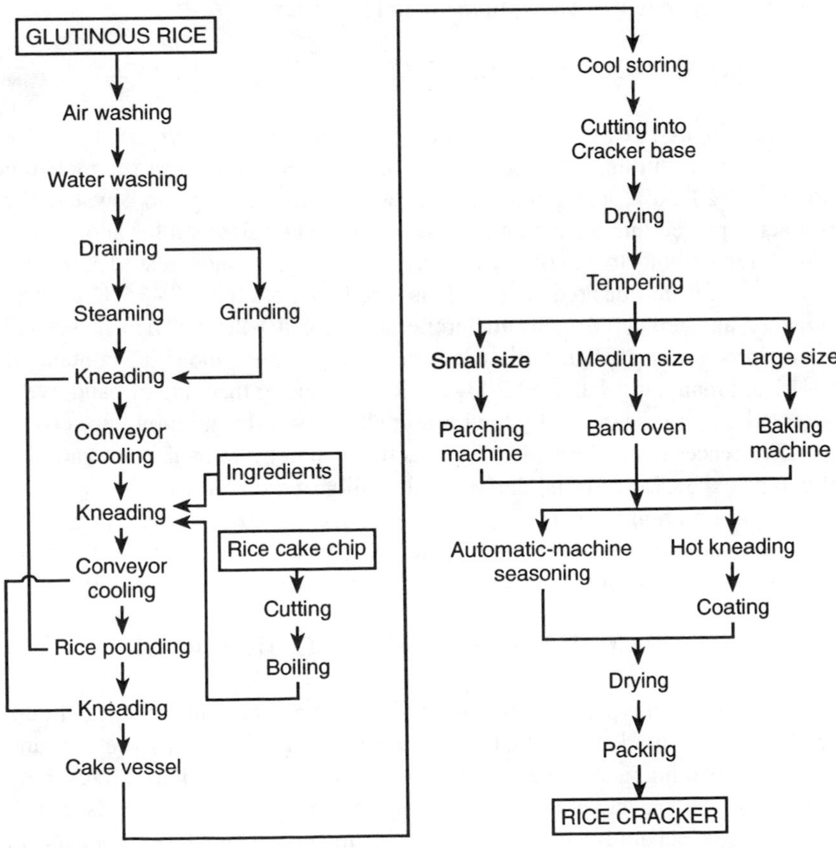

Figure 2.1 Flow chart for making rice crackers *(arare)* (courtesy of TaiAn Food Company, Taiwan).

which contains about 38% moisture, is crushed by rollers into fine powder, passed through an 80-mesh sieve, and steamed for 20 min. After cooling for 2–3 min, it is kneaded three times in a special machine. The kneaded cake is put in a cake vessel, quick-frozen, and kept at 2–5°C for 2–3 days for hardening. The cake is cut into various shapes and dried by hot air at 45–75°C to a final moisture content of 20%. The cake is coated with soy sauce, spices, and other seasoning materials and placed in a continuous baking machine or an oven. After baking, the product is dried in a continuous baking machine at 90°C for 30 min. After cooling, the product is packaged in plastic pouches or aluminum film-plastic pouches to protect it from light and moisture.

Currently, the machines used for manufacturing rice crackers are almost all continuous. With the same machines, one can manufacture both glutinous and nonglutinous rice crackers. It takes 3–4 h for the continuous process, while the conventional method takes 3–4 days. Success in rice cake processing depends on efficient operation of the equipment and procedures.

2.2 NONGLUTINOUS RICE CRACKERS

Nonglutinous rice is used for making *senbei* (*souka* or *nigata* type) rice crackers. After milling, the rice is washed, soaked in water to a moisture content of 20–30%, and ground into powder. After adding more water, the product is placed into a kneading machine where it is steamed for about 5–10 min. After cooling to 60–65°C, the product is rolled and pressed into thin layers and cut into desired shapes. It is dried by hot air at 70–75°C to 20% moisture and tempered at room temperature for 10–20 h. Then, a second drying process is applied until the product contains a moisture content of 10–15%. Finally, it is baked at 200–260°C in a baking machine or band oven. After baking, it is seasoned by the same method used for glutinous rice crackers. Differences in amylose and amylopectin content between the rice varieties affect the expansion rate and the textural quality of the product.

The consumption ratio of *arare* to *senbei* is 100 : 45. *Arare* is more popular to the consumer than the *senbei* type. These products are stable in storage at room temperature because of their low moisture content and low water activity.

3. RICE FLOUR CHARACTERIZATION AND UTILIZATION

Based on starch type, there are two primary types of commercial rice flours available on the market. The first is produced from glutinous or sweet rice and is used for making many types of oriental snack foods, as well as a thickening agent in white sauces, gravies, puddings, and prepared frozen foods. It can prevent liquid separation (syneresis) that occurs when some food products are frozen, stored, and subsequently thawed. The other major type of rice flour is prepared from broken grains of nonglutinous raw or parboiled rice. The flour made from parboiled rice is essentially a precooked rice flour. Rice flour differs from wheat flour in baking properties because rice flour does not contain gluten, and its dough does not retain gases generated during baking. Gluten in wheat flour provides elasticity in the dough to allow retention of CO_2 gas generated during baking. There is, however, a steady demand for more rice flours in making baby foods, breakfast cereals, and snack foods; separating powders for refrigerated, performed, unbaked biscuits, dusting powders, and breading mixes; and for formulating of pancakes and waffles. These industrial applications, combined with household needs, are sufficient to sustain a market for rice flour production.

3.1 PROPERTIES OF RICE FLOURS

Rice flours made from long-, medium-, short-grain and waxy rice are available for commercial and home usage. There are varietal differences in protein, lipid, starch content and the amylose to amylopectin ratio in the starch. Differences in chemical composition between rice varieties contribute to the diversity of chemical and physical properties of various rice products. The characteristics of rice flours can be measured objectively by such properties as their viscosity behavior when heated with water, starch gelatinization temperatures, birefringence endpoint temperature, and water absorption capacity.

Rice flours from each variety, with the exception of the glutinous type (waxy rice), have characteristic viscosity patterns during the heating and cooling cycles when evaluated using a Brabender Visco-Amylograph. Differences in viscosity depend largely on the composition of the starch, grinding method, and particle size and, to a lesser extent, on the protein and oil components. Usually, long-grain rice containing starch with an amylose content of over 22% has a relatively low peak viscosity and forms a rigid gel on cooling (high setback viscosity). Those with starches low in amylose have high peak and low setback viscosities.

3.2 CHARACTERIZATION OF RICE FLOUR FUNCTIONAL PROPERTIES

3.2.1 Brabender Viscosity

A Brabender Visco-Amylograph equipped with a 700-cmg cartridge is very commonly used to evaluate the viscosity behavior of rice flour. A 10% suspension of the sample (50 g) is heated gradually in 450 ml water from 30°C to 94°C at a rate of 1.5°C per min, held at 94°C for 20 min, and then cooled to 50°C at a rate of 1.5°C per min. A test on 20% flour slurries (100 g of rice flour to 400 ml of water) can give estimates of initial gelatinization temperature of rice starch within ±0.5°C, when gelatinization temperature is taken as the point of initial increase in viscosity. These values correlate significantly (coefficient of correlation = 0.94) with birefringence endpoint temperature (BEPT) determined microscopically.

3.2.2 BEPT

BEPT of rice flour can be determined on a 1% slurry using a polarizing microscope equipped with a Kofler hot stage and set at a heating rate of 5°C/min. BEPT is usually recorded at 95–98% loss of birefringence. Good quality glutinous rice flour usually has a BEPT value ranging from 61°C to 62°C.

These techniques are used to characterize rice varieties that give contrasting texture results in yeast-leavened rice breads. Figure 2.2 shows results with four rice flours. The short curves (20% slurries) indicate distinctly different temperatures for initial viscosity increases (i.e., estimate of gelatinization temperature) among the rice flour samples. The full curves (10% slurries) show initial viscosity increases occurring much later in the heating cycle. The 10% slurries provide a complete pasting history that can be correlated with the textural properties of baked products. For example, sample B (IR-8 variety) had a low initial gelatinization temperature, which is favorable for baking. Sample

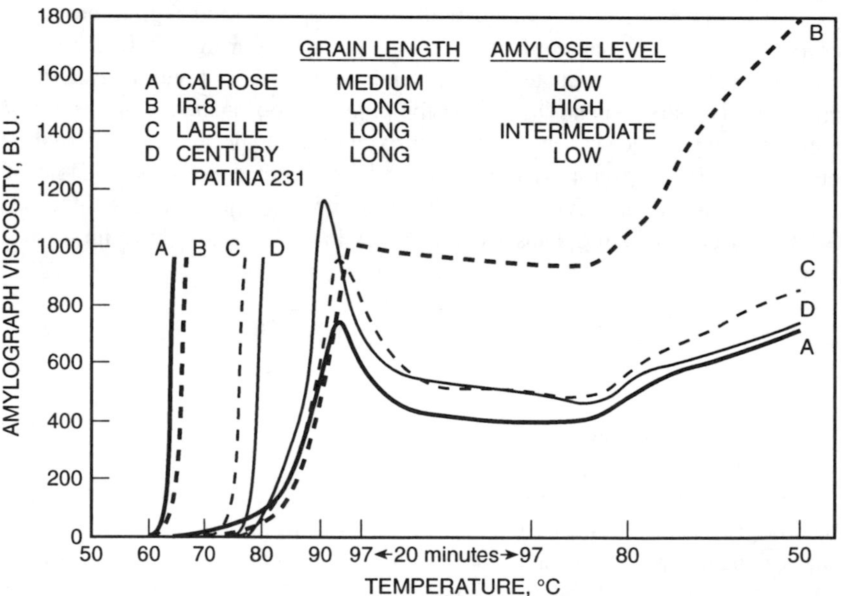

Figure 2.2 Amylograph curves of roller-milled flours from four contrasting rice varieties. Sample B (IR-8) is a Philippine variety; the others are U.S. varieties. Short curves, 20% slurries; full curves, 10% slurries. [Reprinted with permission from Nishita and Bean (1979).]

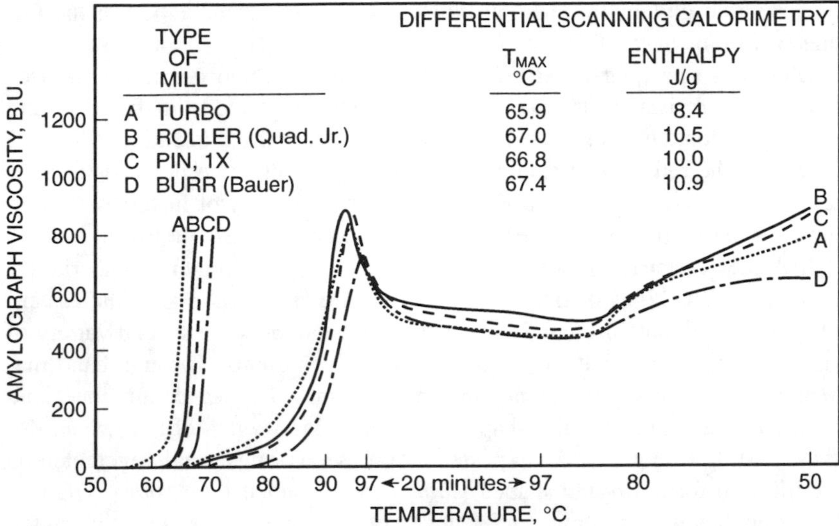

Figure 2.3 Amylograph pasting curves and differential scanning calorimetry data for rice flours ground from the same lot of medium-grain rice on flour different mills. Short curves, 20% slurries; full curves, 10% slurries. [Reprinted with permission from Nishita and Bean (1982).]

B had a high amylose content, which results in increased retrogradation of starch. Bread made from this variety had a harsh, dry, crumbly texture within 24 h.

Amylograph pasting curves and differential scanning calorimetry data for a medium-grain rice on four different mills are presented in Figure 2.3. The curves illustrate the importance of milling on the amylograph pasting curves and differential scanning calorimetry data on the same lot of medium-grain rice. These data can be used to explain the physical properties of rice flour when made into various types of snack foods.

3.3 *Yuan-zi (Tong-Yuan)*

Yuan-zi, or *tong yuan,* is a rice product made from glutinous rice flour and water. The flour is kneaded with water and some cooked rice porridge to form a dough and then kneaded again. The dough is then made into *yuan-zi* by adding fillings through hand molding or machine molding.

There are two types of fillings: (1) sweet and (2) savory. Sweet fillings include sugar, cooked red beans, sesame seed paste, lotus bean paste, or jujube date and nuts. Savory fillings are made of any meat product like pork, chicken, beef, lamb, or shrimp, plus mushroom, dried edible fungus, and Napa cabbage, chopped and cooked with seasonings such as onion, garlic, soy sauce, and monosodium glutamate. There are differences in the composition of the fillings in various parts of China.

Glutinous rice flour differs from those from long- and medium-grain rice flour in its resistance to water separation (syneresis) during freezing and thawing. It contains mostly amylopectin in the starch fraction and less than 2% amylose. The remarkable stability of cooked glutinous rice flour pastes after repeated freeze-thaw cycles was observed during a study of factors determining the stability of white sauces and gravies commonly used in frozen foods.

In Asiatic countries, glutinous rice flour is a very important raw material for making various kinds of rice products. The fillings for *yuan-zi* in the eastern parts of China contain ground pork, Napa cabbage, or some special variety of green vegetable, fat, salt, soy sauce, or other ingredients. Through hand manipulation, the food ingredients are wrapped into the rice dough sheet and form a round- or rectangular-shaped product called *yuan-zi* or *tong yuan.* For sweet tasting fillings, red bean paste, sesame seeds, sugar, and vegetable oils are filled to form a round-shaped *yuan-zi* (also canned *tong yuan*). The rice flour dough acts as a container for the tasty ingredients located at the center. Savory *yuan-zi* is usually elongated in shape (4–5 cm in length and 3–3.5 cm in diameter). Sweet-tasting *yuan-zi* is usually round in shape, 3–4 cm in diameter, but may vary somewhat in size in various parts of China. Success in making *yuan-zi* depends on the high amylopectin content of glutinous rice flour. The taste of the product depends largely on the formula for the filling. The product is served hot after thorough cooking in boiling water for sufficient time to form the cooked *yuan-zi.* If flour from long-grain or medium-grain rice is used, the product will not be satisfactory in texture because of the lack of amylopectin, which is needed for holding the particles together.

The cooked *yuan-zi* can be preserved after packaging in a deep freezer at −18°C for 6 months or longer. The product is consumed hot after thorough cooking in boiling water or as a fried product after heating the cooked *yuan-zi* in a frying pan with vegetable oil. The composition of the fillings may vary with the preference of the cook. In Figure 2.4, a picture of *yuan-zi* samples in various shapes and sizes is presented. There is one sample made of steamed glutinous rice and red beans, which is also consumed as a breakfast food, but it is not the usual form of *yuan-zi,* however.

Glutinous rice is usually milled in a roller mill or impact mill, which can break up the rice kernel to a fine powder. Other milling machines may be used, including pinmill, hammer mill, and Burr mill. The more commonly used mill is the roller mill, which can break up the rice kernel to a fine powder. Rice flour with a finer particle size makes a good *yuan-zi* of fine texture and smooth appearance. Glutinous rice flour is available on the market in a range of particle sizes. The particle size analysis of a typical 66 SS *Mochi* glutinous rice flour is as follows: 0–47 mesh, 3.0%; 47–60 mesh, 21.0%; 60–80 mesh, 33.0%; 80–100 mesh, 10.5%; 100–140 mesh, 11.75%; 140–200 mesh, 7.5%; more

Figure 2.4 A picture of *yuan-zi* in various sizes and forms. They are either elongated or round in shape. The one at the extreme right side is made of glutinous rice and bean kernels, they do not belong to the *yuan-zi* type.

than 200 mesh, 7.5%. The smooth texture of *yuan-zi* after cooking is related to the particle size distribution of the rice flour. The consumer may order a particular size of glutinous rice flour to suit the quality of the desired product.

3.4 COMPOSITE FLOUR FOR BAKING

Composite flours for baking are blends of nonwheat flours with or without wheat. They are blended after milling or during dough preparation at a bakery. Such mixtures are made as adjustments to crop shortages or surpluses. Reports on the use of rice flour in wheat-based dough, baked with 3% shortening in the formula, alleviated most of the problems resulting from substitution of rice for wheat. Use of composite flours in baked products is extensively growing in developing countries. Small amounts of rice flour from Indica or Japonica types of rice can be added to wheat flour if dough improvers are added to compensate for the dilution of the wheat gluten protein. These include ascorbic acid, potassium bromate, lipid-based surfactants, shortenings, and vegetable oils. These additives strengthen and enhance the carrying capacity of the wheat flour for the nongluten flour diluent. Application of 10%, 15%, and 20% substitutions of rice flour pan-bread formula containing 4% shortening is quite common. Rice flour from brown or white milled rice showed a typical dilution effect at 20% substitution. Loaf volumes are lower, and farinograph and extensiograph curves showed slightly less strength. Amylograph viscosity was higher because of less amylase enzyme activity from 80% wheat than from 100% wheat. In Italy, it is very common to add 10% by weight of rice flour to wheat flour in making pastry products.

4. RICE BREAD

The problem associated with rice bread formulation is caused by the absence of gluten in rice flour. The manufacture of Western-style rice pan bread without the addition of gluten presents considerable difficulties because of lack of gluten in rice.

Gluten is the important structure-forming protein. Nishita et al. (1976) and Nishita and Bean (1979) described the development of a yeast-leavened rice bread formula in which the wheat flour is completely replaced by rice flour. They compared the effects of hydroxypropyl-methylcellulose, locust bean and gum, sodium carboxymethyl cellulose, carrageenan, and xanthan gum on the loaf volume of rice bread.

The bread formula consists of 100 parts rice flour, 75 parts water, 7.5 parts sugar, 6 parts oil, 3 parts fresh compressed yeast, 3 parts hydroxypropyl methylcellulose, and 2 parts salt. These ingredients are mixed thoroughly, panned or shaped as rolls, fermented to the desired volume, and baked.

Several gums were tested in the formula presented above as gluten substitutes, including xanthan gum, which was successfully used in wheat starch breads. Hydroxymethyl cellulose provided the proper dough viscosity and film-forming characteristics so that the rice flour dough would retain fermentation gases during proofing and expand during baking to produce a crumb grain similar to that of typical white pan bread. Surfactants that normally improve the texture of wheat breads had negative effects on rice dough and bread because they interfere with the gum-water-rice flour complex such that no fermentation gases were retained and thus no leavening occurred. They reported that the short- and medium-grain rice flour had the necessary physicochemical properties to give the soft-textured bread crumb. The long-grain type yielded sandy, dry crumb characteristics.

There are people who are allergic to eating wheat, oat, barley, and rye breads that contain gluten, which causes them severe abdominal pain, and the rice bread is a good substitute for them. The short- and medium-grain rice kernels have sticky properties when cooked in the traditional manner as raw milled rice. Their starches have a low gelatinization temperature (below

70°C), with amylose contents at 20% or lower. In contrast, the long-grain type produces fluffy cooked rice, with starch that gelatinizes above 70°C, and the amylose content is above 23%. The cooking characteristics of the long-grain types appear to be directly related to the unacceptable crumb grain properties of baked products made from the flour.

Delgado (1977) improved the rice bread by adjusting the level of sucrose, yeast, water, and Methocel, in the formulation. An enriched rice bread was developed by introducing nonfat dry milk. Comparisons of the specific volume, protein, and moisture content were made on the improved and commercial wheat bread. Results showed that the rice bread made from these improved formulas was liked by the sensory evaluation panel.

4.1 STEAMED BREAD

Steamed bread is a staple food in the northern part of China (north of the Yangtze River). Up to 70% of China's bread is consumed in this form. It is made largely from wheat flour. As Asian cuisine continues to become less exotic to mainstream American tastes, steamed breads are gaining popularity as ethnic convenience foods. Steamed bread is defined as the product resulting from steaming fermented dough. In Asia, varieties of steamed bread differ among regions and countries, and regional preferences dictate a product's formulation, processing methods, and final characteristics. The essential ingredients are wheat flour, water, and yeast (Rubenthaler et al., 1990, 1992). Some rice flour, up to 10% by weight may be added to wheat flour in making steamed bread.

There are two types of steamed breads: plain steamed bread, called *mantou* in Chinese. It is prepared daily and generally eaten when it is fresh (Huang et al., 1995). The second type is a filled product called *bao* or *bao-zi.* Filled steam breads or buns are widely preferred in southern China. One of the most familiar varieties of filled steam bread is called *cha shao bao,* which is filled with barbecued roast pork and condiments. An example of a vegetable bun is the Korean-style *kimchee* steamed bread, which is filled with a pickled, spiced cabbage, carrot, and onion mixture. Sweet *bao* may be filled with red bean paste, lotus seed paste, or custard. It is usually eaten as a snack or light meal in southern China (Cantonese dum sum cuisine). These filled products are commonly referred to as "southern-style" steamed bread (Huang et al., 1995).

The desired end-product characteristics of both types of steamed bread dictate the use of appropriate raw materials and ingredients and are important factors in determining consumer acceptance. In general, the steamed bread should have a smooth, shiny surface and a symmetrical, well-rounded shape. In contrast to western-style pan bread, steam bread does not have a crust but, rather, a soft, thin skin. The volume-to-weight ratio should be two and a half to three times the dough weight. Undesirable external characteristics include wrinkles, blisters, dark spots, and a yellow-colored skin after steaming (Yue and Rayas-Duarte, 1997). The quality of steamed bread is heavily influenced by the type of flour. Best results are achieved with the use of a medium- to low-gluten strength flour having a protein content between 10% and 11%.

Steamed bread has been a subject of interest to several rice utilization researchers (Tsen et al., 1982; Rubenthaler et al., 1990; Huang et al., 1995; Wen et al., 1996; Yue and Rayas-Duarte, 1997). They made significant contributions to the knowledge of making steamed rice bread. More detailed information on steamed bread is found in Chapter 4 of this book.

5. RICE CAKES

There are many types of rice cakes available in China, Japan, and other Asian countries. In general, glutinous or nonglutinous rice is milled, refined, and washed. It is then soaked, allowed to drip until the moisture content is lowered considerably, and then steamed. After this, the product is kneaded, packed, and made into various forms. The physical properties of rice cakes are also closely related to the quality and characteristics of the rice variety from which the rice cake is made.

In Taiwan, rice cake is made by a traditional method. The glutinous or nonglutinous rice is soaked overnight in water at temperatures below 25°C. The soaked rice is ground by a stone mill with water to form a slurry, which is then transferred into a cotton cloth bag and sealed with a string. As a means of draining the water out, heavy stones are placed on the bag. The raw rice cake obtained is kneaded with water. For sweetened rice cakes, sugar is added. The product is transferred into a vessel and steamed. For other types of rice cake, ingredients such as salt, monosodium glutamate, crushed radish, crushed mung bean, and crushed taro may be added before steaming.

There is a particular kind of fermented rice cake made in Taiwan called *fakau.* It is made by adding sugar and a leavening agent to a ground, nonglutinous rice slurry. After fermentation, it is steamed and then consumed. In the Philippines there is a similar product called *bibingka,* which is made by the same method. However, salted egg yolk is added on top of the rice cake before steaming.

Some commonly used ingredients for rice cakes are as follows:

(1) Radish rice cake: nonglutinous rice (600 g), water (750 ml), radish (1200 g), salt (20 g), monosodium glutamate (5 g)

(2) Sweetened rice cake: glutinous rice (600 g), white or brown sugar (300 g), water (1000 ml)

(3) Fermented rice cake: nonglutinous rice (300 g), water (500 ml), wheat flour (36 g), white or brown sugar (240 g), leavening (10 g)

5.1 GROUND RICE POWDER CAKE

Glutinous rice is washed, drained, fried in a pan, and ground to a fine powder. Some maltose is added to the rice powder and pressed into the desired mold. If brown color is desired, brown sugar may be added for color and flavor improvement.

Mung bean may be added to the rice product. In this case, the mung bean is washed and cooked in water until the skin has swollen. After removing the seed coat, the bean is cooked to dryness and used as an ingredient. The product made from glutinous rice and mung bean is called *lu dou gao* (mung bean cake) in China.

In the above products, shortening, lard, or cooking oil may be added before pressing into the desired mold. This kind of product has a very attractive taste.

5.2 *ZONG-ZI*

This is a very popular food with glutinous rice wrapped in bamboo leaves. In China it is called *zong-zi,* but in Japan it is called *chimaki.* The bamboo leaves act as a packaging material for the glutinous rice. For example, 600 g of glutinous rice (round grain species) is washed and soaked in water for 1 h, drained, and mixed with 12.5 g soda ash. A 60 g portion of this mixture is wrapped with bamboo leaves to form a tetrahedron and bound with a string. The product is simmered in hot water for 1.5 h. After cooking, the bamboo leaves are removed, and the product is served with honey or sugar. This kind of *zong-zi* is called *chien zong.* The use of soda ash creates a slightly alkaline taste and develops some special aroma. It is possible to make the *tzong-zi* without addition of soda ash. From 600 g of glutinous rice, 20 *chien tzong* can be obtained.

5.3 *ROU ZONG*

Rou zong is another type of *zong-zi* in which pork or ham is used to enrich the nutritive value and sensory quality of the product. Six hundred grams of glutinous rice are washed, soaked in water for 1 h and drained. The bamboo leaves are washed, boiled in water for 5 min, cooled, and taken out. Three hundred grams of pork are cut into 20 small cubes and mixed with 3 Shitake mushrooms cut into small pieces. The product is mixed with 45 ml soy sauce, 1.25 g monosodium glutamate, 1.25 g sugar, 0.75 g black pepper, 2.5 ml sherry wine, and 18 g fried garlic. Sixty milliliters of cooking oil is heated in a frying pan, and 250 g dried shrimp meat is fried until an attractive flavor develops. This is added to the glutinous rice, and the special seasonings are mixed well before frying for several minutes. The special seasonings are 6 g salt, 1.5 g monosodium glutamate, 7.5 ml soy sauce, and 1.25 g black pepper.

Figure 2.5 *Zong-zi* (Bamboo leaf wrapped glutinous rice). Higher portion shows the *zong-zi* with bamboo leaves. The lower portion shows the product after cooking and peeling off the bamboo leaves.

The rice, meat, and seasonings mixture is wrapped with bamboo leaves to form a tetrahedron and bound with string. Finally, the wrapped rice product is placed in hot water and boiled for 20 min or longer. The ingredients mentioned above yield about 20 *rou zong*. The product is served hot with or without soy sauce or chili sauce after removal of the bamboo leaves. A third type of *zong-zi* is made with glutinous rice, red beans, or red bean pastes. It is a very popular sweet rice product wrapped in bamboo leaves (Figure 2.5). *Zong-zi* is served fresh after cooking or preserved in a deep freezer until needed for serving.

6. RICE NOODLES

In China, rice noodle is called *mi fen* and in Japan, *harusame*. There is some difference between the two. *Mi fen* is made from rice only; *harusame* may be made from mung bean, starch, or rice or a mixture of these. In China, Japan, and other Asiatic countries, rice noodles are consumed in the form of soups or snack foods. The rice noodle soups are usually prepared from *mi fen,* water, meat or chicken, green vegetables, soy sauce, and fresh Chinese onion for an attractive aroma. The product is highly attractive in sensory quality because of the special texture, aroma, and taste. In Taiwan, *mi fen* has been manufactured from nonglutinous rice (Taichung No. 1 or similar types). The product can be easily reconstituted in hot water with the addition of some meat or vegetables before serving.

The raw material for making quick cooking *mi fen* rice is nonglutinous rice flour and water. The product is highly elastic and sticky after cooking with water. The use of nonglutinous rice enables easy extrusion of the dough.

The conventional processing method is limited to sun drying. There are a few exceptions using hot air drying or infrared drying methods. The products made by sun drying are less desirable than those made by hot-air drying and infrared light drying. A short description for the manufacture of *mi fen* follows.

Nonglutinous rice is soaked in water for 2–4 h, ground, and mashed into a rice paste, which is then pressed in a bag to force out the excess water. The product is steamed for 50–80 min. The optimum condition is to reach 80% gelatinization of the rice starch. The product is then kneaded in a machine and shaped into a column from which raw rice noodle is extruded. The raw rice noodle is steamed for 30 min, dipped into a seasoning solution, cut, and put into racks for hot-air drying or fried in oil at a temperature higher than 80°C. The final product is then cooled and packaged. *Mi fen* is a dehydrated product. It can be kept for 1–2 years at room temperature after packaging and sealing in plastic films. The product can be reconstituted easily in hot water in a few minutes. To attain the quick reconstitution characteristics, *mi fen* should be dried in a hot air current at 80°C to prevent reverting the gelatinous (alpha) starch to the beta form (Hsieh and Luh, 1991). There are fresh *mi fen* in various forms, packaged in plastic films available in various supermarkets in the United States.

Generally, *mi fen* is a stable product. It can be easily reconstituted when needed for consumption. The rice noodle can be served with the addition of meat, leak, cabbage, soy sauce, and monosodium glutamate. Some fresh green onion pieces may be added to the noodle soup to improve the aroma of the dish before serving.

If mung bean is used in combination with rice, the final product will be more resistant to textural changes during reconstitution. Such product is called *fung-*

shu (or *tong-fun*). Thailand exports large quantities of *fung-shu* to Hong Kong, Korea, Japan, the United States, and other countries.

In the United States, there are several small cereal processors who specialize in making dehydrated noodles sold at supermarkets. It differs from rice noodles in texture, taste, and other characteristics after reconstitution. They are usually made with wheat flour as the raw ingredient.

7. PARBOILED RICE

Parboiling is a hydrothermal procedure in which the crystalline form of starch present in the paddy rice is converted into amorphous form. This is accomplished by soaking the paddy rice in warm water at 65°C for 4–5 h, followed by steaming under pressure in a continuous apparatus, drying, and milling the dried paddy rice. The parboiling process produces physical and chemical changes of the rice kernel, with some economical and nutritional advantages. The process is quite popular in India, Brazil, the United States, and Italy.

The major objectives of parboiling are to (1) increase the total and head yield of the paddy, (2) reduce the loss of nutrients during milling, (3) salvage the wet and damaged paddy, and (4) prepare the rice according to the requirements of the consumers and rice processors.

The changes occurring in the parboiling process are as follows: (1) The starch grains embedded in a proteinaceous matrix are gelatinized and expanded until they fill up the surrounding air spaces. (2) The protein substances are separated and sink into the compact mass of gelatinized starch, becoming less liable to extraction. (3) The water-soluble vitamins and mineral salts are spread throughout the grain. The riboflavin and thiamin content are four times higher in parboiled rice than in milled rice.

The milling yield of parboiled rice is higher than milled rice because there are fewer broken grains. The milled parboiled rice can be kept longer because germination is no longer possible. The enzymes involved in germination were inactivated during the parboiling process. The nutritive value of parboiled rice is richer because of the higher vitamin and mineral content that has diffused into the endosperm. Parboiled and prepared rice are not always favored by some rice consumers because they are prone to oxidative rancidity. Suitable packaging and proper sealing can help to retard rancidity. There are slow oxidative changes of rice lipids during postprocessing storage.

Parboiled rice takes a longer time to cook as compared with ordinary milled rice but is favored by the food industry in canned chicken rice soup and beef rice soup. Parboiled rice is more resistant to textural changes that can occur during the prolonged heating in the canning process.

7.1 THE PARBOILING PROCESSES

(1) Cleaning: The impurities present in paddy rice are weeds, dirt, and extraneous materials, which are usually removed during milling. Some are removed before shelling, and others after polishing along with the broken and damaged grains.

(2) Grading: Sorting on the basis of kernel thickness is essential for good-quality parboiled rice. This is done by means of grading reels fitted with rectangular slots or with wire netting. Further grading may complete the selection according to the length and bulk weight of the grain to obtain a final product of improved and uniform quality. Sorting by bulk weight, if necessary, is done by specific gravity separators.

(3) Steeping: Different varieties of paddy rice have their own rehydration characteristics. An efficient steeping process in medium-temperature water (65°C) is commonly used for the production of parboiled rice. The treatment must be done quickly to avoid fermentation.

The methods used to achieve steeping include the use of high- and medium-temperature water and the addition of wetting agents to the steeping water. Steeping is needed to provide the starch with a sufficient amount of water for gelatinization. A moisture content of 30% or more is required to fully gelatinize the starch in the caryopsis.

In Brazil and in the United States, parboiling of rice has been fully mechanized. The facilities include six or more steeping vessels; rotary drum dryers with built-in steam coils, hot-water tank, and boiler; rotary drum dryers with steam-heat exchangers or husk fired furnaces for drying paddy; mechanical handling equipment; rotary hot-water dryers; bin dryers; milling equipment; and packaging machinery. The steeping vessels are fed with hot water at 80–85°C, and the raw paddy rice is transferred into them from an overhead surge bin. The water is circulated for 15 min and then maintained at 65°C for 4–5 h, after which it is drained off. Steam is let into the built-in steam coils, and the paddy is steamed for 10–20 min and then moved to the dryers through belt conveyors. Discoloration of the parboiled milled rice increases with the duration of steeping and the temperature of the water; subsequent steaming does the same in both cases. The color becomes much deeper once the limit of 70°C is exceeded. The color of the parboiled rice varies with the pH of the steeping water. If the pH is close to 5, discoloration is minimal. The color deepens as the pH rises.

(4) Steaming: The purposes of steaming are to increase the milling yield and to improve storage characteristics and eating quality. Steaming improves the firmness after cooking and achieves better vitamin retention in the milled rice. The quantity of water to be absorbed, the time of exposure to

steam, and the temperature or pressure of the steam itself provide the parameters that will decide the quality of the parboiled rice. Through a variation of these factors, parboiled rice possessing particular characteristics and degree of gelatinization can be obtained.

(5) Drying: The objective of drying parboiled rice is to reduce the moisture content to an optimum level for milling and subsequent storage and to obtain the maximum milling yield. The drying process is done in a horizontal rotary drum dryer to reduce the moisture content to 16%. A part of it was then given a further drying while the rest was put aside, and the second stage began after periods varying from 2 h to 48 h. The optimum temperature and time needed for the second stage drying are related to the temperature of the paddy after conditioning. Generally, slow and prolonged drying in a conventional column dryer with reduced temperature is essential to ensure a maximum yield of whole grains.

(6) Tempering: After drying, the parboiled paddy must be allowed to rest for a time before milling. This time interval is called the tempering period. A tempering period of about 47 h is needed for the product to dissipate the heat it received during drying. Also, the moisture content inside each grain must become uniform throughout. Tempering must be done to ensure dissipation of heat without speeding up the cooling by artificial means. Milling is done only when rice has become stabilized at an ambient level and the grains have hardened and become glassy in texture. The moisture content of parboiled rice may be brought up to 12–14%, even if that of the raw paddy used for the process is below these percentages.

(7) Milling: Parboiled rice, when properly prepared and milled, gives the maximum yield of edible rice with a minimum amount of broken grains. Parboiling gives hardness and seals any cracks in the caryopsis. It is necessary to pass the product through a cone-type whitening machine abrasive or a horizontal cylinder covered with abrasive material in order to remove the pericarp, the perisperm, and the layer of aleuronic cells. Polishing is done in a special machine.

In many cases, parboiled rice is undermilled and still carries most of the aleuronic cells and traces of perisperm, as well as the germ at one end. The bran from parboiled rice has prolonged resistance to the formation of free fatty acids. This makes it better and easier to use for the extraction of edible oil. The bran obtained from raw paddy has a fat content of 12–14%, and the bran from parboiled rice may contain 16–22%. In some modern milling plants, the bran from milled rice is immediately passed through an extruder under 2.8–7.0 kg/sq cm pressure at 138°C for 5–15 s to inactivate the lipase activity, thus preventing the formation of fatty acids caused by enzymatic hydrolysis of the rice oil present in the bran. The bran so treated is more stable during storage.

(8) Color sorting: The parboiled rice must be sorted to remove discolored grains. A flat conveyor belt about 0.9 m wide is used. The speed of the belt is adjustable as desired by the operator.

The rice is spread on the belt in a thin layer and, as it moves along, is inspected by sorters who pick out the discolored ones by suction, using a plastic or rubber tube connected to a centrifugal air pump. The grains thus sucked up are deposited inside a cyclone separator through which the flow of air passes before reaching the pump.

Automatic sorting machines based on photoelectric devices have been used by several rice processors to sort the milled parboiled rice by color. The existence of such machines enables rapid sorting of parboiled rice to separate any discolored grains.

(9) Packaging: Packaging of parboiled rice is an important step in protecting the rice from changes in moisture content during storage. Plastic protective films of desirable chemical and physical properties are employed to make containers for such products. The film must be easy to seal by heating, be resistant to moisture penetration, possess physical strength, and be low in cost.

In recent developments, some research has been published on utilization of microwave energy for parboiling of rice (Velupillai, 1994). A commercial-scale prototype still needs to be developed.

8. QUICK-COOKING RICE

Quick-cooking rice is a popular product in the United States, Japan, and several Western countries. Ordinary milled rice requires 20–30 min to cook to a satisfactory culinary acceptability. In some instances, the rice is soaked, washed, and steamed, requiring an attention time of about 1 h. Thus, effort has been directed by the rice processors toward development of quick-cooking rice.

Up to the present, quick-cooking rice available on the market can be cooked in 5 min or slightly longer to accomplish a satisfactory culinary acceptability. This type of quick-cooking rice enables the housewife, as well as restaurants, to meet the urgent demand for cooked rice when needed. After cooking, the product should match the characteristic flavor, taste, and texture of conventionally cooked rice. The quick-cooking rice must be easily processed in mass quantities and must possess good storage stability for 6–12 months at ambient temperature. The product should be packaged properly in plastic film bags of special design against moisture changes during storage.

To prepare quick-cooking rice, it is necessary to precook the rice and gelatinize the starch to some extent in water, steam, or both. The cooked or partially cooked rice is usually dehydrated in such a manner as to retain the rice grains in a porous and open-structured condition. The finished product should consist of dry, individual kernels, be free of lumps, and have approximately one and a half to three times the bulk volume of the raw rice.

Many quick-cooking rice products, although different in texture, bulk volume, appearance, and performance qualities, are designed specifically for certain consumer niche markets. Some quick-cooking rice for special applications, such as dry soup mixes, casseroles, or other dry food mixtures that have certain dehydration requirements, were designed to be compatible with the ingredients in the mix.

8.1 TYPES OF QUICK-COOKING RICE

Difference in moisture levels, precooking times and temperature, dehydration conditions, and other processing variables results in various types of quick-cooking rice. These range from relatively undercooked rice, requiring 10–15 min of cooking time, to a good quick-cooking rice requiring 5 min preparation time. Minute Rice®, a commercial rice product, can be rehydrated with hot water in a minute or two, which yields a fairly mushy product when boiled with water. Some of these are marketed as ready-to-eat breakfast cereals.

Some consumers prefer long-grain, light, fluffy, or slightly dry individual rice, with typical cooked rice flavor, having no gritty or hard centers. The raw materials used are usually the long-grain, "Indica-type" rice as commonly recognized. A notable exception to this is that people in China, Japan, and Korea prefer "Japonica-type" or short-grain rice, which is somewhat more viscous and pasty when cooked. In fact, as rice becomes "drier" (when cooked) with time in storage after harvesting and milling, the short-grain type may become too dry and nonpasty for textural preference, to the extent that "glutinous" or "sweet rice" may be added in small amounts to increase the pastiness.

Among the types, completely precooked rice should be used in cup and standing form because no further cooking is necessary during preparation, whereas the other types could use either completely or partially cooked rice products. Except for the standing-type commercial product, all others usually contain a seasoning mix. As for quality consistency, cup and standing types give consistent quality in the finished dish because no real cooking is involved, whereas the others may vary, especially simmering and sauté/simmering types. Microwave-type products are most convenient to use because of their short preparation time. Microwave heating time is about one-fourth that of regular stove-top heating.

Practically all quick-cooking rice processes described in patents emphasize the treatment of rice. Efforts have been made by the rice millers to improve milling characteristics and yields, to remove surface fats, to improve storage stability, and to enhance flavor by parching the grain. Some processors use parboiled rice to improve nutritional quality of quick-cooking rice. In the parboiling process, the vitamins and minerals in the bran are infused into the endosperm.

8.2 QUICK-COOKING PROCESSES

Many quick-cooking rice products and processes have been developed and patented during the past decades. Among the processes and products developed, the following are the commercially useful quick-cooking processes (Luh, 1991b):

(1) Soak-boil-steam-dry method: Raw milled white rice is soaked in water to 30% moisture and cooked in boiling water to 50–60% moisture, with or without steam. The product is further boiled or steamed to increase the moisture to 60–70% and then dehydrated to 8–14% moisture to maintain a porous structure. A significant modification of the procedures is a dry-heat treatment to fissure the grains prior to cooking and drying.
(2) Expanded and pregelatinization method: Rice is soaked, boiled, steamed, or pressure-cooked to gelatinize the grain thoroughly, dried at a low temperature to yield fairly dense glassy grains, and then expanded or puffed at a high temperature to produce the desired porous structure in the product.
(3) Rolling or "bumping" method: Rice is pregelatinized as described in the previous section, rolled, or "bumped" to flatten the grains and dried to a relative hard glassy product.
(4) Dry heat treatment method: Rice is exposed to a blast of hot air at 65–82°C for 10–30 min, or at 272°C for 18 s, to dextranize, fissure, or expand the grains. No boiling or steaming is applied. The product cooks in less time than untreated grains.
(5) The freeze-thaw-drying process: Rice is precooked and then frozen, thawed, and dried. The procedure combines the hydration and gelatinization steps 1–3, in addition to the critical steps of freezing and thawing prior to drying.
(6) Gun puffing: This process is a combination of preconditioning the rice to 20–22% moisture followed by steam cooking in a retort at 3.5–5.5 kg/sq cm for 5–10 min. Then the product is puffed to atmospheric pressure or into a vacuum. The optimum terminal condition is at 165°C at 20–25% moisture levels.

9. CANNED, FROZEN, AND FREEZE-DRIED RICE

9.1 CANNED RICE

Various methods have been studied for making canned rice more acceptable. These fall into two categories: wet pack and dry pack. A product in which there is an excess of liquid, such as in soup media, is termed wet pack. Proper density is the prime objective with these types of products. The rice is precooked or blanched sufficiently to promote buoyancy in the product and prevent settling and matting. This washing process removes excess surface starch. The rice is put into cans together with the sauce. The cans are sealed under vacuum and then retorted to sterilize the product. A commercial process has been evaluated for canned white rice packed in 301 × 411 cans with a fill weight of 340 g rice (55–60% moisture) for each can. The initial temperature is 40°C, and the come up time 10–15 min. The recommended processing time at 118.3°C is 55 min. The equivalent sterilization value at 121°C is 13.29 min.

9.2 FROZEN RICE

Frozen cooked rice, like canned rice, is convenient to use since it requires less time to prepare than raw rice. The rice may be frozen plain or in combination with other foods. Rice is an integral part of Chinese frozen dinners. Recently, microwave heating of precooked frozen rice in plastic containers has become a common method for serving frozen cooked rice.

There is a number of excellent frozen rice products available in Japan, Korea, and the United States. Some of them are as combination dishes that can be reheated by the boil-in-bag method or in a microwave oven.

A commonly used method for preparing frozen cooked rice is as follows:

(1) Soak long-grain rice (Indica type) in an abundance of water at 54–60°C, which contains enough citric acid to reach a pH of 4.0–5.5. Sufficient water should be used to cover the rice after soaking for 2 h.
(2) After soaking for 2 h, drain off the excess soaking water, and rinse with more of the same pH water to remove fines.
(3) Drain thoroughly, tapping the screen to shake loose the adhering water.
(4) The soaked, drained rice is placed in layers 5 cm deep or less over a screen supported above the water in the pressure cooker. Place a small volume of water at the bottom of the pressure cooker. Close the vessel, and heat with the vent open until steam is emitted to expel air in the retort. Close the vent, raise the steam pressure to 2.09 kg/sq cm, and hold for 12–15 min. Then blow off steam gradually to prevent violent boiling and flashing.
(5) Place the hot steamed rice in a large amount of water at 93–99°C without stirring. The rice will imbibe water until the grains are large, tender, and quite free. Stirring will cause the rice to become sticky. The rice should be held in a perforated vessel so that water may circulate freely through it.
(6) Cooking by boiling the rice needs only 10–15 min following the method described in step 4. Drain off the hot water, and rinse twice with cold water that has the pH adjustment described in step 1.

(7) Tap and shake to remove the free water, or suck off the free water over a vacuum filter.
(8) Convey the cooked rice on a stainless steel mesh belt through an airblast cooler to reduce it to room temperature, and then package in cartons or plastic pouches. Freeze the rice in air-blast freezers. The rice can also be frozen as individually quick-frozen (IQF) products prior to packaging in a fluidized-bed freezer.

Both boiled and steamed white rice that have been frozen and reheated are virtually indistinguishable from their unfrozen counterparts. Frozen storage at −18.8°C up to 1 year appears to have no deleterious effects on quality.

Prechilling the cooked rice results in removing most of the surface moisture that may be present after cooking and at the same time permits quick-freezing. Before freezing takes place, the individual grains are separated and kept out of contact with one another during the freezing process. The product is then frozen solid. Any appropriate freezing temperature may be used, but excellent results have been obtained by subjecting the rice to a moving air-blast at −34°C. After the grains are solidly frozen, they can be brought together and packed in any desirable manner.

9.3 FREEZE-DRIED RICE

The freeze-drying process consists of removing moisture from foods in a frozen state by sublimation under high vacuum. Because of the low temperature used in the freeze-drying process, undesirable chemical and biological changes in the foods are inhibited. The dried product can be stored in airtight containers for long periods without refrigeration. For institutional feeding and for markets where low-temperature facilities are absent, freeze-dried rice is more desirable than frozen or dehydrated products. In our laboratory we have tried experiments on freeze-dried precooked rice for some of the rice processing industries. The product has excellent eating quality and a long storage life of 1–2 years if packaged properly in containers with proper protection from moisture, light, and oxygen. Because of the high-vacuum and low-temperature requirements in the free-drying process and the higher cost of operation and slower speed in production, the process has not been applied by the industry up to the present time. If special flavor compounds are applied to the prepared rice, it may be possible to apply freeze-drying to special types of ready-to-serve rice products.

10. RICE BREAKFAST CEREALS

Rice breakfast cereals may be divided into two classes: those requiring cooking before serving and those ready to eat directly from the package. Here, we will concentrate on processing of ready-to-eat breakfast rice cereals.

10.1 OVEN-PUFFED RICE

Oven-puffed rice is prepared from short-grain rice (Japonica type). Frequently, the rice is parboiled as described earlier. In a batch process, 635 kg of parboiled rice and 202 L of sugar syrup are mixed. Some salt may be added to improve the taste of the product. The mixture is cooked in a retort for 5 h under 100–150 kPa steam pressure. The cooked rice is broken up and dried to approximately 25–30% moisture in a rotary dryer. The partially dried product is stored in stainless steel bins for 15 h to equilibrate the moisture. This reduces the stickiness and toughens the kernel so that it is in perfect condition for bumping. Lumps may form during the tempering process. They should be broken up, dried to 25–30% moisture in a rotating dryer, and then stored in stainless steel bins for about 15 h to equilibrate the moisture. Lumps may form during the tempering process. They must be broken up before being sent to the flaking rolls. The individual kernels are separated and dried so that a moisture content of 18–20% is reached. The kernels are passed under a radiant heater, which brings the external layers of the rice to 82°C. The outside layers of the kernel are plasticized by the heat so they do not split when the grain is run through the flaking rolls. The rolls used in preparation of oven-puffed rice are set relatively far apart so that they contact only the central part of the rice kernel. The bumped grains are again tempered for about 24 h. To secure the puffing effect, the cooled and tempered rice is passed through a toasting oven that is usually gas-fired. The moist flakes are tumbled through a perforated drum and passed within a few inches of the gas flames. Treatments are at 232°C to 302°C. The transit time is about 30–45 s. The oven-puffed rice emerges from the oven with less than 3% moisture. It is then carried by conveyor belt to expansion bins. The product is cooled to room temperature; sprayed with some vitamins, minerals, and flavor enrichment materials; and then packaged.

The process of making rice cereal such as Special K® is similar to that of oven-puffed rice. The rice kernels are cooked and then, while in a moistened condition, coated with wheat gluten, wheat-germ meal, dried skim milk, brewer's yeast, and other nutritional adjuncts. Following partial drying and tempering, the grain is run through steel flaking cylinders revolving at a speed of about 180–200 rpm. Hydraulic controls maintain a constant pressure at the point of contact of the rolls. The rolls are cooled by internal circulation of water.

The flaking process presses the rice kernels into thin flakes. The product is still rather flexible at this time. The flakes are toasted in the same manner as oven-puffed rice. In addition to being thoroughly dehydrated by the process, the flakes are toasted and blistered. The product may be further enriched by spraying the vitamins and minerals, which is followed by packaging.

10.2 GUN-PUFFED RICE

Rice puffing consists of three steps: heating the cleaned rice, cooking with steam at high pressure in a sealed chamber or gun, and suddenly releasing the pressure. Short-grain rice is preferred for gun puffing. Pearl rice is generally used. The clean milled rice is introduced into the gun manually by a swing spout, and the gun is then closed. With the gun rotating, the heating phase of the process is started. After a period of preheating, superheated steam at 15 kg/sq cm is introduced into the gun. It is important that dry steam is used. Free water will cause clumping, pitting, and uneven expansion. Sufficient time is allowed for the superheated steam to cook the rice to a semiplastic state. Finally, the pressure in the gun is suddenly released by manually triggering the end gate, and the puffed rice is caught in a cage and then dried to 3% moisture before packaging.

Satisfactory puffing depends on attaining grain temperature at which starch exhibits plastic flow characteristics under pressure. The time and temperature at which the rice is preheated are very important to the quality of the final product. In general, the required temperature should be reached as quickly as possible without scorching the grain. The rate of steam flow to the puffing gun is very important and must be controlled precisely. This method is now less popular because of lack of continuity in processing.

10.3 EXTRUDED RICE

Ready-to-eat rice breakfast cereals are being made by extruding superheated and pressured doughs through an orifice into the atmosphere. Either single-screw or twin-screw extruders can be used. The sudden expansion of water vapor in the extrudate as the excess pressure is released results in a volume increase of several times. Apparent specific volumes can reach or exceed those attained by gun puffing, and the process seems to have several advantages over gun puffing, such as high and continuous production rates, greater versatility in product shape, and easier control of product density.

The rice flour mix containing a 60–75% expandable starch base is moistened with water or steam and equilibrated to ensure a uniform supply of extrusion material. The resultant mash is compacted by a screw revolving inside a barrel, which may be heated by steam. The thread of the screw has a progressively closer pitch as it approaches discharge. In some extruder designs, the rice premix is fed directly into the extruder. The water and/or steam are injected into the barrel and mixed with the premix. The pressurizing, shearing, and steam heating bring the dough to a temperature of about 150–175°C and a pressure of 5–10 MPa at the die end. Under these conditions, the dough is quite flexible and easily adapts to complex orifice configurations.

The dough pieces expand very rapidly as they leave the dice orifice, and the expansion may continue for a few seconds since the dough is hot and still flexible and water continues to boil off. The moisture content of pieces is on the order of 10–15% and is too high for satisfactory crispness and stability. Thus, the pieces are flash-dried or dried on vibrating screens in hot-air ovens to a final moisture content of 3–4%. The product may be coated with sugar syrup and flavoring if desired, dried again, cooled, and packaged.

Fortification of ready-to-eat rice breakfast cereals with vitamins, minerals, and flavor compounds is now a very common practice. The usual approach is to add the minerals and more heat-stable vitamins such as niacin, riboflavin, and pyridoxine to the basic formula mix and then spray the more heat-labile vitamins such as vitamin A and thiamin on the product after processing. The nutrients to be added must be stable during processing and storage, and a sufficient amount is added to compensate for losses in processing and storage.

10.4 SHREDDED RICE

Shredded rice is a popular ready-to-eat breakfast cereal. Whole kernel or broken rice can be used as the starting material. The rice is washed and cooked in a rotary cooker with sugar, salt, and malt syrup under 100–150 kPa steam pressure for a period of 1–2 h or until the rice is uniformly cooked throughout, with no white centers. The kernels are soft and pliable but still individual particles. The cooked particles are then discharged at a moisture content of about 40% and partially dried to a moisture content of 25–30%. The dried kernels are tempered to ensure a uniform moisture distribution and form a hard, glazed surface. This process allows the rice kernels to flow freely through the process and reduces hang-up problems.

The shredding rolls are from 15.2 cm to 20.3 cm in diameter and as wide as 60 cm or more. They are much smaller than flaking rolls. On one roll of the pair is a series of about 20 shallow corrugations running around the periphery. In cross section, these corrugations may be square, rectangular, or a combination of these shapes. The other roll of the pair is smooth. Soft and cooked rice is drawn between these rolls as they rotate and issues a continuous strand of dough.

Rice Chex® and Crispix® are made by using two pairs of shredding rolls. Rice Chex® is made with rice as the sole cereal ingredient, while Crispix® is made from rice and corn as the cereal components. The dough sheet formed from the first pair of rolls is placed on a moving belt. The dough sheet from the second pair of rolls is then laid on top of the first sheet on the same moving belt. The layered sheets can be cut by one or two pairs of cutting rolls, which fuse a thin line of the dough sheets into a solid mass at regular intervals to form a continuous matrix of biscuits. The wet biscuits are transferred to a metal belt moving through a gas-fired oven. The shredded rice cereals are toasted, cooled, and broken apart from each other through a vibrator conveyor. Fortification of the shredded rice cereal with vitamins and minerals is a common practice. Packaging of the product in proper containers is a very important step for protection from moisture and atmospheric light changes.

11. BABY FOODS

Rice in the form of rice flour or as granulated rice is used in the formulation of many strained baby foods. Rice flour, glutinous rice flour, and rice polishings are used in preparing baby foods. The largest use of rice in the baby food industry is in the manufacture of precooked infant rice cereals.

11.1 PRECOOKED RICE CEREAL

Precooked rice cereal is frequently prescribed as an infant's first solid food. The cereal must be easily reconstituted with milk or formula, with a minimum of lumps.

The process for making precooked baby foods consists of preparing and cooking a cereal slurry. The objective is to precook the rice and to convert the starch from crystalline to amorphous form. To aid in the digestion of the rice starch, amylase enzymes from an industrial source may be added to the slurry with accurate control of the time and temperature of the process. The rice slurry is first precooked in water to form a slurry and then treated with amylase enzyme to predigest the starch into dextrin and oligosaccharides. Each baby food manufacturer has his own formulation and process for the manufacture of precooked rice cereal. The ingredients used in the formulation of baby foods are rice flour, rice polishings, sugar, dibasic calcium phosphate, glycerol monostearate (emulsifier), rice oil, thiamine, riboflavin, niacin, or niacinamide.

The prepared rice slurry is dried with an atmospheric drum drier. The thickness of the film on the drier surface, the spacing between the drums, the temperature of the drum surface, the drum speed, and the flowing properties of the slurry are controlled with the objective of making an easy-to-digest rice product. The bulk density of the cereal is related to the thickness of the sheet coming off the dryer and the size distribution of the flakes. Because of the high starch content in rice, the apparent viscosity of the cooked rice cereal is markedly affected by a slight variation in the solids content. The solids, drum speed, and the drum temperature are adjusted to obtain a finished product of excellent quality. Packaging of the final product in selected paperboard carton containers appears to be a common practice for markets where extreme climate fluctuations are not common.

Some precooked rice cereals with strawberries or apples appear to be gaining favorable consumer acceptance. These products are prepared in a manner similar to that employed for regular precooked infant cereals. Cereal ingredients, fruit, sugar, oil, vitamins, and minerals are cooked, dried on an atmospheric drum drier, flaked, and packaged. Because of the hygroscopicity of the fruit and sugar, fruit cereals require moisture-proof packages. Some antioxidant may be incorporated in the packaging material to prolong the shelf life of the product. But special permission for such practices from the Food and Drug Administration may be necessary because baby foods require special attention for safety seasons.

Sometimes, some diastase enzymes can be used to lower the liquid requirement for reconstitution. This is a very sophisticated process. The temperature, time of digestion, enzyme activity, and solid content must be closely controlled to obtain a satisfactory product.

Rice cereal may become rancid if packaged in a hermetically sealed container. The package material most suitable for such product is one that allows transmission of both moisture vapor and gas. Most precooked infant cereals are packaged in paperboard cartons. A bleached manila liner on the interior of the carton appears to be very commonly used. The carton is wrapped with a glue-mounted, printed paper label. The tight wrap offers sifting and insect protection to the package.

11.2 EXTRUSION-COOKED BABY FOOD

Besides drum drying of rice baby foods, extrusion cooking is a new method for preparing baby foods. The type of extruders used, the particle size of the rice flour, the moisture content of the rice-water mixture, and the extrusion conditions are some of the important factors influencing the properties of extruded rice baby foods. An example of extrusion cooked rice-based baby foods is the Kasetsart® infant food, developed by Professor Amaran Bhumiratana of the Khasart University in Bangkok, Thailand. It contains 72% rice flour and 13% full-fat soy flour and babymate. A mixture made with 75% milled rice flour and 25% dehulled mung bean is customarily used in the preparation of soups and casserole dishes. Similar products designed for infant feeding also contain these ingredients. Not only is rice a food ingredient, but its use in baby foods has a significant role on the consistency of the product.

11.3 FORMULATED BABY FOODS

Rice cereal products are customarily used in the preparation of formulated baby foods.

Not only is rice a food ingredient, but its use in baby foods has a significant role in the consistency of the product. The variety of rice used in these products is important to the physical properties of the final product. Long-grain rice, because of its higher amylose content, causes the product to thicken during storage caused by starch retrogradation and eventually to produce a very rigid gel and water separation. The presence of free liquid in a product packaged in a glass container is a serious defect. Glutinous-rice flour is a good stabilizer for canned and frozen food products. The stability is achieved as a result of a reduction of amylose/amylopectin rationing the product.

There is a grade of baby food designed for more advanced babies called junior grade. "Junior" baby foods have a coarser texture. They help the babies acquire the mouth feel of solid food. To produce the junior grade baby foods, granulated rice is incorporated into the formulation for many junior vegetable and meat items. In the formulation of junior baby foods, care must be taken to avoid thin consistency after cooking, or particles can settle out to form a mat in the bottom of the jar. To ensure uniform distribution of the junior-sized particles, modified waxy-maize starch is frequently incorporated into the product.

12. SOME POPULAR RICE PRODUCTS

Recent cereal food research has shown consumption of rice or oat bran to be an effective means of reducing cholesterol levels in humans. As a result, cereal processors are using rice and oat brans as ingredients in such snack foods as rice cakes, crackers, cookies, breads, side dishes, hot and cold breakfast cereals, and pancake mixes. The incorporation of bran in cereal foods increases the fiber and nutrient intake of the public. Peoples of Asian origin, as well as the general public, seem to like the various types of rice products shown below.

12.1 PUFFED RICE CAKE

Puffed rice cakes are gaining popularity. In China and the United States, there is a new puffed rice cake industry using medium-grain brown rice as raw material. The puffed rice cake differs from the Japanese rice snack food in composition and taste. It is a disk-shaped puffed product, low in calories (35–40 kcal per cake), and free from cholesterol. Puffed rice cake has become a popular snack food in California. Other minor ingredients, such as sesame seed, millet, and salt, may be used in making the product.

Brown rice is milled from paddy rice by removing the hulls and retaining the bran and polish layers that have higher levels of nutrients and dietary fiber than conventional white rice. Because of consumer interest in low-calorie and dietary fiber-containing foods, puffed rice cakes are gaining widespread consumer acceptance.

The procedures for making puffed rice cake are as follows: Water is added to the medium-grain brown rice to adjust its moisture content to 14–18%. The added water and brown rice are mixed and tempered in a liquid-solids blender and tumbled for a selected time (1–3 h) at room temperature. The moistened rice is then introduced to a rice cake puffing machine that has been preheated to 200°C or higher. An example of the rice cake machine (Figure 2.6) is the Lite Energy® rice cake machine (Real Foods Pty., St. Peters, Australia).

The mold in this type of rice cake machine consists of three parts: a ring-shaped side piece and upper and low platens, which can be moved up or down to adjust the space between them. The rice is then pressed between the movable upper and lower platens. At the end of a prescribed heating time, usually 10–12 s, the upper platen is lifted. The heat-softened rice kernels are

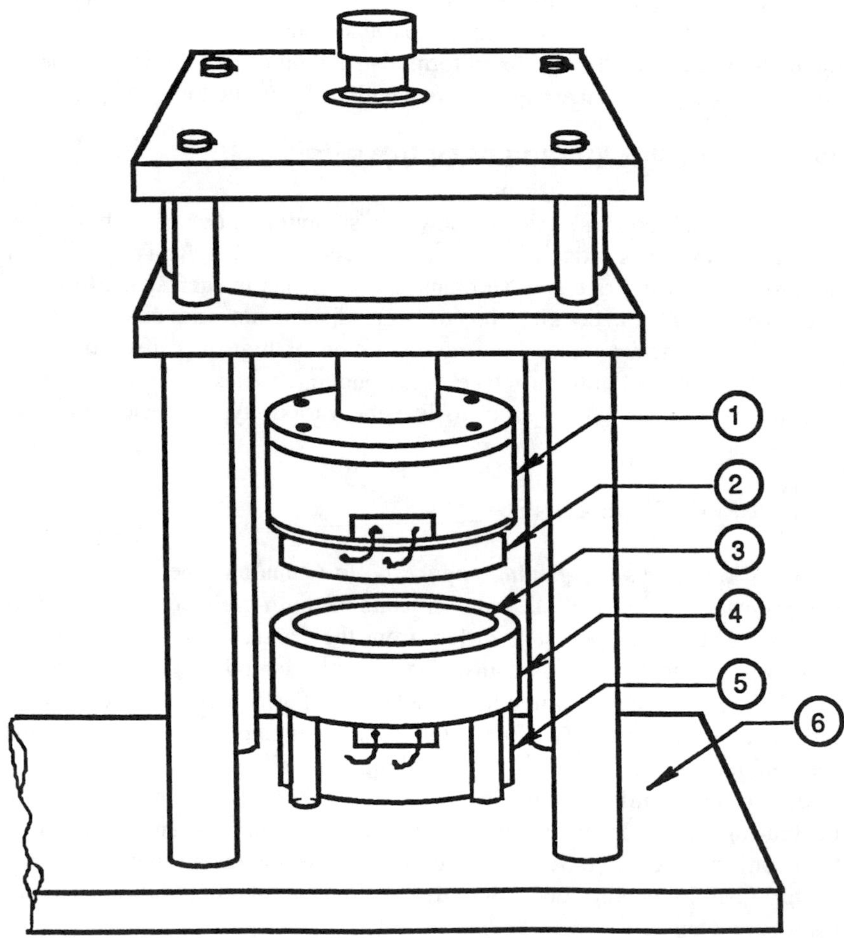

Figure 2.6 Rice puffing mold assemble of the Lite Energy rice-cake machine. (1) Insulation block; (2) upper platen; (3) lower platen; (4) ring; (5) insulation block; (6) base plate. Courtesy: Hsieh and Luh (1991).

puffed because of the sudden release of water vapor as a result of moisture flash vaporization and are fused together to form the rice cake. Each cake is 10 cm in diameter and 1.7 cm high and weighs approximately 10 g. The cakes are then discharged, cooled in air, and packaged and sealed in plastic cylindrical pouches, usually made of polyethylene (PE) or oriented polypropylene (OPP) films.

The influence of moisture content and tempering temperature and time immediately before puffing on rice cake volume has been investigated by Hsieh et al. (1989). In general, a lower moisture level (14% vs. 16–20%) in raw rice and a longer tempering time (5 h vs. 1–3 h) resulted in higher specific volumes in the rice cakes. Higher heating temperature at 230°C versus 200–220°C and an 8-s heating time produces rice cakes of higher specific volumes. Darker colored rice cakes are obtained from combinations of high temperature and longer tempering time.

12.2 EXTRUDED CRISPY RICE CAKES

Another category of rice products is represented by the square crispy cakes, which are similar in shape to the flat bread or crisp bread-type product so popular in Europe. Rice flour is the major ingredient. A typical formula contains 55% rice flour, 42% barley malt flour, 2% malt extract, and 1% salt. Extruded rice cakes are usually produced in self-wiping twin-screw extruders. The moisture content of the preextruding mixture runs about 21–23%, depending on the formulation of the ingredients. The traveling ribbon from the rollers is continuously cut into narrow slit die or dies. Hot and expanded ribbon extrudes are woven to 7.5-cm-square cakes, which are then conveyed to a drying oven. After drying to 2–4% moisture content, the cakes are cooled, stacked, packaged, and heat-sealed in containers made of PE or OPP films. To improve the aroma of the rice cakes, apple, strawberry, and cinnamon extract in vegetable oil are sprayed onto the cakes before packaging.

12.3 RICE FRIES

Rice fries are a snack food using rice as a basic ingredient. The rice is fully cooked in a broth containing butter, salt, and selected seasonings. After cooking, the rice is compressed and pumped or extruded through a 1.3-cm-square die to form ribbons. The ribbons are passed through a cutter, where they are cut into 7.6-cm-long units. The units are then fried in hot (177–204°C) vegetable oil for about 1 min or until a proper crust is formed on their surfaces. Finally, the products are cooled and quick-frozen, packaged, and made ready for shipment. The product has a crisp exterior crust, while retaining the fluffy, light interior. The product may compete with French fried potatoes.

12.4 RICE PUDDING

Rice is frequently used in making puddings. Pastry chefs cook the rice in boiling water and then strain the cooked product. It is then mixed with low-fat milk (2% fat) before cooking is completed. Rice must be handled carefully during cooking to prevent lumps from forming and rice kernel from breaking. Egg yolks, sugar, vanilla, and light cream are other ingredients used in making the dessert. Rice pudding, with a variety of fruit combinations such as banana, pineapple, and strawberry, serves as a popular dessert. It is preserved in 5-oz aluminum cans by the high-temperature, short-time aseptic canning process for travelers, mountain hikers, and parties or gatherings. The flavor and aroma of the pudding is greater in the aseptic canning process, which minimizes the heating process to about 1 min in a Votator-type heat exchanger, followed by rapid cooling and sealing under aseptic conditions.

12.5 RICE KRISPIES®

A very popular rice product called Kellogg's® Rice Krispies® is consumed in large quantities as a snack food, as well as a rice breakfast cereal in America. It is a toasted cereal made of rice, sugar, salt, high fructose corn syrup, and malt flavoring. It is free from fat and cholesterol and serving are in 1-1/4 cup (33 g) portions with 1/2 cup of skim milk. The product is enriched with several vitamins and minerals for improvement of nutritive value. Each serving of rice contains 120 calories, 29 g carbohydrate (10% of recommended daily allowance (RDA), 2 g protein, but no fat. The enriched product contains (in RDA): vitamin A, 15%; vitamin C, 25%; vitamin D, 10%; and thiamin, riboflavin, niacin, vitamin B-6, folate, and vitamin B-12, all 25% of RDA.

The mineral content of rice per serving is as follows: iron, 10% of RDA; phosphorus, magnesium, and zinc, all 4% of RDA; and copper, 2% of RDA.

A Rice Krispies Treats® snack food can be made from the following ingredients:

- margarine — 3 tablespoons
- marshmallows — 10 oz (40 regular size marshmallows)
- Rice Krispies® cereal — 6 cups

The procedure in making the rice treat is as follows:

(1) Melt the margarine in a large saucepan over low heat.

(2) Add marshmallows and stir until completely melted.

(3) Remove from heat.

(4) Add Rice Krispies® cereal; stir until well coated.

(5) Spread it. Use a buttered spatula or wax paper, press mixture evenly into a 13 × 9 × 2-inch pan coated with cooking spray.
(6) Cut the product into 2 × 2-inch squares when cool. The yield is 24 squares.

A microwave oven may be used in making the Rice Krispies Treat®. Heat the margarine and marshmallows for 2 min in a microwave oven in a microwave-safe mixing bowl. Stir the Rice Krispies® into the ingredients. Heat in the microwave oven for 1 min. The temperature and time needs careful observation to get a satisfactory product.

13. SOME TRADITIONAL ASIAN RICE PRODUCTS

There are many rice products that are prepared and consumed for special occasions. The following are several special rice products that are served during some special occasions.

13.1 *NENG GAO*

Neng gao is a special rice cake consumed largely at the end of a year and during celebration of the new year. It differs greatly from the rice snack foods in texture, flavor, and appearance.

The procedure for making *neng gao* (rice cake) consists of simple procedures as shown below.

Glutinous or nonglutinous rice is milled, refined, and washed with water. The product is then soaked, allowed to drip until the moisture is lowered considerably, and then steamed. The cooked rice is kneaded, packed, and made into various forms. Various forms of *neng gao* is presented in Figure 2.7. Some products are sweetened with sugar or enriched with lard and cinnamon flour for improved flavor and aroma. The product is sliced or rehydrated first and then served as a *neng gao* soup after slicing, followed by cooking with some oil, meat, and vegetables. In the sweetened *neng gao,* they are served after steaming to soften the texture. The most famous type is Nin-Poo *neng gao* prevailing along the eastern coast of China, especially in the Jiangsu and Cheking provinces.

The moisture content of *neng gao* is a very important factor influencing the keeping quality and texture of the product. The kneading and steaming process is similar to that described above in the rice cake section. For longer storage, the water activity should be below 0.20, and the product must be stored at temperatures below 10°C. As a means of preventing *neng gao* from mold and microbial spoilage, the steaming and kneading process is carried out with special attention to avoid microbial contamination.

Figure 2.7 Samples of rice nenkau in various forms.

Sometimes, *neng gao* (called *mochi* in Japanese) at an intermediate moisture level is packaged with food grade plastic film. After packing and heat sealing, the product is pasteurized at 80°C for 20 min. Rapid cooling after processing is very important to the quality of the final product. Control of possible microbial contamination as related to storage temperature and moisture content of the rice product is very important to the safety of the consumer.

13.2 *TIANG JIU NIANG*

Tiang jiu niang is a fermented glutinous rice product very popular in China and Japan. The method for making *tiang jiu niang* is presented in Figure 2.8. It is a saccharified and steamed sweet rice made with *jiu zao,* which contains *Rhizopus, Mucor, Monilia,* and *Aspergillus* (Wang, 1991). Some yeast and bacteria may also be involved in the fermenting process.

Tiang jiu niang is a mixture of rice grains and saccharified liquid, which contains 1.5–2.0% alcohol with an acidity of 0.5–0.6% as lactic acid and some glucose, maltose, and oligosaccharides. The product can be kept at 10°C for a

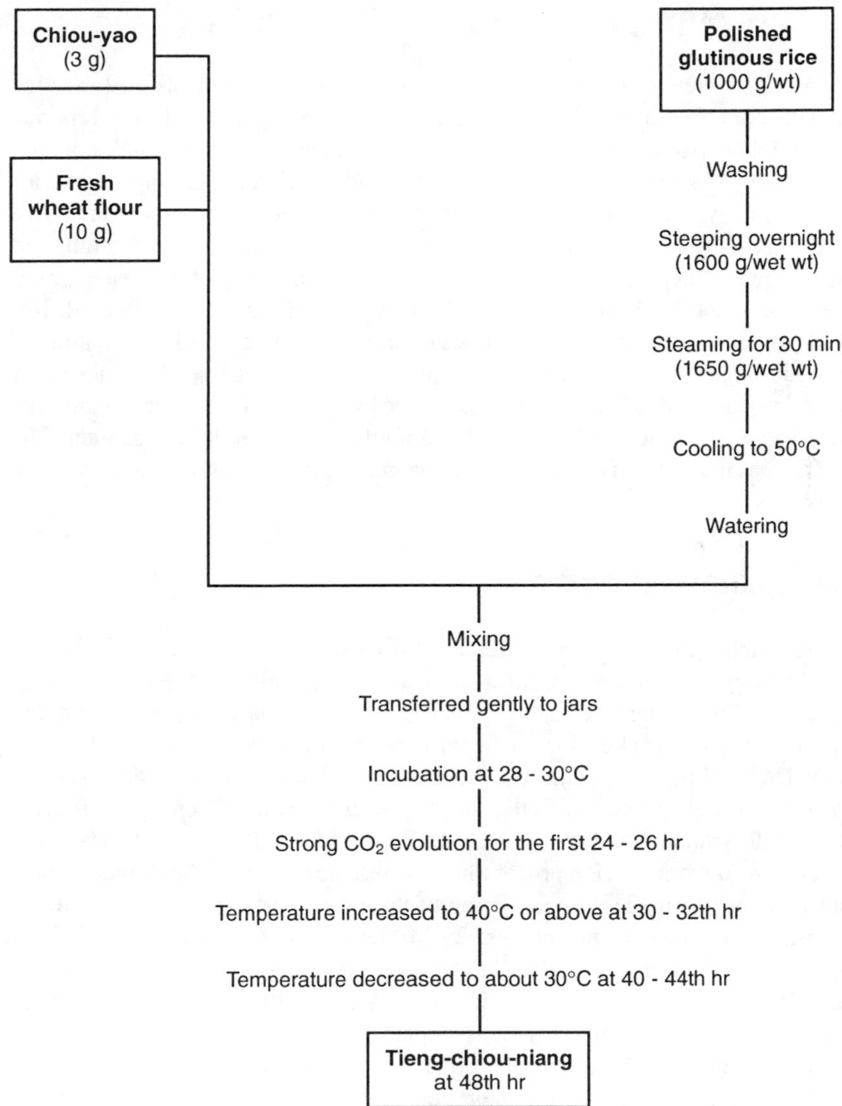

Figure 2.8 Flow sheet for the preparation of *tiang jiu niang* (fermented glutinous rice). Courtesy of Prof. Hsi-Hua Wang, National Taiwan University, Taipei, Taiwan (1991).

time, but those kept at 25°C or above will be further fermented. The acidity may increase to 1% as lactic acid, and the alcohol content may be more than 5%. The product is consumed largely in the winter season and is also used as an ingredient in making special dishes because of its attractive aroma and flavor.

14. SUMMARY

A brief review of the present status of some rice products utilized in some Asiatic countries, as well as in Western countries, was presented. It covers rice snack foods, rice flours, breads, cakes, noodles, parboiled rice, quick-cooking rice, rice preservation by canning, freezing and freeze-drying, breakfast cereals, baby foods, and some specialty rice products. There are other rice products such as rice bran, dextran, maltose syrup, rice oil, rice starch, and beverages, which are also important to the utilization of the rice grain; because of the limitation of space, these latter subjects are not covered. Recent trends on rice research have been shifted partly toward application of genetic engineering to improve the performance of rice seeds in terms of protein quality, starch functions, disease resistance, yield per acre, and nutritive value of the rice grain, which have not been covered in this chapter. The author has made efforts to summarize the more important and well-known rice products.

15. ACKNOWLEDGMENTS

The author thanks Prof. Charles F. Shoemaker, Dept. of Food Science and Technology, University of California, Davis, California; Prof. Hsi-Hua Wang and Prof. Chin-Fung Li of the National Taiwan University, Taipei, Taiwan; Dr. Yuan K. Liu of the Del Monte Corp. Research Center, Walnut Creek, California; Dr. Luping Ning and Alex Balafoutis of Pacific Grain Products, Inc., Woodland, California; Dr. Catharina Ang of the National Center for Toxicological Research, Food and Drug Administration, Jefferson, Arkansas; and friends who work on rice processing and research for their help and encouragement, which made the completion of this chapter possible. The critical and constructive review of this chapter by Mr. Don R. McCaskill, Director of Research Development of the Riceland Foods, Inc. of Stugttgart, Arkansas, and Dr. Wallace A. Yokoyama of the U.S.D.A. Western Regional Research Laboratory, Albany, California, is greatly appreciated. I also thank Pam Garcia and Judy DeStefano of the Department of Food Science and Technology, University of California, Davis, for their help and patience in typing the revised manuscript.

16. REFERENCES

Burns, E. E., and Gerdes, D. L. 1985. Canned rice foods. In: *Rice: Chemistry and Technology,* B. O. Juliano, ed., Amer. Assoc. Cereal Chemists, St. Paul, MN, pp. 557–567.

Chang, T. T., and Luh, B. S. 1991. Overview and prospects of rice production. In: *Rice Production, Vol. 1,* 2nd ed., B. S. Luh, ed., An Avi Book, Van Nostrand Reinhold, New York, pp. 1–11.

Del Gado, C. 1977. Improvement of rice bread. M.S. Thesis, University of California, Davis, CA.

Herdt, R. W., and Palacpac, A. C. 1983. *World Rice Facts and Trends.* International Rice Research Institute, Los Banos, Philippines.

Houston, D. F. 1972. *Rice: Chemistry and Technology.* Amer. Assoc. Cereal Chem., Inc., St. Paul, MN.

Hsieh, F., Huff, H. E., Peng, I. C., and Marek, S. W. 1989. Puffing of rice cakes as influenced by tempering and heating conditions. *J. Food Sci.* 54(5):1310–1312.

Hsieh, F., and Luh, B. S. 1991. In: *Rice Utilization, Vol. 11,* 2nd ed., B. S. Luh, ed., An Avi Book, Van Nostrand Reinhold, New York, pp. 225–249.

Huang, S. D., and Miskelly, D. M. 1991. Steamed bread—A popular food in China. *Food Austral.* 43(8):346.

Huang, S., Quail, K., Moss, R., and Best, J. 1995. Objective methods for the quality assessment of northern-style Chinese steamed bread. *J. Cereal Sci.,* 21:49.

Juliano, B. O. 1985. *Rice: Chemistry and Technology.* Amer. Assoc. Cereal Chemists, Inc., St. Paul, MN.

Kokini, J. L., Ho, C.-T., and Karwe, M. V. 1991. *Food Extrusion Science and Technology.* Marcel Dekker, Inc., New York.

Luh, B. S. 1991a. Canning, freezing and freeze drying. In: *Rice Utilization, Vol 11,* 2nd ed., B. S. Luh, ed. An Avi Book, Van Nostrand Reinhold, New York, pp. 147–175.

Luh, B. S. 1991b. Quick cooking rice. In: *Rice Utilization, Vol. 11,* 2nd ed., B. S. Luh, ed., Van Nostrand Reinhold, New York, pp. 121–146.

Marshall, W. E., and Wadsworth, J. I. 1994. *Rice Science and Technology.* Marcel Dekker, Inc., New York.

Mercier, C., Linko, P., and Harper, J. M. 1989. *Extrusion Cooking,* Amer. Assoc. Cereal Chemists, Inc., St. Paul, MN.

Nishita, K. D., and Bean, M. M. 1979. Physiochemical properties of rice in relation to rice bread. Cereal Chem. 56:185–189.

Nishita, K. D., and Bean, M. M. 1982. Grinding methods: Their impact on rice flour properties. *Cereal Chem.* 59:46–49.

Nishita, K. D., Roberts, R. L., Bean, M. M., and Kennedy, B. M. 1976. Development of a yeast-leavened rice bread formula. *Cereal Chem.* 53:626–635.

Passmore, J. 1991. *The Encyclopedia of Asian Food and Cooking.* Doubleday, a Division of Trans World Publishers (Australia), Pty. Ltd.

Rubenthaler, G. L., Huang, M. L., and Pomeranz, Y. 1990. Steamed bread 1. Chinese steamed bread formulation and interactions. *Cereal Chem.* 67:471.

Rubenthaler, G. L., Pomeranz, Y., and Huang, M. L. 1992. Steamed bread IV. Negative steamer-spring of strong flours. *Cereal Chem.* 69:334.

Tsen, C. C., Ding, W. L., Lin, T. C., and Nassar, R. 1982. Optimizing processing conditions and ingredient formulations for making steamed bread. *Cereal Foods World* 27:451.

Velupillai, L. 1994. Parboiling rice with microwave energy. In: *Rice Science and Technology,* W. Marshall and J. I. Wadsworth, eds., Marcel Dekker, Inc. New York, pp. 263–274.

Wang, H. H. 1991. Fermented rice products. In: *Rice Utilization, Vol. II,* 2nd ed., B. S. Luh, ed., Van Nostrand Reinhold, New York, pp. 195–223.

Webb, B. D., Bollich, C. N., Carnahan, H. L., Kuenzel, K. A., and McKenzie, K. S. 1985. Utilization characteristics and qualities of United States rices. In: *Symposium on Rice Grain Quality and Marketing,* Los Banos Philippines, IRRI, pp. 25–35.

Webb B. D., Bollich C. N., Jackson, B. R., Kanter, D. G., Linscomb, S. D., Moldenhauer, K. A. K., Tseng, S. T., and Peterson, H. D. 1989. Evaluation of rice quality components for named varieties grown in performance trials in Arkansas, Louisiana, Texas, and California 1986–1988. Beaumont, TX: Cooperative Rice Quality Annual Crop Report 1989.

Wen, Q. B., Lroenz, K. J., Stewart, B. G., and Sampson, D. A. 1996. Carbohydrate digestibility and resistant starch of steamed bread. *Stärke* 41(5):80.

Yue, P., and Rayas-Duarte, P. 1997. An Overview of steamed bread. *Cereal Foods World* 42(4):210–215.

CHAPTER 3

Wheat Products: 1. Noodles

HAROLD CORKE
MONISHA BHATTACHARYA

1. INTRODUCTION

NOODLES represent a dominant usage of wheat flour in much of Asia. Many Asian countries, including China, are rapidly growing wheat importers. Exporting nations are increasingly interested in the quality requirements for wheat flour destined for use in noodles. Food scientists are interested in developing new, high-quality, convenient forms of noodles, often developed from traditional formulations and processing methods. A basic understanding of noodles requires an overview of their diversity, an in-depth look at the scientific basis for quality in a few key types, and some guidelines for testing methods and criteria that may aid in product development.

There are a few key points that should be emphasized regarding Asian noodles. In this chapter we will concentrate on wheat flour noodles (rather than starch noodles made from purified starch, generally of other plant sources such as rice, mung bean, or sweet potato). Asian wheat noodles are made from common wheat flour (hexaploid wheat, *Triticum aestivum,* bread wheat), *not* from the durum wheat flour used for European-type pasta. Asian noodles are sheeted and cut from a low-moisture dough, *not* extruded like European pasta. In Asian noodles the key criteria for quality are texture and color. Textural requirements are specific for different types of noodles and according to regional preferences. Discoloration of noodles towards gray is a major negative. In many cases the right color—the right white or yellow—is essential. These basic points should be kept in mind throughout any discussion on Asian noodles. The basic types of wheat or wheat-composite noodle that will be discussed represent four main types: (a) white-salted, consisting of flour, water, and common salt and typified by the Japanese *udon* noodles; (b) yellow-alkaline noodles, consisting of flour, water, and alkali and typified by Cantonese yellow noodles; (c) composite flour noodles where wheat flour is mixed with other starch-based material such as buckwheat flour and typified by the Japanese *soba* noodles; and (d) instant noodles, where the noodle is steamed (for starch gelatinization, i.e., cooking) and fried (for drying) and can be rapidly rehydrated before consumption, typified by the common instant *ramen* packaged noodles. Starch noodles are prepared by mixing purified starch with some pregelatinized starch as a binder, mixing to a "dough," and extruding into boiling water. This type will not be discussed in this chapter.

2. MAKING SENSE OF NOODLES

It is generally not helpful to discuss the immense diversity of noodle types as separate products with a catalog of specific quality traits. The food technologist may often be interested in new product development or in quality control of raw materials, and the wheat scientist or breeder may be interested in defining objectives for targeted selection. To create order out of diversity and to allow flexibility in considering new and unusual noodle variants, we will concentrate on the basic principles of noodle quality. To do this, we consider research results on three significant classes of noodles: the Japanese form of white-salted noodles *(udon),* Cantonese-style yellow-alkaline noodles, and instant noodles. Many other regional or local types of noodles can be shown to be variants of these, with varying formulation, wheat quality demands, and expected texture and color outcomes. In particular, salted noodles other than Japanese *udon,* such as in China, usually have a very different texture and require wheat flour with stronger dough and possibly lower-swelling starch. Starch and protein are the keys to wheat flour quality for noodles. Noodles generally contain salt (NaCl) and/or an alkaline additive (often a mixture of sodium and potassium carbonate). The interaction of starch and protein with these components is important. Understanding how to measure and assess starch and protein quality is essential for any applied research in noodles (Baik et al., 1994; Bhattacharya and Corke, 1996; Crosbie, 1991; Huang and Morrison, 1988).

We should also bear in mind the general wheat situation in Asia. China is by far the biggest producer and consumer of wheat in the world, harvesting nearly 100 million tonnes per year (Figure 3.1). China is also increasingly becoming a major importer of wheat. In Asia, as disposable incomes rise, there is an increase in wheat consumption at the expense of diminishing rice consumption. Projections of Chinese demand for wheat in the coming 5–20 years has led to some dramatic stories in the international news media (Wehrfritz, 1995). Whatever the outcome of China's efforts to increase production, global trade in wheat is set to increase, and prices may be expected to rise in the medium to

Figure 3.1 A wheat field in Hubei Province, China. China, with some 100 m tonnes per year, is by far the biggest wheat producer in the world.

long term. Many tropical Asian countries are also large consumers of wheat products but have essentially no domestic production, such as the Philippines, Malaysia, Indonesia, and Singapore. Although there is significant wheat production in Japan, the agricultural sector is not competitive on price, and Japan has long been a major wheat importer. Thus, Asia seems certain to dominate a fast-increasing global trade in wheat well into the future. Major exporters such as the United States, Canada, and Australia will be exporting into increasingly quality-conscious markets. Scientists and technologists in these countries are fully aware of this significant shift in market development and realize that quality selection will increasingly be focused on Asian end uses such as noodles and steamed bread. Quality criteria for noodles are *not* those for leavened baked bread.

3. HISTORY AND ORIGIN OF NOODLES

Noodles are strips or strands cut from a sheet of dough made of flour, water, and either common salt or a mixture of alkaline salts. They are one of the main staple foods consumed in East and Southeast Asian countries, representing up to 40% of total flour consumption (Miskelly, 1988). It is believed that noodles originated in the north of China as early as 5000 B.C., but their essential modern-day form has developed over the last 2000 years. Present-day noodles *(mian)* were a unique contribution by the *Han* Dynasty (206 B.C. to 220 A.D.) to Chinese culinary art. The development of noodle foods in the *Han* period seemingly can be explained by the fact that techniques for large-scale flour milling were introduced to China from the West during the latter part of the earlier *Han* Dynasty, as a result of the *Han* expansion. *Han* ingenuity in experimenting with such common food materials, combined with the willingness to incorporate technology from other cultures, led to the emergence of an eventually dominant new product in Chinese culinary history (Yu, 1979). The writer *Shu Hsi* in the Western Jin Dynasty (late third and early fourth centuries) noted that the various kinds of noodles "were mainly the invention of the common people, while some of the cooking methods came from foreign lands."

The art of noodlemaking developed further in the *Tang* Dynasty (618–907 A.D.) when the noodles were first cut into long strips. The age-old custom of eating noodles to signify long life is believed to have originated in this era. During the period of the *Song* Dynasty (960–1179 A.D.), the variety of different styles of noodles gradually increased as unique local tastes were developed, and a great variety of popular noodle dishes with inclusion of meat and vegetables came into being (Wang, 1987). The technique of making dried noodles was learned in the *Yuan* Dynasty (1271–1368 A.D.), and in the *Ming* Dynasty (1368–1644 A.D.) noodles were given the name of *mian,* and, since then, have continued to evolve until the present (Huang, 1996).

With increased travel and trade and widespread Chinese migration and emigration, noodles were taken across the country and gradually gained popularity in other countries. Noodles spread from China to Korea, Japan, Thailand, Malaysia, Indonesia, Singapore, the Philippines, Burma, and Vietnam with the many Chinese traders, seafarers, and emigrants who moved into these areas. The similarities of some of the generic words for noodles [*viz., mian, mein,* or *mi* (China); *men* (Japan); *mie* (Indonesia); and *mee* (Thailand, Singapore and Malaysia)], support this common origin (Miskelly and Gore, 1991). In Japan, *udon* noodles were created in the 15th century, and most of the types of noodles found today were available by the 16th century (Nagao, 1979, 1981).

Some sources claim that pasta was invented in China, and the legendary Marco Polo has been credited for the introduction of pasta to Italy, following his travels to China in 1295. However, this is easily refuted, because references to lasagna and pastas as part of the staple Italian diet occur in written records as early as 1279, before Marco Polo set out on his travels (Baroni, 1988; Giese, 1992). Although the terms *pasta* and *noodles* are often used interchangeably, the two products are very different in terms of their raw materials, method of manufacture, the quality attributes of the wheat desired by the breeder, and the textural parameters demanded by the consumer. Pasta usually refers to Italian-style extruded products, such as spaghetti and macaroni, which are made from coarse semolina milled from tetraploid or durum wheat, while noodles are generally of Asian origin and are made from hexaploid or common (bread) wheat. The basic process of manufacture includes mixing the ingredients to a stiff dough of relatively low water content, passing repeatedly between sheeting rolls to get a desired thickness, and finally cutting into strips.

4. DIVERSITY AND GENERAL CLASSIFICATION OF ASIAN NOODLES

There are regional preferences for noodle color, texture, flavor, size and shape, shelflife, and ease of cooking, which in turn depend on the flour characteristics, method of preparation, and the inclusion of other raw materials or chemical additives. The terminology of classifying noodles can be confusing since noodles of almost identical composition have different names in various countries. Moreover, within the same country, noodles of the same formulation are sometimes named to differentiate the manufacturing process and the noodle strand thickness (Dick and Matsuo, 1988). Noodles are commonly classified according to (1) size of the noodle strands, (2) the nature of the raw materials used in their manufacture, (3) the method of preparation, and (4) the form of the product on the market (Crosbie et al., 1990) (Table 3.1). Though a wide diversity exists in the type of noodles, they can be broadly classified into three main groups: white-salted noodles, popular in China, Japan, and Korea; yellow-alkaline noodles, popular in Malaysia, Singapore, Indonesia, Thailand, and southern China; and instant noodles, popular in East and Southeast Asia.

4.1 CLASSIFICATION BY SIZE

White-salted noodles vary widely in the size of the noodle strands. In Japan, the main types classified according to width include very thin noodles *(so-men),* thin noodles *(hiya-mugi),* standard noodles *(udon),* and flat noodles *(kishi-men* or *hira-men)* (Crosbie et al., 1990). *So-men* and *hiya-mugi* are dipped in a cup of cold soy sauce soup and usually served in the summer, while *udon* and *hira-men* are served in hot soy sauce soup in the winter along with boiled noodles, tempura, fried bean curds, boiled fish paste, and/or vegetables (Nagao, 1991). Yellow-alkaline noodles in Japan are thin and rarely more than 1.5 mm wide.

In China, noodles are similarly classified according to the size of the noodle strands but are known by different names than those used in Japan. The main varieties are very thin noodles *(longxu mian),* thin noodles *(xi mian),* flat noodles *(yang chun mian),* and wide flat noodles *(dai mian* or *cu mian).*

4.2 VARIATION BY METHOD OF PREPARATION

The machine-operated process of dough mixing, sheeting, and cutting the sheet into strands is most commonly employed in the manufacture of Oriental

TABLE 3.1 Classification of Asian Noodles.

1. By size:
   - Very thin noodles: *somen* (Japanese), *longxu mian* (Mandarin)
   - Thin noodles: *hiya-mugi* (Japanese), *ximian* (Mandarin)
   - Standard noodles: *udon* (Japanese)
   - Flat noodles: *kishi-men/hira-men* (Japanese), *dai mian/chu mian* (Chinese)
2. Nature of raw materials, pH and organoleptic qualities:
   - White-salted noodles: flour of low to medium protein level (8–10%), water (30–35%), sodium chloride (2–3%). pH 6.5–7; soft elastic texture and a smooth surface
   - Yellow-alkaline noodles: hard wheat flour of medium protein content (10–12%), water (30–35%), alkaline salts, e.g., sodium carbonate, potassium carbonate; pH 9–11; firm, chewy, springy texture and bright yellow appearance
   - Instant noodles: wheat flour of medium protein content (9–11%), water (30–35%), common salt or alkaline salts (2–3%), oil; pH 5.5–9; elastic, chewy texture
   - Buckwheat noodles: wheat flour of high protein content (12–14%), buckwheat flour, water, salt (optional); firm, chewy, tender texture
3. Method of preparation:
   - Machine-operated process: mixing, sheeting and cutting into strips
   - Handmade process: e.g., pulling and stretching of dough to create noodle strands or cutting the dough into strips with a knife
4. Form of product available on market:
   - Uncooked wet noodles
   - Dried noodles
   - Boiled noodles
   - Frozen boiled noodles
   - Steamed-and-fried or steamed-and-dried instant noodles

noodles (Figure 3.2). Handmade noodles have a more fascinating method of preparation and require an experienced hand with considerable skills to produce consistent results (Wang and Ya, 1988). The procedure is much more complex than the sheeting and cutting method and involves stretching and pulling a ball of soft dough repeatedly to create long strands of noodles with uniform thickness (Figure 3.3). The art of making handmade noodles (*tenobe* in Japanese, *la mian* in Chinese) is quickly being replaced by the mechanized process, but the former type of noodles has greater value because of the fine texture and favorable eating quality. Hand-swung noodles are characterized by a soft, smooth, elastic, and chewy texture. Variations of hand-stretched noodles are single hand-stretched noodles, hollow hand-stretched noodles, very thin noodles, flat noodles, noodles stretched in water, oiled stretched noodles, and stuffed hand-stretched noodles (Huang, 1996). The better textural quality of handmade noodles is presumably because of the mode of gluten formation. The gluten strings in machine-made noodles form lines in a direction, whereas those in handmade noodles intertwine lengthwise and crosswise. In

Figure 3.2 Typical small-scale commercial noodle production in Inner Mongolia, China. Large-scale factory production follows essentially the same process.

Japan, a new technique has been employed using high water absorption in dough, to allow dough aging to take place in the machine, thereby making it possible to produce noodles having the palatability of the handmade type on an industrial scale (Nagao, 1991).

Other types of handmade noodles are also found in Asia, which do not require particular skills in preparation and are more frequently prepared on a household level than on a large-scale industrial level. A few variations of the handmade noodles found in China are described here. *Qie-mian* are vermicelli-like fresh noodles made from hand-rolling a dough into a flat square sheet and then cutting the sheet into fine strips with a knife. *Dao-xiao-mian* are ribbon-like, white-salted noodles about ⅓″ wide and have an elastic texture when cooked. They are prepared out of a stiff and dry ball of dough by slicing the dough into ribbon-like strips and directly dropping them into a pot of boiling water. These two types of

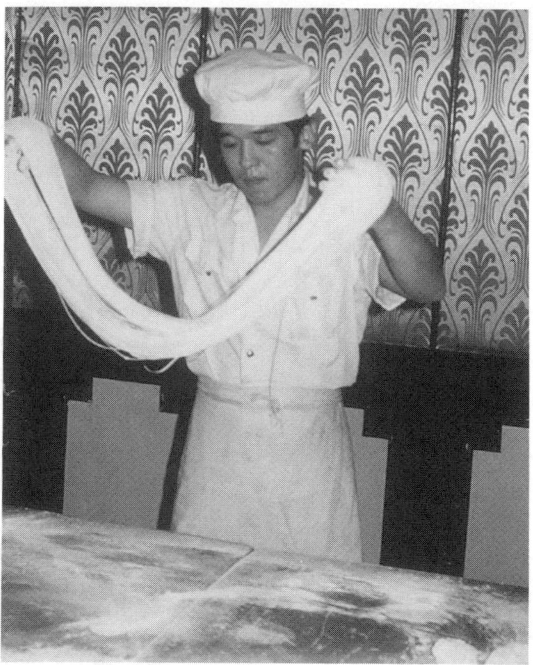

Figure 3.3 Making Chinese *la-mian,* or stretched noodles, in Shanxi Province.

noodles are commonly prepared in Chinese households and are eaten in a soup form along with vegetables, meat, or seafood.

Cat's-ear *(mao-er-dao)* is another unique variety of noodles made by sheeting and cutting a dough and then cutting the strips into 1″ squares. Each square is then pinched into the shape of a cat's ear, boiled, and served in a soup form. Another unique variant is the "flour-fish" noodle *(po-yeu),* where the flour, salt, and water are mixed to form a smooth and grainless paste (batter) instead of a dough. The paste is then added bit by bit to boiling water or soup using chopsticks, and the slices are separated from each other with the help of a spatula. The flour-fish soup is served with vegetables and meat and has an elastic texture.

4.3 VARIATION BY THE FORM OF PRODUCT SOLD IN THE MARKET

Noodles are in one sense a curious product; they are deeply rooted in tradition yet have proliferated to vast diversity. People know intimately the quality attributes they expect of the traditional noodle of their childhood and demand precisely the same attributes from the manufacturer. However, in other ways the consumer may be relatively unbound by tradition. New types of noodles, such as instant noodles, evolve into particular styles over time, but people are more flexible or adventurous in defining their preferences. A look in an Asian supermarket reveals a vast range of products, some quixotic and destined for small-scale and niche sales and others perhaps expanding into the growth categories of the future.

A retail outlet for Park 'n Shop in *Chi Fu Fa Yuen,* Hong Kong, was examined in June 1997. Delicatessen-style food counters gave additional opportunities for production and marketing of premium fresh noodles. Rice noodles were well represented: Vietnamese *mong* noodles (thin, rounded, rice starch noodles, very brittle in texture); Chinese-type *lai* rice noodles (thicker, rounded strands compared to the *mong* type, harder texture); Guangdong Province-type rice noodles (*hofen,* flat, tough texture); steamed rice roll with shrimp [perhaps not a true noodle but a cooked rice sheet embedded with shrimp and onion and rolled to form a thick cylinder (250 mm length, 18 mm diameter)]. Dried noodles included shrimp egg noodles that were oily and very dark yellow; in this case the relatively long shelf-life product was held in the chilled produce section as it was unpackaged. Fresh noodles included the ordinary Chinese-type white-salted; an alkaline type containing shrimp; *dandan* (Sichuanese-type), one of the four famous noodles of China with a bright yellow, soft texture, containing egg and chili; and Rickshaw noodles, an early version of instant noodles, a quick meal for the rickshaw driver, with oiled, yellow, thin flat strands.

Nontraditional types for retail sale in Hong Kong included flat spinach noodles (bright green); fresh carrot noodles (square cut, light orange-brown); and squid ink noodles (black but lightly floured). A processed variant of a traditional type would be exemplified by packaged fresh *udon*-type noodles, which were instant or precooked. Of course, the diversity in dried, packaged noodles is similarly large. We feel that product development should not be constrained by fear of consumer-resistance to change. Perhaps the least prevalent indication in the wide variety of noodles on the market is any clear attempt at health claims. After convenience, health consciousness in dietary selection is the next logical step with increased economic development. We feel that the time is right for nutritional claims to be incorporated into marketing improved protein or reduced-fat versions of particular types of premium noodles.

4.4 NOODLE MANUFACTURE

It is beyond the scope of this review to consider industrial processing in depth; a good discussion of the topic is provided by Huang (1996). A few general points will be made here. Essentially, the noodle-making process is remarkably constant from the smallest scale machine production to the largest

A

B

Figure 3.4 Medium-scale factory processing of dried noodles in China: (a) mixing a low-moisture crumbly dough, (b) sheeting for progressive reduction in sheet thickness and dough development.

scale factory (Figure 3.4). The first step is mixing [Figure 3.4(a)], where a low-moisture crumbly dough is made, mixing the ingredients and allowing some protein hydration. Two types of mixers are in common use: single-shaft mixers and double-shaft mixers with interlinked beaters (Huang, 1996). The moisture content is too low to allow full dough development as in bread making. Further

D

Figure 3.4 ***continued*** (c) cutting into strands, and (d) drying—the critical operation for optimum quality.

dough development to exploit the viscoelastic properties of the wheat gluten takes place during sheeting. "Resting" (actually low-speed mixing) is followed by sheeting [Figure 3.4(b)], which forms a thick sheet of dough and then progressively reduces it in thickness by successive passes through a roller, with reduction in the gap setting to control thickness. Some six or seven reduction passes would be typical, after which the sheet is cut into long strands [Figure 3.4(c)]. Typically, the cut noodle strand will be 0.6–1.4 mm, and the width will be 0.8–6.0 mm (Huang, 1996), depending on the type of noodle being made. After cutting, there is great flexibility in further processing and packaging. The noodle strands could be dried to make packaged dried noodles [Figure 3.4(d)] (the dominant usage in China), or the fresh noodle strands could be formed into wavy blocks, steamed, fried, and packaged (for instant noodles). These variations are discussed in more detail below.

5. WHITE-SALTED NOODLES

To regain some sense of order after consideration of diversity, we will consider in depth several key types of noodle and describe the quality attributes and processing inherent in their production. The Japanese form of white-salted noodles *(udon)* is perhaps the most intensively studied. They are white or creamy white in appearance and have a soft and elastic texture. The boiled forms of Japanese *udon* are the most popular types, which are either sold loose or packed in tightly sealed polyethylene pouches for extended shelf life. They are made from a mixture of flour (100 parts), water (32–35 parts), and salt (2–3 parts). The amount of salt added is based on the type of noodle (boiled or dried), climate, and consumer requirements. The criteria for judging Japanese noodle quality are cooked noodle texture (eating quality), followed by color, taste, surface appearance, cooking loss, and noodle yield (Toyokawa et al., 1989a). The flour used is predominantly made from relatively soft wheat of low to medium protein level (8–10%), low flour ash content (0.36–0.40%), low damaged starch, and a good color grade (Nagao et al., 1977; Crosbie et al., 1990), giving the bright creamy tone and desirable texture to the noodles. The low protein level results in a less tough and lighter colored dough for noodle formation (Oh et al., 1985). Any discoloration, visible specks, or dull gray hue is considered to be of unacceptable quality. Extremely low protein content, abnormally weak gluten, and low amylograph viscosity cause problems in noodle manufacture and reduce the quality and acceptability of the final product (Nagao, 1995).

A number of researchers have described the lab-scale preparation of Japanese white-salted noodles. A typical procedure is described here according to the method of Toyokawa et al. (1989a): 300 g of flour (14% moisture basis), 96 ml of water and 6 g of salt are mixed with a flat paddle in a Hobart mixer for 6 min on slow speed, which helps to distribute water evenly throughout the flour particles. The resulting crumbly dough is sheeted between steel rolls (diameter 180 mm, gap 3 mm) of an Ohtake (Tokyo, Japan) laboratory-type noodle machine; it is then folded end-to-end and put through the sheeting rolls twice. The dough sheet is inactive for 1 h in a plastic bag to allow gluten development and then put through the sheeting rolls three times at progressively smaller gap settings of 2.83, 2.66, and 2.50 mm, respectively, to reduce the sheet to a final thickness of 2.5 mm. All sheeting operations are carried out in the same direction. This allows the gluten fibrils to align in the direction of sheeting, giving more strength to the noodle structure (Hoseney, 1994). The sheet is then cut with a #10 cutting roll into strips approximately 30 cm in length with a 0.30 × 0.25 cm cross section. The noodles are cooked in boiling water until the zone of gelatinization has extended throughout the noodles and then is evaluated for color and textural characteristics.

Oda et al. (1980) reported that starch characteristics are very important to the eating quality of Japanese noodles, while Oh et al. (1983) found that the protein content of flour influenced the chewiness of cooked noodles. Dexter et al. (1979) used a scanning electron microscope to study the dough structure during Japanese noodle preparation. They observed that the starch granules were more loosely held within the gluten protein matrix and the presence of 2% salt in the dough resulted in a smoother and more uniform gluten structure than that observed for unsalted dough. Toyokawa et al. (1989a) carried out a fractionation and reconstitution interchange of gluten, primary starch, tailing starch, and water solubles to investigate the role of each in Japanese noodle quality. The primary and tailing starch fractions were found to be responsible for noodle texture, with the former contributing most to the desirable viscoelasticity in noodle texture. The gluten fraction of the flour affected the noodle color but not texture, while the water-soluble fraction did not have any effect on noodle properties. Toyokawa et al. (1989b) found in a later study that an optimum ratio of amylose to amylopectin is necessary for good noodle quality, and increased levels of amylose reduced the water binding of cooked noodles, resulting in firmer noodles with loss of elasticity. High-quality Japanese noodles have a bright, clean luster; are cream-colored; and have a soft, but elastic, texture and a smooth surface (Crosbie et al., 1990). Wang and Seib (1996), proposing a working model for starch in salted noodles, suggested that low flour protein, along with high-swelling starch, causes extra water to be imbibed inside a strand to produce a noodle with a soft bite. Moreover, low levels of amylose may be leached between granules to produce the desired elasticity.

6. YELLOW-ALKALINE NOODLES

Yellow-alkaline noodles are essentially made from flour (100 parts), water (32–35 parts), and a solution of alkaline salts known as *kansui* or lye water

(1 part). These salts are usually a mixture of sodium and potassium carbonates (typically 9 : 1) or sodium hydroxide in some cases (Moss et al., 1986). The alkaline salts confer a unique flavor and texture to the noodles and are responsible for imparting the typical yellow color by detaching the flavones from starch and allowing their natural color to manifest (Miskelly, 1984; Moss, 1984). Flours from hard wheat, with protein content in the range of 10–12%, with mellow gluten quality, is recommended for fresh alkaline noodles (Moss, 1982). The desired textural characteristics are a bright, even light yellow appearance; free of any darkening or discoloration; a firm clean bite; a chewy and elastic texture with some degree of springiness; and a satisfactory *al dente* reaction on biting (Miskelly and Moss, 1985).

The basic procedure for making alkaline noodles is similar to that of white-salted noodles, and a lab-scale method is described here based on the procedure by Miskelly and Moss (1985). The ingredients [300 g flour (13.5% m.b.), 96 ml water, Na_2CO_3 (2.7 g), and K_2CO_3 (0.3 g)] are mixed in a Hobart mixer for 1 min at slow speed, 1 min at fast speed, and then for 3 min at slow speed. The crumbly dough is sheeted between steel rollers, 2.75 mm apart, in an Ohtake noodle machine. The dough sheet is folded and passed between the rollers twice and then allowed to lie at rest for 30 min in a plastic bag. The single sheet of dough is then passed between the rollers, with the clearance successively reduced to 2.5, 2.0, and 1.5 mm. The sheet is finally cut into strips of cross section 1.5 × 1.5 mm. After standing for 3 h at 25°C, the noodles are placed in wire mesh and cooked in boiling water until the uncooked core has just disappeared (~5 min); the cooked noodles are used for sensory testing.

The principal factors governing the eating quality of yellow-alkaline noodles are protein content, dough strength, and starch paste viscosity (Miskelly and Moss, 1985). In the interior of alkaline noodles, the inherently low-swelling starches of hard wheats are further restricted from swelling by the carbonate salts (Moss et al., 1986). The alkaline salts toughen the dough, affect pasting properties of starch by retarding gelatinization and increasing paste viscosity, inhibit enzyme activity, and suppress enzymatic darkening (Miskelly and Moss, 1985; Moss et al., 1986). The resulting starch gels are hard and strongly elastic, which together with the high protein and relatively small voids (Moss et al., 1987), produces chewy noodles (Wang and Seib, 1996).

Shelke et al. (1990), using a response surface methodology study, suggested that Cantonese wet noodles of excellent quality can be made from wheat flours having a protein content in the range of 10–11.5%, using 1.4–1.7% sodium chloride and 0.7–1.2% sodium carbonate. They observed that sodium carbonate affected all the properties of cooked noodles, protein level affected noodle color and firmness, and sodium chloride had little influence on noodle quality. Sung and Sung (1993) reported that the alkaline reagent increased both the breaking and cutting forces of the noodles, while salt and alkaline reagent together strengthened the dough properties. Lorenz et al. (1994) observed that alkaline salts decreased the dough development time and dough stability and increased the pasting temperature and peak viscosity.

Huang and Morrison (1988) found that sodium dodecyl sulfate (SDS)-sedimentation volume significantly correlated with maximum cutting stress and maximum compression stress and could therefore be used to predict Cantonese noodle quality. Konik et al. (1993, 1994) reported that starch pasting properties correlate well with the smoothness and firmness of yellow-alkaline noodles, and this correlation could be improved further by including protein content, grain hardness, and wheat falling number in a multiple regression equation. Baik et al. (1994) reported that, along with starch pasting properties, protein content and protein quality should also be considered in the evaluation of suitability of flours for making Asian noodles other than Japanese *udon* (white-salted) noodles.

Another typical Cantonese modification of alkaline noodles may add eggs to the formulation, which impart a brighter yellow color and a firmer texture to the product. Such Cantonese-style noodles are sold raw, that is, without any further processing after cutting into strands, and the cooking procedure usually involves frying the noodles thoroughly until both sides are brown (slightly scorched) and crispy and serving with an assortment of meats and vegetables (Shou, 1994).

7. INSTANT NOODLES OR *RAMEN*

Instant noodles *(ramen)* have become one of the most important types of noodles in Southeast Asia and have become a popular convenience food on a global level. They are either steamed and dried or steamed and fried before packaging. Instant noodles are a convenient snackfood because of their ease of cooking, needing only 2–3 min boiling or rehydration with boiling water. The frying process removes the water from the noodle strands, resulting in a porous structure that rehydrates quickly when water is added. When reconstituted in boiling water, no fat should separate in the cooking water, and the noodles should have a strong bite and a firm nonsticky surface. Figure 3.5 shows how instant *ramen* noodles are made in China by a medium-scale commercial process.

Instant noodles are either flavored with seasonings or are plain in taste, accompanied by a separate sachet of soup-base. They are usually packed and sold in a polyethylene bag (Figure 3.6) and have good storage properties, particularly if packed to exclude oxygen and light. *Ramen* may also be packed in a Styrofoam™ cup (Figure 3.7) or bowl with a peelback aluminum cover, along with dried soup-base, vegetables, shrimp, or meat, and can be eaten by pouring hot water into the cup (Nagao, 1996). A survey carried out by the Bread Research Institute of Australia showed that fried instant noodles have,

A

B

Figure 3.5 Medium-scale factory processing of instant ramen noodles in China: (a) cutting into strands and waving into blocks, (b) steaming—the cooking operation.

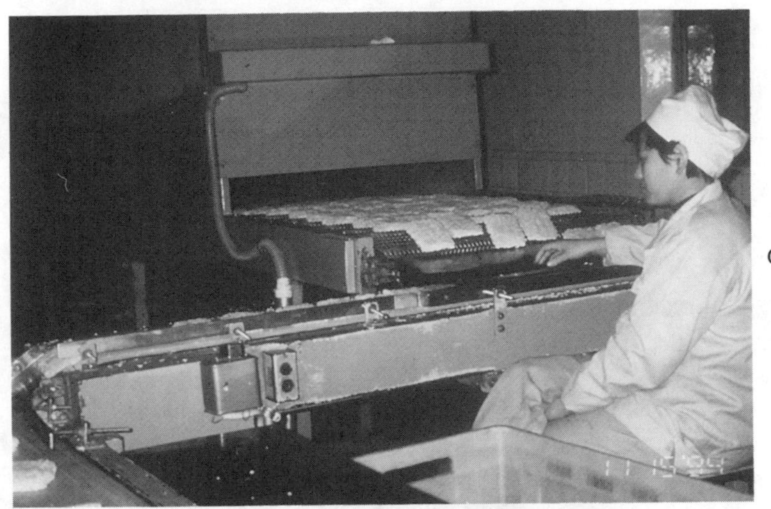

C

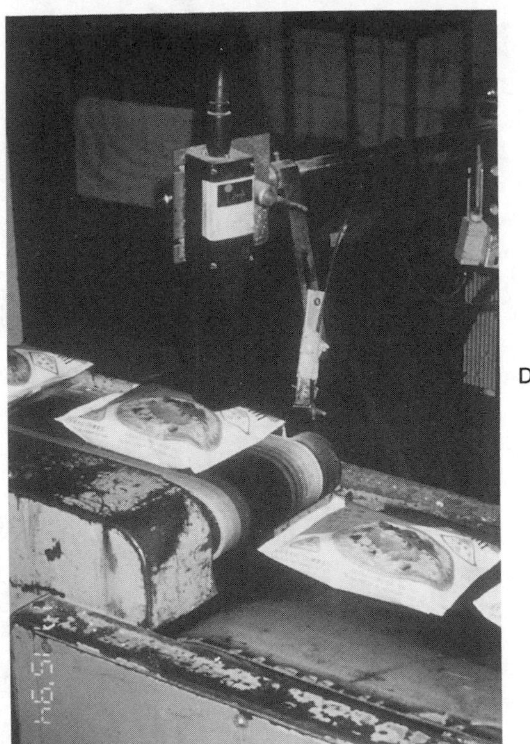

D

Figure 3.4 ***continued*** (c) frying for moisture reduction after dividing into individual portions, and (d) packaging.

Figure 3.6 A typical premium bag package of instant ramen noodles found in Japan.

Figure 3.7 A typical cup package of instant ramen noodles.

on average, 6.6% moisture, 9.5% protein, 21.1% fat, 1.7% ash, and neutral pH. Wheat flour milled from soft wheat with moderate to low protein content is desirable because it facilitates rapid cooking upon addition of hot water. They usually have a shelf life of 4–6 months. For good consumer acceptance, instant-fried, or *ramen,* noodles should be free of rancidity.

A lab-scale preparation of instant fried noodles is described by Kim and Seib (1993). Wheat flour (100 g, 14% m.b.), water (34–36 g), and sodium chloride (2 g) are blended in a Hobart mixer for 1 min at slow speed (speed 1) and 4 min at medium speed (speed 2) with a cake paddle. This formulation should be modified depending on the experimental purpose, taking into account the possible ingredients listed in the next paragraph, in particular the wide usage of alkali in commercial practice. The crumbly mixture is pressed into a thick sheet of 5.5 mm thickness, inactive for 15 min, and then sheeted in seven consecutive steps to a final thickness of 0.8 mm. The dough sheet is cut into noodle strings 0.8 mm wide that are configured in a crinkly or a wavy pattern by moving the collecting plate. Portions of this are then assembled into small blocks and steamed at atmospheric pressure for 3 min to partly gelatinize the noodle strands. The surface of the steamed noodles are dried by forced air at 25°C for 4 min. The noodles are placed in a wire basket fitted with a lid, and the basket is submerged in hot palm oil (177°C) for 50 s. The excess oil is drained of the fried noodles for 2 min, and the noodles are then cooled to room temperature for 2 h and stored in a plastic bag until further testing.

The formulation of instant noodles can be varied and complex. A study in our laboratory (Chen, 1997) on 24 packaged instant products available in the Hong Kong retail market found that flour, oil, salt, and alkali (usually potassium carbonate) were the most common listed ingredients. Others included sodium carbonate, guar gum, carboxymethyl cellulose (CMC), antioxidants, food colorants, egg, potato starch, corn starch, and emulsifiers. Recently, calcium carbonate has become more widely used. One article (Anon., 1995) additionally noted the use of xanthan and alginic acid, phosphates, and sorbitol. The pH of a ground noodle slurry ranged from 5.9 to 8.6, indicating considerable variation in the amount of alkali added. Oil content ranged from 15.0% to 21.5% in fried-type and 1.5–1.8% in two dried-type products. We note that, in order for a laboratory-scale method to mimic these levels of oil content, a typical kitchen-type deep fryer is not suitable, with many types giving far higher oil levels. We find the Rotofryer® (DeLonghi, Italy) to be the most suitable because it has a rotating and angled base in a frying unit, which serves to prevent excess oil uptake. The use of guar gum, CMC, and potato starch is intended to have texture-improving effects on the product. There is now considerable interest and marketing of various alginates (sodium alginate, propylene glycol alginate) by companies such as Nutra Sweet Kelco, a unit of Monsanto (USA) and Pronova (Norway). The alginates have distinct effects on texture and some apparent effects on oil uptake. Oil content is a major cost of the production issue (oil is more expensive than flour or water) and a nutritional concern. Because of the varied formulation of instant noodles (especially the wide range of pH), additive developers should take care to study the customer's product carefully before making recommendations, because some of the materials may respond very differently depending on conditions.

The increased interest in additives to optimize the quality of instant noodles reflects the harshly competitive, and declining, market for commonly packaged *ramen.* However, there is still a healthy and growing market for premium types such as cup noodles.

8. BUCKWHEAT NOODLES, OR *SOBA*

Buckwheat noodles, or *soba,* refer to long and thin brownish noodles made from a composite of buckwheat flour and wheat flour, the ratio of which varies according to the type of the product. They have a unique flavor and texture and are served plain or in soup. The English-language literature on *soba* has been greatly enhanced by the publication of *The Book of Soba* by James Udesky (1988), and we have drawn heavily on that source in the discussion below. *Soba* cuisine has been enjoyed in Japan for more than 400 years because of its light, sweet taste and is considered a perfect food for any season. Quite different from spaghetti, *ramen, udon,* and other wheat noodles, buckwheat flour noodles do not stretch easily, and the dough has to undergo a process of "beating" to form a smooth sheet. This process is known by the traditional Japanese name as *te-uchi,* meaning "beating the buckwheat dough by hand" (Udesky, 1988).

There are many different variations to the process of making *soba,* which have been discussed in detail by Udesky (1988). The most common form of making buckwheat noodles is by combining 60–70 parts buckwheat to 30–40 parts wheat flour, to which 45–48 parts water is added, depending on the protein content of the mixture. The protein content of the wheat flour is generally higher than alkaline noodles (12% minimum), which compensates for the lack of gluten-forming protein in buckwheat flour and helps in binding the flour to form a dough (Crosbie et al., 1990). The flour is mixed together in a circular motion by adding the water little by little. To improve the binding capacity of the flour, boiling water could be used in the initial mixing stages (about 80% of the total water used), which penetrates buckwheat's starch components and activates the binding power of its water-soluble proteins. It is important to add 90% of the water and evenly moisten the flour within the first 30–45 s, taking care to avoid formation of any large clumps. The two most important factors responsible for making successful *soba* noodles are using fresh and well-sifted flour and then performing the initial mixing of the water into the flour in such a way to allow the buckwheat to develop its own viscosity.

The stage when the viscosity becomes apparent is known as the "blossoming" of the dough, and the dough crumbs appear firm, compact, and smooth.

The dough crumbs are then kneaded together, using both the hands, until a firm, smooth, and polished ball of dough is formed, with as few air pockets as possible. The large dough ball is then broken into five or six small portions, covered with a damp cloth, and allowed to develop for 30 min. Each dough portion is then flattened out by using a rolling pin into an oval-shaped sheet with a thickness of approximately 3 mm. The sheet is then sprinkled with buckwheat flour, wrapped with a plastic sheet, and allowed to develop further. This process is repeated with the remaining dough balls, and each new sheet is placed on top of the previous one, with flour sprinkled in between them to prevent them from sticking together. The stacked sheets are then cut into half, folded, and cut into fine strips of noodles.

The *soba* noodles are then dropped into boiling water and stirred gently to prevent them from clumping together. As the starch begins to gelatinize, the cooking water becomes milky and starts to foam. The noodle strands are occasionally lifted from the boiling water and tested for firmness. Once the noodles are done, they are strained and dropped into a bowl of iced water to firm them up and then are stirred briskly to "wash" the starch up. Finally, the noodles are drained thoroughly and served *al dente,* that is, tender, but firm and chewy throughout.

Buckwheat noodles are also widely produced in China. Buckwheat *helao* (Figure 3.8) is a famous local food with more than 300 years of history (Zhang, 1993). Buckwheat flour (500 g) is kneaded to dough with 300 ml water and a little limewash. The water is added in thirds, first kneading the flour, then stirring by hand to form a stiff dough, and then kneading to a soft, smooth dough. The limewash makes the dough turn green. The kneaded buckwheat dough is placed into a *helao* machine to press the dough into a rough noodle shape, and the strands are dropped directly into boiling water to cook. The noodle is eaten with a sauce containing vinegar, salt, chopped onion, pepper powder, gourmet powder and sesame oil. Buckwheat *helao* is light green and is considered to be easily digestible (Zhang, 1993; Lin et al., 1998).

Buckwheat cat's-ear noodles are a traditional food in Shanxi Province (Zhang, 1993). Buckwheat flour (500 g) is kneaded with about 275–300 ml water, forming a dough. The dough is rolled into rounded finger-shaped pieces and coated with flour to prevent them from sticking to each other. They are cut into smaller pieces the size of a fingertip and then placed on the left hand and rubbed with the right thumb to a thin cat's-ear-shaped piece. These pieces are cooked in boiling water. The buckwheat cat's-ear noodles are eaten with a vinegar-based sauce and have an attractive appearance, delicious taste, and smooth mouth feel (Zhang, 1993; Lin et al., 1998).

Figure 3.8 A Chinese type of buckwheat noodles in Shanxi Province. The size and thickness of the strands and the presentation differ from the better known Japanese *soba* buckwheat noodles.

Also in China, dried noodles made of wheat–buckwheat composite flour and even steamed-and-fried instant noodles containing buckwheat are available in the market. We have given details on several variants of buckwheat noodles for several reasons. First, *soba* is a major product category in its own right, and second, the quality of buckwheat noodles depends on factors related to both buckwheat grain quality and to wheat quality where it is present. Relatively little is known about the extent of genetic variation in buckwheat starch, protein, and nonstarch polysaccharide quality that would relate to optimal use in particular types of noodle. A report on starch quality, indicating relatively slight variation and giving a review of prior work, was recently published (Li et al., 1997). Wheat quality for buckwheat noodles should logically emphasize a relatively high protein content and quality in terms of gluten strength. The importance of buckwheat noodles also lies in the fact that they are a major type of composite flour, and lessons can be drawn for technological processing, formulation, and raw material quality for development of new noodle products using wheat mixed with other starch sources. Again, such products are very widespread, both as traditional products and as new products. Almost any grain source can be used to give a unique texture, taste, and nutritional profile to predominantly wheat noodles (e.g., maize, sorghum, millet, *Amaranthus,* and sweetpotato).

9. ADDITIONAL NOODLE TYPES

The following types are generally different forms of the basic types of noodles (white-salted and yellow-alkaline) in the market discussed previously.

9.1 FRESH OR RAW NOODLES (CANTONESE-STYLE)

Fresh or raw (uncooked) noodles are sold without any further processing and contain about 35% moisture. They usually have a thin cross section, which allows them to cook rapidly, and are made from relatively strong flour that imparts a chewy texture when cooked. The most important quality characteristics of these noodles are good color and smooth surface. For good consumer acceptance they must be white or light yellow, depending on preferences in different regions. They have a limited shelf life and are very prone to enzymatic darkening because of the activity of enzyme polyphenoloxidase, unless stored under refrigeration. Only about 2% of white-salted noodles are sold fresh, while a popular form of alkaline noodles (15.5%) is sold in Japan (Crosbie et al., 1990). In Shanghai fresh noodles are the most popular.

9.2 DRIED NOODLES

Dried noodles are raw noodles that have been hung to dry, either in the sun or in a special controlled temperature and humidity drying chamber, to about 8–10% moisture. The dough must have sufficient physical strength to support the noodles as they dry. The noodles should have a uniform shape and cleanly cut sides. Since dull, gray, or brown noodles are considered inferior, highly purified flours are generally used to impart the desired color and opacity to the noodles (Hoseney, 1994). They have an extended shelf life and retain their bright color when stored. Most Japanese-style noodles are sold in this form.

Korean-style dried noodles are another form of popular Asian noodles made from wheat flour, salt and water. A study of the terms describing the textural characteristics of Korean noodle products indicated that cohesiveness, elasticity, and hardness are the most important textural parameters. When cooked, they should have a bright, white appearance, free from any discoloration or darkening, and a soft, elastic eating quality with a smooth surface and clean bite (Lee and Lee, 1985). Flours containing about 10% protein are found to be most suitable, because protein concentration influences many aspects of noodle quality (Lee et al., 1987). Overall, dried white-salted noodles are the most popular noodles in China (the biggest consumer of noodles in the world). The key step in dried noodle production is the controlled drying process, essential for optimum quality. Much of the technology for pasta drying (see Pollini, 1996, and references therein) appears applicable to noodle production.

9.3 BOILED OR WET NOODLES (HOKKIEN-STYLE)

Boiled noodles are essentially fresh noodles that have been partially boiled for 1–2 min until there is only a fine core of dough in the center, surrounded by cooked or gelatinized dough. It is then reboiled prior to consumption, during which time the cooked zone extends to the center (Moss, 1982). They are popular in China, Japan, Singapore, Malaysia, and Indonesia. Hokkien noodles are alkaline noodles that are essentially sold as boiled noodles. After boiling, the noodles are withdrawn from the boiling water, rinsed in cool water, allowed to drain, and finally oiled and sold in plastic bags (Moss, 1984).

The boiled noodles have a moisture content of about 52% and thus have a relatively short shelf life. They do not discolor upon storage because the boiling denatures the polyphenoloxidase enzyme, but they have to be protected from airborne contaminants because their surface provides an ideal substrate for bacterial and moldy growth. Their shelf life can be extended by hygienic manufacturing practices and refrigeration or freezing of the products. In Japan, the shelf life of these products has been increased to several months with the use of retort pouch packaging and partial drying. These noodles are often sold packed in a heatproof cup or bowl with various flavorings and toppings added to it (Miskelly, 1993).

9.4 FROZEN BOILED NOODLES

Frozen boiled noodles are prepared by rapidly freezing the noodles after boiling them. Boiled noodles are washed with water, immersed in cold water below 5°C, and rapidly frozen to a temperature of −30°C. The freezing step preserves the fresh taste of boiled noodles for a long time period and has been applied to *udon* and *ramen.* This type of noodle is mostly sold from a central factory to noodle restaurants, where it is thawed in a specially designed boiling pot and immediately served to guests, thus saving time and labor cost (Nagao, 1996).

10. SUMMARY AND FUTURE DIRECTIONS

Some clues to the future may be gleaned by studying the recent past. The patent literature indicates ongoing work on packaging methods, particularly for the maintenance of freshness and texture in long shelf life "fresh"-type noodles. Ito et al. (1996) described a process for making alkaline double-layered noodles, one layer containing alginic acid or alginate, with steaming, partial drying, and acid treatment steps before thermally sterilized packaging. Other recent developments include use of other starch sources, processed resistant starches, other texturizing ingredients, and processes to reduce discoloration.

There are many problems inherent in studying noodles: product and consumer diversity, the dual demands of tradition and innovation, and the lack of unambiguous quality parameters. The best way to proceed is to understand something of the range of noodle types presently made, to understand something of the starch and protein components of the primary raw material (wheat flour), and to have an open mind about manipulating formulations and processes to achieve interesting and innovative outcomes. There is an expression, "from farm gate to dinner plate," but the implied chain from farmer to miller to technologist and factory to retailer to consumer seems to leave out several key players. The farmer's success is built on the activities of the geneticist and breeder. These are perhaps the most confused at present regarding what the manufacturer really wants in terms of wheat quality. The nutritionist or medical specialist also deserves greater recognition for their potential contribution. A noodle can be a complete meal: protein, carbohydrate, and lipids in widely varying quantities and nutritional qualities. Manipulation of starch digestibility, enhancement of protein quality (amino acid balance) and quantity, reduction of oil content in instant noodles, or increased concern for lipid quality are all possible. These would perhaps represent value-added opportunities for manufacturers while offering useful progress in the healthfulness of our diet. At the practical level, for new food product development, we see several fertile areas for continued research and development work, several of which do not require sophisticated processing technologies. These are

(1) Improvement of testing methods to define instrument–sensory relationships for noodle texture and to develop more detailed sensory profiles and consumer preference data for existing products
(2) Incorporation of research experience from frozen dough technology into a wider range of noodle types
(3) Continued efforts at oil reduction in steamed-and-fried instant noodles, while maintaining the desired flavor and texture of the now-traditional product fried in palm oil
(4) Nutritional/functional food/texture enhancement of noodles by use of protein and starch additives, such as the use of *Amaranthus* or buckwheat protein isolates reported by Bejosano and Corke (1998)
(5) Continued work on the genetic improvement of wheat varieties for noodle production, focusing on starch, protein, and color properties
(6) Development of better predictive methods for the quality of raw materials, particularly wheat flour, intended for noodle production
(7) In general, continued efforts to create diverse, nutritious, palatable noodle products that will continue to be an integral part of convenient and healthy diets worldwide

Noodle scientists and technologists have many interesting years of work ahead to bring their work up to its potential.

11. ACKNOWLEDGMENTS

Special thanks to the Australian noodle group of Graham Crosbie, Di Miskelly, and Sidi Huang for all their generous advice and help over the past 5 years. We also thank all the members of our lab for their comments and input, particularly on Chinese terminology. Financial support was received from the Research Grants Council of Hong Kong.

12. REFERENCES

Anon. 1995. The use of stabilizers in instant wheat noodles. *Asia Pacific Food Industry* (Nov.):80, 82, 84.

Baik, B. K., Czuchajowska, Z., and Pomeranz, Y. 1994. Role and contribution of starch and protein contents and quality to texture profile analysis of Oriental noodles. *Cereal Chem.* 71:315–320.

Baroni, D. 1988. Manufacture of pasta products. In: *Durum Wheat: Chemistry and Technology,* G. Fabriani and C. Lintas, eds., Am. Assoc. of Cereal Chem., St. Paul, MN, pp. 191–216.

Bejosano, F. P., and Corke, H. 1998. Effect of *Amaranthus* and buckwheat proteins on wheat dough properties and noodle quality. *Cereal Chem.* 75:171–176.

Bhattacharya, M., and Corke, H. 1996. Selection of desirable starch pasting properties in wheat for use in white salted or yellow alkaline noodles. *Cereal Chem.* 73:721–728.

Chen, X. F. 1997. The effect of propylene glycol alginates on oil uptake and texture of steamed-and-fried instant noodles. Master's thesis, University of Hong Kong.

Crosbie, G. 1991. The relationship between starch swelling properties, paste viscosity and boiled noodle quality in wheat flours. *J. Cereal Sci.* 13:145–150.

Crosbie, G., Miskelly, D., and Dewan, T. 1990. Wheat quality for the Japanese flour milling and noodle industries. *W. Aust. J. Agric.* 31:83–88.

Dexter, J. E., Matsuo, R. R., and Dronzek, B. L. 1979. A scanning electron microscopy study of Japanese noodles. *Cereal Chem.* 56:202–208.

Dick, J. W., and Matsuo, R. R. 1988. Durum wheat and pasta products. In: *Wheat: Chemistry and Technology, Vol. II,* 3rd ed., Y. Pomeranz, ed., Am. Assoc. Cereal Chem., St. Paul, MN, pp. 507–547.

Giese, J. 1992. Pasta: New twists on an old product. *Food Technology* 46:118–126.

Hoseney, R. C. 1994. Pasta and noodles. In: *Principles of Cereal Science and Technology,* 2nd ed., R. C. Hoseney, ed., Am. Assoc. Cereal Chem., St Paul, MN, pp. 321–334.

Huang, S. 1996. China—The world's largest consumer of paste products. In: *Pasta and Noodle Technology,* J. E. Kruger, R. B. Matsuo, and J. W. Dick, eds., Am. Assoc. of Cereal Chem., St. Paul, MN, pp. 301–325.

Huang, S., and Morrison, W. R. 1988. Aspects of proteins in Chinese and British common (hexaploid) wheats related to quality of white and yellow Chinese noodles. *J. Cereal Sci.* 8:177–187.

Ito, Y., Hanaoka, A., and Tezuka, K. 1996. Method for manufacturing packed wet-type instant noodles. United States Patent 5,534,273.

Kim, W. S., and Seib, P. A. 1993. Apparent restriction of starch swelling in cooked noodles by lipids in some commercial wheat flours. *Cereal Chem.* 70:367–372.

Konik, C. M., Mikkelsen, L. M., Moss, R., and Gore, P. J. 1994. Relationships between physical starch properties and yellow alkaline noodle quality. *Starch/Stärke* 46:292–299.

Konik, C. M., Miskelly, D. M., and Gras, P. W. 1993. Starch swelling power, grain hardness and protein: Relationship to sensory properties of Japanese noodles. *Starch/Stärke* 4:139–144.

Lee, C. H., Gore, P. J., Lee, H. D., Yoo, B. S., and Hong, S. H. 1987. Utilisation of Australian wheat for Korean style dried noodle making. *J. Cereal Sci.* 6:283–297.

Lee, H. D., and Lee, C. H. 1985. The quality of Korean dried noodle made from Australian wheats. *Korean Journal of Food Science and Technology* 17:163–169.

Li, W., Lin, R. and Corke, H. (1997) Physico-chemical properties of common and tartary buckwheat starch. *Cereal Chem.* 74:79–82.

Lin, R., Li, W., and Corke, H. (1998) Spotlight on Shanxi Province, China—minor crops and special foods. *Cereal Foods World* 43:189–192.

Lorenz, K. J., Martin, D. J., and Stewart, B. G. 1994. Chinese-style noodles made from Bühler mill fractions of Australian wheats. *ASEAN Food Journal* 9:156–160.

Miskelly, D. M. 1984. Flour components affecting paste and noodle colour. *J. Sci. Food Agric.* 35:463–471.

Miskelly, D. M. 1988. Noodle and soft wheat quality for South East Asia. In: *Proc. 38th Aust. Cereal Chem. Conf.,* Royal Australian Chemical Institute, Melbourne, Australia, pp. 91–95.

Miskelly, D. M. 1993. Noodles—A new look at an old food. *Food Australia.* 45:496–500.

Miskelly, D. M., and Gore, P. J. 1991. The importance of Asian noodles to the Australian wheat industry. In: *Cereals International,* D.J. Martin and C.W. Wrigley, eds., Proceedings of an International Conference, Royal Australian Chemical Institute, Brisbane, Australia, pp. 271–275.

Miskelly, D. M., and Moss, H. J. 1985. Flour quality requirements for Chinese noodle manufacture. *J. Cereal Sci.* 3:379–387.

Moss, H. J. 1982. Wheat flour quality for Chinese noodle production. *Proceedings Singapore Institute of Food Science and Technology Conference,* Singapore, pp. 234–239.

Moss, H. J. 1984. Ingredient effect in mechanised noodle manufacture. *Proceedings Singapore Institute Food Science and Technology Conference,* Singapore, pp. 71–75.

Moss, H.J., Miskelly, D. M., and Moss, R. 1986. The effect of alkaline conditions on the properties of wheat flour dough and Cantonese-style noodles. *J. Cereal Sci.* 4:261–268.

Moss, R., Gore, P. J., and Murray, I. C. 1987. The influence of ingredients and processing variables on the quality and microstructure of Hokkien, Cantonese and instant noodles. *Food Microstructure* 6:63–74.

Nagao, S. 1979. Wheat—Production and consumption trends. *Cereal Foods World* 24:593–595.

Nagao, S. 1981. Soft wheat uses in the Orient. In: *Soft Wheat: Production, Breeding, Milling and Uses,* W. T. Yamazaki, and C. T. Greenwood, ed., Am. Assoc. of Cereal Chem., St. Paul, MN, pp. 267–304.

Nagao, S. 1991. Noodles and pasta in Japan. In: *Cereals International,* D. J. Martin and C. W. Wrigley, eds., Royal Aust. Chem. Inst., Brisbane, Australia, pp. 22–25.

Nagao, S. 1995. Wheat usage in East Asia. In: *Wheat End Uses Around the World,* H. Faridi and J. M. Faubion, eds., Am. Assoc. of Cereal Chem., St. Paul, MN, pp. 167–189.

Nagao, S. 1996. Processing technology of noodle products in Japan. In: *Pasta and Noodle Technology,* J. E. Kruger, R. B. Matsuo, and J. W. Dick, eds., Am. Assoc. of Cereal Chem., St. Paul, MN, pp. 169–194.

Nagao, S., Ishibashi, S., Imai, S., Sato, T., Kanbe, T., Kaneko, Y., and Otsubo, H. 1977. Quality characteristics of soft wheats and their utilization in Japan. III. Effects of crop year and protein content on product quality. *Cereal Chem.* 54:300–306.

Oda, M., Yasuda, Y., Okazaki, S., Yamauchi, Y., and Yokoyama, Y. 1980. A method of flour quality assessment for Japanese noodles. *Cereal Chem.* 57:253–254.

Oh, N. H., Seib, P. A., Deyoe, C. W., and Ward, A. B. 1983. Noodles. I. Measuring the textural characteristics of cooked noodles. *Cereal Chem.* 60:433–438.

Oh, N. H., Seib, P. A., Ward, A. B., and Deyoe, C. W. 1985. Noodles. IV. Influence of flour protein, extraction rate, particle size, and starch damage on the quality characteristics of dry noodles. *Cereal Chem.* 62:441–446.

Pollini, C. M. 1996. THT technology in the modern industrial pasta drying process. In: *Pasta and Noodle Technology,* J. E. Kruger, R. B. Matsuo, and J. W. Dick, eds., Am. Assoc. of Cereal Chem., St. Paul, MN, pp. 59–74.

Shelke, K., Dick, J. W., Holm, Y. F., and Loo, K. S. 1990. Chinese wet noodle formulation: A response surface methodology study. *Cereal Chem.* 67:338–342.

Shou, T. H. 1994. In: *Chinese Dimsum Recipes,* 6th ed., Hilit Publishing Co. Ltd., pp. 11–23.

Sung, Y. S., and Sung, K. K. 1993. Cooking properties of dry noodles prepared from HRW-WW and HRW-ASW wheat flour blends. *Korean Journal of Food Science and Technology* 25:232–237.

Zhang, B. S., ed. 1993. *Shanxi Traditional Foods,* B. S. Zhang, ed., China Light Industry Publications, Beijing, China.

Toyokawa, H., Rubenthaler, G. L., Powers, J. R., and Schanus, E. G. 1989a. Japanese noodle qualities. I. Flour components. *Cereal Chem.* 66:382–386.

Toyokawa, H., Rubenthaler, G. L., Powers, J. R., and Schanus, E. G. 1989b. Japanese noodle qualities. II. Starch components. *Cereal Chem.* 66:387–391.

Udesky, J. 1988. *The Book of Soba.* Kodansha International Ltd., Tokyo.

Wang, C. X., and Ya, F. 1988. *Flour products in Shanxi Province.* Shanxi Science and Education Publishing House, Taiyuan (in Chinese).

Wang, L., and Seib, P. A. 1996. Australian salt-noodle flours and their starches compared to U.S. wheat flours and their starches. *Cereal Chem.* 73:167–175.

Wang, X. 1987. *One Hundred Varieties of Noodles.* Guangzhou Branch, Publishing House of the Spread of Scientific Knowledge, Guangzhou (in Chinese).

Wehrfritz, G. 1995. Grain drain. *Newsweek* 125(20):8–14.

Yu, Y. S. 1979. Han. In: *Food in Chinese Culture,* K. C. Chang, ed., Yale University Press Ltd., London, pp. 52–83.

CHAPTER 4

Wheat Products: 2. Breads, Cakes, Cookies, Pastries, and Dumplings

SIDI HUANG

1. INTRODUCTION

IN Chapter 3, various types of noodles have been described. In this chapter other popular wheaten foods, including breads, cakes, cookies, pastries, and dumplings, will be discussed. Wheaten foods have had a very important role in the diet and culture of Asian countries since very early times. Today, steamed bread and noodles are very common foods in China and the East and Southeast Asian regions. Over 1.3 billion people consume steamed bread and buns regularly. Dumplings are also a traditional food and are becoming more popular as the general living standard improves. The commercial production of frozen steamed bread, buns, and dumplings, creating more convenience for consumers, has raised their popularity even further.

The inscriptions on bones and tortoise shells dating from the Shang Dynasty (1751–1122 B.C.) indicate that wheat was already widely grown throughout the province of Henan (Fan, 1982). People used stone mortars to grind wheat into flour and made wheaten food by hand. *Bing* was the common name of cooked wheaten foods in ancient times (Wang, 1984). There was further development of wheaten foods during the Han Dynasty (206 B.C. to 220 A.D.). The writer Liu Shi reported on *shou mian* (i.e. fermented dough) in his work *Shi Ming*. This indicated that, at that time, dough fermentation technology was already in use (Huang and Hao, 1994). It was said that, during the "Three Kingdoms" (221–263 A.D.), a well-known statesman and general Geliang Zhu reformed the tribe's dedication ceremony by making steamed stuffed meat buns to replace a human head. Since then, steamed stuffed buns have not only saved many lives, but have also become a popular food in China. Later, these products were introduced to Japan,

Korea, and Southeast Asian countries (Huang and Hao, 1994). Steamed bread and buns have evolved continuously throughout Chinese history so that today there are many styles (or varieties) of steamed bread, buns, and rolls.

2. BREADS

Western-style pan bread was introduced to Asia several centuries ago. A large increase in consumption of this style of bread has been achieved only during the last 20–30 years. However, the traditional wheaten foods, steamed bread, buns, and rolls, are still more popular.

2.1 STEAMED BREAD *(MANTOU)*

2.1.1 Introduction

Steamed bread is a leavened wheat flour product, which is cooked by steaming in a steamer. The most common types of steamed breads, weighing about 130–150 g, are either round or roughly cylindrical in shape, are white in color, and have a smooth, shiny surface devoid of a crust. The crumb texture varies from dense to open, and the flavor varies to suit local tastes (Huang and Miskelly, 1991). Like European bread, one piece of dough can be made into different forms of products such as rolls, buns, and pan bread. The same piece of dough can be used to make different forms of steamed products such as steamed bread, steamed buns, and steamed twisted rolls. Steamed products can be made

Figure 4.1 Steamed bread *(mantou)* of different shapes and sizes.

Figure 4.2 Steamed buns *(bao* or *baozi)* with different fillings.

with or without fillings. The products without filling are called steamed bread, or *mantou* (Figure 4.1), and with fillings are called steamed buns *(baozi* or *bao)* (Figure 4.2). The common fillings for steamed buns are meats, vegetables, and sweet red bean or lotus bean pastes. Other forms of steamed products include twisted rolls with various shapes *(huajuan)* (Figure 4.3).

Figure 4.3 Steamed rolls *(huajuan).*

TABLE. 4.1 The Main Differences of Three Styles of Steamed Bread.

	Northern	Southern	Guangdong
Fat (%)	0	0	up to 10%
Sugar (%)	0	0	up to 25%
Specific volume	~2.5	~3.0	~2.8–3.2
Structure	dense	open	open
Eating quality	elastic, very cohesive	soft, elastic, and cohesive	very soft and elastic but not cohesive
Popular area	northern China	all China	very southern China and southeast Asia

There are three main styles of steamed bread in China and East and Southeast Asian countries: northern, southern, and Guangdong styles (Huang et al., 1995). The differences in quality preference, ingredients, and consumption regions among the three styles of steamed bread are listed in Table 4.1. The northern style (so-called *changmian mantou*), preferred in northern China, has a very cohesive and elastic eating quality, a higher arch domed shape and dense structure. The southern style (so-called *xiaomian mantou,* derived in southern China), now widely popular throughout China, has a soft, elastic, and medium cohesive eating quality, a lower arch domed shape and open structure. Guangdong style, which is popular in the very southern part of China (i.e., Guangdong, Fujian, Hainan, Taiwan, and Hong Kong) and East and Southeast Asian countries, has an open structure, a sweet taste, and a very soft and elastic, but not a cohesive, eating quality. Steamed buns are the most popular product for this style of steamed bread. People usually consume this style of steamed bread as a snack.

Steamed bread is a staple food in the wheat-growing area of northern China, representing approximately 70% of flour produced in this region. In contrast, a lesser proportion is used in the south, where rice and noodles are more popular (Huang and Miskelly, 1991). People in the south often consume steamed bread for breakfast. Steamed buns with a variety of fillings are popular throughout China. Steamed bread dough for northern- and southern-style steamed bread is made of flour, water, and yeast (1.5% fresh yeast or about 1% of dry yeast). In Guangdong-style steamed bread, up to 25% sugar, 10% fat, and 1.2% salt are added.

2.1.2 Steamed Bread Preparation

The sponge and dough process is a traditional procedure for steamed bread making, but the "no time" dough process is increasingly used in many commercial steamed bread factories. The traditional sponge and dough procedure includes mixing of the dough, fermentation, neutralization, remixing and molding, proofing, and steaming (Huang and Miskelly, 1991).

2.1.2.1 Mixing of Dough

Dough is mixed in factories using mixers operating at slow speed or by hand in the home. In steamed bread making, the dough-mixing process aims at two major objectives: (a) the thorough and uniform dispersion of ingredients to form a homogeneous mixture and (b) bringing about the physical development of gluten in the dough into a uniform structure. Such a structure possesses an optimum degree of plasticity, elasticity, and viscous flow. In the traditional sponge and dough process, the first of these objectives is attained during mixing. The physical development of gluten in the dough is achieved by the mixing, remixing, and fermentation processes. These operations result in a continuous three-dimensional network of thin, hydrated protein films that enclose the starch granules and flour particles together with occluded air bubbles.

2.1.2.2 Fermentation

Proper dough fermentation is very important for the quality of steamed bread. In China most steamed bread is fermented using either freshly prepared "starter" dough or sourdough kept from the previous day. Sourdough is cheap and readily available. When starter dough or sourdough is used, the fermentation is called sourdough fermentation. In recent years, modern yeast factories have been established in China, and the use of yeast (fresh or dry) for steamed bread fermentation is increasing, particularly in medium-sized steamed bread factories.

When sourdough fermentation is used, the fermented dough is very sour because of fermentation by-products of the *Lactobacillus* spp. It needs to be neutralized (see Section 2.1.2.3).

Fully fermented starter dough is suitable for steamed bread making, whereas overfermented or old starter dough is not. Baking powder is occasionally used for steamed bread made at home. When starter dough is used, fermentation of steamed bread dough includes three steps: the preparation of starter dough, dough fermentation, and neutralization. The starter dough is prepared fresh by dispersing the sourdough remaining at the end of the day into water and adding additional flour and mixing well. The dough is then usually fermented overnight for use on the following day.

There are four types of fermentation processes that are used for different types of steamed bread.

(1) *Full fermentation* (1–3 h, depending on the season) is used to make dough for steamed bread of southern style, steamed rolls, and steamed buns *(da bao).*

(2) *Partial fermentation* (0.5–1.5 h) is used to make dough suitable for steamed buns with juicy fillings *(xiao bao).*

(3) *No-time fermentation* is increasingly used in commercial production of steamed bread. Flour, water, starter dough, and sodium carbonate are mixed, and the dough is then molded, proofed, and steamed. The amount of sodium carbonate added is variable (about 6 g per kilogram of flour) and depends on both the temperature in proofing cabinet and proof time.

(4) *Remixed fermentation dough* is often used to make northern-style steamed bread. After the dough is fully fermented, more flour is mixed in at a ratio of additional flour to fermented dough up to 40% by weight. Another use of this dough is for making a type of steamed bread with a cross cut on the top surface *(kaihua mantou)* (Huang and Miskelly, 1991).

2.1.2.3 Neutralization and Remixing

Because the full fermented dough is acidic, it needs to be neutralized with an alkali such as a 40% sodium carbonate solution. It is critical to control this stage for good quality steamed bread. If the dough is overneutralized, the steamed bread will become yellow or dark in color with a strong alkaline flavor and an objectionable, bitter taste caused by hydrolysis of protein to small peptides. Steamed bread made from underneutralized dough has a sour smell and taste, smaller volume, poor appearance, and dense structure. The pH of dough before neutralization is 3.7–4.0 and after neutralization, 6.4–6.7 (Huang and Miskelly, 1991).

Addition of alkali assists fermentation as it neutralizes acids produced by lactic acid bacteria during fermentation and allows production of carbon dioxide and water by yeast. However, it does destroy some vitamins in the dough. The amount of alkali required varies with the extent of fermentation, temperature, and ambient conditions. Experience is required both in judging the correct level of alkali and in the technique of alkali addition. The key point is how to make even and correct neutralization of dough. This is achieved by using a special kneading technique by hands or the combination of mixing and sheeting by machine.

Several sensory methods are used to check the degree of neutralization. These include the smell, taste, and structure of the cut dough section; the sound made when striking the dough with the palm of the hand; or cooking a test piece of dough (Anon., 1989). The neutralization point can also be determined by checking the pH of the dough. When the neutralization is complete, the dough may require further mixing, depending on the strength of the flour used and the style of steamed bread made.

2.1.2.4 Molding

The dough is divided into pieces, usually weighing 130–150 g. The dough can be molded into a long cylindrical piece and cut into smaller pieces or shaped into round domes, either by hand or a molding machine.

2.1.2.5 Proofing

Dough pieces are placed on a tray made of bamboo or aluminum. In home, proofing is carried out at room temperature. In factories proofing is carried in a proof cabinet, where temperature and humidity are controlled, for 30–40 min (depending on levels of sugar added) at about 40°C when the no-time dough procedure is used.

2.1.2.6 Steaming

The tray containing the dough pieces is transferred to a steamer and is steamed for about 20 min at the appropriate steaming rate. Optimum steaming has been achieved if the bread recovers to its original state after the surface of the steamed bread is gently pressed with an index finger.

2.2 STEAMED BUNS *(BAOZI)*

2.2.1 Variety of Buns

There are numerous varieties of steamed buns with differences in size, texture, shape, and filling. Despite this wide range, there are really two major types of steamed buns: *da bao* (large buns) (50 g of flour for one or two buns) and *xiao bao* (small buns) (50 g of flour for three to five pieces of steamed buns). The dough for *da bao* is fully fermented, while the dough for *xiao bao* is partially fermented, typically using half or one-third of the time required for full fermentation. The partially fermented dough is tender, yet still has the strength to hold the juicy fillings. The texture of *da bao* is open and *xiao bao* is dense. The range of fillings is diverse (sweet and savory) for *da bao,* and usually only juicy filling (adding chicken or pork broth) is used for *xiao bao* (keeping the filling in the refrigerator for 1–2 h can make it firmer for use). Meanwhile, *xiao bao* is more delicate in shape and filling, and it is served immediately after steaming in a small steamer (Huang and Hao, 1994).

There are two types of fillings for steamed buns: sweet and savory. Sweet fillings include sugar, cooked beans, red bean paste, lotus bean paste, red (or black) jujube date, chopped leaf lard mixed with sugar, and a mixture of various nuts with sugar. Other common sweet fillings include sesame seed, dried fruits, preserved mixture of Chinese wisteria flower, preserved mixture of rose flower and sugar, and taro paste.

Savory fillings include chopped vegetables, meat, or a combination of both.

(1) Vegetables
   (a) Raw vegetables: Fresh vegetables are cleaned and chopped, and extra juice is squeezed out. This is mixed with seasonings and other ingredients such as chopped and cooked bean curd or bamboo shoot or some hydrated dried vegetables such as Chinese mushroom.
   (b) Cooked vegetables: Dried and preserved vegetables such as dried day lily, dried Chinese mushroom, dried edible fungus, or the tip of bamboo shoots are mostly used in combination with starch noodles, bean curd, and some fresh vegetables. Preserved vegetables and/or dried vegetables are hydrated, chopped, and cooked with seasonings.

(2) Meat
   (a) Raw meat: Raw meat is widely used for steamed bun fillings. Pork, beef, and lamb (particularly pork) alone or combined with poultry meat and seafood such as prawn and crab meat are used as the bases for numerous varieties of fillings. There are several steps to make raw meat fillings. The first is to select tender and lean pork with a small portion of fat and then mince the meat. The second is to season the minced meat. In northern China, the typical formula is: meat (100 parts), soy sauce (20 parts), salt (2 parts), sesame oil (10 parts), and a small amount of chopped ginger and shallots. In southern China sugar is always added to the filling. The third step is to add either water or gel solution. The addition of water is a way to make the filling moist and tender. The amount of water added depends on the proportion of lean meat to fat. If the proportion is high, more water is needed. Approximately 1 kg of mince pork with 60% of lean meat requires 400 g of water. After adding seasonings, water is gradually added as the meat is stirred in one direction until the filling becomes quite sticky. The mixture should be refrigerated for 1–2 h before use. This is very popular in northern China. A very well known steamed bun called *gou bu li baozi* uses this type of filling (Anon., 1989).
   (b) Addition of gel solution: Addition of a gel solution is the way to improve the palatability and water-holding capacity of the fillings. This is widely used in southern China. The gel is made of finely chopped cooked pork skin. Broth made from ham, chicken, or dried scallops may be mixed with the gel solution. It is cooled and added to the mince-meat filling. The amount of the gel added to meat depends on how well the dough is fermented. If the dough is fully fermented, less gel should be added; if dough is partially fermented, then more gel can be added.

(c) Cooked meat fillings: There are two ways to make cooked meat fillings: (1) Chop the raw materials into small pieces and cook them with seasonings, or (2) chop cooked materials into small pieces and mix them with sauce or other seasonings. The key for the first method is to cook the raw materials separately according to their tolerance to cooking and to use a proper amount of sauce, seasonings, and thickening (starch solution). A well-known filling of this type is Chinese roasted pork filling *(chasao xian).*

(3) Mixture of meat and vegetables: This type of filling is made by mixing processed raw or cooked vegetables and meat. The most popular fillings are a mixture of chopped pork and fresh vegetables *(chai rou bao xian)* and a mixture of chopped meat and preserved vegetable *(maiganchai rou xian* and *dongchai rou xian).*

2.2.2 Steamed Bun Preparation

Transfer the fermented dough (described above) to a lightly floured work surface and knead it a few times, and then roll it by hand into a cylinder 40 cm long and 4 cm wide. Cut the cylinder into 2.5-cm pieces. Turn the pieces on their flat ends and press with the palm to flatten them. Roll each one with a rolling pin into a 7.5-cm circular-shaped piece, rotating the dough as you roll and rolling the edges thinner than the center.

Hold one round with the palm of your left hand and add about 1.5 tablespoons of filling to the center of the round. Use the thumb of the left hand to push the filling down while pinching and pleating the edge of the round with the thumb, forefinger, and the middle finger of your right hand to enclose the filling. In China and other East and Southeast Asian countries, buns with sweet filling are usually steamed upside down to identify them as sweet ones. First, pinch the edge and twist the bun to make sure it is tightly closed, and then put the closed end down. After steaming, these sweet buns may be marked with a spot of red food coloring to further identify them (Lin, 1986).

2.2.3 Proofing and Steaming

Each filled bun is placed on a 7.5-cm square of nonstick baking paper and then placed on a baking sheet with a small space between them. They are left in a warm place for about 20–30 min (depending on temperature), or until they almost double in size.

The buns (on the paper) are then transferred and placed about 2.5 cm apart on the steamer rack. The buns are steamed over high heat for 15–20 min. The best way to serve the buns is directly from the steamer; otherwise, transfer them to a plate to cool. In a plastic bag they will keep for 4–5 days in a refrigerator or for up to 2 months in a freezer. Refrigerated buns can be reheated by steaming for about 5 min and frozen buns for 7 min. Reheating by microwave oven takes about 1 min.

2.3 STEAMED ROLLS *(HUAJUAN)*

Steamed rolls *(huajuan)* are another common form of steamed product. Within this category there are a variety of shapes and flavors. Steamed rolls have the following method of preparation in common: fermented and neutralized dough is rolled out into a large thin flat shape (may be round, square, or rectangular in shape). A layer of a particular condiment is spread over this dough. It is then rolled and cut into pieces, shaped, and then steamed. In general, steamed rolls can be divided into three groups: rolled rolls, folded rolls, and stretched rolls. Details are discussed in the following sections.

2.3.1 Rolled Rolls *(Juan Huajuan)*

The fully fermented dough is neutralized and kneaded until smooth. The dough is rolled into a thin flat rectangular piece. It is brushed with vegetable oil and sprinkled with salt and flour (or fillings such as sugar, sweet bean paste, jujube date, sesame paste, chopped shallots, five-spice powder, Sichuan peppercorn powder, or other savory fillings). The dough is rolled tightly from one end to the other (or from both sides towards the middle) to form a long cylinder. The cylinder is cut crosswise into small pieces (about 35–70 g of dough) and formed into different shapes, proofed, and steamed for about 20 min. The variety of fillings and shapes makes these types of rolls very attractive.

2.3.2 Folded Rolls *(Zhedie Juan)*

A well-known variety of folded rolls is "one thousand layer" rolls. The processing procedure is as follows: The fully fermented dough is neutralized and kneaded until smooth. The dough is rolled into a 27 × 27 cm square piece and cut down the center into two equal pieces. One piece is brushed with vegetable oil and sprinkled with flour, and the other piece is placed on top. Repeating this process, the dough piece has 2, 4, 8, 16, and 32 layers after folding five times. Finally, roll the dough piece into a 4-cm-thick cylinder, proof, and steam; then the product is cut into 5-cm-thick pieces and served. Another well-known roll of this style is "Lotus leaf" roll (Anon., 1989).

2.3.3 Stretched Rolls *(Chen Huajuan)*

This type of roll is made from sheets of stretched dough. Two popular varieties are "silver thread" rolls (Figure 4.4) and "golden thread" rolls. The "silver thread" rolls are processed as follows:

(1) Mix 4 kg of flour and 1.6 kg of water with 4 kg of sourdough; allow this mixture to ferment.

(2) In a separate bowl, mix 1 kg of flour with 500 g of water into dough.

(3) Neutralize the fermented dough with alkali (usually 40% of sodium carbonate solution). Mix two-thirds of the neutralized dough with one-third of the water dough prepared in step 2. Add 1 kg of white sugar, mix well, and let rest for a while. Stretch the dough into a string of 4–5 mm in diameter, brush it with oil (a combination of lard and sesame oil), and cut the string into 7-cm-long pieces.

(4) Divide the remaining third of neutralized dough into 60-g pieces. Roll each piece into a rectangular sheet with the edge thinner than the center. Place the cut strings on the middle of the sheet. The dough sheet is folded lengthways over the strings, leaving a part of the sheet still flat on the preparation surface. The ends of the folded dough sheet are then folded towards the center, sealing the strings inside. The folded sheet is then completely rolled up.

Figure 4.4 Silver thread rolls *(yinsi juan)*.

(5) Proof the folded pieces for approximately 5 min and then steam for 20 min (Anon., 1992).

Another variation of making "silver thread" rolls is to cut the oiled and stretched string into 16-cm-long pieces. Combine 25 of the cut pieces together, hold each end of the combined strings with each hand, and twist and fold back on it to form a secure shape. Proof, steam, and serve (Figure 4.4).

2.4 FLAT BREAD

Flat bread is a baked, leavened wheat flour product with low specific volume. It can be further classified according to whether it has a single layer or two layers. Flat bread is a staple food for many people (Quail, 1996). It is particularly important for people in the Middle East region and the Indian subcontinent. The following discusses three types of flat bread popular in the Indian subcontinent and the northwest part of China.

2.4.1 Chapati

In India, Pakistan, and Bangladesh wheat is consumed mainly in the form of *chapati*—unleavened flat bread.

Traditionally, wheat is milled into whole meal flour *(atta)* on a power-driven stone mill *(chakki)*. About 95–97% extraction *atta* is used for the preparation of various whole meal products as described below.

In India, *chapati* is generally consumed at every meal of the day. The two most important quality parameters of *chapati* are softness and flexibility. Dough is made by hand mixing the flour with an optimum amount of water. Then it is let alone for 15–20 min at room temperature. The dough is divided into about 60- to 100-g pieces, and each is rounded between the palms of the hands and sheeted into a disk 2–3 mm thick using a wooden rolling pin. It is usually immediately baked on a preheated iron griddle *(tawa)* at 230°C for 1 min on each side (Sidhu et al., 1988). Because *chapati* is quite thin and lacks sugar, it is very susceptible to moisture loss and staling after baking; therefore, it is usually prepared fresh for each meal. Since consumers prefer hot, fresh *chapati,* it is usually only kept for 8–10 hours after baking.

Chapati is also known by other names, such as *phulka* and *roti.* The main variations of the *chapati* are the thickness and size. *Phulka* is about 12–15 cm in diameter and 1–2 mm in thickness (each weighs about 30–50 g). *Roti* is normally 2–3 mm thick and 20–25 cm in diameter (each dough piece weighs about 130–160 g).

The first criterion for acceptability of *chapati* from a consumer is color. A creamy yellow is the most desirable color for this product (Sidhu et al., 1988). It is affected by the wheat cultivar, flour extraction rate, and the processing treatments the flour has undergone. In urban Punjab, there is an increasing trend towards the consumption of whiter *chapati* made from wheat flour of lower (80–86%) extraction. Although the color of the product is improved considerably by extensive debranning before milling into *atta,* the textural characteristics are affected adversely. Certain fungal diseases such as Karnal bunt also affect the color of the *chapati.*

Although at present more *chapati* is prepared fresh at home for the main meals, considerable demand for factory-made products is steadily developing because of urbanization of the country. It is very likely that significant commercial production of *chapati* in India may start in the near future (Sidhu, 1995).

2.4.2 *NAAN*

In Northwest China, the Xinjian Autonomous region, the major flat bread known as *naan* is a single layered product. It is usually a circular-shaped loaf with a 15- to 20-cm diameter. The edge is approximately 3 cm thick and the center 1 cm thick. Flour is mixed with salt, water, and sourdough to form stiff bread dough. The dough is fermented before sheeting to form a thin dough piece, and then it is proofed for a short period, after which the center of the loaf is pressed flat to create a thinner center. Just prior to baking, the dough is docked with a spiked roller. It is then baked on the wall of a beehive oven. This is similar to the popular tannor-style bread common in the Middle East (Huang and Quail, 1995).

2.4.3 Chinese-Style Flat Bread *(Shaobing)*

Shaobing is a popular baked wheaten product in China and other Asian countries. It is said that *shaobing* originated earlier than steamed bread in China (Wang and Li, 1994). This hot layered bread is similar to Middle East flat bread. Different ways of cooking are used: cooking in a pan, baking in an oven, cooking in a suspended round cooking plate over hot charcoal, or baking in a large barrel-shaped oven with a hot charcoal fire inside (Anon., 1989). Slit horizontally almost all the way around, *shaobing* becomes the traditional "pocket" for *you tiao* (i.e., deep fried devils—described in Section 4.1.1), as well as for numerous other fillings. A very well known variety is *huangchao shaobing.* Following is a description of ingredients, processing, and quality characteristics:

(1) Making the roux (mixture of wheat flour and vegetable oil or lard): Add 0.5 kg of peanut oil or 0.7 kg of lard to 1 kg of flour and mix well.
(2) Fillings: There are many fillings, such as crab meat, eggs, prawn meat, ham, jujube paste, bean paste, Chinese preserved vegetables, mushroom, dried meat floss, mixed nuts, and more.

(3) Making the dough: The dough is made over 2 days. On the first day, flour and water (mixture of 70% of boiling water and 30% of cold water) but no yeast are mixed together to form a crumble. It is then kneaded and divided into several pieces and allowed to reach a warm temperature. The small dough pieces are combined into one piece and kneaded until smooth; then sourdough is added and mixed well, and the dough is covered and left to ferment. On the second day, the fermented dough is mixed with another piece of dough, which is made from a mixture of flour and cold water. The dough is neutralized and is ready for making *shaobing* (Anon., 1989).

(4) Forming the bread: The dough is kneaded and rolled into a long strip and then cut into 30-g pieces. Each piece is flattened by the palm of the hand, and a small piece of roux is placed on the center. The roux is completely sealed by the dough layer by pinching the edges of the dough together. The dough is rolled out into a rectangular sheet and is then rolled tightly from one end to the other. This coiled dough is then flattened and rolled out into a thin circle. The filling is enclosed by this circular sheet of dough, and the dough is flattened with the sealed side down. It is finally rolled into a round piece (6–7 cm in diameter). Sugar solution is brushed on the smooth side, and some baked sesame seeds are sprinkled on it.

(5) Baking: Water is spread on the other side of the dough piece. The dough piece is slapped with this watered side against the inside wall of a long barrel-shaped oven with a hot charcoal fire inside at the bottom. It puffs up in 4–5 min and then one at a time is slightly loosened with a long spatula and scooped up swiftly with the other hand.

(6) Serving and storing the bread: *Shaobing* is best eaten hot when it has just been made, but it can be covered and eaten later. Reheat it in a preheated 200°C oven for about 5 min. *Huangchao shaobing* has a multilayered structure and a crisp and delicious taste (Anon., 1989).

2.5 FLOUR QUALITY REQUIREMENTS

As mentioned above, there are three main styles of steamed bread in China and East and Southeast Asian countries. They have different formulas and different quality characteristics. Therefore, separate investigations were carried out on flour quality requirements for the three styles of steamed bread.

2.5.1 Four Quality Requirements for Northern-Style Steamed Bread

In a study of forty-three Australian and six Chinese wheat flours, significant correlation was found between flour quality factors and steamed bread quality (Huang and Quail, 1996). Protein quality and quantity and starch quality were important factors determining steamed bread volume. Dough strength was the

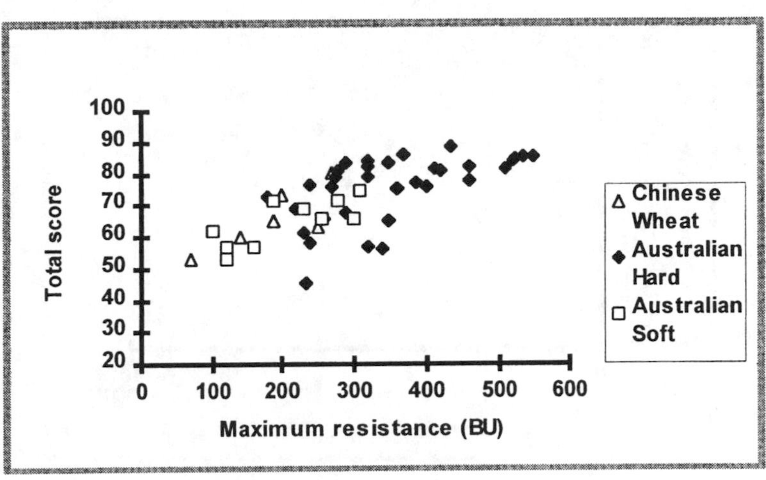

Figure 4.5 Total score vs. maximum resistance, $r = 0.7, p < 0.001$.

major determinant of overall steamed bread quality (Figure 4.5). Recent work (Zhu et al., 1997) indicated that medium to strong flours with high protein content were suitable for this style of steamed bread.

Rapid Visco Analyzer (RVA) (Newport Scientific, Australia) viscosity parameters were also found to be significantly correlated with steamed bread quality. Generally, flour samples showed better correlation with steamed bread quality than starches extracted from the flours. Of the flour RVA viscosity parameters, peak viscosity showed the best correlation with steamed bread quality, with a higher viscosity associated with higher bread scores. However, no one test clearly defined the flour quality requirements. Hard wheat flour was found superior to soft wheat flour for this product. Medium to strong dough strength is recommended for the production of good quality steamed bread.

2.5.2 Flour Quality Requirements for Southern-Style Steamed Bread, Buns, and Rolls

Fifty Australian and seven Chinese wheat flours were used in a study of flour quality requirements for southern-style Chinese steamed bread, buns, and rolls (Huang and Quail, 1996). Again, both protein quantity and quality were identified as major determinants for a specific volume of steamed bread. Protein content had a more significant effect on the specific volume of this style of steamed bread than for northern-style steamed bread. Dough strength plays an important role in determining the overall quality of steamed bread (Figure 4.6), but to a lesser extent than for northern style. There was significant

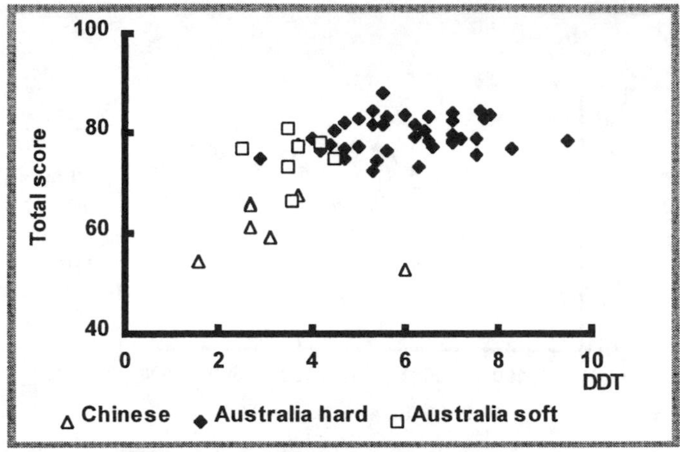

Figure 4.6 Total score (TS) vs. dough development time (DDT), $r = 0.57$, $p < 0.001$.

negative correlation between external smoothness of steamed bread and propein content and between smoothness and flour dough strength.

Rain damage has a significant effect on the quality of steamed bread. Flour from slightly damaged wheat (falling number = 230 s; falling number is a simple viscosity measurement to check if wheat has been subjected to weather-damage) resulted in an unacceptable quality of steamed bread only in surface smoothness. However, in the case of severe rain damage (falling number = 62 s), the resulting steamed bread had a very dull and rough surface, with big blisters and some brown spots, a flat shape, and a very sticky and inelastic eating quality.

Flour with medium protein content and dough strength, a medium to high falling number, and a low ash content was recommended for this product. No significant correlation was found between flour RVA viscosity parameters and steamed bread quality. No single test can clearly define the flour quality requirements for southern-style steamed breads. Hard wheat was found superior to soft wheat for this product.

The flour quality requirements for steamed buns and rolls are the same as for southern-style steamed bread.

2.5.3 For Guangdong-Style Steamed Bread, Buns, and Rolls

The majority of Guangdong-style steamed bread is prepared with fillings. Preferences in sweetness, cohesiveness, and structure range throughout different regions and countries. There are few reports about flour quality requirements for this style of steamed bread. Flour quality specifications from several modern mills in Hong Kong, Shenzhen, and Guangzhou indicate that flours with a protein content of 7.5–9.0%, wet gluten of 19–22%, and ash of 0.45–0.55% is used for this product. However, some steamed bread manufacturers from Taiwan use flour with a protein content up to 12.8% and wet gluten up to 34%. Although there is no agreeable explanation for such a large difference in flour quality requirement, it may have resulted from different recipes used in different regions. There is a need to investigate these differences and the flour quality requirements for this style of steamed bread.

2.6 INNOVATION

Over the past 2 decades, the rapidly growing economies in China and East and Southeast Asian countries have led to an improvement of living standards. The demand for convenience and quality steamed products is increasing. Many innovative products have been developed, particularly among those distributed to supermarket chains. Sold both fresh and frozen, an enormous variety of types is available. For example, layered steamed breads with chocolate, pandan, or taro colorings have been widely marketed. Whole meal steamed bread has also recently appeared in markets. In addition, there are some new types of steamed breads made from mixtures of wheat flour with other flours such as buckwheat, millet, sorghum, black rice, or maize flour. These new products are marketed as health foods and are sold in grain stores in northern China.

There is increasing production of steamed bread, buns, and rolls in factories equipped with modern machines. An automatic production line for steamed bread was established in China in the 1980s (Huang and Hao, 1994). There are several manufacturers who produce modern steamed bread and bun machines, such as Rheon (Japan) and Yang Zheng (Taiwan).

2.6.1 Frozen Products—Convenience Food

Steamed bread, buns, and rolls are traditionally made at home; however, as living standards improve and people have less time, commercially manufactured products are a fast-growing market. The freezing, chilling, and microwave technologies have made these traditional foods adapt easily into modern lifestyles. The freezing process greatly extends the shelf life of these products. Adding to this convenience, frozen products can be reheated directly in a microwave in less than a minute. Now people can have hot, delicious steamed buns for breakfast almost as simply as having corn flakes.

There are still many new challenges facing the manufacturer to improve the quality of steamed products. For example, what is the optimum procedure for freezing and microwave reheating processes? More research is needed to resolve some adverse effects of freezing and microwave reheating processes on the quality of steamed products.

2.6.2 Current and Potential Market Status

Fresh and frozen steamed buns are very popular in Japan, Korea, China, and Southeast Asian countries. Currently in China, people in the city still buy more unpacked fresh steamed bread, buns, and rolls than the packed and chilled or frozen ones simply because the packed ones are more expensive. However, as the living standard continues to improve, the consumption of packed chilled or frozen steamed bread, buns, and rolls will continue to increase steadily.

3. CAKES

Baked and steamed sponge cakes are popular in Asian countries, and traditional unleavened pancakes are also popular. The following is a review of some important types of sponge cakes and pancakes.

3.1 JAPANESE CAKE—CASTILLA

The Portuguese introduced *castilla* to Japan in the 16th century, and modifications to the original formula, baking process, and flavor have led to what is known as Japanese cake (Nagao, 1995).

Ingredients for *castilla* include flour (100 units), eggs (200–210 units), sugar (180–210 units), starch syrup (15–30 units), water (10–20 units), and sweet sake. The whole eggs are beaten and sugar is added; the mixture is blended at low speed. Then all the other ingredients are gently added to it. The batter is then poured into a wooden frame placed on a steel plate covered with a sheet of kraft paper, and the frame is covered with a second heat-intercepting plate. The initial baking is carried out for 25 min at 180°C; then the temperature is reduced to 150°C, and it is baked for a further 30 to 35 min (Nagao, 1995).

Castilla has a higher proportion of sugar, egg and honey than sponge cake, which is essential to maintain the soft, moist, and masticator characteristics of this product. Flour quality requirements for *castilla* are similar to that for sponge cake (Nagao, 1995).

3.2 STEAMED CAKES

There are many varieties of steamed cakes, differing in shapes and minor ingredients, but the main ingredients are same.

3.2.1 Thousand Layered Cake

Flour is mixed with water and sourdough and is left for full fermentation. The risen dough is removed, neutralized, and kneaded until smooth. The dough is rolled into a long roll and cut into smaller pieces. Each piece is rolled into a rectangle; a mixture of flaked coconut, sugar, and melted butter (filling) is spread in the middle of the rectangle. One-third of the rectangle sheet is folded over the center and the surface is spread with filling; then the remaining side is folded over the filling, and the strip is turned to horizontal. This process is repeated three times to create a layered square. The square is placed on a piece of nonstick baking paper and shredded papaya is sprinkled. It is allowed to rise for 30 min and then placed in a steamer and steamed for 30 min over high heat. After steaming, it is cut into slices and served (Huang, 1974; Anon., 1989). This type of cake has a soft and multilayered structure and a sweet taste.

3.2.2 *Fonggao*

Flour is mixed with sourdough and water (100:20:60) and fully fermented. The dough is neutralized, and then castor sugar (50) and some baking powder are added and then it is fermented again. The fermented dough is transferred into a steamer lined with nonstick baking paper, and some chopped dried fruit and papaya threads are spread on top. It is then steamed for 20 min over high heat. The product is cut into pieces and served (Wang and Ya, 1988).

3.2.3 Steamed Sponge Cake

A tube pan is greased or lined with nonstick baking paper. Eggs are beaten with brown sugar until thick and lemon-colored. Then evaporated milk, vanilla extract, baking soda, and some melted shortening are added, and the mixture is then beaten for 1 min. Flour and baking powder are sifted and folded into the egg mixture. The mixture is poured into a lined steamer and steamed for 30 min over high heat. It is then removed and cut into slices and served (Huang, 1974).

3.3 PANCAKES AND PAN-FRIED CAKES

These are unleavened wheat flour products and baked in a pan (without adding oils in the pan) or pan fried (adding cooking oil to the pan). There are six main types of pancakes with many more varieties. The main technical points for making these products are mixing the dough to a proper extent that the product requires, baking the pancake at a proper temperature, and turning over the pancake at the proper time. The following introduces the methods for preparing the six main types of pancakes.

3.3.1 *Da Bing*

Mix flour (100 portions) with water (50 portions) and a little salt; knead it until soft and smooth and let the dough sit for 1 h. Divide the dough into 800-g (or 500 g and 250-g) pieces and roll each dough piece into round thin pieces. Brush the surface of the dough piece with vegetable oil and roll up into a cylinder; then fold the cylinder, roll it into a round shape, and brush it with vegetable oil again. Bake the round dough piece in a hot pan; when the bottom side is browned, turn it over and bake the other side until both sides have a golden brown color. It is then ready to serve.

Da bing is a layered product with a cohesive eating quality. It can also be cut into pieces and stir-fried or cooked with vegetables and meat (Anon., 1989).

3.3.2 *Gu Bing*

Mix flour, sourdough, and water (100:75:50) into dough and neutralize the dough with an alkali solution. Mix additional flour (100) into the dough by continuously kneading the dough until uniform and smooth. Sheet the dough between the rollers of a noodle machine repeatedly until very smooth. Then roll into a 5-cm-thick round pieces and bake in a flat pan with moderate heat for 10 min; it is then docked using chopsticks, turned over, and baked for 20 min. Turn it over again and bake it for another 15 min.

Gu bing has a very cohesive eating quality and a sweet taste (Anon., 1989).

3.3.3 *Jiachang Bing*

Mix flour with some boiling water, let it cool slightly and add some cold water, and mix to a smooth dough. Roll the dough into a cylinder and cut into 100-g pieces. Roll each piece into a rectangle, and brush the surface with cooking oil evenly and roll up the piece; flatten tightly and fold the piece sideways into a round. Flatten again and roll each piece into an 18-cm circle. Lay one cake flat in a hot pan (with a small amount of oil). Cover and cook for 1.5 min over low heat; flip over cake and cook another 1.5 min; remove and cut into small slices (Figure 4.7), and serve with stir-fried meats and vegetables.

Jiachang bing can be made into many varieties by adding some special ingredients such as chopped shallots; sesame paste; and mixture of sugar, sesame oil, and flour or others. (Anon., 1989).

3.3.4 *Bo Bing* ("Lotus Pad" Pancake)

Place flour in a mixing bowl; add some boiling water and mix until smooth, and then add cold water; mix again until smooth. Roll into a long roll and cut into 50-g pieces. Roll each dough piece into a 10-cm pancake; spread the surface lightly with vegetable oil. Place another pancake on top of the oiled one. Press together and roll again into a 30-cm paper-thin pancake. Bake both sides of the pancake in a pan. Remove and separate the pancakes; wrap into fourths and arrange on plate (Figure 4.8). These pancakes may be served with stir-fried pork and eggs *(mosiu rou)* or with other dishes. It also is served with

Figure 4.7 *Jiachang bing*—a pancake with a layered structure.

Figure 4.8 *Bo bing* (lotus pad pancake).

famous Beijing roasted duck and spring onions with smaller sized pancakes (Huang, 1974; Anon., 1989).

3.3.5 *Chinyu Bing*

This kind of pancake is made in a special way. Flour is mixed with salt, water, and vegetable oil. The dough is left alone for a time and then stretched into a roll, cut into pieces, and brushed with vegetable oil. Each piece is coined, flattened, and baked on both sides on a hot oiled pan. It is crispy and tasty (Anon., 1989).

3.3.6 *Xian Bing* (Chinese Meat Pie)

Xian bings are pancakes with meat fillings. Method: Mix flour (100 portions) with warm water (60 portions); then knead it into smooth dough. Roll the dough into a long roll and cut into pieces. Flatten each piece by hand and wrap with a portion of seasoned minced meat. Bake on both sides in a pan and serve (Anon., 1989).

4. COOKIES

These products are usually prepared by deep frying a wheat flour dough, in which special chemicals or shortening were added. They are served as breakfast or a snack.

4.1 *GUZI* AND *MAHUA*

Traditionally, *you tiao* and *you bing* are called *guzi.* These products are made from dough prepared by mixing flour with water, salt, and two additives: alum and sodium carbonate. Alum makes dough harder and therefore very crisp when fried. The sodium carbonate gives the dough aeration, making it puff up. The function of salt is to increase the strength of dough and to improve its gas-holding ability (Lin, 1986).

There is a special technique to make this kind of dough. The ratio of flour to alum to sodium carbonate to sodium chloride, and to water varies with temperature. It is about 100:2.4:1.2:2.4:60 in spring and autumn; 100:3.4:1.3:3.4:60 in summer; and 100:2.2:1.1:2.2:60 in winter. Grind alum, sodium carbonate, and salt into powders separately and mix them according to the rates above; then dissolve them in water. Then mix medium-strength flour with the solution and knead it until uniform. Punch the dough from the left side to the right side; fold the dough around the side of container to the middle and punch and fold. Wait for 20–30 min and repeat this punch and fold procedure another two times (the more the better). Brush some vegetable oil on the surface of the dough; cover the dough with a clean cloth and wait for more than 2 h in summer and 3–4 h in winter. The preparation of dough is quite critical to the quality of the final product. These products are cooked by frying although the temperature used for frying varies with different products (Anon., 1989).

4.1.1 Deep-Fried Devils *(You Tiao* or *You Zha Gui)*

You tiao are deep-fried strips of dough that are puffy and crisp on the outside and soft on the inside. They are popular throughout China as a breakfast or as a snack (Figure 4.9). They are traditionally served wrapped inside baked sesame seed bread *(shaobing)* and accompanied by soymilk. The preparation is as follows.

A piece of fully rested dough (described above) is flattened and left alone for a while. Then it is cut into 16-cm-wide pieces, and each is rolled into a 7-mm-thick rectangular piece, which is cut into several 1.7-cm-wide pieces. Brush the top of two with water, and then lay the other two on top. Press a chopstick lengthways on top of a pair of strips—this makes the two stick together. Hold the ends of the two pieces and gently pull until about 30 cm long. Fry them in hot oil and immediately turn them, so that all sides become lightly browned (Lin, 1986).

Figure 4.9 *You tiao*—deep-fried devils.

4.1.2 Deep-Fried Pancakes *(You Bing)*

You bing is very popular in northern China. Preparation of dough and frying conditions are similar to *you tiao.* A piece of fully rested dough is rolled into a cylinder and cut into pieces. Each piece is rolled into a thin round or oval piece. Two parallel cuts are made down the middle of piece, which is fried in hot oil until golden brown.

There are also some other varieties, such as *tang bing* (deep-fried pancake with sweet filling in the middle), *jiao chuan* (fried dough ring), and *bo chui* (deep-fried thin pancake) (Anon., 1989).

4.1.3 Round Pretzels *(Mahua)*

There are two types of *mahua:* sweet and savory. They are very crisp and tasty. The most famous ones are made in Tianjin. The ingredients for dough preparation of *mahua* are similar to *you tiao* and *you bing,* except that no sodium chloride is added and sourdough is used. In some cases alum is not used, and other ingredients such as eggs, vegetable oil (or lard), and sugar are added (Anon., 1989). The processing procedures are as follows. Dissolve sugar, alum, and sodium carbonate in warm water, and add some vegetable oil to the solution; stir it into a suspension. Add flour to the suspension and then combine with a piece of fully fermented dough that is neutralized with alkali; knead it until even and smooth and let rest for a short while. Roll the dough into a paper-thin sheet (13 cm × 40 cm × 0.4 cm), and cut the sheet into 48 pieces. Lightly stretch each piece by the left hand while the other hand keeps the other end of the dough firmly pressed on the table; twist the dough between fingers of the left hand; bring both ends of the piece together, and twist so that the ends are connected into a circle. Deep-fry round pretzels until golden brown (Huang, 1974; Anon., 1989). *Mahua* is characterized by its crispiness, good taste, and attractive shapes.

4.2 TRADITIONAL SWEET AND SAVORY PRODUCTS IN THE INDIA SUBCONTINENT

There are many varieties of sweet and savory wheat-based snacks in the Indian subcontinent. Products like *samosa, mathi, shakarpara, gujiya, holige, modaka, kachori, sev,* and *mattar* are popular snacks. They are made by deep-frying a stiff flour dough after molding. Flour is mixed with shortening (3–5%), salt and sugar (2–3% each), and sufficient water to produce a stiff dough. The dough is then flattened to a suitable thickness, cut into desired shapes, and deep-fried at about 130–170°C. The crispness is the most desirable attribute of all these snacks for consumer acceptance. The amount of shortening added to the dough, moisture content of the dough, and frying temperature are the determinants for the crispness of the snack. Among these snacks, *samosa* and *kachori* are the most popular savory products. Both have a moist filling, usually consisting of sliced potato, peas, pulse (beans), and spices; therefore, these products have a very short shelf life (Sidhu, 1995).

5. PASTRY PRODUCTS

There are two ways that pastry products are cooked. The first is deep-frying in mildly hot oil at first and finishing under higher heat. The color of deep-fried pastry is pale and has a very flaky texture and a rich flavor. The second method is oven baking. Baked pastry is also flaky but drier and not as rich as the deep-fried pastry.

5.1 FRIED CHINESE FLAKY PASTRY PRODUCTS *(SU BING)*

Su bing is another specialty that is usually made for special occasions. There are many traditional fillings such as red bean paste, lotus bean paste, meat, Chinese roast pork, nuts, white turnips with ham, and jujube red date fillings.

Mix flour, lard (or vegetable shortening), and water (100:2:40) into a water dough (dough A), and knead until smooth. Mix flour with lard (or vegetable shortening) (100:50) to make an oil dough (dough B), and knead it until smooth. Flatten dough A to a circle shape and place dough B in the center of the dough A circle and wrap edges around dough B to completely enclose; pinch edges to seal. Roll the dough into a thin rectangular sheet and fold one-third of the dough from right to left and then the remaining one-third from left to right, placing it on top of the first third. Roll the dough into a thin rectangular sheet again, and roll the piece tightly from one side to the other into a roll of about 33 mm in diameter. Cut it into 60-g pieces and divide each piece into two half pieces; roll each half into a circle shape and put in about 15 g of red bean paste filling. Using index finger and thumb, fold outer edge into pleats. Fry them over low heat for 6 min and then over high heat until golden brown; remove, drain, and serve (Anon., 1989).

There are other varieties such as *su jiao* (shaped similar to Chinese dumplings), *su he,* and other pastries with various shapes.

5.2 BAKED FLAKY PASTRY PRODUCTS

The main varieties of baked pastry products are *pan xiang su, tao su* (Figure 4.10), *meigui su,* curry puffs (Figure 4.11) and flaky pastry with associated fillings. The ingredients and processing methods are similar to the fried flaky pastry products, but they are baked in an oven. The most well known variety is Chinese moon cake *(yue bing),* which is very popular in China and other East and Southeast Asian countries and is served at the Moon Festival of the Chinese calendar

Figure 4.10 *Tao su*—popular baked pastry.

Figure 4.11 Curry puff *(gali jiao).*

(August 15). There are two styles of moon cake: Guangdong-style moon cake (*guangshi yu bing,* Figure 4.12) and flaky moon cake *(sushi yu bing,* Figure 4.13). The Guangdong-style moon cake has a smooth crust decorated with embossed figures of rabbits, pagodas, and more, while *sushi* moon cake has a flaky skin.

A typical Guangdong (Cantonese)-style *yu bing* dough formula contains flour (100 portions), water (16 portions), white sugar (40 portions), maltose (8 portions), peanut oil (22 portions), eggs (12 portions) for glazing the surface of *yu bing,* and filling *(wuren xian)* (240 portions). One very common filling consists of a mixture of chopped almonds, peanuts, watermelon seed nuts, peanut oil, maltose, white sesame seeds, steamed wheat flour, rice flour, lard, and glacé winter melons, sugar, rose wine, and rose flower. Another common filling is lotus bean paste containing a whole salted egg yolk. Other fillings include ham, roast duck, and dates (Anon., 1992).

Dissolve white sugar in boiling water and then add the maltose. Mix the flour with the sugar solution and then with the peanut oil. Knead the dough until even and smooth. Divide the dough (which has been at rest for about an hour) into small balls and flatten them. Place the filling in the cavity and fold the edges to seal. Mold the filled pieces in a wooden molder. Brush the top lightly with a beaten egg yolk and bake for about 15 min at 190°C.

Short potent all-purpose flour milled from soft wheat or a mixture of soft and hard wheat flour (80:20) is suitable for this product. The baked product should have a smooth, fine texture that melts in the mouth (Nagao, 1995).

The processing method of flaky moon cake *(sushi yu bing)* is similar to the flaky pastry products described in Section 4.1.

5.3 FRIED SESAME BALL *(KAI KOU XIAO)*

Medium protein content, semi-hard wheat flour mixed with soft flour is used for the manufacture of fried sesame balls in Taiwan. Its ingredients include flour, sugar, lard or margarine, egg, baking powder, and water (Nagao, 1995).

Mix sugar, lard, egg, and water until thoroughly combined; sift flour and baking powder, and add to the sugar mixture. Knead to a soft dough and cut into 2 cm × 5 cm pieces. Roll each piece into a ball and dip in water. Roll in sesame seeds; deep-fry the balls over low heat until they expand and open, and then turn heat to high and deep-fry until golden brown; remove, drain and let cool, and serve.

5.4 SPRING ROLL *(CHUNJUAN)*

Spring rolls are not only popular in Asian countries, but also in other countries around the world. There are two main styles of spring rolls: Shanghai and Guangdong styles. The Shanghai-style spring roll wrapper is made from wheat flour dough that is briefly cooked on a griddle. The Guangdong-style spring

Figure 4.12 Guangdong-style moon cake *(guangshi yu bing).*

Figure 4.13 Flaky moon cake *(sushi yu bing).*

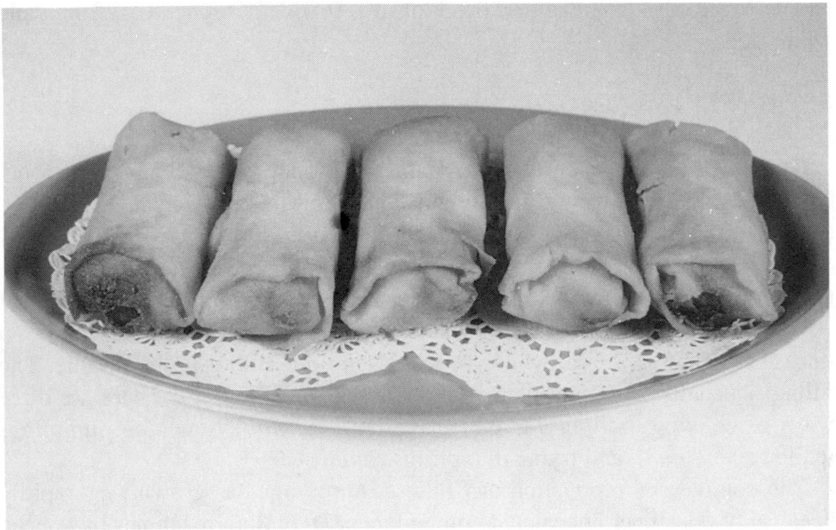

Figure 4.14 Spring rolls *(chuanjuan)*.

roll wrapper is made of wheat flour dough containing egg. Typical fillings for Shanghai-style spring rolls are cooked seasoned pork mixed with cooked vegetables such as Chinese chives, hotbed chives, spinach, bean sprouts, shepherd's purse, and bamboo shoots. The fillings for Guangdong-style spring rolls are a mixture of cooked shredded pork, vegetables, and shrimp. Shanghai-style spring rolls are thin and brittle after deep-frying (Figure 4.14); Guangdong-style spring rolls are thick and crunchy.

The wrapper for Shanghai-style spring rolls is formed when dough (very soft) is pressed in a circular motion onto a hot and ungreased surface of an electrical skillet or griddle. The temperature is important, because if the pan is too cold, a thick layer of dough will stick to it, and if the pan is too hot, the dough will not stick at all. Quickly pull the dough back; a very thin film of dough will form on the hot surface. It will start to dry at the edges in a few seconds and can then be peeled off ready for use (Lin, 1986). Machine-made commercial wrappers for both styles of spring rolls are available in Chinese grocery stores and supermarkets.

Put about two spoonfuls of filling onto the lower half of the wrapper and spread it into a 10-cm line. Then fold the edge over the filling and roll the wrapper to the center. Bring the two end flaps over the top of the enclosed filling. Brush the edge of the wrapper with beaten egg or starch solution and roll the cylinder into 12- to 13-cm-long spring rolls. Deep-fry the spring rolls in hot oil (180°C) until golden brown and serve hot with (or without) Chinese vinegar and chili sauce.

There are also some other fried pastry products such as *sachima* and *chiao guo*.

6. DUMPLINGS *(JIAOZI)*

Chinese dumplings are another main traditional wheat flour-based food. They are popular not only throughout China, but also in Japan, Korea, and Southeast Asian countries. Chinese dumplings used to be the food for special occasions only, such as festivals and for entertaining guests; however, they are becoming more generally consumed as living standards continue to improve.

Chinese dumplings are usually made from wheat flour dough sheeted to a thickness of 1.5–2.0 mm, which wraps a delicious filling in the middle. The filling is usually fine mincemeat, vegetables, and seasonings. There are three ways of cooking: boiling dumplings (called *shui jiao*), steaming dumplings (called *zheng jiao*), and frying dumplings *(guotie).*

The commercial production of Chinese dumplings has grown very rapidly because of consumer appeal and convenience. Dumpling machines have been widely adopted for commercial production, although they are still often made at home by hand.

Dough prepared for dumplings should be white, bright, transparent, and smooth and have cooking resistance (does not break up during boiling). These features are largely dependent on the flour quality used for production.

6.1 BOILED DUMPLINGS *(SHUI JIAO)*

Boiling is the most common way of cooking dumplings. Many people learn how to make dumplings in their childhood in northern China. Flour is mixed with water (about 40%) to get a mixture that just holds together, which is similar to noodle dough. Knead the dough until even and smooth and let it sit at rest for 30 min. Roll the dough into a cylinder and cut into 15-g pieces. Roll each piece into an 80-mm round piece with the edge thinner than the center. Put a tablespoon of filling in the center of the wrapper and fold the edge over to make a half moon shape; then with the forefinger and thumb of the right hand, press and twist the edge into small pleats. Cook the dumplings in boiling water over high heat. It is served hot (Figure 4.15) with a simple dip of Chinese vinegar and hot chili sauce. There are many varieties of fillings for dumplings. Most of the savory fillings for steamed buns (Section 2.2.1) can also be used for dumplings, but not the juicy fillings with gel.

6.2 FRIED DUMPLINGS *(GUOTIE)*

Guotie are fried *jiaozi.* They are first pan-fried and then steamed with a little water in the pan. They take on a crisp golden brown texture at the bottom and a soft one on top, bursting with juice inside (Figure 4.16).

Figure 4.15 Boiled dumplings *(shui jiao).*

Figure 4.16 Fried dumplings *(guotie).*

The amount of water used for the dough for *guotie* is the same as *shui jiao* (about 30–40% of flour weight), but 60% is boiling water and 40% is cold water. Mix the flour with boiling water first, and then gradually add cold water and knead it until smooth. The way to make the wrapper, fillings, and dumplings is the same as for *shui jiao*. Heat a pan until hot and add 2 tablespoons of vegetable oil. Place the dumplings in a winding circle in the pan. When the bottoms of the dumplings turn to light brown, add some water. Cover and cook until the water evaporates. Add some more water and cook again. Uncover the pan and let the dumplings fry a little more until a golden brown crust forms on the bottom and the dumplings start to puff up. Transfer the dumplings to a plate with brown sides up and serve hot with Chinese vinegar or chili sauce (Lin, 1986; Anon., 1988).

6.3 STEAMED DUMPLINGS *(ZHENG JIAO)*

These delicate steamed dumplings are usually served hot right from a steamer.

The fillings for steamed dumplings are similar to *shui jiao*. Raw meat fillings are very popular for this type of dumplings. The way to make steamed dumpling wrappers is also similar to *shui jiao*, except that there is a difference in dough making. The amount of water used for steamed dumpling dough is the same as *shui jiao*, but 70% is boiling water and 30% is cold water. Mix the flour with boiling water first, and then gradually add cold water and knead it until smooth.

Figure 4.17 Popular steamed dumplings—*har gaw*.

Steam the raw dumplings over medium to high heat for about 8–10 min and serve immediately. Steamed dumplings contain a lot of juice. The best way to eat them is to pick up with chopsticks and immediately transfer it to a small plate or a soup spoon, so that when you bite into it no juice is lost. Serve the dumplings hot with Chinese vinegar (Lin, 1986).

Steamed dumplings can be made into many shapes with many varieties of fillings. This makes the steamed dumplings very attractive and popular.

Guangdong-style steamed dumplings are very popular not only in China, but also in other countries around world (one of the main varieties for *yum cha*) (Figure 4.17). *Har gaw* (steamed prawn dumplings) and *fun gaw* (steamed meat dumplings) have a very delicate taste. The wrappers, made from a mixture of wheat starch and tapioca flour, are extremely white and semitransparent after cooking. The prawn dumplings are always made in the shape of a pouch, and the meat dumplings are always crescent-shaped. This style of steamed dumplings is served as a snack. For detailed methods of preparation of these steamed dumplings, refer to Florence Lin's (1986) book.

6.4 WONTON

While people in northern China enjoy delicious *jiaozi,* a dumpling with a thick wrapper, people in southern China have *wonton* (also called *chaosou*), a dumpling with a light, thin wrapper, as their favorite food (Figure 4.18). The fillings for *wonton* are usually a mixture of mincemeat (particularly pork), vegetables, and seasonings or are vegetables along (or mincemeat) with seasonings.

Figure 4.18 *Wonton*—a favorite food in southern China.

Mix flour with water (30–40%) and some salt (or alkali as well) to form a crumbly mixture and let it rest for 10 to 15 min. Sheet the dough in the rollers of a noodle machine, in which the gap is gradually reduced. Finally, it is cut into 7-cm-square, 1-mm-thick sheets. Place 2 teaspoons of filling in the center of the wrapper. Fold the wrapper in half from the bottom edge, over the filling, and then fold in half again in the same direction, encasing the filling. Bring the two top inner corners together, leaving the outer flap, and seal the wonton by pressing the ends firmly together using moistened fingers. Cook the wontons in boiling water and serve with broth (chicken broth with seasonings). Wontons can also be shallow fried or deep-fried, served hot, and dipped in Chinese vinegar and chili pepper sauce.

6.5 OPEN-FACED STEAMED WONTON OR DUMPLING *(SAOMAI)*

Saomai is a similar product to dumpling, but the wrapper is much thinner than dumplings (Figure 4.19). They usually serve as a snack instead of as a meal. There are different fillings for *saomai.* The most famous ones are those with glutinous rice *(nuomi saomai)* and vegetable fillings *(feichui saomai)* made in Jiangsu Province (Anon., 1989).

(1) Preparation of filling: Mince pork is stir-fried with chopped bamboo shoot, Chinese mushroom, and seasonings and then mixed with cooked glutinous rice.

Figure 4.19 *Saomai*—a delicious steamed snack.

(2) Preparation of wrapper: Flour is mixed with boiling water (100:30) (or mixed with 60% boiling water first and then with 40% cold water). The dough is left to cool and then kneaded, rolled, and cut into 10-g pieces; each piece is rolled into a circle shape with a waved edge using a special roller.
(3) Making *saomai:* Place 1 tablespoon of filling in the center of the wrapper. Taking the wrapper between the index finger and thumb of the left hand, and gather edges together to make a waist shape; use the thumb of the right hand to press down the filling so that it is compact and has a smooth top. Steam 5 min over high heat; remove and serve hot (Anon., 1989).

6.6 FROZEN DUMPLINGS

Frozen dumplings have been available since the mid-1970s in East Asia. In 1988 Beijing Eastern Food Company produced frozen dumplings for the first time in China and immediately had a major success. Since then, more and more frozen food factories have been established, and the dumpling machine manufactured in Harbin City in Heilongjiang Province has predominantly been used throughout China. The frozen Chinese dumplings, wonton, *saomai,* and so on are all available in city supermarkets in China. The price of the frozen dumplings is more acceptable than previously, and the quality of the frozen products has improved. However, there are still some quality problems from two main sources: processing technology and the quality of the flour used.

6.6.1 Flour Quality Requirements

There are some recommendations on flour quality requirements for Chinese dumplings. They indicated that the flours with medium dough strength, flour particle size 7–12 xx, low starch damage, and falling number above 230 seconds are suitable for Chinese dumplings (Li, 1994).

6.6.2 Current and Potential Market Status

Chinese dumplings are very delicious and healthy. Since the introduction of frozen dumplings to supermarkets, dumpling popularity has greatly increased because of their convenience and acceptable price. They may become more popular than instant noodles in the future because of further improvements in living standards and in the quality of frozen dumplings.

7. MEAT ALTERNATIVES

Wheat gluten has been used in Chinese and Japanese cooking for a long time. Wet gluten separated from flour is processed into different dishes such as fried gluten balls and *kaofu.*

7.1 FRIED GLUTEN BALL *(YU MIANJIN)*

Wet gluten is divided into small pieces and deep-fried in hot vegetable oil. It expands into a hollow ball (Figure 4.20). It can be added as a high-protein ingredient with a light chewy texture to soups, stews, and many other dishes (Passmore, 1991). Gluten balls can also be stuffed with fine-minced meat and seasonings. Fried gluten pieces can also be cooked with vegetables and seasonings and served as a dish (Figure 4.21).

7.2 *KAOFU*

Kaofu is another popular gluten product in China. It is prepared by fermenting wet gluten and then steaming. It is a sponge-like product and is cut into pieces and cooked with seasoning and meat or sauce and other vegetables. It has little taste, but its sponge-like structure absorbs sauce and seasonings, giving it a delicious taste (Figure 4.22).

7.3 CURRENT AND POTENTIAL MARKET STATUS

As a substitute for meat, gluten dishes are very popular in Asian countries. Meat consumption is increasing steadily as living standards improve; however, overconsumption of animal meat causes health problems for

Figure 4.20 Fried gluten ball *(yu mianjin).*

Figure 4.21 A fried gluten dish.

Figure 4.22 A tasty fermented gluten dish *(kaotu)*.

humans. For good health, people are seeking alternative foods such as gluten foods and soybean products. These will be more popular in the future.

8. ACKNOWLEDGMENT

The author is very grateful to Mrs. Sandi Ormston for her excellent photographic work. Thanks also go to Miss Susan To, Mrs. Miriam Andrade, Mr. David Mugford, and Dr. Peter Gras for their helpful suggestions.

9. REFERENCES

Anon., 1989. *Techniques to Making Wheaten Foods.* Chinese Commercial Publishing House, Bejing (in Chinese).

Anon., 1992. *Four Hundred Varieties of Delicious Wheaten Foods.* Jin Dun Publishing House, Beijing. (in Chinese).

Fan, C. Y. 1982. The initial approach to several issues of agricultural development in ancient China. *J. Chinese Agricultural History* 1 (in Chinese).

Huang, S. H. 1974. *Chinese Snacks.* Dept. of Home Economics, Wei-Chuan Foods Corp., Taipei, Taiwan.

Huang, S. D. and Miskelly, D. M. 1991. Steamed bread—A popular food in China. *Food Australia* 43(8):346–47, 350–51.

Huang, S. and Hao, Q. 1994. Steamed bread processing. In: *Food Processing and Wheat Quality Improvement,* Z. Lin, ed., Chinese Agricultural Publishing House, Beijing, pp. 359–388 (in Chinese).

Huang, S. and Quail, K. 1995. Wheat based foods. *Cereal International,* autumn, pp. 6–9.

Huang, S., Yun, S., Quail, K. and Moss, R. 1996. Establishment of flour quality guidelines for northern style Chinese steamed bread. *J. Cereal Science* 24:179–185.

Huang, S. and Quail, K. 1996. In: *Proceedings of the 46th Australian Cereal Chemistry Conference,* September 1–6, 1996, Sydney, pp. 315–318.

Li, Y. 1994. In: *Proceedings of '94 International Symposium & Exhibition on New Approaches in the Production of Food Stuffs and Intermediate Products from Cereal Grains and Oil Seeds,* November 16–19, 1994, Beijing, pp. 415–421.

Lin, F. 1986. *Florence Lin's Complete Book of Chinese Noodles, Dumplings and Breads.* Willia Morrow and Company, Inc., New York.

Nagao, S. 1995. Wheat usage in East Asia. In: *Wheat End Uses around the World,* H. Faridi, and J. Faubion, eds., American Association of Cereal Chemists, St. Paul, MN, pp. 167–189.

Passmore, J. 1991. *The Encyclopaedia of Asian Food & Cooking.* Doubleday (Australia) Pty Ltd.

Quail, J. K. 1996. *Arabic Bread Production.* American Association of Cereal Chemists, Inc., St. Paul, MN.

Sidhu, J. S., Seibel, W., Bruemmer, J. M., and Zwingelberg, H. 1988. Effect of flour milling conditions on the quality of India unleavened flat bread (chapati), *J. Fd. Sci.* 53(5):1563–1565.

Sidhu, J. S. 1995. Wheat usage in the Indian subcontinent. In: *Wheat End Uses around the World,* H. Faridi, and J. Faubion, eds., American Association of Cereal Chemists, St. Paul, MN, pp. 191–213.

Wang, C. and Ya, F. 1988. *Wheaten Foods in Shanxi.* Shanxi Science-Education Publishing House, Taiyuan, Shanxi. (in Chinese).

Wang, G. and Li, Y. 1994. Bread processing. In: *Food Processing and Wheat Quality Improvements,* Z. Lin, ed., Chinese Agricultural Publishing House, Beijing, pp. 84–159. (in Chinese).

Wang, H. 1984. *Mozhijiaoshi.* Zhejiang Art Publishing House, Hangzhou. (in Chinese).

Zhu, J., Huang, S., O'Brien, L., Wei, X. and Mares, D.J., 1997. In: *Proceedings of the 47th Australian Cereal Chemistry Conference,* September 14–19, 1996, Perth, pp. 272–275.

CHAPTER 5

Foods from Other Grains and Starchy Materials

SEIICHI NAGAO

1. INTRODUCTION

GRAINS and starchy materials are the primary caloric source for the people in Asia, and their cultivation and use have developed with the culture. In most nations of Asia, rice or wheat is the staple food, but other grains and starchy materials are also eaten in many ways. Rice and wheat products have been covered in previous chapters, and this chapter covers food products derived from other grains and starchy materials, including amaranth, arrowroot, barley, buckwheat, corn, grain sorghum, Job's tears, millet, oat, potato, rye, sweet potato, and yam. The proximate composition of grains, starchy materials, and their products described in this chapter is listed in Table 5.1.

2. AMARANTH

The seed of amaranth *(Amaranthus)* was a very important cereal for people of Aztec, Maya, and Inca but was forgotten after the defeat by Spaniards in the 16th century. Later, it was introduced to Asia and is now grown in northern and southern India, Nepal, northeastern Pakistan, southwestern China, and Japan. Though it is not widely used today, it is expected to be a promising crop to solve the problem of food shortages in the 21st century. In Japan, the plant breeding of amaranth has been done recently by the National Agriculture Research Center.

Among about 60 varieties known in the world, *Amaranthus hypochondriacus* and *Amaranthus caudatus* are the main varieties grown in Asia (Sakamoto, 1997). The protein content of the seed depends on variety and growing condi-

TABLE 5.1. Proximate Composition of Other Grains, Starchy Materials, and Their Products

Grains, Starchy Materials, and Their Products	Moisture (%)	Protein (%)	Lipid (%)	Carbohydrates Non-fibrous (%)	Carbohydrates Fiber (%)	Ash (%)
Arrowroot,						
starch	13.9	0.2	0.2	85.6	0	0.1
Barley,						
covered grain	14.0	10.0	2.8	66.9	3.9	2.4
naked grain	14.0	10.6	2.8	69.4	1.4	1.8
pearled grain	14.0	8.8	2.1	73.5	0.7	0.9
roasted flour	4.5	13.2	3.6	71.9	4.3	2.5
Buckwheat,						
grain with husk	14.5	10.8	2.8	61.0	9.0	1.9
straight grade flour	13.5	12.1	3.1	68.5	1.0	1.8
1st grade flour	14.0	6.1	1.6	77.2	0.3	0.8
2nd grade flour	13.5	10.3	2.7	71.6	0.3	1.6
3rd grade flour	13.0	15.1	3.6	65.2	0.7	2.4
groats	12.8	9.7	2.5	73.1	0.5	1.4
raw noodle	33.0	9.8	1.9	54.2	0.3	0.8
boiled noodle	68.0	4.8	1.0	25.8	0.2	0.2
dried noodle	12.0	13.6	2.6	70.2	0.4	1.2
Corn,						
grain	14.5	8.6	5.0	68.6	2.0	1.3
boiled sweet corn	74.7	3.3	0.4	19.7	1.2	0.7
corn meal, degermed	14.0	8.3	4.0	71.1	1.4	1.2
corn grits	14.0	8.2	1.0	75.9	0.5	0.4
corn flour	14.0	6.6	2.8	75.3	0.7	0.6
popcorn, popped	4.0	10.2	22.8	58.2	1.4	3.4
cornflakes	4.5	7.8	1.7	83.2	0.4	2.4
starch	12.8	0.1	0.7	86.3	0	0.1
Grain sorghum,						
grain	12.0	10.3	4.7	69.5	1.7	1.8
pearled grain	12.5	9.5	2.6	73.9	0.4	1.1
Job's tears,						
pearled grain	13.0	14.2	5.9	64.8	0.8	1.3
Millet, Barnyard grass,						
grain	13.0	9.3	4.8	61.3	8.3	3.3
pearled grain	12.0	9.8	3.7	72.4	0.8	1.3
Foxtail millet,						
grain	13.0	9.9	3.7	63.5	7.0	2.9
pearled grain	12.5	10.5	2.7	72.4	0.5	1.4
glutinous cake	47.9	4.8	1.2	45.7	0.2	0.2

(continued)

TABLE 5.1. (continued)

Grains, Starchy Materials, and Their Products	Moisture (%)	Protein (%)	Lipid (%)	Carbohydrates Non-fibrous (%)	Carbohydrates Fiber (%)	Ash (%)
Proso millet,						
grain	13.5	12.7	3.8	57.1	9.1	3.8
pearled grain	14.0	10.6	1.7	71.7	0.8	1.2
Oat,						
grain	12.5	13.0	6.2	54.7	10.6	3.0
oatmeal	11.5	13.5	5.6	66.5	1.1	1.8
Potato,						
raw	79.5	2.0	0.2	16.8	0.4	1.1
mashed, dehydrated	7.5	6.6	0.6	81.2	1.6	2.5
chips, fried	2.5	4.7	35.0	52.6	1.8	3.4
starch	18.0	0.1	0.1	81.6	0	0.2
Rye,						
grain	12.5	12.7	2.7	68.5	1.9	1.7
flour	13.5	8.5	1.6	75.0	0.7	0.7
bread	35.0	12.2	2.5	48.0	0.4	1.9
Sweet potato,						
raw	68.2	1.2	0.2	28.7	0.7	1.0
baked	62.0	1.3	0.2	34.6	0.8	1.1
starch	17.5	0.1	0.2	82.0	0	0.2
Yam,						
Chinese yam	82.6	2.2	0.4	13.5	0.3	1.0
Japanese yam	69.6	2.8	0.7	25.4	0.5	1.0

Source: Standard Tables of Food Composition in Japan, Fourth revised edition, edited by Resources Council, Science and Technology Agency, Japan.

tions, but the average figure is estimated to be about 15%. Its protein is rich in lysine and tryptophan compared to most cereals. Leucine is the limiting amino acid. It was reported that lectin, enzyme inhibitor, antibacterial peptide, microcomponents to lower cholesterol, and more were found in the seed of amaranth (Konishi, 1997).

Laddoos in India is a traditional confection produced by puffing the amaranthus seed. Pearled whole grain mixed with pearled rice is boiled or cooked to a porridge and pilaf.

It is possible to blend amaranthus with wheat flour to make noodles, pasta, breads, cookies, and a variety of snack foods. Since amaranth is not a crop of the grass family, it can be used as a substitutive food for allergy therapy. The antihistamine action and cholesterol lowering action of amaranth are worthy of notice, but its ingestion is advised under a doctor's supervision (Chiba, 1997).

In Japan, these are trials to use the flour in the production of tempura mix, *manju* (steamed or baked bun), candy, and *miso*. Biscuits and wafers made from amaranth flour were put on the market by a large biscuit company and were well received by some consumers. In China, it is used as a material to produce local liquor and vinegar. The color of bran is utilized to give color to soy sauce, and the flower of amaranth is also used to produce honey in China.

There are some waxy varieties. New foods utilizing amaranth's starch property are expected to be developed in the near future.

3. ARROWROOT

Arrowroot *(Maranta arundinacea)* is a perennial plant of legume. Starch of characteristic quality is taken from its smashed root. Because it is rich in amylose of low molecular weight, arrowroot starch is easily gelatinized by pouring hot water over it, which makes it turn to clear paste. *Kuzuneri,* which is made by dissolving the starch and sugar in hot water, is a digestible food for infants and patients (Okada, 1996).

4. BARLEY

Barley *(Hordeum vulgare* L.) is a crop in the grass family. In covered barley, the husks adhere to the kernel and remain attached after threshing. The main types of cultivated covered barley, depending on the arrangement of grains in the ear, are two-rowed and six-rowed. There is also naked barley from which husks can be easily removed from the kernel. Two-rowed covered barley is used for malting and brewing. Both covered and naked six-rowed barleys are processed to a product that can be blended with pearled rice to prepare a staple food in Japan. Possible uses of both pearled barley and barley flour have been extensively explored and promoted by a group of people headed by S. Hayashi (Mitsunage and Hayashi, 1993; Mitsunaga et al., 1994).

4.1 PEARLED BARLEY

In Asian countries, the percentage of food use in barley consumption is higher than that in other continents. A large amount of pearled, six-rowed barley was consumed by boiling with pearled rice in Japan. However, the consumption of pearled barley has gradually decreased to less than 30,000 tons per year because of the rapid increase in rice production and the diversification in modern dietary life. However, pearled barley taken with pearled rice as a staple food is now drawing some people's attention as a possible source of dietary fiber. Cooking it with beans is popular in China.

Figure 5.1 Flaked barley.

Pearling of barley uses more technique than that of brown rice. Pearled barley is manufactured by gradually removing the hull and outer portions of barley kernels by adhesive action. Barley of high protein content is generally low in pearling yield and unfavorable in the color of pearled product. A soft and chalky kernel has better pearling characteristics than a hard and vitreous kernel.

Pearled barley is processed into three types of products, which are flaked (Figure 5.1), white, and round. As shown in Figure 5.2, the manufacturing process of flaked barley comprises damping, attrition polishing, steaming, flaking, cooling, and drying of polished barley, and that of white barley consists of damping, attrition polishing, breaking into two pieces along a crease, removing of the crease, and steaming of pearled barley. Both products are blended and boiled with pearled rice to be eaten as a healthy staple food in Japan today. As for the quality of flaked barley, the extent of pearling, moisture content, and flaked shape is very important. Size distribution, gloss, the width of remaining crease, and the content of cracked kernels are the other quality criterion to be considered. The recommended blending ratio of flaked or white barley to pearled rice in modern dietary life in Japan is about 10–20%, though the figure has been up to 30–70% in the past. Since the blending of barley products with pearled rice tends to decrease the pasty texture of the boiled dish easily, adding 5% more water

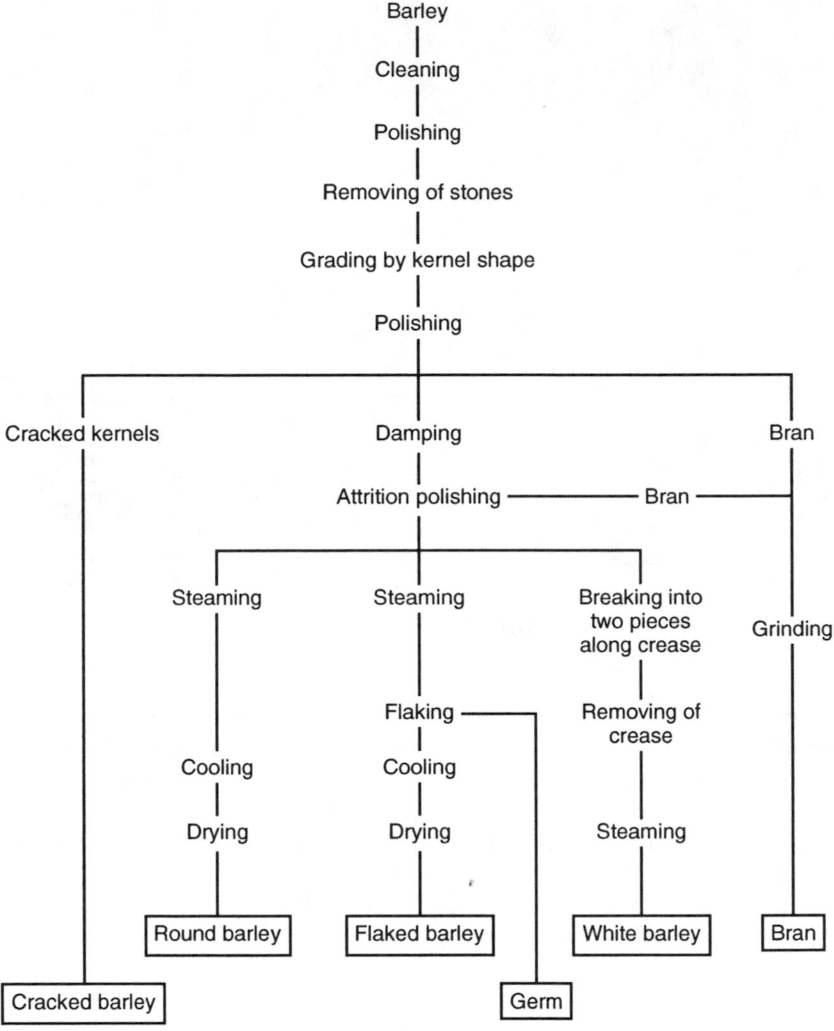

Figure 5.2 Manufacturing process of pearled barley products.

than usual to the pearled grains at boiling is advised. Round barley is processed by damping, attrition polishing, steaming, cooling, and drying.

Recently, a new product made by blending flaked barley with the pearled products of waxy foxtail millet, waxy proso millet, and amaranth together with white sesame was put on the market by a company specializing in barley products. The blending of this product with pearled rice is recommended by the company to prepare a healthy and tasty boiled dish.

4.2 BARLEY FLOUR

Flour milling of barley was studied by Mitsunaga et al. (Mitsunaga et al., 1994). Barley grains were polished and milled with a machine used in the polishing of brewer's rice and modified for this purpose. Grains were polished to several products, of which extraction rates were 85 to 30%, and were milled into flour. About half of the particles in each milled product was 10–30 μm in diameter.

They found that a barley flour of low extraction rate could be mixed with wheat flour to be used in the production of noodles, breads, rolls, cakes, and cookies. The eating quality of those products was unique and a little different from that of 100% wheat flour products. Noodles containing 50–70% barley flour of high quality are light in taste and different from traditional noodles. Barley flour can be used for cooking in many ways, too. Like noodles, starch in barley flour generally gives a light taste to the cooked dishes. Barley flour of low extraction rate is possible for use as a material in the production of boiled fish paste.

4.3 BARLEY TEA

Barley tea stored and served cold is very popular during a hot Japanese summer. To prepare barley tea at home, in the office, or at a restaurant, roasted

Figure 5.3 A new product of barley tea.

barley with hulls is decocted in boiling water, and the tea grounds are removed. It is most common to drink it cold, but some people prefer to drink it hot.

The recent advance of roasting technology made it possible to develop and market a new product (shown in Figure 5.3) whose essential components dissolve even in cold water. When this new product is used, barley tea can be easily prepared just by steeping a tea bag of 11.5 g in 1 L of water kept in a refrigerator for 1–2 h. Blending of well- and lightly roasted barley kernels is another newly developed technique. Thoroughly roasted barley kernels give decocted tea a natural color and flavor, while the lightly roasted barley kernels create a favorable flavor.

4.4 OTHER USES

Though soybean is the main material in *miso* (fermented soybean paste) production, round barley can be used to make a variety of local *miso* such as *inaka miso* and *kinzanji miso. Miso* fermented from barley is unique in its appearance because of its black creases remaining in the final product. Barley is a good material to make a plain type of soy sauce.

Instant barley *risotto* was developed and put on the market (Sugiyama, 1994). To make the product, pearled barley kernels are steamed, freeze-dried, and heated for 5 min. *Mugi shochu* (distilled spirits from barley) is a traditional and popular product in some local areas in Japan. To produce it, steamed round or flaked barley is added to unrefined spirits fermented by three steps and is distilled.

Mugi-rakugan is a traditional cookie made from barley flour. Barley flour (100 parts), sugar (60 parts), and water are mixed and put into a wooden mold. After taking it out of the mold, the dough is heated and dried (Sugiyama, 1994). Its light and crispy taste was popular long ago in Japan. *Mugikogashi* (roasted barley flour) is flour milled from roasted barley, which is also popular under the names of *kousen* or *hattaiko* in Japan.

Barley meal is used to make breakfast cereals and baby food. It is also blended with a kind of milk to make a dumpling in Tibet (Leonard and Martin, 1963).

5. BUCKWHEAT

Buckwheat *(Fagopyrum esculentum* Moench) is grown in China, Nepal, Japan, and other countries in Asia, where it has been traditionally eaten in many ways. The most popular way of eating buckwheat in those countries is in the form of flour, but some people eat it in the form of grain. Buckwheat noodle is a very popular food in Japan, which is unique among the many uses of buckwheat in the world.

5.1 BUCKWHEAT MILLING

Stone mills have been traditionally used in buckwheat milling. Buckwheat kernels are ground through a pair of stone mills, and husks are removed by sifting. Product milled by this method is straight grade flour.

Though stone mills are still used to extract a special high-grade flour, most buckwheat flour mills in Japan have been equipped with a continuous milling system using roller mills. Continuous milling systems consist of three processes: cleaning, hulling, and milling. In the cleaning process, dockage and foreign materials are removed by using separators, and brush machines clean the surface of grains. Several methods are applied to hulling. In the process of using stone mills, cleaned buckwheat kernels are graded to several kernel sizes by using wire sieves, and the graded kernels are separately hulled by using stone mills, the clearance being adjusted to the size of kernels. Hulling has to be done very carefully so as not to destroy the structure of the endosperm, and the quality of hulled kernels directly affects the quality of the end products. A small amount of lower grade flour is obtained in the hulling process.

Hulled kernels are ground by the first roller mills and are sifted by a sifter. A flour obtained at this stage is called first-grade flour, which is white in color, high in starch content, and poor in aroma and flavor. The leftover is ground by the second roller mills and is sifted again. Flour that passes through the sifter is called second-grade flour, which is rich in aroma and flavor, light yellow in color, and high in nutritive value. The leftover at this stage is ground by the next roller mills and is sifted again. Flour passed through the sifter now is called third-grade flour, which is rich in flavor, slightly green in color right after milling, high in nutritive value, and poor in taste. Because the third-grade flour contains a high amount of protein, lipid, and fiber (as shown in Table 5.1), it is not easy to grind into fine particles. Reducing the particle size of the third-grade flour and how it is blended to the first- and second-grade flour decide the quality of the finished products.

5.2 BUCKWHEAT FLOUR

The use of quality buckwheat is the first and most important step in milling to get quality flour. Complete hulling is also essential to the quality of the flour. Buckwheat kernels are still alive even after harvesting, and the chemical change in the components such as lipid and chlorophyll during storage may easily deteriorate the quality of buckwheat. Therefore, storage at a low and fixed temperature is necessary to keep the freshness in the flavor and color of buckwheat. Since buckwheat is very sensitive to high temperatures, careful attention has to be paid not to overheat at roller milling and dough mixing in noodle making.

Commercial buckwheat flours vary in freshness and ash content because of the blending ratio of the third-grade flour to the first- and second-grade flours.

A special-grade flour of approximately 10% extraction rate was formerly obtained by a special grinding using stone mills, but it has generally been replaced by the first-grade flour.

In Japan, buckwheat consumption has been gradually increasing yearly and was about 120,000 tons in 1996. Domestic production remains at the level of approximately 20,000 tons, and the remainder is imported from China, the United States, Canada, and other countries.

5.3 BUCKWHEAT NOODLE

Soba, or buckwheat noodles, became popular in the *Edo* period (1665–1885) in Japan. Since then, *soba* noodle has been a very popular snack food for the Japanese. Even today, many office workers eat *soba* noodle for lunch. Various kinds of *soba* noodle dishes are available. Among them, the most popular items are fresh *soba* noodles such as *mori soba* (a cold noodle eaten with a dip) and *kake soba* (soup noodle). Boiled *soba* noodles packed in polyethylene bags are sold at supermarkets, grocery stores, and noodle shops. A dried type is also available.

The basic process for the production of fresh *soba* noodle by machines comprises mixing of the raw materials, dough sheeting, combining of two sheets, rolling, and cutting. Buckwheat flour (30 parts or more) and hard wheat flour (70 parts or less) are mixed with water (about 28 parts) in the manufacture of typical *soba* noodle. The ratio of wheat flour to buckwheat flour varies according to the type of product and the quality of buckwheat flour. Wheat flour used for *soba* noodle production is 0.50–0.70% in ash and 12–14% in protein content. Yam or marine algae may be added to the ingredients to increase adhesiveness in the finished product. Salt is not used in the production of *soba* noodle lest the binding capacity of wheat flour and the fresh flavor of buckwheat flour should be lost. Mixers varying in type and size are used in the industry. A typical mixer has a rotary shaft with special blades that contribute to uniform mixing and to a little gluten development through the beating action of the blades. However, in the manufacture of *soba* noodle, mixing should be done quickly, preferably within 10 min.

The stiff and crumbly dough pieces are divided into two, and each portion is passed through a pair of sheeting rolls to form a noodle sheet. The diameter of the sheeting roll is usually 180 mm. The roll gap at this stage should be narrow to help the development of gluten by the pressure at the moment of passing. The two sheets are then combined and passed through a second set of rolls, where the gap is adjusted to that of the finished product. The sheet is then cut into strands of the desired width. The width of a *soba* noodle strand is determined by the groove of cutting rolls. Cutting rolls of 1.4–1.6 mm in width are usually used in *soba* noodle production. *Soba* noodle strands are finally cut into the proper lengths by using a cutter (Nagao, 1981, 1996).

Figure 5.4 *Zaru soba,* cold noodle served on a bamboo basket on which thinly cut, dried seaweed is sprinkled.

At *soba* noodle shops or restaurants, fresh *soba* noodles are cooked soon after making them or are kept in a refrigerator to be consumed within a few hours. Boiled *soba* noodles are prepared by cooking uncooked wet noodles in boiling water for 1–3 min, with the timing adjusted precisely to give optimal textural characteristics to the product. There are various types and sizes of boiling pots. Since the volume of boiling water in the pots is desirable to be 10–20 times that of the uncooked wet noodles, a proper size of boiling pot should be chosen. Immediately after removal from the boiling pot, they are washed, cooled in running water, and drained quickly.

Cooked *soba* noodles are served in many ways. *Mori soba* and *kake soba* are the most popular among them. *Mori soba,* cold noodles served on a bamboo basket, is eaten with a special dip made from soy sauce and broth. *Zaru soba,* as shown in Figure 5.4, is a variety of *mori soba* on which thinly cut, dried seaweed is sprinkled. *Kake soba* is the basic type of boiled noodle served in hot soy sauce soup and has many variations. On the top of boiled noodles in a bowl, a fresh egg, shrimp or vegetable tempura, fried *tofu* (bean curd), boiled fish pastes, vegetables, or boiled duck, are placed. The dish is eaten with a pair of chopsticks.

Figure 5.5 Dried *soba* (buckwheat) noodle.

Figure 5.6 Buckwheat groats called buckwheat rice in Japan.

For the manufacture of dried *soba* noodles, uncooked wet noodle strands of several meters in length are hung on rods in a special drying chamber equipped with devices to remove humid air and to introduce fresh air. The rate of drying and humidity in the drying chamber must be closely controlled. Some examples of dried *soba* noodles are shown in Figure 5.5.

5.4 BUCKWHEAT GROATS

Buckwheat groats (as shown in Figure 5.6), called buckwheat rice in Japan, are made by the hulling of buckwheat grain. In a hulling method developed in 1940, buckwheat grains soaked in water for 4 h are drained, steamed, dried, and hulled. By gelatinization during steaming, starch in the endosperm shrinks and helps to make a gap between endosperm and hull. It is boiled and eaten by itself or with pearled rice.

A cup of buckwheat groats is added to six glasses of boiled milk little by little and is mixed. After adding a little salt and butter to it, it is boiled for 10–12 min with occasional stirring. It can be eaten hot or cold, and sugar may be added to it. Buckwheat groats are possible to be cooked as soups or fried.

Figure 5.7 *Soba boro,* cookies made from the mixture of buckwheat flour and wheat flour.

5.5 OTHER USES

Buckwheat tea is prepared by decocting roasted hulled buckwheat kernels in boiled water, and the tea grounds are removed. Some people in Japan like it as a healthy beverage.

Soba manju, steamed buns made from buckwheat flour, is a traditional and popular confection in some mountain areas in Japan. The dough, made from buckwheat flour, wheat flour, sugar, water, ammonium carbonate, and sodium bicarbonate, is divided and rounded into small balls, which are flattened. Sweet bean jam filling is placed in the center of the flattened dough piece and enclosed by sealing the edges of the dough piece. The finished dough pieces are placed on racks and steamed.

Soba boro (shown in Figure 5.7) is a traditional cookie made from a mixture of buckwheat flour and wheat flour in Japan, with a crispy texture and peculiar taste.

Soba gaki is a convenient food easily prepared at home. Buckwheat flour and hot water are hand-mixed in a bowl, using a pair of chopsticks, just prior to eating. It is eaten as is or is soaked in soy sauce.

6. CORN

Corn (*Zea mays* L.) for human consumption is processed in several ways. It is (a) ground to make hominy, corn grits, corn meal, or corn flour; (b) processed for a variety of breakfast foods; (c) steamed, boiled, or parched to be eaten as whole grains; (d) processed for many kinds of foods; and (e) treated to extract corn starch.

Corn is an important material in the brewing and alcohol industry and is a quality material for extracting oil. Powdery corn fiber—developed by Takeuchi—is 80.8% in dietary fiber content (Takeuchi et al., 1990). When 30 g of cookies made from wheat flour (95 parts) and corn fiber (5 parts) were eaten every day for a week, it was proven to be good for constipation.

6.1 HOMINY AND CORN FLOUR

The consumption of corn meal ground dent or flint corn by millstones is very limited in Asia. Because it is a whole grain product including germ portion, it has a rich and oily flavor and stores very poorly. Corn meal produced by this method is generally mixed with wheat flour to make baked or steamed flour foods at home.

Roller machines are commonly used to make hominy or corn grits, corn meal, and corn flour. Generally, flint or dent corn is the material for making corn grits and coarse meal, while white corn or dent corn is used for making fine meal and corn flour. After adding sugar and salt, hominy is steamed, dried, flaked, and cooked to make corn crisp and corn flakes. Corn flour is used for making snacks and cereals, mixed with wheat flour in the production of bread and cake mixes, and utilized as a binder in sausage production.

6.2 CORN STARCH

The process of starch extraction from corn comprises steeping, degermination, grinding by friction, washing, shifting, centrifuging, filtering and drying. Dent corn is most commonly used for starch extraction, but the use of waxy or high amylose corn is increasing because of its high extraction rate and peculiar quality.

The use of corn starch is broad. It is widely used in the production of (a) sauce, soup, and pie to increase viscosity; (b) gravy and dressing as a stabilizer; (c) chewing gum and drops as a gelling agent; (d) icing of cakes as a moistening agent; (e) ice-cream cones as a binder; and (f) baked products as dust.

6.3 SWEET CORN

Canned products of corn are made from sweet corn. They are in whole style and cream style. The uniformity of corn grain in maturity and shape is a very important quality criterion for the whole-style product. The processing process comprises cutting, creaming, seasoning, canning, and sterilizing. Cream-style product is made by cutting the head portion of corn grain a little, grinding into small particles, adding salt and sugar syrup, steaming at 82–93°C, blending with starch, canning, and sterilizing.

7. GRAIN SORGHUM

Grain sorghum (*Sorghum vulgare* Pers.) grown in dry land is a very important food material in some countries in Asia. Porridge and dumplings are the most common uses of grain sorghum.

7.1 PORRIDGE

Pearling is the first and most important step in preparing to eat grain sorghum. In former times it was steeped in water and pearled by stone mills in China and Mongolia. As the testa of grain sorghum cemented to the endosperm, tannin and pigments remained in the pearled product, which made it difficult to store for a long time.

In 1930s, new pearling machines of an attrition type were developed one after another in northeastern China, and they are still used. Gradual pearling consisting of three or four steps is used in the pearling of grain sorghum. The pearling yield is 75–85% and is influenced by the variety, moisture content, and hardness of grain sorghum. The quality of the product pearled by those developed machines is much better than that of the product pearled by stone mills. The pearled grain sorghum is cooked into porridge or boiled with pearled rice or barley in China.

7.2 DUMPLING

In former days, pearled grain sorghum was milled into flour by using roller mills or impact grinding machines in Japan. The quality of grain sorghum flour depended on the milling yield. Grain sorghum flour was kneaded with water, rounded to a small and round shape, and steamed to make *dango,* which is a round flour dumpling. It is coated with sweet red bean paste or dipped in soy sauce. Three or four balls of *dango* are skewered on a bamboo stick and are eaten.

7.3 OTHER USES

In China, *kaoliang chiu* (grain sorghum spirits) is made from *kaoliang,* a variety of grain sorghum (Miki, 1935). *Kaoliang chiu* is colorless and transparent and has a peculiar flavor. The alcohol content of *kaoliang chiu* is about 60%. Cracked *kaoliang* steamed and cooled to a warm temperature is mixed well with fermented powder and water and fermented for 8–10 days in a fermentation chamber. After fermentation, it is distilled to get *kaoliang chiu.* Grain sorghum was also used in the manufacture of *miso* and soy sauce in former days.

8. JOB'S TEARS

8.1 JOB'S TEARS GRAIN

Job's tears (*Coix lacryma-jobi* L.) is a plant in the grass family, which is produced in tropical Asia, China, and Japan. It originated in Indonesia in ancient times, but most of its production was replaced by rice in southeastern Asia. It is covered with a hard husk, but its husk can be removed easily by a thresher.

From early times, Job's tears has been believed to be effective for removing warts, promoting urination, strengthening the stomach, and promoting energy. From 1981, the production of Job's tears on paddy fields has been promoted as a converted crop from surplus rice in Japan. Yields of Job's tears are high even in paddy fields. In order to increase its production, plant breeding to develop high-yielding and high-quality varieties is a pressing need (Ono, 1992).

It is rich in protein and lipids, as shown in Table 5.1, and is processed to flaked product, *miso, shouchu* (distilled spirits), and vinegar. A new use of Job's tears as a health food is becoming popular in Japan.

8.2 JOB'S TEARS FLOUR

After removing the hard husk by using a thresher, Job's tears is gradually pearled by a pearler to get a white hulled product, as shown in Figure 5.8, which is blended in pearled rice to make boiled food.

Roller mills are used to extract the white Job's tears flour of a fine particle size from a hulled product. Since it is thought to be a healthy food, it can be blended to any wheat flour products such as breads, cookies, noodles, rice crackers, and so on. Several unique ways of eating Job's tears flour are being promoted and are becoming popular. In one promotion, it is recommended mixing a small quantity of it with milk, tea, water, or yogurt and drinking the mixture every day. Eating it with parched bean flour is also advised.

Figure 5.8 Hulled adley.

8.3 JOB'S TEARS TEA

Tea decocted from roasted Job's tears is believed by some people to be useful to maintain and promote health. It is produced by roasting Job's tears with the husk.

9. MILLET

Millet is a generic term for five genera *(panicum, setaria, echinochloa, pennisetum,* and *paspalum)* in the Panicea family and *eleusine* genus in the Chlorideae family. Barnyard grass (*Echinochloa crus-galli* L.), finger millet (*Eleusine coracana* Gaertn.), foxtail millet (*Setaria italica* Beauv.), koda or ditch millet *(Paspalum serobiculatum),* pearl or cattail millet (*Pennisetum typhoideum* Rich.), and Proso millet (*Panicum miliaceum* L.) are grown in India, China, and other Asian countries.

They can grow on sterile soil and in severe climates. Rather high yields are generally expected in spite of their short growing period. They have been used for food since primitive ages, but rice, wheat, barley, corn, and potato replaced most of their uses over time. However, some people in Asia are still eating them in many traditional ways.

Recently, the consumption of millets as a substitutive cereal for allergy therapy is increasing little by little in Japan. Nishizawa et al. (1997) studied the effect of dietary protein of proso, foxtail, and barnyard grass (Japanese millet) on the cholesterol and triglyceride levels in plasma of rats. When rats were fed with diets containing 20% millet protein for 21 days, high-density lipoprotein cholesterol in plasma was significantly higher than that of rats given 20% casein or soybean protein diets (Nishizawa et al., 1997).

9.1 BARNYARD GRASS

Barnyard grass (*Echinochloa crus-galli* L.), which is called *sawa* millet in India and Japanese millet in Japan, is grown in many Asian countries.

Pearling is the first step in the utilization of barnyard grass, but it was not easy because of the close adhesion of bran coat to endosperm. To make its pearling easy, three conditioning methods for industrial use were developed by Obara in the 1930s (Obara, 1938, 1940), and the basic concept of their development is still used. They are (a) drying only; (b) soaking in water, steaming, and drying; and (c) steaming and drying. In the case of method (b), kernels of barnyard grass are washed in water, soaked in water for about 2 h, and dehydrated. Then the kernels are steamed, cooled down gradually, and dried. Too long a steaming may deteriorate the pearling property of barnyard grass kernels.

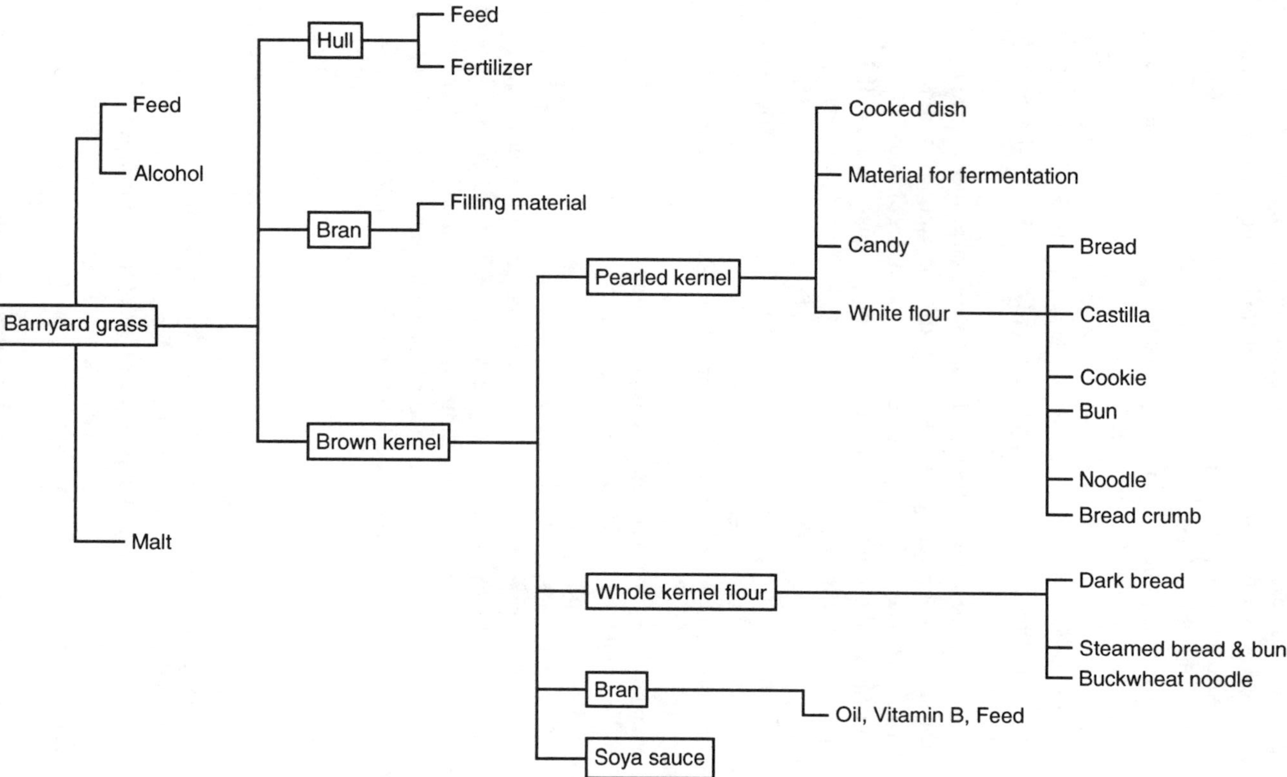

Figure 5.9 Possible uses of barnyard grass.

After conditioning, pearling is done in three steps. At the first step, a rice-polishing machine is used to remove the outer bran layer. Second, the inner bran layer is separated from the brown kernels by a roller machine. Finally, brown kernels, after mixing with a portion of inner bran, are polished by using a rice-pounding machine three times. The pearled product produced by conditioning method (a) is preferable in quality, including color and chemical stability, to that produced by other methods.

In India, people in lower income brackets eat barnyard grass. Flour milled from pearled barnyard grass is kneaded with water by hand and baked on a plate. Pasty food prepared by blending barnyard grass flour with milk and sugar is also eaten.

People living in some mountain areas in Japan have eaten pearled barnyard grass in many ways: boiled barnyard grass from pearled kernels, porridge from coarse ground flour, and steamed or baked glutinous cake from flour (Masuda, 1990).

Today, in Japan, those foods made from barnyard grass are regarded as a kind of health food. Figure 5.9 shows the possible uses of barnyard grass, which were summarized by Obara. Whole kernel flour from pearled barnyard grass can be mixed in the production of dark breads, steamed breads, steamed buns, and noodles. In the same way, fine flour milled from pearled kernels can be used as a material in the formula of breads, *castillas,* cookies, steamed buns, noodles, and bread crumbs.

Diastatic activity of barnyard grass sprouts is almost equal to that of barley malt after digesting for 17 h. Therefore, it can be used as a material for the production of starch syrup.

Since the starch property of barnyard grass is similar to that of rice, its malt can be used for the fermentation in *miso* production. The color, gloss, and taste of *miso* made from barnyard grass malt are desirable. The chemical change of *miso* made from barnyard grass malt during storage is similar to that made from rice malt. Soy sauce fermented and extracted from barnyard grass has a little lighter color than that from wheat, but it is desirable in taste.

9.2 FINGER MILLET (AFRICAN MILLET)

In some regions of India, flour milled from finger millet (*Eleusine coracana* Gaertn.) is eaten as a staple food. It is also used in the production of confectioneries. Leavening made from wheat flour is called *ragi* in Indonesia, which corresponds to the *alias* of finger millet in India. This fact suggests that leavening was made from finger millet in former times (Taira, 1968).

9.3 FOXTAIL MILLET (ITALIAN MILLET)

Foxtail millet (*Setaria italica* Beauv.) is grown widely in China, but very little in Japan. Hulled foxtail millet of an approximately 80% extraction rate is obtained by using a roller-type dehuller or a threshing machine. Then, it is pearled to the product of 75% extraction rate by using a pearler, which is used for food. Hulled foxtail millet has been used for some confectionery production as a preferable material for a long period of time. Sprouted foxtail millet may replace malt in some uses. Maltose forming power in sprouted foxtail millet is weaker than that in sprouted barley and rye, but it is stronger than that in sprouted oat (Obara, 1949). Amylase in sprouted foxtail millet shows a stronger liquefying ability in thinner starch solution, and its most suitable pH is 5.0.

Foxtail millet has been used for the production of *awamori* (Okinawan rice brandy) in Okinawa, Japan, and *shochu* (distilled spirits) in Kagoshima, Japan. In China, foxtail millet is a widely used material in the brewing of local wine (Taira, 1969). Waxy foxtail millet is soaked in water and steamed to make it porridge-like. Then leavening powder and properly fermented, unrefined wine are mixed with the steamed waxy foxtail millet. After fermenting for 5–8 days, a proper amount of water is added, and local wine is squeezed through cotton sacks.

9.4 PEARL MILLET (CATTAIL MILLET)

Pearl millet (*Pennisetum typhoideum* Rich.) has been grown in Asia since prehistoric times. In dry and warm areas such as some parts of India, it is an important crop. Milling is the first step to utilize the pearl millet. Flour is extracted by using stone mills or roller mills. Wet milling is used to separate pearl millet flour into components such as starch and protein. Pearl millet is used for the production of malt, too.

9.5 PROSO MILLET (COMMON MILLET)

Proso millet (*Panicum miliaceum* L.) can be grown in a rather dry and northern area like central Asia because of its short growing period. Since the kernel is soft and easy to break, polishing has to be done lightly to remove the outer hull. A rice-polishing machine of horizontal and cylindrical type is used to get polished proso millet of an approximately 80% extraction rate.

Both roller mills and impact grinders are used for the milling of polished proso millet. The quality of flour milled by roller mills is better than that by impact grinding. Polished proso millet has been eaten in the form of a boiled dish and porridge for many years. Flour from waxy proso millet is a preferable material for the production of dumplings and several traditional confections in Japan. Popular proso millet dumplings called *kibi-dango* are shown in Figure 5.10.

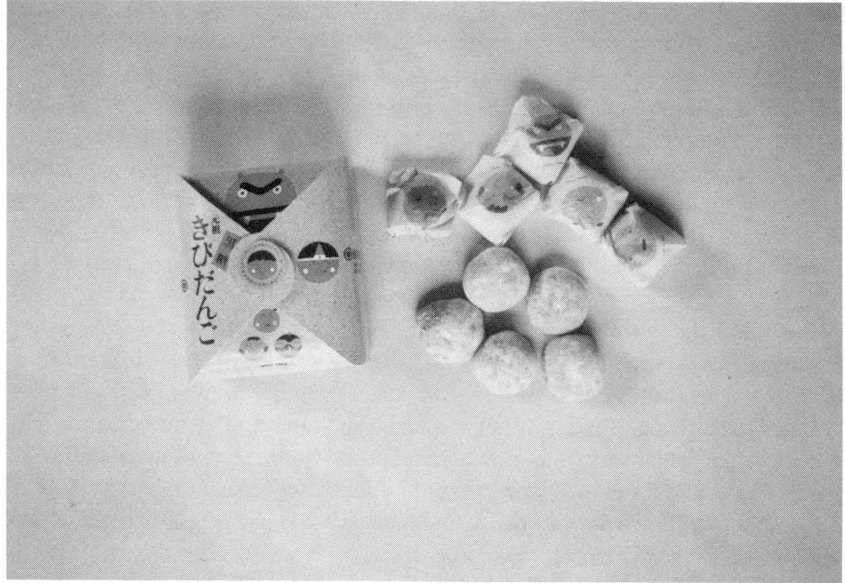

Figure 5.10 Proso millet dumplings called *kibi-dango.*

10. OAT

Oat (*Avena sative* L.) is not very popular as a food material in Asia, but it is valued as a feed for racehorses. The production of oat in Asia is very limited, so the majority is imported from Australia, the United States, and other countries.

Oat is rich in protein and lipid compared to other cereals, as shown in Table 5.1. Lipid in oat contains a large amount of lecithin in addition to olein, stearin, and palmitin. The favorable taste of oat porridge is thought to come from the lipid components. Rolled oats and oatmeal are the main products processed from oat. Breakfast cereals containing oatmeal are becoming popular in Japan.

11. POTATO

Most of the potato (*Solanum tuberosum* L.) consumed in Asia is processed to make starch, alcohol, starch syrup, glucose, and so on, but some are used for cooking in many ways. Both traditional and Western cooking of potato coexist in Japan. In cooking, the potato is used in the form of tissue as it is, cell unit and destroyed cell. The varieties of potato dishes are listed in Table 5.2 (Yamazaki and Shimada, 1983).

TABLE 5.2. Varieties of Potato Dishes.

Classification	Dishes
Boiled dish	Boiled potato, mashed potato
Soup	Potage, vegetable soup, chowder, *miso* soup
Cooked dish	Stew, potato cooked with milk, butter, soysauce, and seasoned vinegar
Panfried dish	Fresh potato
Fried dish	Potato chip, fried potato, straw potato, ribbon potato, German fried potato, *Croquette*
Baked dish	Baked potato, roast potato, *Glatin*
Dressed dish	Potato dressed with French sauce, mayonnaise, or white sauce

There are two types of potato varieties. One is a powdery type such as *Norin* 1 and *Dansyaku* varieties in Japan, which are good for *Pomme de terre nature* and mashed potato, and the other is a viscous type such as *Benimaru* and *Mayqueen,* which are suitable for boiled, roasted, and fried dishes and for thin slicing for salad.

12. RYE

Though a large amount of rye (*Secale cereale* L.) has been consumed for food in northeastern Europe and the former Soviet Union, its consumption in Asia is very limited. A variety of rye breads and their baking technology were introduced to Japan mainly from Germany and the United States, and they are becoming popular among Japanese consumers. However, it will take some time for rye bread to permeate among a majority of consumers because people are not accustomed to its unique flavor—very different from that of other grains.

Whole rye meal, white rye flour, and rye flakes are produced by some large flour milling companies in Japan. Several German types of rye breads are on the market. Their quality is very high and similar to the quality products in Germany, but most of them are made from a mixture of rye products and wheat flour. Rye breads are still not very common in other Asian countries.

13. SWEET POTATO

Sweet potato (*Ipomoea batatas* Poiret) is a parennial plant in the Calystegia family. The inner portion of sweet potato is white, yellowish white, or yellow in color, which is an inherent characteristic for all varieties. Because its active β-amylase reacts on starch components to make maltose during storage and cooking, the cooked product tastes sweet.

Figure 5.11 Sweet potato jelly, a kind of *youkan* (sweet beans jelly).

About 7% of sweet potato production is used as a material for cooking in Japan, and the balance is processed to make starch, alcohol, starch syrup, and the like. Sweet potato is eaten in many ways. In Japan, steaming is the most popular way of cooking sweet potato, while baking under far-infrared rays is also popular, especially during the winter season. Sweet potato boiled in starch syrup *(ba-si-di-gua)* is popular in Chinese dishes. Peeled sweet potato is cut into about 15 g each, washed in water, dried of surface water, and fried. Frying temperature is 150°C, but it is raised to 180°C just prior to taking out. The fried sweet potato is put into a mixture of oil, sugar, and vinegar heated to 140°C and is mixed up quickly. The finished product is so sticky that it has to be dipped in water right before serving.

To make a cake called *sweet potato* a whole sweet potato with skin is baked in an oven or is steamed. It is cut into two pieces lengthwise, and the inner portion is taken out. The inner portion is strained and mixed with milk, sugar, and butter; returned to the original skins; brushed with egg yolk on the surface; and baked for a short time in an oven to give a "burned" color.

Sweet potato is a popular material to make *kinton* (mash). Chestnut *kinton* is a typical dish prepared for the New Year's ceremony. Peeled sweet potato washed in water is boiled quickly in a 0.5% alum solution, and water is removed. Certain spices, such as gardenia, may be added. Chestnuts cooked with sugar are mixed with the strained sweet potato.

Figure 5.12 Sliced sweet potato boiled with sugar, a kind of sweet.

Figure 5.13 A unique product produced by the cooking and drying of the sliced skin of sweet potato.

Sweet potato jelly (as shown in Figure 5.11) is a variety of *youkan* (sweet bean jelly). Sliced sweet potato boiled with sugar (as shown in Figure 5.12) is loved by many as a kind of sweet in Japan. Recently, a confectionery shop has put a unique product on the market (see Figure 5.13), which is produced by the cooking and drying of the sliced skin of sweet potato. In the promotion, it is mentioned that the product contains a lot of dietary fiber and the essential nutrients of sweet potato.

14. YAM

Yam is a root grass with runner. Both cultivated and wild species are utilized. Chinese yam (*Discorea batatas* Decne) is a long, fan-shaped, massive type grown widely in China and Japan. A Japanese yam (*Discorea japonica* Thunb.) is the most popular wild species in Japan. The viscous substance peculiar to yam is a glycoprotein composed of mucin protein and mannan.

Yam soup is the most popular and delicious way of eating yams. Peeled yam is grated by using a wooden pestle in an earthenware mortar to make it viscous. Cooled broth with seasonings is added to the grated yam little by little and then grated. Chopped Welsh onion and sliced seaweed are sprinkled on the top of this dish and then eaten. A massive type of yam is most suitable for making grated yam due to its high viscosity. The viscosity drops rapidly at a temperature over 60°C. Because the cell membrane of yam is thin and contains little cellulose, uncooked yam can be eaten. Since the viscous substance of yam has a foaming ability, it is used in the production of fluffy fish paste and *manju* (bean-jam bun) as a foaming agent. It is also a good binder in the *soba* (buckwheat) noodle production.

15. CONCLUDING REMARKS

Buckwheat, corn, potato, and sweet potato are very important food materials in Asia, while some grains and starchy materials have been switched over to other crops over time. Each crop is characteristic in its nutritive value and as a food material. Some of them may be promising as crops to make up for the food shortage in the coming age of overpopulation. Intensive research and development in the field of breeding, cultivation, nutritive components, processing, and cooking are expected.

16. REFERENCES

Chiba, T. 1997. Application of amaranth to a substitutive food for allergy therapy. *Science of Food* 238:32–35 (in Japanese).

Konishi, Y. 1997. Nutritional characteristics of amaranth. *Science of Food* 238:23–31 (in Japanese).

Leonard, W. and Martin, J. 1963. *Cereal Crops.* The Macmillan Corp., New York.

Masuda, A. 1990. *Food Culture of Foxtail Millet and Barnyard Grass.* Miyai Shoten, Tokyo, Japan (in Japanese).

Miki, S. 1935. Food and beverage industry in Manchuria. *Food Research* 105:167 (in Japanese).

Mitsunaga, T. and Hayashi, Y. 1993. Food property of barley. *Science of Cookery* 26(3):236–243 (in Japanese).

Mitsunaga, T., Shimizu, M., Inaba, K., Yoshida, T., and Hayashi, S. 1994. Polishing and milling of barley grain for wide use in foodstuffs. *Research Bulletin of Agricultural Department, Kinki University* 27:55–62.

Nagao, S. 1981. Soft wheat uses in the Orient. In: *Soft Wheat: Production, Breeding, Milling, and Uses,* W. T. Yamazaki and C. T. Greenwood, eds., American Association of Cereal Chemists, Inc., St. Paul, MN, pp. 267–304.

Nagao, S. 1996. Processing technology of noodle products in Japan. In: *Pasta and Noodle Technology,* J. E. Kruger, R. B. Matsuo and J. W. Dick, eds., American Association of Cereal Chemists, Inc., St. Paul, MN, pp. 169–194.

Nishizawa, N., Fudamoto, Y., Shimanuki, S., Nagasama, T., and Abe, Y. 1997. Food functionality of millets: effects of dietary protein from proso, foxtail, and Japanese millets on cholesterol and triglyceride metabolism, especially plasma HDL levels. Abstracts of the Annual Meeting of the American Assiciation of Cereal Chemists, October 12–16, 1997, San Diego, CA, p. 222.

Obara, T. 1938. Study on the pearling of barnyard grass. *Journal of Brewing Science* 16:387 (in Japanese).

Obara, T. 1940. Malt and *miso* of barnyard grass. *Journal of Brewing Science* 18:341–347 (in Japanese).

Obara, T. 1949. *Science and Utilization of Cereals.* Kawade Shobo, Tokyo, Japan (in Japanese).

Okada, T. 1996. *Encyclopedia for Exploring Japanese Taste.* Tokyodou Shuppan, Tokyo, Japan (in Japanese).

Ono, S. 1992. *Dictionary of Japanese Foods.* Noubunnkyo, Tokyo, Japan (in Japanese).

Ootomo, T. 1976. *Consideration on Buckwheat.* Shibata Shoten, Tokyo, Japan (in Japanese).

Sakamoto, S. 1997. Cultivation and utilization of amaranth. *Science of Food* 238:36–42 (in Japanese).

Sugiyama, N. 1994. Other grains. In: *Vegetable Foods I,* M. Shimada and M. Shimomura, eds., Asakura Shoten, Tokyo, Japan, pp. 176–204 (in Japanese).

Taira, H. 1968. Finger millet. *The Torrid Zone* 3:41–44 (in Japanese).

Taira, H. 1969. Foxtail millet, barnyard grass and proso millet. *Journal of Japanese Association of Tropical Medicine* 4(2):25–30 (in Japanese).

Takeuchi, M., Sigawara, M., and Kawamura, S. 1990. Dietary fiber of corn. *Starch Science,* 37:115–120 (in Japanese).

Yamazki, K. and Shimada, K. 1983. *Cookery and Its Theory.* Doubun Shoin, Tokyo, Japan (in Japanese).

CHAPTER 6

Oriental Soyfoods

KESHUN LIU

1. INTRODUCTION

1.1 SOYBEANS AS A CROP

BASED on historical and geographical evidence and according to Hymowitz (1970), the soybean *(Glycine max)* first emerged as a domesticated crop in the eastern half of North China around the 11th century B.C. of the *Zhou* Dynasty. Because it is easily grown and adaptable to a wide range of soils and climates and because it contains high protein and oil, the soybean was considered one of the five sacred grains (or *wu gu* in Chinese), along with rice, wheat, barley, and millet, which are essential to the Chinese diet and civilization.

As the Zhou Dynasty expanded and trade increased, the soybean slowly migrated to South China. Later, it spread to Korea, Japan, and Southeast Asia. It was not until the 18th century that the soybean reached Europe and then North America. Today, soybeans have become one of the most economical and valuable agricultural crops, not only in the Far East region but also in other parts of the world. As in the past, China produces a large quantity of soybeans, with annual production in the region of 14 million metric tons. However, on a global basis, estimated at 140 million metric tons of production annually, China ranks third, after the United States and Brazil (*Soya Bluebook Plus,* 1998).

1.2 COMPOSITION AND NUTRITIONAL QUALITY OF SOYBEANS

Soybeans are known as "yellow jewel" or "miracle beans" because of their unique chemical composition. Among cereal and other legume species, the soybean has the highest protein content (around 40% on a moisture-free basis); other legumes have a protein content between 20–30%, whereas cereals have a protein content in the range of 8–15%. The soybean also contains about 20% oil, the second highest content among all food legumes (peanuts contain about 48% oil, the highest amount). Because of this, the soybean is considered to be an oilseed. In addition to protein and oil, the third major component is carbohydrate (about 35%), which is the major source of dietary fiber. Other valuable components found in soybeans include phospholipids, vitamins, minerals, and such phytochemicals as isoflavones.

In addition to their large quantity, soybean oil and protein are also of high quality. The fatty acid with the highest percentage in soybean oil is linoleic acid (about 53%). It is followed in decreasing order by oleic (about 23%), palmitic (about 11%), linolenic (about 8%), and stearic acid (about 4%). Therefore, soybean oil is a healthy oil, containing higher percentages of mono- and polyunsaturated fatty acids. It not only provides us with calories and essential fatty acids, but also helps reduce the risk of heart disease by lowering serum cholesterol levels.

Soy protein contains all the essential amino acids, namely, isoleucine, leucine, lysine, methionine, cysteine, phenylalanine, tyrosine, threonine, tryptophan, valine, and histidine. Most of these amino acids are present in amounts that closely match those required for humans or animals. The only limiting amino acids are sulfer-containing ones, mainly methionine and cysteine. However, such a deficiency is increasingly seen as less problematic, particularly for human nutrition. One reason is that protein quality depends on many factors, including amino acid composition of the protein, amino acid requirements for humans or animals fed the protein, digestibility of the protein, amounts of naturally occurring antinutritional factors, and the method for assaying protein quality. In the earlier days of protein assay, the protein efficiency ratio (PER) was used, which is measured by feeding the protein of interest to rats. Since rats have a much higher need for sulfur-containing amino acids than humans do, soy protein quality was underestimated. In recent years, a new method of protein assay has been adopted by the Food and Drug Administration. It is called the protein digestibility corrected amino acid score (PDCAAS), which compares the pattern of essential amino acids in a protein with the requirements that humans have for essential amino acids and then adjusts for how well that protein is digested. In other words, PDCAAS measures how limiting the limiting amino acid is in a protein. Under this new assay method, soy protein has now had a score very close to 1, the highest rating possible, the same rating for such animal proteins as egg white and casein (FAO/WHO, 1990).

Another reason is that the deficiency in sulfur-containing amino acids can readily be corrected by a simple change of dietary patterns. For example, since cereal grains are normally rich in sulfur-containing amino acids, one easy solution to the problem of deficiency in soy has been to combine soyfoods with cereals. This dietary pattern has in fact been practiced in the Orient throughout the history of soyfood consumption, even without much understanding of food chemistry and human nutrition. Another solution is to increase the volume of soy consumption in a single meal. Since the so-called "limiting" for certain amino acids is rather relative, by increasing the serving size, any requirement for an amino acid can be met regardless of how limiting it is.

1.3 SOYFOODS: FROM THE EAST TO THE WEST

During the course of soybean cultivation, the Chinese had gradually transformed soybeans into various forms of soyfoods. *Tofu,* soy sauce, *jiang* (soy paste), and soy sprouts are among the popular ones. This transformation makes soybeans as a food more versatile, more tasteful, and more digestible. Because of their high contents of protein and oil, with a fairly balanced amino acid profile and abundant essential fatty acids, the traditional soyfoods have nourished Chinese peoples throughout history to the present.

Along with the method of soybean cultivation, the ways of preparing soyfoods and eating soybeans were gradually introduced to Japan, Korea, and other Far East regions. People in these regions not only accepted soyfoods, but also modified them and even created their own types to suit for their local taste. Japanese *natto* and Indonesia *tempeh* are just two examples. Within the last century, the art of preparing soyfoods has now spread to the rest of the world, thanks to agricultural and processing innovation, cultural exchanges, and the influence of Chinese and other Asian immigrants. More importantly, many modern processing technologies have found applications in making traditional soyfoods. This has led to their large-scale production in many regions of the world, with improved and consistent quality. Most recently, new medical research has unveiled the role of soyfoods in preventing and treating chronic diseases such as cancers and heart diseases (Messina, 1997). There is no doubt that soyfoods are gaining popularity throughout the world.

1.4 SOYFOOD CLASSIFICATION

In general, traditional soyfoods are classified as nonfermented and fermented. Nonfermented soyfoods include soymilk, tofu, soy sprouts, *yuba* (soymilk film), *okara* (soy pulp), immature soybeans, soynuts, and toasted soy flour, whereas fermented soyfoods include soy sauce, soy paste, *natto, tempeh, sufu* (fermented tofu), and soy nuggets (fermented whole soybeans). Tables 6.1 and 6.2 list English names, local names, and a general description and the uses of nonfermented and fermented soyfoods, respectively.

Most nonfermented soyfoods are consumed mainly for nourishment. In contrast, many fermented soyfoods are generally served as seasonings or

TABLE 6.1. Names, General Description, and Utilization of Nonfermented Oriental Soyfoods.

Soyfoods	Native Names: Chinese[a]	Japanese	Korean	Others[b]	General Description and Uses
Soymilk	*Doujiang,* *Dou nai,* *Dou ru*	*Tonyu*	*Kong kook* *Doo goo*		Heated water extract of soybeans after grinding and filtering. Resemble dairy milk. Served hot or cold as breakfast, beverage, or with other foods.
Tofu (Soybean curd)	*Doufu* *(Toufu)*	*Tofu*	*Doo bu*	*Tahu* (In.) *Tau foo* (Ma.) *Tokua* (Ph.)	White protein curd precipitated from soymilk with a salt or acid. Bland taste. Cooked with or without meat, vegetable, and seasonings. Served as a main dish or soup. Can also be fried, grilled, fermented, or frozen and then dried.
Soy sprouts	*Huang dou ya* *Dao dou ya*	*Daizu no moyashi*	*Kong na mool*		Germinated soybeans in dark. Yellow cotyledons with white sprouts. Cooked and served as vegetable or in soup.
Yuba (Soymilk film)	*Dou fu pi,* *Fu zhu*	*Yuba*	*Kong kook*	*Fu chok* (Ph.)	Creamy, yellowish protein-lipid film formed from the surface of boiling soymilk. Sheets, sticks, or flakes. As a food wrapper or cooked with meat or vegetable or in soup.

(continued)

TABLE 6.1. (continued)

Soyfoods	Native Names				General Description and Uses
	Chinese[a]	Japanese	Korean	Others[b]	
Vegetable soybean (Immature soybeans)	*Qing dou, Mao dou*	*Edamame*	*Put kong*		Fresh green soybeans harvested at 80% maturity. Cooked in pods or pod removed and served as snack or vegetable.
Okara (Soy pulp)	*Dou zha*	*Okara*	*Bejee*	*Tempeh gembus* (In.)	Insoluble residue after filtration of soy slurry into soymilk. Made into dish, salted as pickle, or fermented into *tempeh.*
Roasted soybeans (Soynuts)	*Chao da dou*	*Iri-mame*			Dry or oil-roasted soybeans, seasoned or nonseasoned. Nutty flavor. Served as snacks or made into powder.
Roasted soy powder	*Chao dou fen, Dou fen*	*Kinako*	*Kong ka au*	*Bubuk kedelai* (In.)	Yellowish powder by dry roasting and then grinding soybeans. Nutty flavor. As coating for rice cake or sprinkled over cooked rice.

[a]Mandarin Chinese (or Cantonese).
[b]In. = Indonesian, Ma. = Malaysian, and Ph. = Philippine.

TABLE 6.2. Names, General Description, and Utilization of Fermented Oriental Soyfoods.

	Native Names				
Soyfoods	Chinese[a]	Japanese	Korean	Others[b]	General Description and Uses
Jiang or *miso* (Fermented soy paste)	*Dou jiang* *Jiang* *(Chiang)*	*Miso*		*Tauco* (In., Ma.) *Tao si* (Ph.)	Whole soybeans with wheat flour, rice, or barley, fermented with *Aspergillus*, *Pediococcus*, *Zygosaccharomyces*, *Torulopisis*, and *Streptococcus*. Light yellow to dark paste, salty and meaty. Served as all-purpose seasoning for dishes or soups.
Soy sauce	*Jiang you* *(Chiang yu)*	*Shoyu*	*Kang jang*	*Kecap* (In., Ma.) *Tayo* (Ph.)	Whole soybeans (or defatted soy flake) and wheat fermented with *Aspergillus*, *Pediococcus*, *Torulopsis*, and *Zygosaccharomyces*. Dark brown liquid, salty and meaty. Served as all-purpose seasoning for dishes and soups.
Natto	*Na dou* *Shui dou chi*	*Natto*			Cooked soybeans fermented with *Bacillus natto*. Soft whole beans covered with viscous and sticky polymer and have a distinct aroma. Seasoned and served with cooked rice.

(continued)

TABLE 6.2. (continued)

Soyfoods	Native Names: Chinese[a]	Japanese	Korean	Others[b]	General Description and Uses
Tempeh	*Tian bei*	*Tempe*		*Tempeh kedelai* (In.) *Tempeh* (Ma.)	Cooked then dehulled soybeans fermented with *Rhizopus*. Soft cotyledons bound by white mycelia. Cake-like and nutty flavor. Fried or cooked and served as a part of meal, snack, or in soups.
Sufu (Chinese cheese)	*Dou fu ru, fu ru (Tou fu ju)*	*Sufu*			Firm tofu fermented with *Actinomucor* or *Mucor*. Creamy cheese-like, salty, distinct aroma. As a condiment, served with or without further cooking.
Soy nuggets (Black beans)	*Dou chi (Toushih)*	*Hamanatto*		*Tao si* (Ph.)	Whole soybeans and wheat flour fermented with *Aspergillus*. Soft whole beans with black color, salty and meaty taste. Cooked with vegetable or meat, or served as a seasoning.

[a]Mandarin Chinese (or Cantonese).
[b]In. = Indonesian, Ma. = Malaysian, and Ph. = Philippine.

condiments in cooking or making soups. They contribute more in flavor than in nutrition to the diet. The two exceptions for fermented ones are *tempeh* and *natto,* which are consumed as part of the main meal and contribute protein and oil to the diet, in addition to their characteristic flavor. The proximate chemical composition of selected Oriental soyfoods is listed in Table 6.3.

In this chapter, various traditional soyfoods are discussed with respect to their preparation methods, principles, and utilization. Because of the diversity of ethnic soyfoods and their preparation methods, it is impossible to cover each soyfood in sufficient detail. With regard to modern ways of using soybeans, such as processing into soy oil and soy protein ingredients, no discussion is given here because they are beyond the scope of this chapter. Additional information on the subject can be found in Shurtleff and Aoyagi (1975, 1976, 1979), Watanabe and Kishi (1984), Applewhite (1989), Shi and Ren (1993), Ning (1995), Fang (1997), and Liu (1997).

TABLE 6.3. Proximate Composition (g/100 g fresh weight) of Some Traditional Soyfoods.

Soyfood	Moisture	Protein	Fat	Carbohydrate	Ash
Soybeans (mature)[a]	12.0	34.3	17.5	31.2	5.0
Soybeans (green vegetable)[b]	60.4	14.3	8.2	14.9	2.2
Soymilk[c]	90.8	3.6	2.0	2.9	0.5
Soft tofu[d]	90.3	5.3	0.9	2.6	0.9
Firm tofu[d]	84.0	10.7	2.1	2.3	0.9
Deep-fried tofu[d]	45.2	24.6	20.8	7.9	1.5
Okara[d]	87.0	2.6	0.3	9.4	0.7
Yuba (dried)[d]	7.1	50.5	23.7	15.6	3.1
Sufu (red)[d]	55.5	14.6	5.7	6.4	17.8
Sufu (white)[d]	56.5	14.4	11.2	5.5	12.4
Natto[a]	58.5	16.5	10.0	12.4	2.6
Rice *miso* (light)[a]	50.0	12.6	3.4	21.2	12.8
Rice *miso* (red)[a]	50.0	14.0	5.0	16.2	14.8
Soybean miso[a]	47.5	16.8	6.9	15.9	13.0
Jiang (chunky)[e]	48.6	11.6	5.2	29.3	7.4
Tempeh[f]	64.0	18.3	4.0	12.7	1.0

[a]From Hesseltine and Wang (1972).
[b]From Liu (1996).
[c]From Chen (1989).
[d]From Shi and Ren (1993).
[e]From Shurtleff and Aoyagi (1983).
[f]From Winarno (1989).

2. SOYMILK

Based on the method of preparation, soymilk is generally divided into traditional soymilk and modern soymilk. Traditional soymilk, known as *dou jiang* in Chinese, is made by a traditional method in the home or on the village level. Being considered as an intermediate product during *tofu* production, *dou jiang* is generally served fresh and hot during breakfast. The product not only has a limited shelf life, but also possesses a characteristic beany flavor and bitter or astringent taste, with all nutrients coming solely from original soybeans.

In contrast, modern soymilk, sometimes referred to as soy beverage or soy drink, is produced by the use of modern technology and equipment. Known as *dou ru* or *dou nai* in Chinese, the product has a relatively bland taste with its own commercial identity and standards. In most cases it is flavored, sweetened, and/or fortified for better taste and better nutrition and packed for longer shelf life, as compared with traditional soymilk. It may also be in a powdered or condensed form. Sold as a milk substitute, a healthful soft drink, or an infant formula (with fortification), soymilk is particularly important to those infants who suffer from malnutrition caused by lack of a dairy milk supply in certain regions of the world and to individuals with allergies and diseases associated with dairy milk consumption. Thus, for the past several decades, soymilk has been produced on a large commercial scale, not only in the Orient, but also in Europe, North America, and Latin America.

2.1 TRADITIONAL SOYMILK

In China, traditionally, soymilk is made by soaking, rinsing, and grinding soybeans into a slurry. This is followed by filtering the slurry to separate the residue and cooking the soy extract to become edible (Figure 6.1). The method is basically the same as the one used originally for making soymilk during the second century B.C.:

- *soaking:* Dry whole soybeans, preferably beans with large seed size and light hilum, are cleaned, measured (or weighed), and then soaked in water overnight. The volume of water is normally about two to three times the bean volume.
- *draining and rinsing:* The soaked beans are drained and rinsed with fresh water two to three times.
- *grinding:* The wet, clean soybeans are now ground in a mill (previously a stone mill or hammermill, now mostly an electrical mill) with the addition of fresh water. The water:bean ratio is normally in the range of 6:1–10:1. The slurry is collected in a large container.
- *filtering:* The bean slurry is filtered through a screen, cloth, or pressing sack, with or without a wooden level press. The residue, known as soy

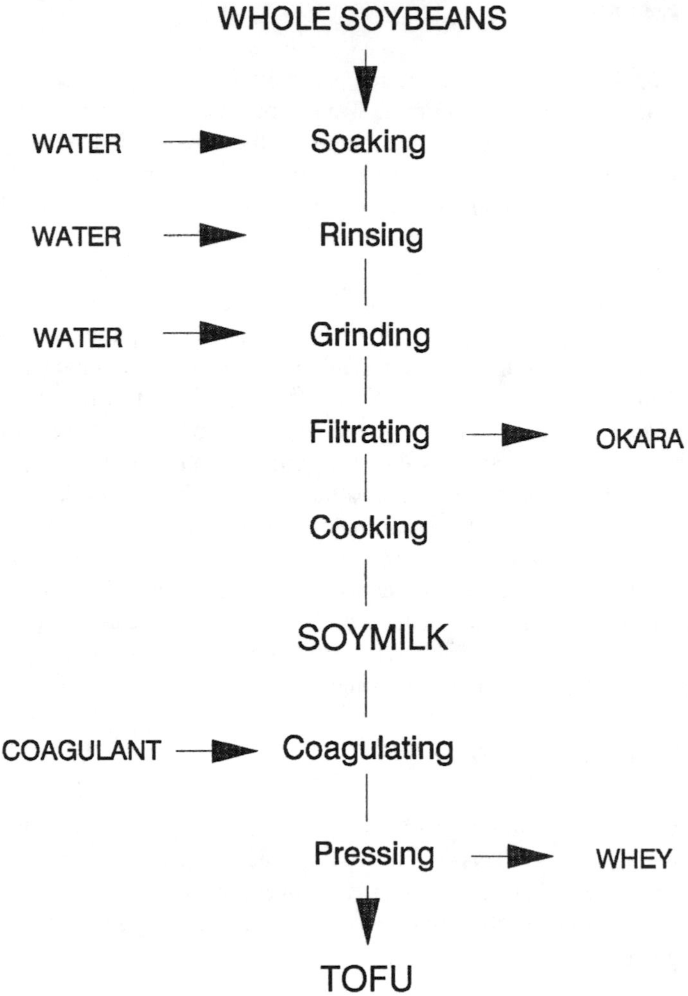

Figure 6.1 A traditional Chinese method for making soymilk and tofu.

pulp in English, *okara* in Japanese, and *dou zha* in Chinese, is removed. It is normally washed once or twice with water (cold or hot), stirred, and pressed again to maximize milk yield. The total volume of the combined filtrate (raw soymilk) is about six to ten times the original bean volume.

- *cooking:* The raw milk is transferred to a big wok or pot and then heated until boiling. To avoid burning, slow heating with frequent stirring is necessary. After boiling for about 10 min, the hot soymilk is ready to serve or is transferred to another container for later consumption. Hot milk can now be further processed into tofu by adding a coagulant, which will be discussed shortly.

Alternatively, soy slurry may be first heated before filtering into soymilk. This procedure is particularly popular in Japan. Regardless of which comes first—heating or filtering—soymilk or soy slurry must be heated to become edible. This is because the heat treatment plays multiple roles: (1) inactivating such antinutritional factors as trypsin inhibitors and lectins, which are naturally present in soybeans; (2) denaturing soy proteins to some extent so that they become more digestible; (3) blanching off objectionable raw bean flavors present in the raw milk; (4) inactivating residual lipoxygenases so that lipid oxidation and its resultant beany flavor are minimized; and (5) increasing storage life by killing microbes carried from the raw materials or from the processing equipment (Kwok and Niranjan, 1995).

2.2 MODERN SOYMILK

The basic principle for making modern soymilk is very similar to the traditional Chinese method. The procedure also includes selecting and cleaning raw soybeans, incorporating water, grinding, filtering, and heating. However, in modern methods, several key steps are modified or added in such ways that reduce beany flavor, increase production yield, and improve overall product quality and consumer acceptance (Chen, 1989; Wilson, 1989; Liu, 1997).

2.2.1 Techniques to Reduce Beany Flavors

To reduce or eliminate the beany flavor, many techniques have been used. Basically, there are three strategies, including (1) preventing beany flavor formation, (2) stripping off the responsible volatiles once they are formed, and (3) masking the residual off-flavor with flavorings.

During the 1960s and 1970s, several studies were conducted to elucidate the chemistry of beany flavor formation during the preparation of soymilk (Wilkens et al., 1967; Nelson et al., 1976). It was found that the beany flavor of soymilk in particular and soy products in general results mainly from peroxidation of polyunsaturated fatty acids or esters catalyzed by an enzyme known as lipoxygenase. The reaction produces many volatile compounds, including ketones, aldehydes, and alcohols, most of which impart undesirable flavors.

Based on this finding, two procedures have been developed to prevent beany flavor formation by inactivating the enzyme. One is a hot-grinding method and the other is a preblanching method. In the hot-grinding method, also known as the Cornell method, unsoaked, dehulled soybeans are ground in a preheated grinder with hot water. The slurry is maintained at temperatures between 80°C and 100°C in the grinder to completely inactivate the lipoxygenase, and then it is boiled in a steam-jacketed kettle with constant stirring for 10 min. The heated slurry is filtered by a centrifuge or a filter press. The resulting soymilk is formulated, bottled, sealed, and finally sterilized at 121°C for 12 min (Wilkens et al., 1967).

The preblanch method, known also as the Illinois method, starts with blanching presoaked soybeans in boiling water for 10 min or by placing dry beans directly into hot water for 20 min. Either of the procedures will hydrate soybeans and inactivate enzymes. The beans are then drained and ground with sufficient cold water to make 12% bean solids. The slurry is first heated to about 93°C and then homogenized. The product may be formulated, pasteurized, homogenized again, and bottled. Soymilk thus produced has a high yield since all original materials end in the final product (Nelson et al., 1976).

Once off-flavors are formed, the only way to eliminate them is to strip off the responsible volatile. To accomplish this task, a deodorization process is available for soymilk production. The process involves passing cooked soymilk through a vacuum pan at a high temperature and high vacuum condition. The method is fairly complex and expensive and is used by a number of large soymilk manufacturers only in conjunction with other techniques.

2.2.2 Commercial Methods

Most commercial methods employ multiple techniques to improve both soymilk quality and yield (Shurtleff and Aoyagi, 1979; Chen, 1989; Wilson, 1989). A representative procedure of commercial soymilk processing is the Tetra Alwin Soy® process line, designed and marketed by Alfa Laval (Lund, Sweden). As shown in Figure 6.2, the production line is aimed for continuous extraction of soybase from soybeans. Cleaned soybeans are fed from a bean feeder into a grinder funnel, where hot water is added. The hot water inactivates lipoxygenases and at the same time forms an air lock to minimize air in the soy slurry. Lipid oxidation is thus effectively prevented. As an option, sodium bicarbonate can also be dosed into the funnel in order to improve the flavor. The beans are ground, and the slurry is pumped to a decanter centrifuge for *okara* separation. Before removal by a pump, *okara* is blended with water and decanted once again to increase the efficiency of extraction. The clarified soy extract is pumped to a steam injection head where live steam is injected into the extract. The temperature increases instantly, and the product is held for a certain time in a holding tube to inactivate trypsin inhibitors and other antinutrients naturally present in soybeans. The time/temperature requirement is based upon 85% inactivation of the inhibitors. This step is followed by flash cooling in a vacuum chamber, which also removes volatile off-flavors and air in the extract. The cooled soybase is ready for blending with other ingredients to formulate various products, including soymilk, *tofu,* soy ice cream, and soy yoghurt. In making soymilk, pasteurization is needed before packaging, storage, and/or distribution.

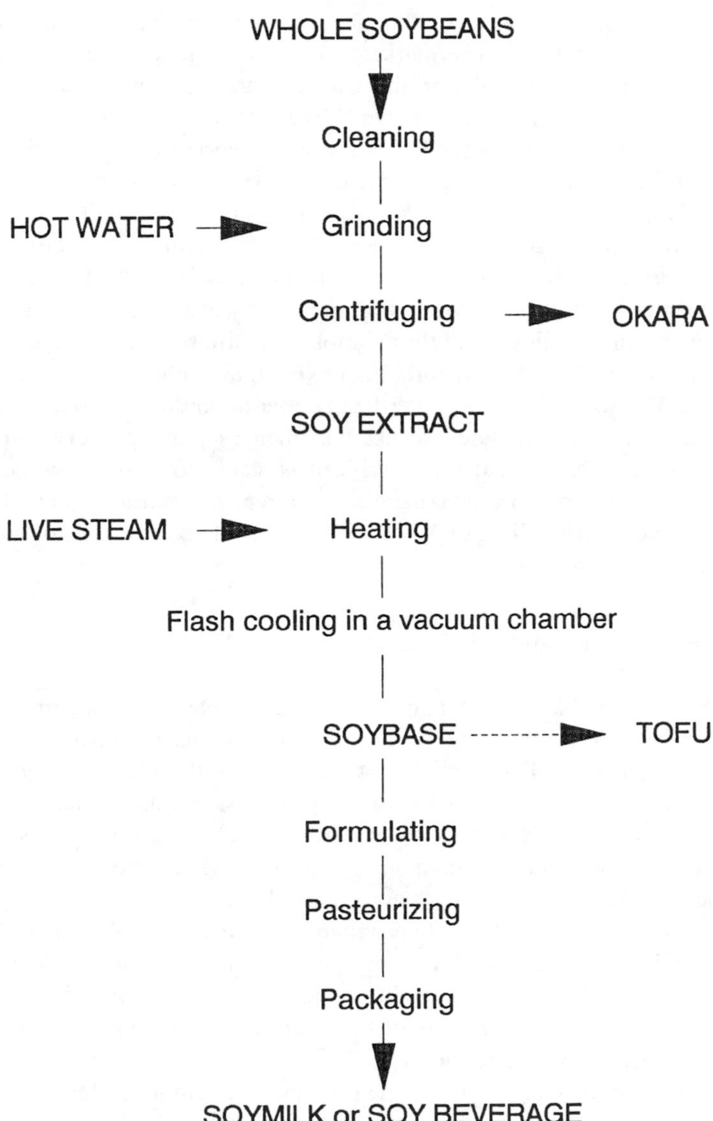

Figure 6.2 A commercial processing method (Alfa Laval) for making soybase and a subsequent product—soymilk.

There are many commercial soymilk manufacturers around the world. Among them, Hong Kong-based Vitasoy International Holdings Limited is a leading one. The company mass produces a range of soy drinks under a brand name of Vitasoy®, along with a range of *tofu* products and other noncarbonated beverages and food products. Vitasoy® soymilk comes in different flavors, including regular, malt, chocolate, melon, vanilla, and cocoa, and is distributed in mainstream supermarkets and/or health food outlets, not only in the Far East regions, but also in the United States, Canada, Australia, New Zealand, and other European countries. More recently, the company launched a new variant, Vitasoy Calci-Plus®, a soymilk enriched with calcium. Although the detailed procedure for producing Vitasoy® soymilk has been kept secret, the basic step has been disclosed to outsiders (Shurtleff and Aoyagi, 1979). Briefly, whole soybeans are subject to an optimum treatment of dry heat and pressure before being dehulled and milled to flour. After the flour is soaked and boiled in a cooker, the soymilk is extracted in a centrifuge, all on a continuous basis. This is followed by formulation with different ingredients, such as vitamins, sugar (about 5%), natural malt extract, and others, to have different products. The soymilk is finally sterilized either in bottles and sold as such or by an ultra-high-temperature process and then aseptically packed in Tetra Brik® cartons. The product thus sterilized is stable for months without the need for refrigeration. In China, there are also many commercial soymilk producers. Among them, Jiangsu VV Group is considered the king of soymilk production in China.

2.2.3 Formulation and Fortification

One of the key factors to obtain widespread acceptance of modern soymilk is to formulate soymilk with different types of sweeteners, flavoring agents, and other materials. Formulation not only masks the characteristic beany flavor and bitter taste associated with soy products, but also imparts different types of flavors and tastes that suit various customers. It also improves the nutritional value of soymilk. A common list of ingredients used for formulation includes honey or maple syrup, sugars, vanilla extract, locust bean, cocoa powder, orange juice, and salt. Other ingredients include malt or malt flavor, coffee, and almond extract. Mixing soymilk with non-fat cow's milk solids or coconut milk is also popular. In addition, fat and/or lecithin are often added to increase richness and creaminess of the final product although such practice generally requires homogenization.

Besides formulation, soymilk made for babies or sold for general consumption in developing countries is normally fortified with vitamins, minerals, and in some cases, amino acids. The most widely used nutrients for fortifying soymilk are vitamin B_{12}, calcium, and methionine.

2.2.4 Homogenization, Thermal Processing, and Packaging

Final processing for modern soymilk may include homogenization, pasteurization, or sterilization, and packaging. Homogenization is normally done after formulation and/or fortification as a way to mix all ingredients well. It also further stabilizes soymilk emulsion, reducing the tendency of chalkiness. Pasteurization extends the shelf life of the final product by destroying the vegetative bacteria. The product is then packaged in glass bottles, plastic pouches, or gable-top cartons and must be stored under refrigeration. The shelf life of pasteurized products is usually within 7 days. Because the product undergoes a low temperature and short time treatment, the essential nutrients and the original flavor are generally preserved. Alternatively, soymilk may be subjected to sterilization under a higher and prolonged heat treatment. The resulting product has a longer stable shelf life even without refrigeration.

Aseptic packaging is a modern technology that not only extends product shelf life, but also maximally maintains original quality. In this process, soymilk and the packaging material are sterilized separately. Soymilk is first subjected to 140–150°C for 2–8 s to inactivate microbial spores and then to a vacuum system by spraying for flash-cooling to 60–75°C to get rid of undesirable flavor compounds. The product is finally filled into a sterilized package in a sterile environment and sealed hermetically to prevent recontamination. The technique has a number of advantages compared with conventional packaging methods, including lower cost, lighter weight, easy handling and stocking, longer shelf life, and favorable consumer acceptance. Thus, it is widely used for modern soymilk production (Sizer, 1989).

2.3 CONCENTRATED AND POWDERED SOYMILK

Soymilk may be concentrated and dried to become concentrated or powdered products. Such products have the advantages of reduced bulk volume and extended shelf life, yet their disadvantages are the inevitable decline in flavor and the added equipment and processing costs. There are four drying methods that can be applied to soymilk, including spray-drying, freeze-drying, vacuum roll-drying, and atmospheric roll-drying (also known as drum-drying). Spray-drying is the most widely used commercial method for drying milks because the very short time of heat contact and the high rate of evaporation give a high quality of the product, with a relatively lower cost. A number of spray-dried soymilks are now available on the market; some are sold together with a coagulant to make instant homemade tofu.

2.4 FERMENTED SOYMILK

Certain lactic acid bacteria have the ability to grow in dairy milk to produce various types of fermented dairy products, including acidophilus milk (sour milk), cultured buttermilk, yogurt, cheese, and other cultured milk products. The microbial action not only increases the shelf life and nutritional value of these products, but also makes them more pleasant to eat or drink.

Soymilk resembles dairy milk in composition, so it can also be fermented by lactic acid bacteria to produce such products as sour soymilk and soy yoghurt. Fermentation of soymilk offers not only a means of preserving soymilk, but also a possibility for modifying or improving its flavor and texture so that it becomes more acceptable to Westerners. It also leads to new types of soy products that resemble cultured dairy products but are at a low cost.

Certain lactic acid bacteria such as *L. acidophilus, S. thermophilus, L. cellobiosis, L. plantarum,* and *L. lactis* have been shown to grow well in soymilk, but they produce less acid in soymilk than in cow's milk (Wang et al., 1974; Mital and Steinkraus, 1979). The major reason is that soymilk lacks monosaccharides and the disaccharide lactose. Instead, it contains such sugars as sucrose, raffinose, and stachyose, which are not readily digestible by many lactic starters because of the lack of α-galactosidase in these organisms. Therefore, numerous efforts have been made to increase acid production during lactic fermentation of soymilk. These include selection of culture strain, alteration of processing conditions during soymilk preparation, addition of fermentable sugars, and/or enrichment with dairy ingredients (Mital and Steinkraus, 1979; Chopra and Prasad, 1990).

2.5 SOYMILK COMPOSITION AND STANDARDIZATION

On average, soymilk has total solids at 8–12%, depending on the water: bean ratio in its processing. Among them, protein is about 3.6%; fat, 2.0–3.2%; carbohydrates, 2.9–3.9%; and ash, about 0.5%. Such a composition of soymilk compares favorably with those of cow's milk and human milk (Chen, 1989). In addition, soymilk is lactose-free and contains higher levels of protein, iron, unsaturated fatty acids, and niacin. However, it has lower amounts of fat, carbohydrates, calcium, riboflavin, thiamine, methionine, and lysine (Kosikowski, 1971). Therefore, many commercial soymilks are fortified with these nutrients.

In order to ensure that consumers enjoy quality soymilk and also to discourage production of a very diluted soymilk, many countries have established soymilk quality standards, including Japan, China, Singapore, and Thailand (Chen, 1989). It is important for soymilk producers to meet these standards when seeking both domestic and oversea markets.

3. TOFU

Tofu is water-extracted and salt- or acid-precipitated soybase in the form of a curd, resembling a soft white cheese (Figure 6.3) or a very firm yoghurt. On a wet basis, a typical pressed tofu with a moisture content in the range of 85% contains about 7.8% protein, 4.2% lipid, and 2 mg/g calcium. On a dry basis, it

Figure 6.3 Homemade firm tofu.

contains about 50% protein and 27% oil, the remaining components being carbohydrates and minerals (Wang et al., 1983).

Tofu is inexpensive, nutritious, and versatile. It can be served as a meat or cheese substitute. Yet compared with meat or cheese, it has much lower calories because of its higher protein/fat ratio. It is also cholesterol-free, lactose-free, and lower in saturated fat. Because of its bland taste and porous texture, tofu can be prepared with virtually any other foods. Most popularly, it is served in soups or separate dishes stir-fried with meat and/or vegetables. It can also be further processed into such secondary tofu products such as deep-fried tofu, dried-frozen tofu, and fermented tofu *(sufu)*.

Historians believe that the method for preparing both soymilk and tofu was invented by Liu An of the Han Dynasty in China, in about 164 B.C. About 900 years later, it spread to Japan and then to other Far East countries. Since then, tofu has been the most popular way to serve soybeans as a food in the Far East. Even today, there are thousands of tofu shops throughout China, Japan, and other Southeast Asian countries or regions, where many types of tofu are produced daily for local consumption. In recent years, tofu has become increasingly popular throughout the world, as increased numbers of consumers are looking for health foods of plant origins. This has led to increasing development of an infrastructure for large-scale commercial tofu production and distribution.

3.1 METHODS OF TOFU PREPARATION

Many methods are available for making tofu today, and all of them are derived from the traditional Chinese method developed some 2000 years ago. Basically, the procedure starts with preparation of soymilk (see Section 2.1 and Figure 6.1). After the milk is boiled for about 10 min, it is transferred to another container, usually a wooden barrel or a pottery vat, and allowed to cool down. At the same time, a coagulant suspension is prepared by mixing a powdered coagulant with some hot water well. Traditionally, either powdered gypsum or nigari is commonly used. When soymilk cools to about 78°C, the coagulant solution is added with rigorous stirring. When tiny curds appears (normally in less than 30 s), the container is covered, and coagulation is allowed to complete for about 30 min.

The soy curd thus formed is now ready for molding. It is first broken by stirring and then transferred to a shallow forming box lined with cloths at each edge. The four ends of the forming box cloth is pulled up and folded over curds in the box. The box is now covered with a wooden lid that is smaller than the box size and on the top of the lip, a few bricks or stones are placed. For about 30 min, whey is pressed out and tofu becomes firm. Cooled tofu is finally cut into cakes that are ready to be served or immersed in cold water for short storage or for sales at local markets.

Today, the traditional method is still popular at home and village levels. However, based on the same principle, many new methods have been created to suit for making different types of tofu products and for using different types of equipment for varying scales of production. This variation in tofu making will be discussed in the following sections.

3.2 FACTORS INVOLVED IN TOFU MAKING

In the Orient, tofu making has long been considered an art. Without undertaking some forms of apprenticeship, not everyone can make tofu or tofu with good quality. Even until today, with our great understanding of protein chemistry, it has been a difficult task to make tofu with consistent quality and yield even under a well-controlled processing condition. The major reason for this difficulty arises from the fact that there is a complex interaction of many factors that are involved in the making of tofu (Watanabe et al., 1964; Bourne, 1970; Saio, 1979; Skurray et al., 1980; Wang and Hesseltine, 1982; deMan et al., 1986; Beddows and Wong, 1987a; Ohara et al., 1992; Shi and Ren, 1993; Ning, 1995;

Cai and Chang, 1997; Shih et al., 1997; Evans et al., 1997). These factors (or variants) mostly center around three key areas: the way soymilk is prepared, the way soy protein is coagulated, and the way tofu is pressed and packaged.

In general, factors affecting soymilk preparation include soybean varieties, whether soybeans are ground hot or cold, whether soy slurry is heated before or after filtration, the water to bean ratio, and the extent of heat applied to soymilk. Factors affecting coagulation include the temperature at which a coagulant is added, the type and concentration of coagulants, the mode of adding coagulants, and the duration of coagulation. Factors involved in the molding step include whether curds are broken and then pressed to separate whey and the pressure and time applied to press curds. To be a master in tofu making, one needs constantly to make a choice or keep these factors in balance to maximize both tofu yield and quality. A few major ones are briefly discussed below; for a detailed discussion, refer to Liu (1997).

3.2.1 Soybean Varieties

Traditionally, soymilk and tofu manufacturers prefer large-seeded soybeans with clear hilum and high protein content. These beans are now specially bred in North America, China, and Japan (Liu et al., 1995). Such beans, known as tofu beans, are believed to produce tofu with whiter color, higher yield, and better overall qulaity, compared to regular field beans known as oil beans.

3.2.2 Concentration of Soymilk

The concentration of soymilk or total solids in soymilk is closely related to the water: bean ratio. It can be measured easily with a refractometer and expressed as °Brix. In general, higher solids content in soymilk correlates with a harder texture and lower yield of tofu (Beddows and Wong, 1987a; Ohara et al., 1992; Cai and Chang, 1997; Shih et al., 1997). In general, tofu manufacturers use soymilk with a solids content ranging between 5% and 12%.

3.2.3 Heat Process of Soymilk

Heating soymilk after filtering (or in some processes, soy slurry before filtering) is essential, not only for improving nutritional quality and reducing beany flavor, but also for denaturing proteins so that they can coagulate into curds in the presence of a coagulant for tofu making. However, extended heat treatment should also be avoided since it not only leads to destruction of such nutrients as essential amino acids and vitamins, Maillard browning, and development of cooked flavor, but also produces a tofu with reduced yield and poor quality. Regardless of whether soymilk or soy slurry is heated, it is recommended that 10 min at near 100°C is optimum for heat treatment (Saio, 1979).

3.2.4 Type of Coagulants

Coagulants that are widely used for tofu making are calcium sulfate, nigari, and glucono-δ-lactone (GDL, or simply known as lactone). Calcium sulfate is the most widely used tofu coagulant. It is also the oldest one used in China, with over 2000 years of history. It comes from a translucent, crystalline, white stone named gypsum, a dihydrate form of calcium sulfate, in mountains. The stone is baked and crushed before being used as a coagulant.

Nigari is a by-product during production of table salt from seawater. It is a mixture of mineral compounds naturally found in and sedimented from seawater, from which table salt has been mostly removed. It consists primarily of magnesium chloride plus all of the other salts and trace minerals in seawater.

GDL is a fine, white, odorless, crystalline powder with a sour taste. An oxidation product of glucose, it is made industrially from corn starch, followed by a fermentation process. When dissolved in water, it is slowly hydrolyzed to gluconic acid by water. First used in Japan as a tofu coagulant during the 1960s, lactone is fundamentally different from nigari- and gypsum-types of coagulants since coagulation by lactone results from action of an acid rather than a salt. The great advantage of lactone as a tofu coagulant is that, by controlling temperatures, it allows completion of mixing and packaging before coagulation. Thus, it is particularly suitable for automation and aseptic packaging. Since each type of coagulant has advantages and disadvantages, in commercial processes, a mixture of coagulants, such as that of GDL and calcium sulfate or magnesium chloride, is sometimes used.

3.2.5 Concentration of Coagulants

When a proper amount of coagulant is used, whey becomes transparent, with amber or a pale yellow color and sweet taste. No uncoagulated soymilk remains. However, if too much coagulant is added, whey has a slightly bitter taste, its color turns more yellowish, and curds have a coarse and hard texture. In contrast, if too little coagulant is used, whey is cloudy, and there may be some uncoagulated soymilk remaining. Within a certain concentration range, as the coagulant concentration increases, tofu bulk yield and protein recovery decrease while tofu hardness, fracturability, and elasticity increase. Furthermore, this changing pattern varies with soybean varieties (Sun and Breene, 1991; Shih et al., 1997).

3.2.6 Coagulation Temperature

The temperature of soymilk at the time of adding a coagulant affects the coagulation rate as well as the tofu texture and yield. At a high temperature, proteins possess high active energy, leading to fast coagulation. The resulting tofu tends to have a low water-holding capacity, a hard and coarse texture, and therefore a low bulk yield. When the coagulation temperature is low, the effect is just the opposite. However, if the temperature is too low (below 60°C), coagulation becomes incomplete, and tofu contains too much water and is too soft to retain its shape. Generally, the temperature falls in the range of 70–80°C (Beddows and Wong, 1987b).

3.2.7 Coagulation Time

After adding a coagulant, it is desirable to let the soymilk-coagulant system stand still for a while after a coagulant is added, since completion of coagulation requires a certain period of time. If the time is too short, coagulation is incomplete. If too long, the temperature of the system decreases to such an extent that the subsequent molding step becomes difficult. In general, for silken tofu, the standing time should be about 30 min; for regular tofu, 20–25 min; and for firm tofu, 10–15 min (Shi and Ren, 1993).

3.2.8 Process Automation

During commercial production of tofu, coagulation has been the most difficult step for automation. However, since the 1980s, different types of tofu coagulation machines have emerged in Japan, China, and South Asia. The machine automatically adds the calculated amount of coagulant to the heated soymilk with a stirring device on both batch and continuous bases. It also allows the coagulation system to stand for a certain period of time before molding. Therefore, the finished product tends to have uniform quality. In some large commercial processing plants, tofu is now made continously and automatically from the raw bean cleaning stage to the final stage of packaging.

3.2.9 Packaging

In rural areas, tofu is sold fresh in the local market in cakes covered with water in a wooden bail or box. However, for commercial distribution, most tofu products are packaged in polyethylene containers, filled with water, sealed, pasteurized, and finally chilled before distribution. They need to be maintained refrigerated before consumption and have a shelf life of about 2–3 weeks. In addition, there is a special type of tofu products under the brand name of Mori-Nu®. It is processed under a patented technology and aseptically packaged in a tetrahedral box made of laminated carton paper. It has a shelf life up to a year even without refrigeration.

3.3 VARIETIES OF TOFU

There are many different types of tofu on the market. Based on water content and textural properties, tofu is generally classified as soft (silken), regular, or firm tofu. Basically, all tofu is made in a similar fashion, except for variations in the water:bean ratio, the type and concentration of coagulants, the way a coagulant is added, and the amount of whey being pressed out.

3.3.1 Silken Tofu

Soft or silken tofu has a soft cheeselike texture but is firm enough to retain its shape after slicing. Silken tofu is normally made from rich soymilk containing 10–12% solids. After being finely filtered, the soymilk is allowed to cool to 65–70°C. It is then mixed with a relatively low concentration of calcium sulfate. Over a period of 30–60 min, a fine, smooth, yet firm curd forms. The curd is neither broken nor pressed. The entire box-sized block of silken tofu is then cut into cakes, removed from the box under water, and cooled. In Japan, tofu made without removing the whey is known as *kinugoshi* tofu.

In China, soft tofu is known as *shui doufu,* meaning water tofu, or *neng doufu,* meaning tender tofu. It is normally made by pressing unbroken bean curds to remove some whey. The product has a soft texture but is firm enough to be cut into cakes and packed in containers with water. It is usually served in soups and delicate sauteed dishes. Also, in China, there exists another tofu product with a texture even softer than silken tofu. Known as *doufu-nao* or *dou-nao* (means tofu brain) or *dou hua* (tofu flower), the product is specially made by using thin soymilk and a reduced amount of coagulant. Smooth and tender tofu curds are consumed fresh after scooping into a deep bowl topped with a little soy sauce or sugar.

3.3.2 Regular and Firm Tofu

Regular and firm tofus are mostly pressed tofus, which are known as *momen* tofu in Japan. The difference between the two is that firm tofu is harder than regular tofu. Sometimes, the term *extra firm* is used by some manufacturers. In any terms, the textural difference among silken, regular, firm, or extra firm is rather relative. There are no standards in absolute textural values for them, and the texture of the same type of tofu may vary with manufacturers, seasons, and even batches.

There are two basic features in making pressed tofu. First, the coagulant is stirred into hot soymilk, and curds are thus formed upon cooling. Second, the curds are broken and then pressed while they are still warm. The smaller the particles of broken curds, the firmer the tofu. Also, the heavier the weight or the higher the pressure applied during pressing, the firmer is the tofu. As the texture becomes firmer, the water content becomes less. Firmer texture and reduced water content make tofu easier to handle and more similar to meat or cheese; they also help tofu keep its shape well during cooking. Therefore, pressed tofu is ideal for use in pan-frying, deep-frying, grilling, freezing and then drying, and dicing into an ingredient for other foods or soups.

3.3.3 Chinese Semidry Tofu

Doufu gan, or semidry tofu, is one of the firmest varieties of tofu found in China. It results from heavily pressing tofu curds during molding. The product contains about 62% water or less and has a chewy, meaty texture. It is served as a meat substitute for a main dish or as a starting material for making Chinese savory tofu (see Section 3.4.5).

3.3.4 Chinese Tofu Sheets and Tofu Noodles

The most unusual variety of pressed, semidry tofu is Chinese tofu sheets, known as *qian zhang,* meaning 1000 sheets. With a soft, flexible texture, each sheet looks like a 6 to 12-in square of canvas with a clothlike pattern imprinted on both sides. They are prepared by ladling a thin layer of firm curds onto about 20–100 individual pieces of cloth stacked consecutively in a tall wooden frame. The alternate layers of cloth and curds are then pressed for several hours beneath a very heavy weight. The finished products are sold in market-places for use as a food wrapper. Alternatively, they are cut into strips to become pressed tofu noodles or bean curd shreds, which are served in soups or special dishes.

3.3.5 Lacton Tofu

When a tofu is made mostly with GDL, a modern coagulant, it is known as lactone tofu. The process starts with the preparation of rich soymilk (10–12% solid content). The milk is heated and deaerated before cooling to about 30°C. Air bubbles form during milk production, and their presence affects tofu quality. This is particularly true for lactone tofu, which is characterized by a fine and smooth texture. Deaeration is carried out by passing heated milk to a vacuum evaporator, which not only reduces air bubbles, but also strips off some off-flavor volatiles. The cold soymilk is mixed with GDL at a concentration of 0.25–0.30% (based on the milk volume). The mixture is immediately run into a container and sealed. The container is then immersed in hot water (85–90°C) for 30–50 min for coagulation to occur. The resulting curd is cooled in the container by immersing in cold water. It is then refrigerated. Lactone tofu also has three types based on texture: soft, regular, and firm. In most cases, lactone tofu is not made by GDL alone, but by a combination of GDL and calcium sulfate or magnesium chloride.

3.4 VARIETIES OF TOFU PRODUCTS

Tofu is very versatile as a food; it can be served fresh or cooked with vegetables and/or meat in thousands of dishes and soups. It can also be further processed into various secondary tofu products, including deep-fried tofu, grilled tofu, frozen tofu, dried-frozen tofu, fermented tofu, and more. In most cases, these processed tofu products have different characteristics, end uses, and commercial identities than the original plain tofu discussed above.

3.4.1 Deep-Fried Tofu

When various types of tofu, preferably those with a firm texture, are deep-fried, they become deep-fried tofu. Available in various shapes (cubes, triangles, flat cakes, etc.), the product is known as *youzha doufu* in Chinese and *aburra-age* or *mama-age* in Japanese (Figure 6.4). The process of deep-frying improves tofu quality in a number of ways. In addition to imparting a meaty texture, flavor, color, and aroma, it extends the shelf life by destroying microorganisms and lowering moisture content. It also facilitates transportation by

Figure 6.4 Dish made of tofu and mushroom.

lowering the weight, eliminating the need for packaging in water, and increasing cohesiveness of the product.

Deep-fried tofu can easily be prepared at home from regular tofu. It is also commercially produced on a large scale. Like most deep-fried foods, deep-fried tofu is at its very best just after being prepared, while still crisp and sizzling. Of course, it can also be stored for later consumption. Commercially made products may be served as is without reheating and adding seasonings. Alternatively, they may be seasoned and further cooked before serving or are used as an ingredient for other dishes.

Deep-fried tofu comprises a relatively large proportion of the tofu prepared in the Far East region. In Japan, it is estimated that one-third of all tofu served is deep-fried. Furthermore, deep-fried tofu appears to satisfy Western tastes best because of its meaty texture and lack of beany flavor. It is also suited for Western cooking styles and serves readily as a meat substitute in a wide variety of recipes.

3.4.2 Japanese Grilled Tofu

Grilled tofu may be one of the earliest ways of eating tofu in Japan. The product is not often found in China. To prepare grilled tofu, one needs first to make a tofu that has a strong, cohesive structure. Such firm tofu is prepared from the same curds used to make regular tofu. However, the curds are broken into fine particles and pressed with a heavy weight for a relatively long time (30–50 min), so that more whey is ladled out. Finally, the pressed tofu is cut into cakes about 5 in. long, 3 in. wide, and 2 in. thick, which are further pressed for about 1 h between alternate layers of bamboo mats sandwiched between large wooden boards. Tofu thus prepared becomes so compact that it will hold together when skewered and grilled over a charcoal fire.

In Japan, freshly grilled tofu is either served sizzling hot, seasoned with a little *miso* or soy sauce, or used as a basic ingredient in *miso* soups, simmering tofu, and other dishes. Because it contains relatively less water, it readily absorbs the flavor of seasoned broths, soups, or casseroles. It also has longer shelf life than regular tofu. Traditionally, grilled tofu is one of the featured ingredients in Japan's New Year's cuisine.

3.4.3 Frozen Tofu

When fresh tofu is frozen, the structure and basic characteristics of the tofu undergo a radical transformation. All the water in the tofu turns into ice, whereas protein and other solids congeal into a firm network. As soon as the frozen tofu is placed in warm water, the ice thaws, leaving a network of protein and solids. Like a sponge, the tofu becomes resilient, highly absorbent, and cohesive and has a tough chewy texture. It also has a characteristic flavor. Good quality frozen tofu can be prepared with less than 12 h of freezing and may be used immediately or stored under refrigeration for later use. It makes an excellent, low-cost replacement for gluten meat or cutlets in vegetarian menus. Made almost exclusively in rural farmhouses and temples by putting tofu outdoors in the cold winter, frozen tofu was never sold as a commodity in the past. Today, many people can make it at home by placing fresh tofu into their own freezer. There is also a limited amount of commercial production.

3.4.4 Japanese Dried-Frozen Tofu

Freezing is one of the best ways to preserve tofu; in particular, frozen tofu, when dried, becomes a highly concentrated source of protein and energy, with an extended shelf life (up to 12 months) and easy handling and storage. The product, known as dried-frozen tofu, is very popular in Japan. The estimated commercial production volume is nearly one-tenth that of fresh and deep-fried tofu products.

Originally, dried-frozen tofu resulted from letting ordinary tofu freeze at night in winter and thaw and be pressed the next day. As this process was repeated over and over, the water content of the tofu drained off and diminished. Fresh tofu is difficult to press and heat directly because heating causes reduction in volume, case-hardening, toughening, and browning. But freezing and storing tofu at a temperature below 0°C changes the texture of the tofu, which becomes sponge-like and permits pressing and drying out water. Using the same principle, the product is now produced in large quantities by large manufacturers. The commercial process of dried-frozen tofu starts with freezing regular or firm tofu in a very cold freezing room with strong winds provided by huge electric fans. This makes it possible to reduce tofu temperature to −10°C within 1–2 h. The frozen tofu is then stored in a freezing warehouse (−1 to −3°C) for about 20 days before being placed on a wide conveyor belt, thawed under a spray of warm water, pressed between heavy rollers to expel the water, and dried in a tunnel dryer. The product is finally permeated with ammonia in large vacuum chambers before being packed in airtight cellophane bags. Readily reconstituted by placing in cold or warm water, dried-frozen tofu requires only a few minutes to cook and is a versatile protein ingredient for many dishes.

The presence of ammonia gas helps the finished product to swell and soften properly during rehydration and cooking because it is retained in the openings of the spongy tofu but is driven out at this stage. Alternatively, instead of using ammonia gas, which has a distinctive odor and other disadvantages, phosphate and carbonate (or baking soda) solutions are introduced at the thawing stage and allowed to penetrate into the thawed tofu just before drying. The chemicals have the same function as ammonia gas in helping dried-frozen tofu swell poorly and become soft when rehydrated and cooked (Watanabe and Kishi, 1984).

3.4.5 Chinese Savory Tofu

In China, there are many specially made tofu varieties known as savory tofu. Some are made by simmering pressed tofu in a soy sauce and water until they turn dark brown. The product is known as soy sauce tofu. Others are prepared by simmering in a mixture of soy sauce, oil, and five fragrant spices. They are known as *wu xiang* (five fragrant spices) tofu. Still others are simmered in a chicken or pork broth seasoned with soy sauce, red peppers, or other flavorings. These savory tofu products are served as hors d'oeuvres, thinly sliced like cold cuts and often accompanied by drinks. They are also used as a side dish or topping for a main meal.

3.4.6 Fermented Tofu (*Sufu* or Chinese Cheese)

When fresh tofu is fermented with a strain of certain fungi, such as *Mucor hiemalis* or *Actinomucor elegans,* it becomes a new product known as *sufu* or Chinese cheese. The product, known as *doufu ru* or *furu* in mandarin Chinese and *toufu ju* or *fuju* in Cantonese, consists of tofu cubes covered with white or yellowish-white fungus mycelia, having a firm texture, salty taste, and characteristic flavor (Figure 6.5). Although relatively unknown in some adjunctive countries such as Japan and Korea, *sufu* was produced in China long before the

Figure 6.5 *Sufu* (Chinese cheese).

Ching Dynasty and is consumed mainly as an appetizer or relish by all segments of the Chinese people, including those living overseas.

3.4.6.1 Varieties of Sufu

Various types of *sufu* are produced in China, depending on the type of microorganisms used during fermentation and the type of flavorings and coloring used in the brine or added later. Flavorings commonly used include sugar, wine, chilies, soy sauce, sesame oil, rose essence, and more. These additives impart either flavor or color to the finished product, resulting in different types of products. For example, *sufu* is normally white in color, but the addition of red rice and soy mash to the brine results in a red product known as "red *sufu.*" Other types of *sufu* found in China include *sufu* with chili, *sufu* with chili and sesame oil, *sufu* with five fragant spices, *sufu* in *jiang,* and *sufu* in rice wine or with fermented rice mash, which imparts alcoholic fragrance.

Sufu may also be named after the region where it is made. Some famous products in China include *Shaoxing sufu, Guiling sufu, Kedong sufu* of Helongjiang Province, *Jiajiang sufu* of Sichuan Province, *Tangchang sufu,* also from Sichuan Province, all being named after their production regions (Shi and Ren, 1993). Most of these products are made solely based on secret recipes that have been passed down through the generations. Therefore, their preparation methods are very traditional, complex, and unique.

3.4.6.2 Preparation Methods

Preparation methods vary with types of *sufu* and regions, but all involve three basic steps: preparing tofu, molding (first fermentation), and brining (second fermentation) (Wai, 1968; Shi and Ren, 1993). A common method is outlined in Figure 6.6. Firm tofu is prepared and then diced into uniform pieces, with sizes ranging from 1 in. to 3 in. In a natural fermentation, the diced pieces are arranged on woven bamboo trays or on rice straw in some places. The trays are placed in direct sunlight for at least several hours to let solar radiation naturally kill many unwanted microorganisms and then are stacked on shelves in an incubation room at 20–35°C. The molds already inhabiting the trays or rice straw begin to inoculate the tofu naturally. In a pure culture fermentation, freshly cut tofu cubes are first immersed in an acid-saline solution (6% sodium chloride plus 2.5% citric acid) for 1 h and then subjected to sterilization at 100°C for 15 min. The cubes are separated from one another in a tray with small openings in the bottom and top to facilitate the circulation of air. This helps mycelium development on all sides of the cubes. After cooling, the cubes are then inoculated over their surface with a suspension or dried powder of spores of a selected, pure-cultured microorganism.

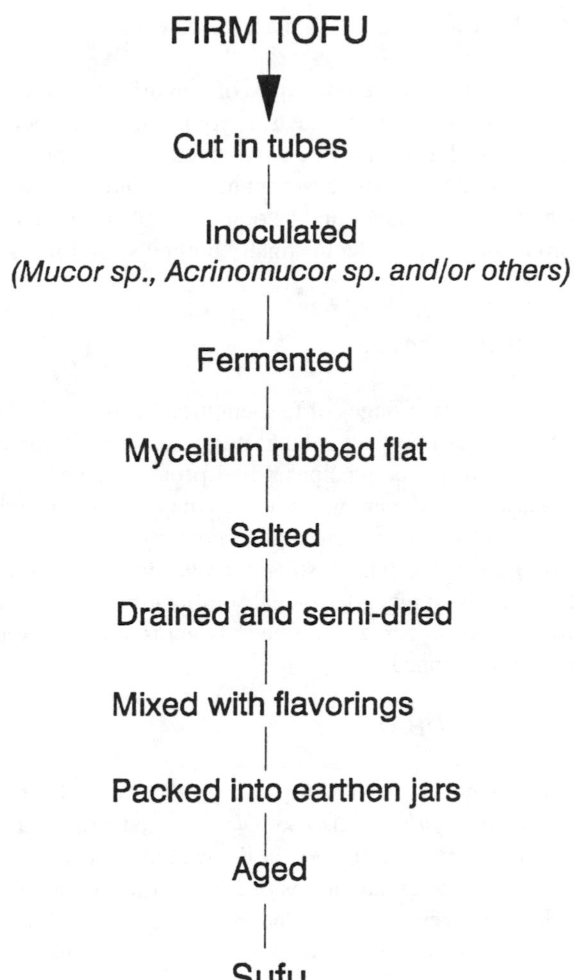

Figure 6.6 A flowchart for making *sufu* out of firm tofu.

The inoculated tofu pieces, either naturally or with a pure culture, are now transferred to an incubation room. The recommended temperature and relative humidity are 25–30°C and about 97%, respectively. After several days of incubation, depending on the temperature and the type of culture, each cube is covered with a fragrant cottony mycelium. This intermediate product is known as *pehtze.* To ensure good formation of a dense and thick texture of mycelial mat, the mycelium on the *pehtze* is normally rubbed flat before the *pehtze* is salted, seasoned, and aged in an earthware crock with a brine. The crock is tightly sealed and put in a cool, dark place for several weeks or even several months.

The brine may contain different types of flavorings and colorings. This results in different types of *sufu* products. A common brine would be one containing 12% sodium chloride and rice wine amounting to about 10% ethanol. The end product has a characteristic flavor and color and very salty taste. Some varieties may have a putrid flavor and be objectionable to some individuals. In commercial production, the product is finally bottled with brine, sterilized, and marketed.

3.4.6.3 Preparation Principle

Making *sufu* requires two stages of fermentation (Wai, 1968). The first stage allows molds to grow as many mycelia as possible. At the same time, various enzymes are produced, including lipases and proteases. The second stage of fermentation, known as brine aging, promotes major biological changes since many enzymes produced during the first fermentation are now released into the brine and become active. This results in flavor development and compositional changes. There is an increase in total soluble nitrogen, a decrease in total insoluble nitrogen, and an increase in free fatty acids, although total lipids and nitrogen remained unchanged.

4. SOYMILK FILM *(YUBA)*

Yuba is another soyfood derived from soymilk. Named after a Japanese word for soymilk film, *yuba* is also known as dried bean curd in English, *doufupi* or *fuzhu* in Chinese, *kong kook* in Korean, and *fuchok* in Malaysian.

Yuba has a very chewy texture and is one of the oldest "texturized" protein foods (Figure 6.7). On average, *yuba* contains 55% protein, 26% neutral lipids, 12% carbohydrate, 2% phospholipids, 2% ash, and 9% moisture (Wu and Bate, 1972). Among several Oriental soyfoods, soy protein in *yuba* has the highest percentage of digestibility (almost 100%) (Ikeda et al., 1995). Because of limited production and high cost, *yuba* is considered a delicacy.

To make *yuba,* one needs to make a rich soymilk first. The soymilk is then heated in a flat, open pan to near boiling (about 80–90°C), and a film gradually forms on the liquid surface because of surface dehydration. After the film becomes toughened, it can be lifted with two sticks or by passing a rod underneath it. The film is hung on a line or spread on a galvanized wire mesh for drying. In this manner, films are continuously formed and removed from the surface of the soymilk until no further film formation occurs. Generally, 10 to 20 such sheets can be made before it is necessary to refill the pan with fresh soymilk.

Figure 6.7 *Yuba* (soymilk film).

During *yuba* production, the protein and lipid contents in the film formed successively decrease, while the carbohydrate and ash contents gradually increase; therefore, the first several pieces of *yuba* to be lifted off the heated soymilk are considered premium products. They have a creamy white color, mild flavor, and less sweet taste. They stay relatively soft and flexible, even when dried. The later-formed products are regarded as second grade, because they become sweeter with a faintly reddish tinge, lack internal cohesiveness, tear more easily, and become brittle when dried.

Yuba is commonly sold in three different states: fresh, semidried, and dried. Fresh *yuba* needs to be served as soon as possible after it has been made since it is highly perishable. It is usually thought to be the most delicious type. Semidried *yuba* has a longer shelf life than the fresh one but not nearly as long as dried *yuba.* Dried *yuba* is quite brittle and has a relative long shelf life. It is the most common of the three forms sold in the market. It has different shapes: flat sheets, long rolls, small rolls, U-shaped rolls, and large spirals. Immediately before cooking, semidry or dry *yuba* needs to be soaked until fully hydrated.

In spite of its nutritional excellence, *yuba* is appreciated primarily for its unique flavor and texture. There are various ways of using *yuba.* It can be used as a wrapper for other foods or used in soups or cooked with other food materials. Fresh *yuba* can also be made into meat analogs by pressing into hinged molds and then steaming to create a number of forms resembling such animals as chickens, fish, and ducks. The product is widely served in vegetarian restaurants under such names as vegetarian chicken, vegetarian fish, or Buddha's chicken. It may be packed or canned and is sold in marketplaces. Because *yuba* is a delicacy, a food containing *yuba* is commonly considered special.

5. SOYBEAN SPROUTS

Bean sprouts have been used as a food in the Orient since ancient times. They are made either from soybeans or mung beans *(Phaseolus aureus)* through germination under a dark condition. Soybean sprouts are more popular in Korea and South China, but less popular than mung bean sprouts in most other parts of the world.

To produce soybean sprouts, soybeans, preferably freshly harvested, small- to medium-seeded beans with good vigor, are first soaked in warm water (40–50°C) for 3–4 h, washed well, and then spread in thin layers in a deep container (or bucket) with holes at the bottom for water draining. A cloth is normally placed over the bottom of the container to prevent passage of the beans. The container is covered with hay, rice straw, or other material to screen from light but allow air exchange, and then it is put in a place where the temperature is kept around 23°C.

During germination, heat is built up because of active seed metabolism. Thus, it is necessary to sprinkle fresh water over the beans in the container three to four times a day. Addition of water not only provides moisture for seeds to germinate and for new seedlings to grow, but also helps to reduce metabolic heat. However, excessive moisture is unfavorable for rapid sprouting because it tends to limit the oxygen supply and leads to decay by mold infection. Also, light should always be avoided during the process because it causes sprouts to develop roots and turn green, both of which are undesirable.

In less than a week, when the majority of sprouts reach a length of about 8 cm, they are washed, dehulled, and ready for serving or transporting to a sale market. The finished product is crispy and has a distinct taste and is composed of yellowish cotyledons and long, bright white sprouts (Figure 6.8). In a typical germination process, 1 lb of dry soybeans can produce 7–9 lb of fresh bean sprouts.

Compared with original dry soybeans, soy sprouts offer several nutritional advantages. These include increased contents of vitamin C and β-carotene (Bates and Matthews, 1975) and reduced levels of such antinutrients as the flatulence-causing oligosaccharides and mineral-binding phytic acid. Serving as a vegetable throughout the year, soybean sprouts are used in soups, salads, and side dishes. During cooking, it is desirable to minimize heating to maintain the inherent crisp texture and distinct taste and to minimize destruction of vitamins.

Figure 6.8 Soy sprouts.

6. GREEN VEGETABLE SOYBEANS

With a green or green-yellow color, soft texture, and large seed size (resulting from high moisture content and specially selected varieties), green vegetable soybeans are normally hand-picked at about 80% maturity in the green-yellow pod from the field. Therefore, they are also known as immature soybeans or fresh green soybeans. Direct consumption of green vegetable soybeans is very popular in China, Japan, and some other Far East countries or regions. Steamed or boiled in water before or after shelling, normally for less than 20 min, and lightly salted or spiced, these immature beans can be served either as a delicious green vegetable with a main meal or as a tasty hors d'oeuvre, often with beer or other alcoholic drinks. In Japan, immature soybeans are known as *edamame* and are sold fresh or frozen in the market. They may also be made into roasted beans, which have a crunchy texture and greenish-beige color, and sold as *irori mame*. In the West, frozen and canned immature soybeans have appeared on the market. There is at least one U.S. company, known as SunRich, Inc., Hope, Minnesota, which is currently marketing immature soybeans under the brand name of "Sweet Beans®."

Depending on stages of immaturity, green vegetable soybeans contain protein in the range of 11–16% and oil in the range of 8–11%, whereas on a moisture-free basis, they have protein and oil compositions very close to that of matured soybeans. However, compared with field-dried mature soybeans, fresh immature soybeans offer several distinct features, including a less beany flavor, higher vitamin content, lower levels of antinutrients, tender texture, and less cooking time (Liu, 1996). In addition, green color and soft texture enhance their appeal as a vegetable.

7. OTHER NON FERMENTED SOYFOODS

Besides soymilk, *tofu, yuba,* soy sprouts, and fresh green beans, other non-fermented soyfoods include *okara,* roasted soynuts, boiled or steamed whole soybeans, and roasted soy grits or powder.

7.1 OKARA

Okara, also known as soy pulp in English, and *doufu zha* or *dou zha* in Chinese, is the insoluble residue after filtration of soy slurry into soymilk. Therefore, it is considered a by-product of soymilk and tofu preparation, yet for every pound of dry soybeans made into soymilk or tofu, about 1 lb of *okara* is generated. More specifically, on an average, 53% of the initial soybean dry mass is recovered in tofu, 34% in *okara,* and 16% in whey. About 72% of soy protein is recovered in tofu, 23% in *okara,* and 8% in whey; the respective average percentages for soybean oil recovery are 82, 16, and <1 (van der Riet et al., 1989).

Although the actual composition depends on the specific process, as well as soybean variety, according to van der Riet et al. (1989), fresh *okara* contains 76–80% moisture, 2.6–4.0% protein, and the remaining percentage for other solids. When dried, it contains 25.4–28.4% protein, 9.3–10.9% oil, 40.2–43.6% insoluble fiber, 12.6–14.6% soluble fiber, and 3.8–5.3% soluble carbohydrates. Therefore, *okara* contains high fiber content and appreciable amounts of protein and oil.

The major use of *okara* is as feed. However, there are various ways of using *okara* as food. For example, in some parts of China, *okara* is salted and spiced and serves as a pickle, or it is simply made into a dish with a meat or vegetable. In other parts of China, *okara* is pressed into cakes and allowed to ferment for 10–15 days until each is covered with a white mycelium of *Rhizopus* mold. The cakes are dried in the sun and then deep-fried or cooked with vegetables. A similar product is also popular in Indonesia, which is known as *tempeh gembus.* Sometimes, *okara* may be mixed with soybeans before fermentation. With a growing awareness of dietary fiber in human health, there is an increasing interest in using *okara* as a food ingredient. For example, plain or flavored *okara* can easily be dehydrated on a drum dryer to make a dry staple, which might be further milled into a flour before being used as a high-fiber food ingredient.

7.2 ROASTED OR COOKED WHOLE SOYBEANS

When clean, whole soybeans are roasted for about 30 min, they become brown and acquire a characteristic toasted flavor. Upon cooling, the roasted beans, known as soynuts, can be eaten like roasted peanuts as a snack or used as an ingredient to add a crunchy texture and nutlike flavor to a wide variety of salads, sauces, casseroles, and *miso* preparations. Besides dry-roasting, whole soybeans may be oil-roasted.

Roasted beans may be covered with various flavoring and coating materials, including sugar, chocolate, onion, and garlic. Such coated soynuts are now seen in the Western market. Compared with roasted peanuts, soynuts provide a higher protein content with lower cost.

In the Far East, whole soybeans are sometimes consumed directly after soaking and cooking (steaming or boiling) until their texture becomes tender. Salt, oil, soy sauce, and other spices and seasonings may be added during cooking. When meat is also added, the dish becomes a tasteful and Oriental version of "pork and beans."

7.3 ROASTED SOY POWDER

When roasted soybeans are ground into powder, they become roasted soy powder, which is similar to modern full-fat soy flour, except that it contains the seed coat and has a nutty flavor. The product is known as *doufen* in Chinese and *kinako* in Japanese. Like roasted soynuts, roasted soy flour is an inexpensive source of good-quality protein for supplementing Oriental diets. In China, roasted soy flour may be mixed with lard and sugar and used as a filling or coating material for pastry. In Japan, one of the favorite ways of using roasted soy flour is to spread it on rice or rice cakes. In Indonesia, the flour is mixed with spices, such as garlic and chili powders, and served with *longtong,* which is boiled rice wrapped in banana leaves.

8. FERMENTED SOY PASTE *(JIANG* AND *MISO)*

Soy paste is an important fermented soyfood in the Far East. It has a color varying from a light, bright yellow to very blackish brown; a distinctively pleasant aroma; and a salty taste (Figure 6.9). Soy paste is commonly known as *jiang* (Mandarin) or *chiang* (Cantonese) in China, *miso* in Japan, *jang* in Korea, *Taucho* in Indonesia, and *taotsi* in the Philippines.

Developed in China some 2500 years ago, *jiang* was the progenitor of the many varieties of soy paste and soy sauce that are now used throughout the world (Shurtleff and Aoyagi, 1983). At present, Chinese *jiang* and Japanese *miso* are the two most popular types of soy paste. Although sharing the same progenitor, the two differ in many aspects. Chinese *jiang* is made from soy-

Figure 6.9 Chinese *jiang* (top) and Japanese red and white *miso.*

beans and wheat flour. The finished product may not be ground so that individual particles of soybeans are present. It is used mainly as an all-purpose seasoning for dishes and soups. However, Japanese *miso* is made from soybeans mixed with rice or barley or from soybeans alone. The finished product is a paste resembling peanut butter in consistency and may have a sweet taste. It is mainly dissolved in water as a base for various types of soups in Japan (Fang, 1997).

8.1 *KOJI, KOJI* STARTER, AND INOCULUM

Before describing the methods of preparing soy paste, as well as some other fermented soy products, it is necessary to discuss some terms first.

8.1.1 *Koji*

The Chinese counterpart for the word *koji* is *qu,* meaning bloom of mold. Made by growing molds on rice, barley, wheat, soybeans, or a combination, *koji* contains a great variety of enzymes that digest starch, protein, and lipid components in raw materials. It is an intermediate product for making not only *jiang* and *miso,* but also some other fermented products such as soy sauce, soy nuggets, and Japanese *sake.*

The microorganisms found in *koji* almost always belong to fungi species, *Aspergillus oryzae* and/or *A. sojae. A. oryzae* molds reproduce only asexually and have the ability to utilize starch, oligosaccharides, simple sugars, organic acids, and alcohols as carbon sources, and protein, amino acids, and urea as nitrogen sources. The mold is aerobic, with growth optima generally at a pH of 6.0, a temperature of 37°C, and a water content of 50% in a medium. When the air supply is limited or water content of the medium is below 30%, its growth slows down. When a temperature is below 28°C, its growth also becomes slow, but enzymatic activities remain high.

8.1.2 *Koji* Starter

Koji starter, also known as seed *koji, koji* seeds, or *tane-koji,* provides spores of microorganisms to make *koji.* Preparation of *koji* starter is essentially done the same way as making regular *koji* for soy paste and soy sauce, except that in making *koji* starter, pure culture and different raw materials are used and longer fermentation is needed to produce abundant spores. In addition, a sterile condition is needed to avoid contamination.

Since many molds, including *A. oryzae,* are ubiquitous, up until several decades ago, wild spores of the species were used as the starter for *jiang, miso,* or soy sauce preparation. However, the modern process for making *koji* starter begins with growing a selected *A. oryzae* strain on an agar slant in pure culture. The strain is selected for its special abilities by natural selection or by induced mutation to give a desirable *koji* for a particular fermentation. Therefore, there are many varieties of commercial *tane-koji,* each having a different capacity to break down protein, carbohydrate, and lipid in raw materials. It is very important to select a suitable variety for making a particular product. For example, for salty rice *miso* rich in protein, a *tane-koji* of high proteolytic activity is suitable, whereas for sweet rice *miso* rich in starch, a *koji* starter with high amylolytic activity is preferable.

8.1.3 Inoculum

In addition to the *koji* mold, halophilic yeasts and lactic acid bacteria play an important role in developing flavors during fermentation of soy paste and some other fermented soy products. This is particularly true in the later stage of fermentation. Traditionally, in making *jiang, miso,* or soy sauce, a soundly fermenting product from a previous batch was used as an inoculum to be mixed with salted *koji* and cooked soybeans. The inoculum generally contains a selected flora of salt-tolerant yeasts and bacteria capable of growing under anaerobic conditions. The dominant organisms are yeast *Zygosaccharomyces rouxii* and *Torulopsis* sp. and certain lactic acid bacteria such as *Pediococcus halophilus* and *Streptococcus faecalis* (Hesseltine and Wang, 1972). Since pure culture of these organisms speeds up fermentation and reduces the influence of weed yeasts and bacteria, its use in the commercial preparation of certain fermented soy products has been popular in recent years.

8.2 CHINESE *JIANG*

The Chinese word *jiang* has a broader meaning than the Japanese word *miso.* In addition to fermented soy paste, *jiang* refers to almost all other types of fermented paste made with such materials as wheat, peanuts, peppers, sesame, or seafoods. The most common types of *jiang* made from soybeans are chunky *jiang,* hot chunky *jiang,* Sichuan red-pepper *jiang,* Cantonese red *jiang,* and soy nugget *jiang.* Chunky *jiang* (or *dou-ban jiang*) is a variety with a chunky texture caused by the presence of unmashed beans. Hot chunky *jiang* is spicy hot because of the addition of hot peppers. Sichuan red-pepper *jiang* is prepared in Sichuan Province and renowned for its abundant use of red peppers. Soy nugget *jiang* is made from soy nuggets known as *douchi* in Chinese, another fermented soyfood to be discussed shortly.

There are at least three methods being used in China to prepare *jiang:* traditional household method, pure culture method, and enzymatic method (Shi and Ren, 1995; Fang, 1997).

8.2.1 Traditional Household Method

Koji preparation usually starts in March and April. Soybeans are boiled, drained, and then mixed with a small amount of wheat flour. The resulting mixture is spread onto a bamboo tray and kept in an incubation room for a week. The inoculation of the mold is usually done spontaneously in the incubation room or by being mixed with a previous batch of *koji.* In some parts of China, the mixture is covered with a type of wormwood plant, which presumably provides *koji* molds. Since pure culture is not used, the resulting *koji* normally consists of more than one kind of mold. The *koji* thus prepared is mixed with salt and a small amount of water and packed into large earthenware jars or crocks. The container is then placed in a courtyard, on rooftops, or on apartment balconies. The mixture is stirred daily with the top content exposed to the sun. To avoid dust and rain, a transparent plastic film may be used to cover the jar. Insomuch as fermentation continues over all the summer months, the *jiang* becomes darker and thicker. The finished product has a strong aroma, high salt content, dark color, and thick consistency. This method, which has been handed down from generation to generation, is still very popular in many rural areas of China.

8.2.2 Pure Culture Method

The traditional household method for preparing *jiang* is a natural phenomenon and may result in a product with the richest flavor, yet it has disadvantages of a long preparation time, strong influence by seasons, inconsistent product quality, lack of hygiene, and easy contamination with unwanted microorganisms. Therefore, many plants in China have now adopted a pure culture method under controlled conditions, such as the one shown in Figure 6.9. Whole soybeans are cleaned, soaked for 3–5 h, and drained. They are then steamed for 40 min and cooled to 80°C before being mixed with wheat flour, which may or may not have been steamed or roasted. The mixture is allowed to cool to 38–40°C and then is inoculated with a *koji* starter (0.3–0.5%). The inoculated mixture is now packed in a wooden box or bamboo tray and put in an incubation room. The temperature is controlled under 36°C with plenty of air flow. After 30–36 h of incubation with occasional stirring, the *koji* mold has already sporulated, and the mature *koji* turns olive green. The *koji* is now mixed with warm brine; packed into vats, jars, or concrete pools; and allowed to ferment. After at least 15 days, depending on the fermentation temperature, it becomes *jiang*.

8.2.3 Enzymatic Method

In this method, soybeans and wheat flour, after proper heat treatment, are first mixed with brine, *koji* enzyme powder, and an inoculum. *Koji* enzyme powder is made in a way similar to that of making *koji* starter, except that the mature *koji* is finally dried and made into powder. After 15 days of fermentation, the mixture turns into *jiang*. By eliminating the step of *koji* making, the new method shortens total production time and reduces labor and cost.

8.3 JAPANESE *MISO*

There are many varieties of *miso* in Japan based on the type and ratio of raw materials, salt concentration, and the length of fermentation and aging. On the basis of the raw materials used, *miso* is commonly classified into rice *miso,* barley *miso,* and soybean *miso.* Although all three types of *miso* contain salt and water, rice *miso* is made from rice and soybeans, barley *miso* from barley and soybeans, and soybean *miso* from soybeans only. Each type may be further classified by taste into three groups: sweet, semisweet (medium salty), and salty *miso.* Each group may then be further divided by color into white, yellow, red, or brown *miso.* Among these varieties, rice *miso* is the most popular one in Japan, being about 80% of the total *miso* consumption (Watanabe and Kishi, 1984).

The method for making *miso* may vary with variety, but the basic process is essentially the same as making Chinese *jiang*. For example, rice *miso* is made in five distinct steps: preparation of rice *koji,* treatment of soybeans, mixing of all ingredients, fermentation, and pasteurization and packaging (Figure 6.10).

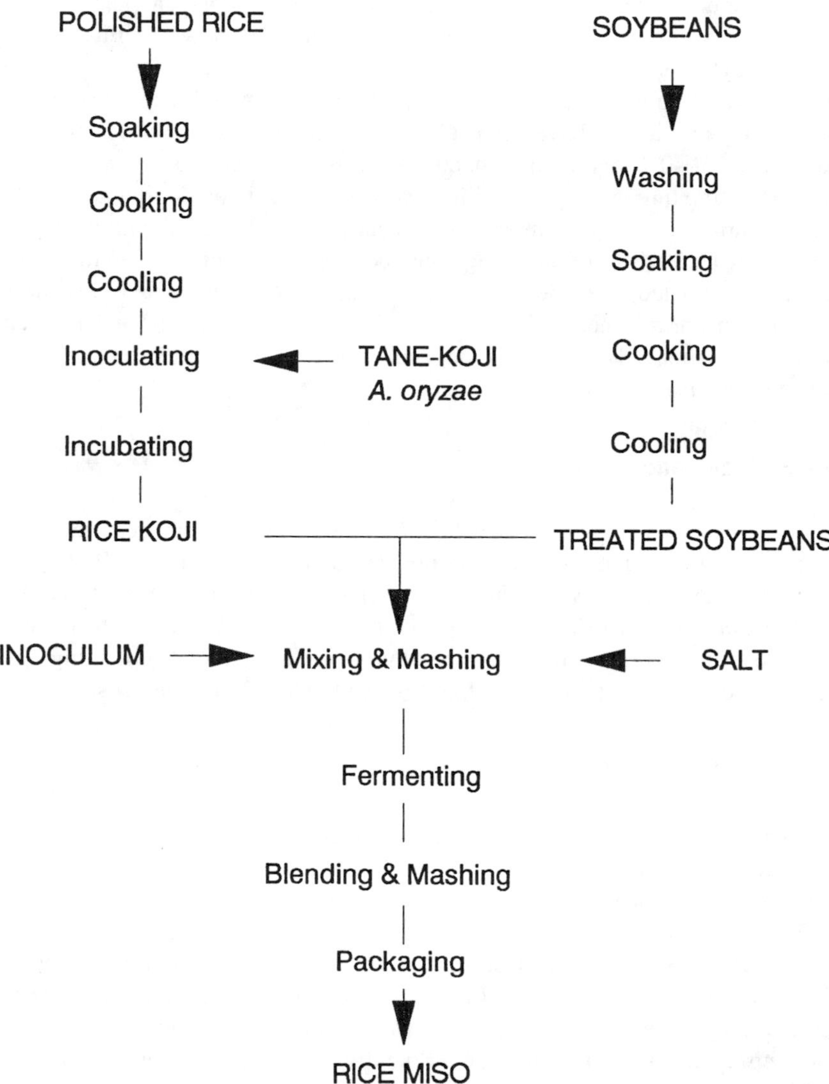

Figure 6.10 A common method for making Japanese rice *miso.*

8.3.1 Preparing Rice *Koji*

Nonglutinous, polished rice is cleaned, washed, and soaked in water at about 15°C overnight or until the moisture content increases to about 35%. After draining, the rice is steamed for about 40 min. When cooled to 35°C, the cooked rice is inoculated with *koji* starter containing *A. oryzae* spores, at a concentration of about 0.1% of the rice. In modern plants, the cooking and inoculating process are carried out continuously.

After 15 h of incubation in a box at a temperature of 30–35°C and a relative humidity higher than 90%, the incubated rice is transferred to the *koji* room where it is spread to a 4-cm depth on the *koji* trays and allowed to ferment. When the temperature goes up to 35°C or more, the young *koji* is turned over and stirred for good aeration for further fermentation. After about 40 h of inoculation, when the cooked rice is completely covered with white mycelium, *koji* becomes mature and ready to be harvested.

A good *koji* has a pleasant smell, lacks any musty or moldy odor, and is quite sweet in taste. The mature *koji* is taken out of the *koji* room to be mixed with salt to halt any further mold development. Recently, there has emerged a mechanical *koji* fermenter that is equipped with an air conditioner and a mechanical stirrer to break the sponge-like lumps that develop during cultivation.

8.3.2 Treating Soybeans

Concurrent with the *koji* preparation, whole soybeans are cleaned, washed, and soaked in water overnight. It is possible to reduce the soaking time by increasing the soaking temperature. The soaked beans are steamed under a pressure of 0.7–1.0 kg/cm^2 for 20–30 min or until they are tender enough to be mashed between the fingers. Soybeans, preferably with a light yellow coat and a clear hilum for white or light-colored *miso* types, are generally cooked in boiling water with a bean : water ratio of 1 : 4, to prevent browning from steaming. Although batch-type cookers are widely used, continuous cookers have been used recently in large factories. Cooked beans are cooled to room temperature with a belt conveyor-type cooler. Sometimes, they may be mashed with a chopper while hot.

8.3.3 Mixing and Mashing

After being cooled to room temperature, the cooked soybeans are mixed with salted rice *koji* and water containing inoculum, which may come from a previous batch or pure culture. The mixed materials are roughly mashed by passing them through a motor-driven chopper with 5-mm perforations. Homogeneous mixing is important to maintain normal fermentation.

The proportions of ingredients at the time of mixing determine the type, flavor, and appearance of the final product. These include water content, the proportion of rice *koji* to soybeans, and the amount of salt. Water content affects, not only the rate of fermentation, but also the consistency of the final product. For *miso,* the final mixture should contain water in the range of 48–52%. In addition, darker types of *miso* require a relatively higher portion of soybeans (in the range of 50–90%) than rice as compared with whiter *miso.* Red or brown *miso* contains 11–13% salt, whereas white *miso* has 4–8% salt. The lower content of salt in white *miso* permits more rapid fermentation but gives the product a shorter shelf life.

8.3.4 Fermenting

After mixing and mashing, the mixture is packed tightly into open tanks or vats. The containers are traditionally made of wood or concrete; however, some modern plants use steel vats coated with epoxy resin, stainless steel vats, or glass-lined resin vats. Weights equivalent to 5–10% of the total *miso* are placed on the sheet to force liquid to the surface, ensuring anaerobic conditions. The young *miso* is allowed to ferment in a controlled temperature, normally in the range of 30–38°C for a period up to 6 months, depending on the type of *miso* to be made. To retain the homogeneity of *miso,* at certain stage of fermentation, the *miso* is transferred from the original vat to another.

8.3.5 Pasteurizing and Packaging

After ripening, *miso* is blended, if necessary, and mashed again through a chopper with a plate cutter having perforations of 1–2 mm. The mashed *miso* is then packaged in a resin bag or cubic container for marketing after being pasteurized with a steam jacket or mixing with preservatives such as 2% ethyl alcohol or 0.1% sorbic acid.

8.4 PRINCIPLES OF MAKING *JIANG* OR *MISO*

Regardless of variations in preparing Chinese *jiang* and Japanese *miso,* the fermentation principle of these products is very similar to each other. In general, all preparations involve treatment of raw materials, preparation of *koji,* mixing of ingredients, and fermentation.

Treatments of raw materials include soaking and heating. Soaking hydrates proteins and other grain components; therefore, it facilitates the effects of subsequent heating. Insufficient hydration leads to insufficient denaturation of soy protein, as well as softening of soybean texture during subsequent heat treatment. The effects of heating include denaturing proteins so that they can be utilized either by *koji* mold or hydrolyzed by *koji* enzymes, inactivating naturally present toxins, such as trypsin inhibitors and lectins, softening soybeans, sterilizing soybeans, and removing the unpleasant bean odor (Nikkuni et al., 1988).

During *koji* making, as spores of the *koji* mold germinate and grow into mycelia and eventually sporulate, various enzymes are produced. Major enzymes include protease, amylase, glutaminase, lipase, hemicellulase, pectinase, and esterase. These enzymes become active during the subsequent stage of fermentation. As a result, components from the original materials are degraded. For example, starch from rice, barley, or wheat is converted by amylase into dextrin, maltose, and glucose, which contribute the sweet taste to *miso.* Protein from soybeans is converted by proteases and peptidases into water-soluble nitrogen compounds containing mainly oligopeptides and amino acids. The amino acids, particularly glutamic acid, contribute the delicious taste to *miso* (Hondo and Mochizuki, 1968). At the same time, oil from soybeans is hydrolyzed in part by lipase to free fatty acids and glycerol.

As the enzymatic digestion progresses, it generates fermentable substances such as simple sugars for the growth of yeasts and lactic bacteria that come from the inoculum. Under anaerobic conditions, these organisms further break down sugars to acid, alcohols, and other substances. As fermentation and aging continue, there are complex chemical and biological interactions among various components in *miso* (Shibasaki and Hesseltine, 1962). In general, acids react with alcohols to produce esters, which contribute a distinct aroma to *miso.* Amino acids and sugars interact to produce browning substances, which contribute in part to the color of *miso.* Since amino acids play the dual role of enhancing flavor and darking color, *miso* with the richest color is often considered to be richest in flavor.

9. SOY SAUCE *(JIANGYOU* OR *SHOYU)*

Soy sauce is a dark brown liquid extracted from a fermented mixture of soybeans and wheat (Figure 6.11). With a salty taste and sharp flavor, it has been served as an all-purpose seasoning for thousands of years. The product is known as *jiangyou* (Mandarin) or *chiangyu* (Cantonese) in China, meaning oil from jiang; *shoyu* in Japan; *tao-yu* in Indonesia; and *tayo* in the Philippines. Among all fermented soyfoods, soy sauce is now the widest accepted product, not only in Far East but also in Western countries.

There are many types of soy sauce. Based on preparation principles, soy sauce is divided into three groups: fermented soy sauce, chemical soy sauce, and semichemical soy sauce. Based on geographical location, there are Chinese and Japanese soy sauces. Under each ethnic group, soy sauce is further divided based on differences in raw ingredients, methods of prepara-

Figure 6.11 Soy sauce.

tion or duration of aging. With Chinese soy sauce, there are traditional soy sauce (*Lao chou,* meaning old extract), modern soy sauce (*sheng chou,* meaning raw extract), regular soy sauce, which is made of soybeans and wheat, and mushroom soy sauce, which is made of soybeans, wheat, and mushroom.

Within the Japanese soy sauce category, there are five main types that are officially recognized. *Koikuchi shoyu* is made from equal amounts of wheat and soybeans in the *koji* and serves as an all-purpose seasoning. It represents about 85% of total soy sauce production in Japan. *Usukuchi shoyu* is the second most popular type of soy sauce in Japan and is commonly used as a seasoning for such food of which the original flavor and color need to be preserved. The remaining three types of soy sauce are produced and consumed only in isolated localities for special uses in Japan. Among them, *tamari shoyu* is very similar to the traditional Chinese type of soy sauce. It is made by a *koji* containing a large proportion of soybeans over wheat. In contrast to *tamari shoyu, shiro shoyu* is made by using a very high ratio of wheat to soybeans in the *koji* and further by being fermented under conditions that prevent color development. *Saishikomi shoyu* is produced by using equal amounts of wheat and soybeans in the *koji.* However, raw soy sauce, instead of a brine solution, is mixed with the *koji* before the second fermentation.

9.1 CHINESE *JIANGYOU*

Among the Chinese soy sauces, there are traditional soy sauce and modern soy sauce. Traditional soy sauce is made by a method thousands of years old, whereas modern Chinese soy sauce is made by a new method developed not long ago.

9.1.1 Traditional Household Method

This is essentially the same method for making Chinese *jiang* described earlier, except for the final stages of preparation, where a brine solution is added to the mature *koji* in a large earthenware jar. The jar is left outside, and the mash is exposed to the sun for several months. At the end of the brine fermentation period, a bamboo basket is dipped in the mash, and the liquid accumulated in the basket is transferred to another earthenware jar and exposed to the sun for several weeks more. The premium grade of soy sauce is thus made. Fresh brine is then added to the residue mash two more times to extract the second and third grade of soy sauce. This is also followed by several weeks of exposure to the sun.

Although centuries old, the traditional method for making soy sauce is still popular today, not only in China, but also in Southeast Asia, where small-scale commercial production prevails. The manufacturers in these regions may add caramel and monosodium glutamate to the second and third extracts. They may also pasteurize their soy sauce before bottling.

9.1.2 Modern Methods

Although the traditional method is still used at a domestic level, many commercial soy sauce producers in China have now adopted a new method developed recently (Fukushima, 1981). Briefly, defatted soymeal and wheat bran are used in a ratio of 60:40. After treatment, their mixture is inoculated with an improved strain of *Aspergillus.* As a result, the time for making *koji* is reduced from 48 h to 24 h. The matured *koji* is then mixed with a small volume of the brine (about half the usual amount) that has about three-quarters as much salt and then is allowed to ferment at a relatively higher temperature (40–45°C). Under such low-water, low-salt, and high-temperature conditions, it takes only 3 weeks to complete fermentation. The fermented mash is transferred to another tank, mixed with additional brine, and then heated to more than 80°C. This is followed by the separation of the liquid in the tank through gravity. The soy sauce may be pasteurized or mixed with benzoic acid before finally being clarified, bottled, and shipped to the market. Compared with the traditional method, the new method is very economical because of the significant reduction in processing time. However, the ratio of amino acid nitrogen to total nitrogen in the final product is not as high.

9.2 JAPANESE *SHOYU*

Although there are some variations in making different types of Japanese soy sauce, their basic steps are the same, including treatment of raw materials, *koji* making, brine fermentation, pressing, and refining (Fukushima, 1979; Beuchat, 1984; Watanabe and Kishi, 1984; Yokotsuka, 1986). A typical process for *koikuchi shoyu,* the representative Japanese type of soy sauce, is outlined in Figure 6.12.

9.2.1 Treatment of Raw Materials

The initial step is to treat soybeans and the wheat simultaneously. Whole soybeans are soaked overnight at an ambient temperature. The soaked soybeans are cooked under a steam pressure or in an open pan until very soft. If defatted soy products are used, which has been popular, they are first moistened by spraying with water in an amount equal to 30% of their own weight. This is followed by pressure-steaming for 45 min. The heated soybeans or soy grits are allowed to cool to less than 40°C within a short period of time.

Concurrent with the treatment of soybeans, whole kernel wheat is roasted and cracked in rollers into four or five pieces. When wheat flour and wheat bran are used, they are steamed after being moisturized.

9.2.2 *Koji* Making

The treated soybeans and wheat are mixed in a certain proportion, depending on the type of end product to be made; for example, for *koikuchi shoyu,* the ratio of soybean (or defatted soy meal) : wheat is about 1 : 1, whereas for *tamari shoyu,* the ratio is 9:1. The mixture is inoculated with seed *koji* or a pure culture containing *Aspergillus oryzae* and/or *A. sojae* at a concentration of 0.1–0.2%.

In traditional *koji* making, the inoculated mixture is put into small wooden trays and kept for 3–4 days in a koji-making room. During the mold growth, the temperature and moisture are controlled by manual stirring. In modern koji making, however, the cultured mixture is put into the shallow, perforated vat and kept in a *koji* room where forced air is circulated, and temperature and humidity may thus be controllable (as in the case with an automatic *koji*-making system). After about 3–4 days, when the mixture turns green-yellow as a result of sporulation of the inoculated mold, it becomes mature *koji.*

In the early stage of *koji* making, temperatures as high as 30–35°C are preferable for mycelium growth and the prevention of *Bacillus* as a contami-

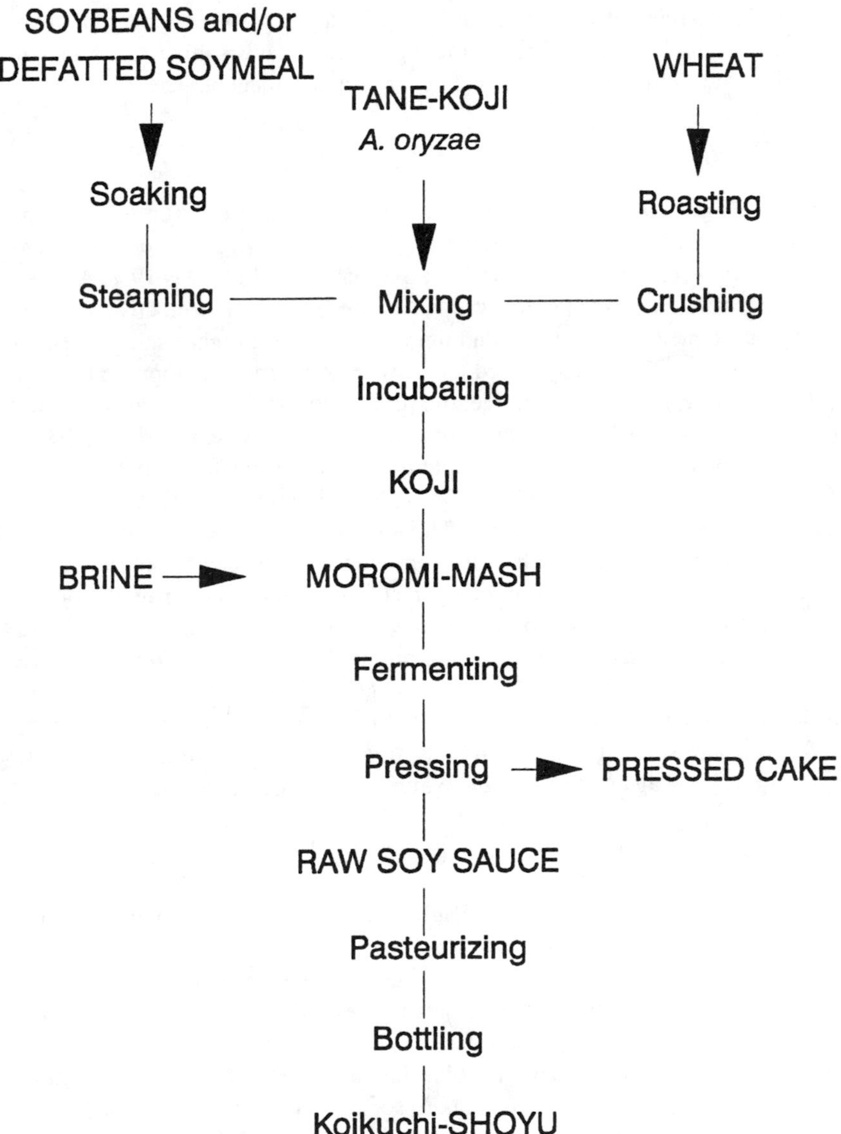

Figure 6.12 A common method for making Japanese *koikuchi shoyu* (soy sauce).

nant. In the later stage, just before spore formation or after the second cooling, a lower temperature (20–25°C) is necessary to allow maximum production of enzymes. Alternatively, *koji* may be prepared at a constant low temperature of 23–25°C for a relatively longer time (66 h). In any case, when the temperature rises above 35°C because of active mold growth, it is advisable to cool the *koji* material twice, either by hand-mixing or by a mechanical device.

9.2.3 Brine Fermentation

Mature *koji* is now mixed with an equal amount or more (up to 120% by volume) of a salt solution to form the liquid mash known as *moromi* in Japanese. The final concentration of NaCl in the mash should be 17–19%. A lower salt concentration promotes the growth of undesirable putrefactive bacteria during subsequent fermentation and aging. However, a higher salt concentration (in excess of 23%) may retard the growth of desirable halophilic bacteria and osmophilic yeasts. In the home, the mash is put into an earthen crock, and the fermentation is under ambient temperatures. In this case, a period of 10–12 months may be necessary for completion of the brine fermentation stage. However, on an industrial level, the mash is kept in large wooden containers or concrete vats with aeration devices. The temperature of the surroundings can be mechanically controlled. Thus, the fermentation time can be shortened.

Temperature is also an important factor during brine fermentation. In general, the higher the temperature, the shorter is the fermentation time. However, lower temperature fermentation results in better products because the rate of enzyme inactivation is slow. A good quality of soy sauce can be made by a 6-month fermentation when the temperature of mash is controlled as follows: starting at 15°C for 1 month, followed by 28°C for 4 months, and finishing at 15°C again for 1 month (Watanabe and Kishi, 1984).

9.2.4 Pressing

After months of fermentation and aging, the mash becomes matured. In the case of home processing, raw sauce may be removed from the mash simply by siphoning off from the top or filtering through cloth under a simple mechanical press. In commercial operations, a batch-type hydraulic press is commonly used. Recently, automatic loading of the mash into a filter cloth or continuous pressing by a diaphragm-type machine has emerged for effective filtration. The filtrate obtained is stored in a tank to separate the sediments at the bottom and the floating oil on the top.

9.2.5 Refining

Raw soy sauce may be adjusted to the standard salt and nitrogen concentrations. It is then pasteurized at 70–80°C to inactivate enzymes and microorganisms, enhances the unique product aroma, darkens color, and induces formation of flocs, which facilitates clarification. After heating, the soy sauce is clarified by either sedimentation or filtration. Kaolin, diatomite, or alum may be added to enhance clarification before filtration. The clear supernatant is packed immediately into cans or bottles. In some cases, preservatives such as sodium benzoate and paraoxy-benzoate may be used.

9.3 PRINCIPLES OF MAKING SOY SAUCE

Just like making *jiang* or *miso,* there are two stages of fermentation that occur in soy sauce preparation. The first fermentation is solid state and occurs during *koji* making where various enzymes are produced under aerobic conditions. The second fermentation occurs after the addition of brine, known as brine fermentation. It is mainly anaerobic. At the earlier stage of brine fermentation, enzymes from *koji* hydrolyze proteins to yield peptides and free amino acids. Starch is converted to simple sugars, which in turn serve as substrates for growth of various types of salt-resistant bacteria and yeasts. These organisms become dominant in sequence as fermentation progresses. All these enzymatic and biological reactions, together with concurrent chemical reactions, lead to the formation of many new volatile and nonvolatile substances that contribute to the characteristic color, flavor, and taste of soy sauce (Fukushima, 1981; Yokotsuka, 1986).

9.4 CHEMICAL SOY SAUCE

Traditionally, soy sauce is made by fermentation as described. However, soy sauce can also be made by acid hydrolysis. The resulting product is known as chemical soy sauce or protein chemical hydrolyzate. Briefly, defatted soy products or other proteinaceous materials are first hydrolyzed by heating with 18% hydrochloric acid for 8–12 h. After hydrolysis, the hydrolyzate is neutralized with sodium carbonate and filtered to remove the insoluble materials. The resulting product is a clear dark-brown liquid, known as chemical soy sauce; however, chemical soy sauce does not possess the flavor and odor of fermented *shoyu.* Therefore, to improve its quality, chemical soy sauce is often blended with fermented *shoyu* to become semichemical products before being sold.

9.5 PROXIMATE COMPOSITION OF SOY SAUCE

The chemical composition of soy sauce is rather complex and varies with types and even batches. In a typical Japanese fermented soy sauce, the soluble solids are divided almost equally between inorganic (46%) and organic components (47%). Sodium and chlorine are the principal inorganic constituents.

Amino acids are the principal organic components, comprising almost 25% of the total soluble solids. They are followed by carbohydrates, 13%; polyalcohols, 5%; and organic acids, nearly 3%. Of the total nitrogen, about 40–50% are amino acids, 40–50% peptides and peptones, 10–15% ammonia, and less than 1% protein. There are 18 amino acids present, and glutamic acid and its salts are the principle flavoring agents. Sugars present are glucose, arabinose, xylose, maltose, and galactose, whereas sugar alcohols are glycerol and mannitol. Organic acids found in *shoyu* are lactic, acetic, succinic, citric, formic, and pyroglutamic. In addition, there exist trace amounts of organic bases, such as ardenine, hypoxanthine, xanthine, quanine, cytosine, and uracil, all of which are believed to be metabolites of nucleic acids (Yokotsuka, 1986).

9.6 QUALITY ATTRIBUTES AND GRADES

In general, a good *shoyu* has a salt content of about 18% and a pH value between 4.6 and 4.8. A product with a pH below this range is considered too acidic, suggesting acid production by undesirable bacteria. Other quality factors include nitrogen yield, total soluble nitrogen, and the ratio of amino nitrogen to total soluble nitrogen. The nitrogen yield is the percentage of nitrogen of raw materials converted to soluble nitrogen in the finished product, showing the efficiency of enzymatic conversion. The total soluble nitrogen is a measure of the concentration of nitrogenous material in the *shoyu* indicating a standard of quality. The ratio of amino acid nitrogen to total nitrogen is an accepted standard for overall quality of a soy sauce. The higher the ratio value, the better is the quality. The normal range is 50–60%. All these quality attributes are affected by factors related to almost every step of processing, including choice of raw materials, steaming conditions, selection of a *koji* starter, and conditions for both *koji* making and subsequent brine fermentation.

In Japan, for each of the five types of soy sauce, there are three grades—special, upper, and standard—based on organoleptic evaluation, total nitrogen content, soluble solids other than sodium chloride, and color. Since the quality of chemical soy sauce is generally considered inferior to fermented soy sauce, semichemical and chemical soy sauce cannot be graded as special.

10. JAPANESE *NATTO*

Originating in the northern part of Japan about 1000 years ago, *natto* is one of the few products in which bacteria predominate during fermentation. When properly prepared, it has a slimy appearance, sweet taste, and a characteristic aroma (Figure 6.13). In Japan, *natto* is often eaten with soy sauce and/or mustard and served for breakfast and dinner along with rice. Similar products are also found in Indonesia and Thailand.

Figure 6.13 Japanese *natto.*

10.1 METHODS OF PREPARATION

As outlined in Figure 6.14, to make *natto,* soybeans, preferably small seeded, are washed and soaked overnight. The soaked beans are then steamed for about 30 min, drained, and cooled to about 40°C. Traditionally, the treated soybeans are wrapped with rice straw and set in a warm place for 1–2 days. Rice straw is credited for not only supplying the fermenting microorganism, *Bacillus natto,* but also for absorbing the unpleasant odor of ammonia released from *natto* and imparting the aroma of straw to the product.

However, since there is a great chance of contamination by unwanted microorganisms from rice straw, the quality of the product is rather difficult to control. Ever since the isolation of the responsible *Bacillus natto,* the old straw method has been largely abandoned in favor of pure culture fermentation. Instead of wrapping with rice straw, the treated beans are inoculated with a pure culture suspension of *B. natto* at a concentration of about 15 ml/100 kg raw beans and thoroughly mixed before being packed into wooden boxes or perforated polyethylene bags. The packages are put into shallow sliced-wood or polystyrene trays and set in a warm, thermostatic chamber with a controlled

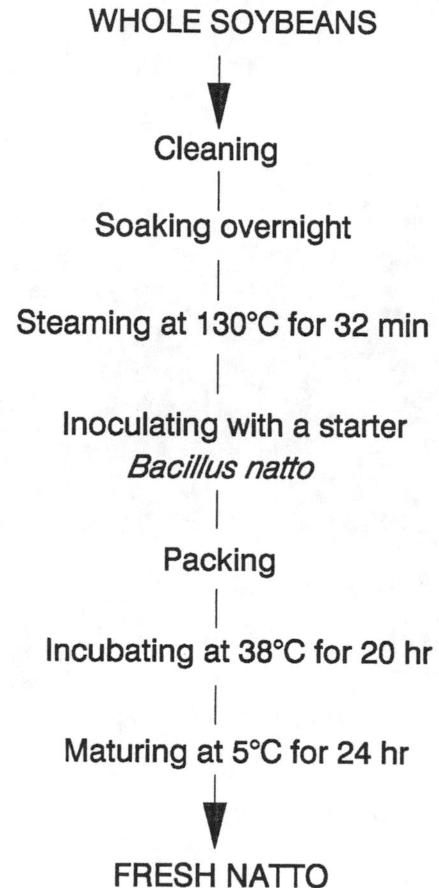

Figure 6.14 A common method for making Japanese *natto.*

temperature of 38°C. After 16–20 h of fermentation, the bacteria will have covered the beans with a white sticky coating, indicating the time for harvesting. For better quality, the package may be kept at a refrigerating temperature for 1–2 days to allow maturation and then taken out for consumption or retailing as needed. The practice is most common in large plants.

Good-quality *natto* should have a characteristic flavor, intact beans with a viscous appearance, and soft texture. There are many factors that affect the quality of *natto.* A key factor is selecting the right soybeans. Unlike for other fermented soyfoods, for production of *natto,* round, small seeded soybeans with a high soluble sugar content, yellowish seed coat, and a clear hilum are preferred. Small beans have a larger surface than large or medium beans and thus absorb water faster, require a shorter steaming time, and allow faster growth of the *natto* organism. Because soluble sugars serve as an initial carbon and energy source for natto organisms, their high content promotes microbial growth. It also makes the finished product taste sweeter. Since *B. natto* is aerobic, good aeration with a sufficient supply of oxygen is required during the entire fermentation. Also, temperature should be controlled below 40°C, and overfermentation should be avoided since it leads to release of ammonia, which not only spoils the *natto* flavor, but also destroys the *B. natto* and promotes spoilage by other organisms.

10.2 PRINCIPLES OF PREPARATION

During fermentation, *B. natto* bacteria grow, multiply, and sporulate. At the same time, the bacteria secrete various extracellular enzymes, including protease, amylase, γ-glutamyltranspeptidase (GTP), levansucrase, and phytase. As *natto* bacilli grow, the enzymes they secreted or produced catalyze many chemical and enzymatic reactions, which lead to formation of the characteristic sticky material, as well as the characteristic aroma and flavor. The viscous material consists of polysaccharide (a levan-form fructan) and γ-polyglutamic acid. The latter contains D- and L-glutamate in varying proportions, depending on the amount of manganese ion and the type of amino acids in the media.

During *natto* fermentation, there are no significant changes in fat and fiber contents of soybeans and in the fatty acid composition of soy lipids; however, the soluble carbohydrates, such as sucrose, raffinose, and stachyose, almost completely disappear. So does citric acid, the major organic acid in steamed soybeans. At the same time, many volatile components, which contribute the characteristic aroma and flavor of *natto,* are produced by the *natto* bacteria (Kanno and Takamatsu, 1987).

11. INDONESIA *TEMPEH*

Tempeh, or *tempe* in some literature, is made by fermenting dehulled and cooked soybeans with mold, *Rhizopus* sp. Freshly prepared *tempeh* is a cake-like product, covered and penetrated completely by white mycelium, and has a clean, yeasty odor (Figure 6.15). When sliced and then deep-fat fried, it has a nutty flavor, pleasant aroma, and crunchy texture, serving as a main dish or meat substitute.

Tempeh is widely believed to have originated in Indonesia centuries ago. Although relatively unknown in the surrounding countries such as Thailand, China, and Japan, where soybeans form an important part of the diet, *tempeh* continues to be one of the most popular fermented foods in Indonesia. Because of its meat-like texture and mushroom flavor, *tempeh* is well suited to Western tastes. It is becoming a popular food for a number of vegetarians in the United States and other parts of the world.

Figure 6.15 Indonesian *tempeh* (courtesy of Mr. Seth Tibbott, Turtle Island Foods, Inc., Hood River, OR).

There are many methods available for making *tempeh* (Winarno, 1989; Hachmeister and Fung, 1993). Among them, the two most representative ones are traditional and pilot plant methods.

11.1 TRADITIONAL METHOD

Traditionally, making *tempeh* is a household art in Indonesia. The method of preparation varies from one household to another, but the principle steps are basically the same (Figure 6.16). Soybeans are cleaned and then boiled in water for 30 min before hand-dehulling. The dehulled beans are soaked overnight to allow full hydration and lactic acid fermentation. The soaked, dehulled beans are cooked again for 60 min, drained using woven bamboo baskets, and spread on a flat surface for cooling to room temperature. In certain places, soybeans are soaked in water until the hulls can be easily removed by hand or feet and washed away with water, and then boiled until soft, normally for at least 30 min. This avoids double cooking procedures.

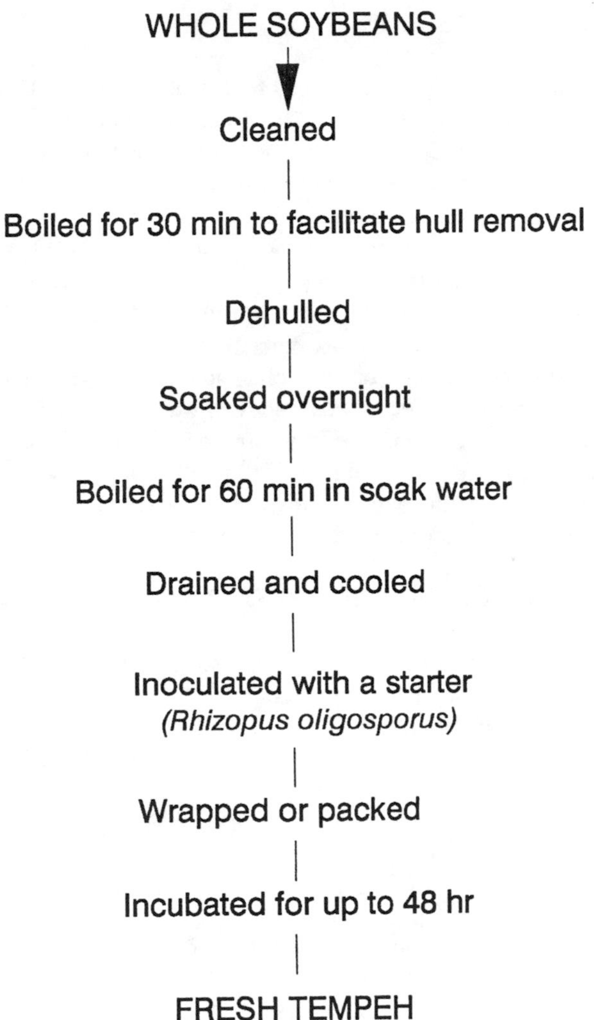

Figure 6.16 A traditional Indonesian method for making *tempeh*.

The treated beans are now inoculated with a traditional starter known as *usar* or an inoculum from a previous batch, both containing *R. oligosporus* spores. The mixture is wrapped in banana leaves or perforated plastic bags, approximately a quarter pound per package. Fermentation is allowed to occur at room temperature for up to 18 h or until the beans are bound by white mycelium. Alternatively, inoculated beans are spread on shallow aluminum foil or metal trays with perforated bottoms and covered with layers of banana leaves, waxed paper, or plastic films that are also perforated.

11.2 PILOT PLANT METHOD

The pilot plant method was developed by Steinkraus et al. (1965). Briefly, the graded soybeans are heated at 93°C for 10 min to loosen hulls, cooled, and then passed through a properly spaced Burr mill to crack the beans. The hulls are separated by a gravity separator or an aspirator. The dehulled beans are hydrated in water or 0.85% lactic acid solution for 2 h at 25°C or 30 min at 100°C. Addition of an acid results in a drop of pH to 4.3–5.3, which is intended to control bacterial growth that might otherwise contaminate *tempeh.* The hydrated beans are boiled for 90 min in the soaking water. After being drained and cooled, the beans are inoculated with pulverized pure *R. oligosporus* starter containing both mycelia and spores. The inoculated beans are spread out on perforated trays, covered with waxed paper to prevent dehydration and excessive aeration. The trays are now put in an incubation room, which is maintained at 35–37°C and 75–78% relative humidity. After about 18 h, when the mold covers the entire beans, *tempeh* is ready for harvesting. It is then dehydrated and packaged in plastic bags before shipment and storage.

11.3 PRINCIPLES OF *TEMPEH* PREPARATION

There are many aspects of changes during *tempeh* fermentation, including temperature, pH, and chemical composition of the soybean substrate. All of these are brought about by microbial growth and enzyme actions.

According to Steinkraus et al. (1960), during the first 20 h at 37°C, the mold spores germinate, and the temperature of the mass rises gradually. During the following 5 h, the mold grows rapidly, reaches a peak and then gradually subsides. Concomitant with mold growth is the rise and fall of the temperature. At the peak, the temperature may reach as high as 43–44°C. By this time, the beans are already knitted into a compact mass by mold mycelia, and the *tempeh* is ready to be harvested. Beyond this stage, the mold sporulates and NH_3 is produced because of protein breakdown.

As microorganisms grow, they produce various enzymes that break down soybean components. This leads to some compositional changes. Compared with *miso* and soy sauce, these changes are much less vigorous because of limited production of enzymes by the *tempeh* mold. In general, in comparison of tempeh and unfermented dehulled soybeans, there are some increases in levels of free amino acids and free fatty acids, a slight decrease in oil content, and no significant changes in protein and ash contents. In addition, there are significant increases in levels of several vitamins, including riboflavin, vitamin B_6, nicotinic acid, pantothenic acid, biotin, and folacin. Some of these increases are severalfold, although thiamin was found to change little (Murata et al., 1967). Furthermore, in some *tempeh,* there is formation of vitamin B_{12}, which is widely attributed to the presence of certain contaminating bacteria, mainly *Klebsiella* (Liem et al., 1977).

12. SOY NUGGETS *(DOUCHI* OR *HAMANATTO)*

Soy nuggets, or *douchi* in mandarin Chinese, *toushih* in Cantonese, and *hamanatto* in Japanese, is made by fermenting whole soybeans with strains of *Aspergillus oryzae,* although some other strains of fungi or bacteria may also be responsible. The finished product (Figure 6.17) has a flavor similar to *jiang* or soy sauce. Because of its black color, it is also known as salted black beans in the West. Soy nuggets are commonly used as an appetizer to be consumed with bland food or as a flavoring agent to be cooked with vegetables, meats, and seafoods.

Originating in China before the *Han* Dynasty (206 B.C.), the soy nugget is considered to be the progenitor of many types of fermented soy paste and soy sauce. It is the first soyfood to be described in written records. The product continues to be popular in China and certain regions of Japan. Similar products are also produced in the Philippines (known as *tao-si*) and East India (known as *tao-tjo*).

Figure 6.17 Chinese *douchi* (soy nuggests).

Because of variations in preparation methods, soy nuggets vary in texture, taste, salt, and moisture content from country to country. For example, Japanese *hamanatto* is softer in texture and higher in moisture content, as compared with the Chinese counterpart, *douchi. Tao-tjo* in India tends to have a sweet taste because sugar is often added to the brine. Even among the Chinese soy nuggets, there exist many varieties. Based on raw material, soy nuggets are classified as those made of yellow soybeans and those of black beans. Based on taste, there are plain (less salty), salty, and wine types. Based on microorganisms involved, there are *Mucor* type, *Aspergillus* type, and bacterial type (Shi and Ren, 1993).

12.1 CHINESE *DOUCHI*

The methods of preparing soy nuggets may vary with regions, but the essential features are similar to a traditional method that has been handed down from generation to generation. Whole soybeans are soaked for 5–6 h and then steamed or boiled in water until soft. The cooked beans are inoculated either naturally or with a *koji* starter. Under natural inoculation, predominance of a specific microorganism depends on incubation conditions. When air exchange is sufficient, incubation at temperatures of 5–10°C for 15–20 days helps the growth of *Mucor* sp. whereas incubation at 26–30°C for 5–6 days is suitable for growth of *Aspergillus* sp., and incubation at 20°C for 3–4 days with coverage of rice stock or pumpkin leaves promotes growth of *Bacillus* sp. (Shi and Ren, 1993).

After *koji* is matured, as evidenced by the appearance of abundant mycelium and spores, it is washed with water. Washing helps remove extra mycelium, spores, and contaminants and ensures that the finished product is shinning and free of mold odor and bitter taste. After washing, *koji* is mixed with 40–47% water, salt, and spices before being put into a jar for fermentation. Fermentation normally is carried out under natural conditions for several months. Sometimes, soy sauce is used in place of brine. In other regions, spices, wine, or sugar may be added at this stage. The aged beans become wet soy nuggets, which may be further dried to make dry soy nuggets. The finished product consists of intact beans with blackish color and has a salty taste and soy sauce flavor. Because of relatively high salt and low water contents, the product can be kept for a long time.

12.2 JAPANESE *HAMANATTO*

In Japan, a similar product known as *hamanatto* is produced, especially in the vicinity of Hamanatsu, Shizuoka Prefecture, from which the name of the product was perhaps derived. Soybeans are soaked and steamed until soft. They are drained and cooled before mixed with parched wheat flour in a soybean:wheat ratio of 2:1. The mixture is inoculated with a strain of *A.*

oryzae (*koji* starter) and then distributed among shallow wooden boxes. After fermentation at 30–35°C for up to 50 h, the beans are dried in the sun until the moisture content of the mixture decreases from the original 30–35% to 20–25%. They are now covered with a brine (15° Baume) and allowed to age under pressure in a tank for several weeks or months. The aged beans are dried in the sun. Sometimes, ginger pickled in soy sauce may be added after drying. The finished product normally contains about 10% salt and 38% water and can be stored for a long time. An apparent difference in making Chinese *douchi* and Japanese *hamanatto* is that the *koji* for the former is made from cooked soybeans only while the *koji* for the latter is made from a mixture of cooked soybeans and parched wheat flour.

13. REFERENCES

Applewhite, T. H. 1989. *Proceedings of the World Congress on Vegetable Protein Utilization in Human Foods and Animal Feedstuffs.* Congress held in Singapore, Oct. 1988. American Oil Chemists' Society, Champaign, IL.

Bates, R. P. and Matthews, R. F. 1975. Ascorbic acid and β-carotene in soybeans as influenced by maturity, sprouting, processing and storage. *Proc. Fla. State Hort. Soc.* 88:266–271.

Beddows, C. G. and Wong, J. 1987a. Optimization of yield and properties of silken tofu from soybeans. I. The water:bean ratio. *Int'l. J. Food Sci. Technol.* 22:15–21.

Beddows, C. G. and Wong, J. 1987b. Optimization of yield and properties of silken tofu from soybeans. II. Heating processing. *Int'l. J. Food Sci. Technol.* 22:23–27.

Beuchat, L. R. 1984. Fermented soybean foods. *Food Technol.* 38(6):64.

Bourne, M. C. 1970. Recent advances in soybean milk processing technology. *FAO/WHO/UNICEF Protein Advisory Group Bulletin,* No. 10. United Nations, New York.

Cai, T. D. and Chang, K. C. 1997. Dry tofu characteristics affected by soymilk solid content and coagulation time. *J. Food Quality,* 20:391–402.

Chen, S. 1989. Preparation of fluid soymilk. In: *Proceedings of the World Congress on Vegetable Protein Utilization in Human Foods and Animal Feedstuffs,* T. H. Applewhite, ed., Am. Oil Chem. Soc., Champaign, IL, pp. 341–351.

Chopra, R. and Prasad, D. N. 1990. Soymilk and lactic fermentation products—A review. *Microbiologie Aliments Nutrition* 8:1–13.

deMan, J. M., deMan, L., and Gupta, S. 1986. Texture and microstructure of soybean curd (tofu) as affected by different coagulants. *Food Microstruct.* 5:83–89.

Evans, D. E., Tsukamoto, C., and Nielsen, N. C. 1997. A small scale method for the production of soymilk and silken tofu. *Crop Sci.* 37:1463–1471.

Fang, J. G. 1997. *Production Technology of Jiang and Related Products.* China's Light Industry Publisher, Beijing (in Chinese).

FAO/WHO. 1990. *Protein Quality Evaluation.* FAO/WHO Nutrition Meetings, Report Series 51. Food and Agriculture Organization/World Health Organization, Rome.

Fukushima, D. 1979. Fermented vegetable (soybean) protein and related foods of Japan and China. *J. Am. Oil Chem. Soc.* 56:357.

Fukushima, D. 1981. Soy proteins for foods centering around soy sauce and tofu. *J. Am. Oil Chem. Soc.* 58(3):346.

Hachmeister, K. A. and Fung, D. Y .C. 1993. Tempeh: A mold-modified indigenous fermented food made from soybeans and/or cereal grains. *Crit. Rev. Microbiol.* 19(3):137–188.

Hesseltine, C. W. and Wang, H. L. 1972. Fermented soybean food products. In: *Soybeans: Chemistry and Technology,* A. K. Smith and S. J. Circle, eds., AVI Publishing Co. Westport, CT, p. 389.

Hondo, S. and Mochizuki, T. 1968. Studies on the degradation process of soybean protein during *miso* making. Part II. Changes in peptides during *miso* making. *Nippon Shokuhin Kogyo Gakkaishi* 15:414.

Hymowitz, T. 1970. On the domestication of the soybean. *Economic Botany,* 24(4):408–421.

Ikeda, K. Matsuda, Y., Katsumaru, A., Teranishi, M., Yamamoto, T. and Kishida, M. 1995. Factors affecting protein digestibility in soybean foods. *Cereal Chem.* 72(4):401.

Kanno, A. and Takamatsu, H. 1987. Changes in the volatile components of *"Natto"* during manufacturing and storage. *Nippon Shokunhin Kogyo Gakkaishi* 34(5):330–335.

Kosikowski, F. V. 1971. Nutritive and organoleptic characteristics of non-dairy imitation milks. *J. Food Sci.* 36:1021.

Kwok, K. C. and Niranjan, K. 1995. Effect of thermal processing on soymilk. *Intl. J. Food Sci. Technol.* 30(3):263–265.

Liem, I. T. H., Steinkraus, K. H., and Cronk, T. C. 1977. Production of vitamin B_{12} in tempeh, a fermented soybean food. *Appl. & Environ. Microbiol.* 34:773.

Liu, K.-S. 1996. Immature soybeans, nutritional quality and organoleptic feature. *INFORM,* 7(11):1217–1223.

Liu, K.-S. 1997. *Soybeans: Chemistry, Technology, and Utilization.* Chapman & Hall, New York. (now acquired by Aspen Publishers, Inc., Gaithersburg, MD)

Liu, K.-S. Orthoefer, F., and Thompson K. 1995. The case for food-grade soybean varieties. *INFORM* 6(5):593.

Messina, M. 1997. Soyfoods: Their role in disease prevention and treatment. In: *Soybeans: Chemistry, Technology, and Utilization,* K.-S. Liu, ed., Chapman & Hall, New York, (now acquired by Aspen Publishers, Inc., Gaithersburg, MD) pp. 442–478.

Mital, B. K. and Steinkraus, K. H. 1979. Fermentation of soymilk by lactic acid bacteria, a review. *J. Food Prot.* 42(11):895–899.

Murata, K., Ikehata, H., and Miyamoto, T. 1967. Studies on the nutritional value of tempeh. *J. Food Sci.* 32:580.

Nelson, A. I., Steinberg, M. P. and Wei, L. S. 1976. Illinois process for separation of soymilk. *J. Food Sci.* 41:57–61.

Nikkuni, S., Okada, N., and Itoh, H. 1988. Effect of soybean cooking temperature on the texture and protein digestibility of *miso. J. Food Sci.* 53(2):445.

Ning, S. S. 1995. Soybeans and soy products. In: *Cereal and Soybean Processing and Technology,* S. S. Wu, ed., China's Light Industry Publisher, Beijing, pp. 411–462 (in Chinese).

Ohara, T., Karasawa, H., and Matsuhashi, T. 1992. Relationships of coagulation characteristics and properties of "Kori-tofu" in a controlled soymilk coagulation system. *J. Jap. Soc. Food Sci. Technol.* 39(6):543–554.

Saio, K. 1979. Tofu—Relationships between texture and fine structure. *Cereal Foods World* 24(8):343.

Shi, Y. G. and Ren, L. 1993. *Soybean Products Technology.* China's Light Industry Publisher, Beijing, China (in Chinese).

Shibasaki, K. and Hesseltine, C. W. 1962. *Miso* fermentation. *Economic Botany* 16:180.

Shih, M. C., Hou, H. J., and Chang, K. C. 1997. Process optimization of soft tofu. *J. Food Sci.* 62(4):833–837.

Shurtleff, W. and Aoyagi, A. 1975. *The Book of Tofu.* Ten speed press, Berkeley, CA.

Shurtleff, W. and Aoyagi, A. 1979. *Tofu and Soymilk Production: The Book of Tofu, Vol. II.* The Soyfoods Center, Lafayette, CA.

Shurtleff, W. and Aoyagi, A. 1983. *The Book of Miso.* Ten speed press, Berkeley, CA.

Sizer, C. E. 1989. Aseptic packaging of soymilk. In: *Food Uses of Whole Oil and Protein Seeds,* E. W. Lusas, D. R. Erickson, and W. K. Nip, eds., Am. Oil Chem. Soc., Champaign, IL, pp. 98–101.

Skurray, G., Cunich, J., and Carter, O. 1980. The effect of different varieties of soybean and calcium ion concentration on the quality of tofu. *Food Chem.* 6:89–95.

Soya Bluebook Plus, 1998. Soyatech, Inc., Bar Harbor, ME.

Steinkraus, K. H., Hwa, Y .B. van Buren, J. P., Hackler, L. R., and Hand, D. B. 1960. Studies on tempeh—An Indonesian fermented soybean food. *Food Research* 25(6):777–788.

Steinkraus, K. H., van Buren, J. P., Hackler, L. R., and Hand, D. B. 1965. A pilot-plant process for the production of dehydrated *tempeh. Food Technol.* 19(1):63.

Sun, N., and Breene, W. M. 1991. Calcium sulfate concentration influence on yield and quality of tofu from five soybean varieties. *J. Food Sci.* 56(6):1604.

Tsai, S.-J., Lan, C. Y., Kao, C. S. and Chen, S. C. 1981. Studies on the yield and quality characteristics of tofu. *J. Food Sci.* 46:1734.

van der Riet, W. B., Wight, A. W., Cilliers, J. J., and Datel, J. M. 1989. Food chemical investigation of tofu and its byproduct *okara. Food Chem.* 34:193–202.

Wai, N. S. 1968. Investigation of the various processes used in preparing Chinese cheese by the fermentation of soybean curd with mucor and other fungi. Final Technical Report. Institute of Chemistry, Academia Sinica, Taiwan.

Wang, H. L. and Hesseltine, C. W. 1982. Coagulation conditions in tofu processing. *Process Biochem.* 17(Jan./Feb.):7–12.

Wang, H. L., Swain, E. W., and Kwolek, W. F. 1983. Effect of soybean varieties on the yield and quality of tofu. *Cereal Chem.* 60:245.

Wang, H. L., Kraidej, L., and Hesseltine, C. W. 1974. Lactic acid fermentation of soybean milk. *J. Milk Food Technol.* 37(2):71.

Watanabe, T., Fukamachi, C., Nakayama, O., Teramachi, Y., Abe, K., Suruga, S., and Miyanaga, S. 1964. *Research into the Standardization of the Tofu Making Process.* National Food Research Institute, Japan (in Japanese).

Watanabe, T. and Kishi, A. 1984. *Nature's Miracle Protein: The Book of Soybeans.* Japan Publications. Inc., Tokyo and New York.

Wilkens, W. F., Mattick, L. R., and Hand, D. B. 1967. Effect of processing method on oxidative off-flavor of soybean milk. *Food Technol.* 21:86.

Wilson, J. C. 1989. The commercial utilization of soybeans, soymilk and soymilk derivatives. In: *Proceedings of World Soybean Research Conference,* pp. 1750–1766. Buenos Aires, Argentina, March 5–9, Esta Publicación Contiene los Trabajos, Buenos Aires.

Winarno, F. G. 1989. Production and utilization of tempeh in Indonesian foods. In: *Proceedings of the World Congress: Vegetable Protein Utilization in Human Foods and Animal Feedstuffs,* T. H. Applewhite, ed., Am. Oil Chemists' Soc. Champaign, IL, p. 363.

Wu, L. C. and Bates, R. P. 1972. Soy protein-lipid films. 1. Studies on the film formation phenomenon. *J. Food Sci.* 37:36.

Yokotsuka, T. 1986. Soy sauce biochemistry. *Adv. Food Res.* 30:196–329.

CHAPTER 7

Chinese Meat Products

YOULING L. XIONG
FANG-QI YANG
XINGQIU LOU

I. INTRODUCTION

MEAT can be defined as the edible portion of postmortem muscle tissue from domestic mammalian species, including pigs, cattle, sheep, and rabbits; poultry species, including chickens, turkeys, ducks, geese, and ostriches; marine species, including fish and crustaceans; and a variety of game animals, such as deer and elk. Traditionally, however, the word *meat* implies muscle tissue from the so-called red meat animals: beef, pork, lamb/mutton. The edible muscle tissue from other species is usually excluded from the traditional definition of meat. To clarify the confusion, the term *muscle foods* is now accepted by food scientists, meat processors, and consumers to refer to all edible muscle tissues, regardless of species (Hedrick et al., 1994). In this chapter, only meats from three major mammalian animals (pork, beef, and lamb) will be discussed. Poultry and fish products are dealt with in other chapters of the book.

Meats are highly nutritious foods that contain high-quality proteins and essential fatty acids and are an excellent source of minerals and vitamins. They have been an important part of the human diet since the prehistoric age, not just because of their nutritional value, but also because of unique texture and flavor characteristics. In most developed countries, including the United States, Canada, United Kingdom, France, Germany, and Australia, meat consumption (red meat) has declined overall in recent years (AMI, 1997), largely resulting from consumers' health and nutritional concerns. In contrast, meat consumption in China has increased steadily over the past 10 years because of drastic improvement in the standard of living (Table 7.1). Currently, per capita

TABLE 7.1. Per Capita Meat Consumption (Disappearance) in Selected Countries in 1996.

Country	Beef and Veal (kg)	Pork (kg)	Lamb[a] (kg)	Broiler Chicken (kg)	Turkey (kg)
Argentina	60.8	n/a[b]	1.7	19.0	n/a
Australia	37.3	18.5	16.6	24.4	n/a
Brazil	29.3	9.0	n/a	22.1	0.5
Bulgaria	12.2	28.7	6.2	n/a	n/a
Canada	34.6	33.2	n/a	25.9	4.5
China	3.6	30.0	1.8	3.9	n/a
Denmark	20.2	67.6	n/a	11.6	2.9
Egypt	8.5	n/a	1.4	5.4	n/a
France	25.0	36.9	5.2	11.7	6.3
Germany	14.5	46.2	1.1	7.1	4.0
Hong Kong	15.2	27.6	n/a	36.6	n/a
Ireland	14.9	38.8	9.5	18.9	6.2
Italy	20.5	35.1	1.7	11.7	4.8
Japan	12.3	16.7	0.6	13.4	n/a
Korea (South)	10.0	19.2	n/a	n/a	n/a
Mexico	19.6	9.6	1.6	12.0	0.7
New Zealand	30.3	n/a	25.1	n/a	n/a
Philippine	3.0	11.0	n/a	n/a	n/a
Poland	10.5	40.0	0.1	5.5	1.4
Romania	8.1	18.0	2.7	n/a	n/a
Saudi Arabia	2.7	n/a	13.0	31.0	n/a
Singapore	5.9	30.0	n/a	30.4	n/a
South Africa	14.7	n/a	4.5	16.5	n/a
Spain	11.0	52.3	6.2[a]	22.6	0.8
Taiwan	3.4	43.0	n/a	n/a	n/a
Turkey	10.1	n/a	6.1	n/a	n/a
United Kingdom	14.4	23.7	6.3	19.2	4.8
United States	31.3	22.3	0.5	32.5	8.2

[a]Includes mutton and goat.
[b]n/a: not available.
Source: Adapted from AMI (1997).

red meat consumption in China has reached 35.4 kg, with pork being the predominant species (84.7%). While most meat is consumed fresh, that is, prepared in the kitchen, the Chinese meat processing industry has experienced a rapid development in recent years because of increased consumer demands and availability of new processing technologies.

1.1 AN OVERVIEW OF THE CHINESE MEAT INDUSTRY

Meat processing in China has a long history. Most of the Chinese commercial meat processing methods today have been developed from the country's several thousand years of culinary arts. Thus, it is not surprising that the principle of processing is, by and large, the same between the two, except, of course, that commercial processing is done on a disproportionately larger scale with a higher degree of automation. Indeed, many of the traditional Chinese meats can be prepared either at home, in a restaurant, or in a commercial meat plant. In many cases, it is difficult to distinguish between products made at different places in terms of their palatability characteristics. The early Chinese literature on meat processing and meat products dates back to 1000–1500 B.C. Among the best known publications were *A Consumer's Essay, Food and Medicine,* and *Meat Regulations.* This ancient literature provided some detailed accounts for meat processing in that era. In the late A.D. 600 (exact date not known), a scientist named Jia gathered information on most meat processing and culinary technologies available at the time and compiled them into a book titled *All People's Needs.* In this encyclopedic documentation, meat products are divided into several categories, including roasted, sausage, marinated, braised, and other types based on products' unique formulations and the specific cooking or processing methods employed. The comprehensiveness of this book lent itself to the evolution of the Chinese meat industry to its current large-scale commercialization.

The processing technology for Chinese meat products experienced some rapid advancements during the Tang (A.D. 618–907) and Song (960–1279) Dynasties, which were marked by prosperity of the Chinese culture. Among the new products developed was *Jinhua* ham, a premium meat product that has become internationally popular today. In the last dynasty (Qing, 1616–1911) of the Imperial China, many other ethnic meat products had been developed, for example, Canton *cha shao* meat, Wuxi pickled pork ribs, and Zhenjiang braised pork, which are still popular foods in Chinese households. In the late 19th century, over 50 kinds of meat products were described in the book *Menu in the Garden,* for example, breaded steamed pork and soy sauced meat. Both products are enjoyed by today's consumers.

Western meat processing technologies were introduced to China in the 1840s, starting in Shanghai, Canton, and several other coastal cities or regions. There were two main varieties: bacon and hotdogs or frankfurters. Since 1950, many Chinese meat processing plants have adopted foreign meat animal slaughter and processing production lines. Among the most rapid technological developments in the Chinese meat industry are meat product refrigeration, freezing, and packaging. Also markedly improved are processing equipment and sanitation conditions. The introduction of modern sausage and ham production lines in the mid-1980s further stimulated the development of the

Chinese meat processing industry. Today, there are a number of joint ventures between Chinese and the United States or European countries. More and more, meat products are currently being produced and exported with a combination of Chinese and Western processing technologies. Recent new products or product forms include quick-frozen meatballs and traditional Chinese luncheon meat cooked at low temperatures. These meats, available to domestic families, are now also exported to international markets.

The meat industry in China has undergone a drastic expansion over the last 50 years. From the mid-century to 1994, the total production of meat and meat products (excluding poultry and fish) had increased from 2.2 million tons to 37.0 million tons (CASTP, 1993; USDA, 1995), leading to an increase in per capita meat consumption from 4.1 kg to 35.4 kg, to which pork, beef, and lamb contributed 84.7, 10.2, and 5.1%, respectively (AMI, 1997). Today, China leads the world in total pork production with an annual yield of 32 million tons (USDA, 1995). By the year 2000, the total pork production will likely reach 40 million tons, based on the current production rate. Because of its large population (1.2 billion), however, China remains low in per capita meat consumption compared to many Western nations (Table 7.1).

1.2 VARIETIES OF CHINESE MEAT PRODUCTS

Over 100 kinds of meat products are available in the Chinese food market today. Most are traditional types characteristic of Chinese dietary custom and culture. A small percentage belongs to nontraditional types developed in recent years using the processing technology introduced from the United States or Europe. The classification of meat products is based on animal species, the product formulation, processing methods, and product characteristics. Many processed Chinese meat products are named after a specific geographic location or a specific place, very similar to many sausage products produced in Germany and countries in the Middle East. The vast majority of meat products processed in China are prepared from pork, followed by beef, sheep, chicken, duck, fish, rabbit, and game animals.

On the basis of processing methods and product characteristics, Chinese meat products can be divided into six major groups: (1) cured meats, (2) braised and seasoned meats, (3) smoked and baked meats, (4) sausages, (5) canned meats, and (6) dried meats (Table 7.2).

2. PRINCIPLES OF CHINESE MEAT PROCESSING

2.1 GENERAL PRODUCT FORMULATIONS AND PROCESSING

Pork is the prevalent raw material used for the production of Chinese meats. Raw meat materials destined for processing must be inspected for whole-

TABLE 7.2. Classification of Chinese Meat Products.

Group	Type	Example
Cured	Salted meats	*Zhejiang* salted pork
	Winterized meats	*Sichuan* winterized meat
	Meats pickled in soy sauce	*Beijing* soy sauce pickled pork
	Hams	*Jinhua* ham
	Bacon	Smoked pork belly
Braised and seasoned	Seasoning marinated meats	*Taiyuan* six-taste meat
	Five-spice meats	*Shanghai* five-spice pork
	Pork ribs braised in soy sauce	*Wuxi* pickled rib
Smoked and baked	Smoked, fully cooked	*Guangdong* stick cooked pork
	Smoked, partially cooked	Cured and smoked pork loin
Sausage	Traditional sausages	Fresh pork sausage
	Starch-type sausages	Starch pork sausage
	Emulsion-type sausages	Ham sausage
Canned	Fresh processed meats	Steamed pork
	Flavored meats	Pork with pickled black bean
	Further processed meats	Luncheon meat
Dried	Shredded meat	*Tai cang* shredded pork
	Beef and pork jerkies	Curry pork jerky
Other products	Pork and beef tendon	Export pork tendon
	Meatballs	Steamed beef ball
	Variety of meats	Pork liver sausage

someness and compliance with sanitation requirements. The fat level is not a particular concern in most products, and therefore, it is not uncommon that a Chinese meat product contains 40% or more fat. To enhance the product flavor characteristics and appearance, a variety of additives are incorporated into the product formulations. Based on their specific functions, additives can be grouped into (1) taste enhancers—salt, sugar, soy sauce (either light- or dark-colored), wine (liquor), vinegar, and monosodium glutamate (MSG); (2) aroma enhancers—prickly ash, bark cassia, clove, fennel seed, green onion, garlic, and numerous other natural ingredients derived from fruits, seeds, flowers, flower buds, stems, leaves, and roots of plants; and (3) pigments or colorants—various natural pigments, such as carmine (a red pigment), and color precursors, such as glucose and ascorbic acid, which produce brown pigments via browning reactions. Many of the natural flavor substances have been used for hundreds of years by the Chinese people for treatment of lost appetite, digestive imbalances, and other diet-related disorders. Hence, the use of these natural ingredients may be of phamaceutical and nutraceutical advantages as well. In recent years, new technologies have been developed to concentrate the specific flavoring compounds through distillation and chemical extraction. The application of these condensed flavorings can be more accurately controlled. As of today, the use of red/pink pigments extracted from plants and vegetables (e.g., carmine) is permitted to enhance the red color of meat product. However, this practice, considered adulteration, is prohibited in the United States and in most other countries. Legume, tuber, and cereal starches are commonly used to bind water and modify product texture.

2.2 DIFFERENCES BETWEEN CHINESE AND WESTERN MEATS

Chinese meat products differ considerably from those produced in the United States, Germany, and other European nations in many ways, including raw material selection, product formulation, and the processing procedure (Table 7.3). While most Chinese meats are made from pork, products made in Western countries are comprised of beef and poultry (particularly turkey), in addition to pork. For example, many frankfurters and luncheon meats are pre-

TABLE 7.3. Major Differences between Chinese and Western Meat Products and Processing.

Parameter	Chinese Meats	Western Meats
Raw materials	Predominantly pork	Beef, pork, chicken, and turkey are all major species
Meat/fat particle size	Large	Small
Fat content	Wide range	Emphasize low
Processed flavor additives	Soy sauce, grain alcohol, sesame oil	Soy flour, milk solids
Natural flavorings	Huge selection of spices and seasonings (star anise, cassia bark, clove, prickly ash, fennel, nutmeg, ginger root, orange peel, etc.)	Limited and different variety paprika, (pepper, cumin, etc.)
Cooking methods	Braise, bake	Rarely braise
Processing time	Long	Short
Processing automation	Low	High
Regulations	Few and some tolerance; red pigments permitted	Numerous and strict

pared using a mixture of beef, pork, and chicken or turkey. In most Chinese meat products, meat is usually either kept intact to preserve the characteristics and identity of the whole cuts, or coarsely ground or cut into particles and cubes substantially larger than those found in Western meat products. A good example is sausage. Chinese-style sausages contain distinctly large (e.g., 0.7–1.0 cm) meat and fat particles, while sausage products from European countries are usually prepared using more finely ground (typically <0.5 cm diameter) or finely chopped (emulsified) meat and fat. Soy sauce is commonly added to Chinese meats to enhance flavor and appearance, but it is never an ingredient in Western meat products. Wine or liquor is formulated into many Chinese sausages, but it is rarely used in Western products. The type of spices/seasonings used also differs for Chinese and Western products. Many dry or semidry meats made in China have a relatively high salt content (>4%), differing from Western counterparts, which typically contain only 2.5–3% salt. Because of the relatively low water activity, a greater degree of dehydration, and high temperature of cooking, most cooked Chinese products (as indicated in Table 7.2) are marketed unrefrigerated. The relatively short distribution (i.e., product usually sold within a close vicinity, with a high turnover rate) and the lack of cooling facilities are also reasons that refrigeration is not commonly applied to these meats. Cooked Chinese sausages and a variety of other high-salt meats are usually vacuum packaged to extend shelf life when stored at room temperature.

3. CURED MEATS

3.1 PRINCIPLE OF PROCESSING

This group of Chinese food encompasses a wide range of salted meats cured with or without added nitrite or nitrate. The main products in this group of meats are salted meats, winterized meats (semidry), meats pickled in soy sauce, ham, and bacon. Pig head, pig tail, pig feet, corned beef, and salted lamb are some of the most popular salted products.

Technologically, curing is done by four different methods: (1) dry rubbing, (2) soaking, (3) a combination of the above two methods, and (4) brine injection. In dry rubbing, salt, nitrate or nitrite, and other ingredients are incorporated into meat through rubbing or massaging the meat cuts. Soaking is a common practice for small cuts in which meat is submerged in a brine. Brine injection is a relatively new technique introduced from Europe. In this method, a specific amount of brine is stitch pumped into meat. Because of the high efficiency of brine distribution in muscle tissue, this method of curing is gaining popularity for large cuts, such as hams and bacon. Traditionally, crude salt is used. Because the salt is contaminated with nitrate, no additional nitrate or nitrite compounds are needed to initiate the curing reaction.

3.2. SALTED MEATS *(YAN ROU)*

Essentially, all parts of the animal carcass could be processed by salting. The processing involves three major steps: (1) raw material selection and sorting, (2) trimming, and (3) salting. For large pieces of raw materials such as sectioned loin, it is often desirable to surgically cut open the meat to facilitate salt penetration. Salted meats, called *yan rou* in Chinese, are usually prepared by soaking meat in a concentrated brine, and the salting time varies, depending on the size of the meat (i.e., approximately 3 days per kilogram of meat). A typical brine is composed of salt and other functional and flavor ingredients. Not only does salt preserve meat by increasing the osmotic pressure of meat and causing dehydration, but it also contributes to the flavor and firmness of meat.

If meat is dry rubbed, salt will be applied on two or three different days that are approximately 4–5 days apart, depending on the size of the product. For example, for salted pork legs, about 18 kg of salt (per 100 kg of meat) is divided into three portions that are applied on different days within a 10-day period. After final salting, meat is overhauled and stored in the brine extract for two more weeks to allow salt equilibrium. To extend shelf life, salted meat should be stored in a concentrated brine (about 24–25° Brix) and kept in a cooler.

3.3 WINTERIZED MEATS *(LA ROU)*

The name *winterized* is derived because these products are preserved in brine and kept through winter to allow meat to dehydrate and develop sharp flavors under natural, low-temperature conditions. It is believed that the long-time, low-temperature processing produces the best aroma, taste, and texture of cured meat. Traditionally, winterized meats are processed in the 12th month on the lunar calendar, known as *La Yue* (about late January or early February on the solar calendar). The first part of the processing is salting, as described above, which takes about 3 weeks. Salted meat will be soaked in water to remove excess salt and subsequently dehydrated in a well-ventilated area under the sun when the ambient temperature is relatively low. This aging process typically takes about 3 months to complete. Finished products will have a low water activity, firm texture, a bright pinkish-red color caused by the formation of nitrosylmyoglobin, and a strong flavor attributed to chemical and enzymic changes in lipids, proteins, and carbohydrates.

Today, most commercial meat processing plants are replacing the lengthy natural aging and dehydration process. Instead of natural drying, salted meats are briefly dried (for 3–4 h only) under the sun before they are placed in an oven to allow accelerated dehydration and flavor development. The oven temperature should be controlled within 45–50°C, and total baking time should not exceed 4–5 days. During baking, meat may be smoked, depending on the specification of the finished product. Finished winterized meats should appear oily and shining on the surface; the fat tissue should look somewhat golden-yellow and translucent, and the lean should be pinkish-red, resulting from nitrite-based curing reactions. A typical winterized pork product will have a 72–75% yield.

3.4 MEATS PICKLED IN SOY SAUCE *(JIANG ROU)*

These products have unique characteristics because a special processing protocol is required. Raw meat is initially soaked in a brine or curing solution for 6–8 h and then partially dehydrated by baking in a 45–50°C oven for 10 h. Soy sauce is then brushed onto the surface of the meat. Because of partial dehydration, the meat picks up soy sauce easily. The meat coated with soy sauce is wrapped with a paper towel and baked for another 15–20 h. After removing the paper by soaking in warm water, the meat is returned to the oven and baked for 24 h. The finished product typically has a 100–105% yield and has a strong soy sauce flavor.

3.5 HAMS

Hams are made from the hind legs of the pig. Processing of Chinese-style hams involves curing, soaking, sun drying, and aging with numerous detailed steps. Unlike other cured meats, ham requires a long time to process, that is, 8–10 months. Among the most famous ham products in China are *Jinhua* ham (Figure 7.1), *Rugao* ham, and *Yiwei* ham, all named after their geographic locations. Despite variations in specifics, the basic steps involved in the processing of these ham products are the same. Overall, Chinese-style hams differ sharply from brine-injected hams manufactured in the United States but are similar in many aspects to the country ham (e.g., Kentucky country ham). Figure 7.2 provides the major steps in the manufacture of *Jinhua* ham using the traditional processing method.

In the traditional method, *Jinhua* ham is cured in the winter season. The fresh hind leg is strictly controlled within 5–7.5 kg to ensure product uniformity. For each 100 kg of fresh leg weight, 7–8 kg salt is used. Under normal conditions, a 5-kg pork leg will be cured by applying salt on days 1, 2, 7, 13, 20, and 27, with the first four applications using 75, 200, 75, and 25 g salt, respectively (SSFI, 1961; CSTIB/CFRI, 1985). The last two applications require only a small amount of salt (<20 g) to ensure that the ham is completely dusted with salt. Salted hams are cured at 0–10°C for 33–40 days. After curing, excess salt on the ham will be removed by washing in water. Washed, clean ham is subsequently dried in the sun for 5–6 days, after which the skin should have turned brownish tan and oil on the surface should start melting.

Figure 7.1 *Jinhua* ham, one of the most famous ham products in China, is named after its geographic location.

Subsequently, hams are hung on racks and allowed to ferment for 2–3 months, during which the unharmful green molds should gradually grow on the ham surface, an indication of proper progression of fermentation. Fermented hams are cleaned by brushing off the molds and are trimmed of irregular pieces to smooth out the surface. Cleaned hams are then aged in a dry room for 3–4 months, making sure they are well spaced to allow proper air flow and dehydration. Aging normally lasts about 9 months. Finished *Jinhua* ham (as well as most other Chinese-style hams) has a 55–60% yield.

With new processing technologies developed in recent years, *Jinhua* ham, as well as other Chinese-style hams, can now be produced in a much shorter period of time. In particular, the fermentation and aging processes used in the traditional method can be accelerated by placing the cured hams in a temperature-

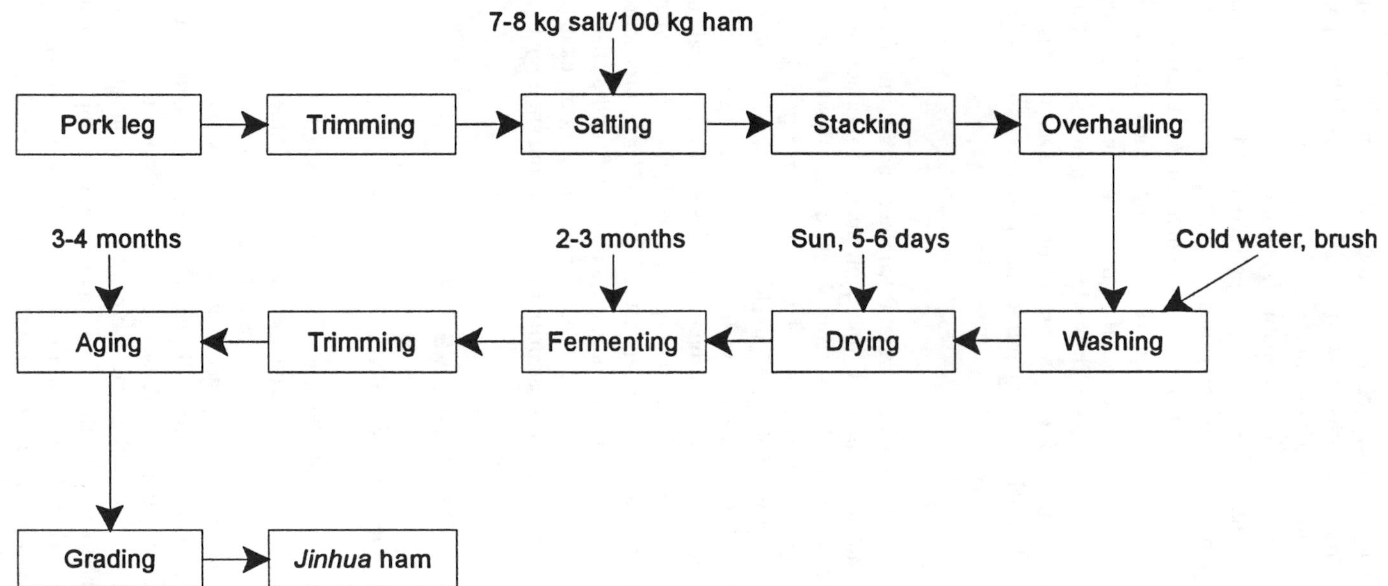

Figure 7.2 Schematic representation of *Jinhua* ham processing.

programmable aging room with 60% humidity, and the entire processing time can be cut down from 9 months to as short as 1–2 months. Other processing steps, however, remain largely the same.

As the first prize winner in the Panama International Merchandise Exhibition held in 1915, *Jinhua* ham has long been recognized as a premium meat product. Today, joining several other meat products, *Jinhua* ham is routinely exported to meet the world's demand. It is sold in South Asia, Japan, Australia, and Europe. Recently, several countries (e.g., Australia and Germany) have reportedly experimented with some modified manufacturing processes to make *Jinhua* ham. Despite the similarity in product characteristics, it is difficult to reproduce all the flavor and texture attributes of *Jinhua* ham that is manufactured using the traditional slow fermenting and aging processes.

3.6 BACON

Bacon is prepared from pork belly that is trimmed and cut into a rectangular shape. Bacon is not a traditional Chinese product; its production in China began only in recent years, adopted from Western bacon-processing technology. Bacon is produced by injection of brine into fresh pork belly through stitch pumping, or by dry rubbing with salt, nitrite, and other curing adjuncts such as erythorbic acid and phosphate. The curing process for bacon usually takes 2–4 weeks, much shorter than the curing process for a typical cured and winterized Chinese meat product. It is expected that in the next 10 years or so, the typical Western bacon-processing method (formulation, procedures) may be modified to better suit Chinese consumers and to reflect the characteristics of traditional Chinese meat products (e.g., spice use). As an example, sun-drying could be added to the process prior to cooking and smoking to improve the product flavor and texture.

4. BRAISED AND SEASONED MEATS

4.1 PRINCIPLE OF PROCESSING

This group of traditional Chinese products is prepared by cooking meat marinaded with various plant-derived seasonings and other flavor-eliciting substances. Soy sauce is the basic ingredient in these products. Many of the seasonings are known for their pharmaceutical functions and have been used as Chinese medicines for hundreds of years. Hence, these plant flavor materials may be of certain clinical or disease-preventative value to meat products. Because cooked products will contain a considerable amount of marinade, it is quite troublesome to package them in a way suitable for shipping unless they are shipped frozen. For this reason, marinated and cooked meats are usually sold locally. However, with new packaging technology developed in Western countries (e.g., cryogenic freezing and vacuum packaging), together with the increased efficiency in railway or highway transportation systems, it is becoming possible to market these products as convenient, ready-to-eat foods throughout the country.

Marination and cooking are two major steps in the processing of braised and seasoned meats. The formulation of marinade varies greatly with geographic locations. For example, people from the Deep South such as Guangdong province prefer sweet meat; those from the Southwest such as Sichuan prefer hot, heavily spiced meat, while consumers from the Northwest such as *Shanxi* like marinades containing some vinegar. Despite variations in product formulation, the basic ingredients used to prepare braised and seasoned meats are more or less the same. They include soy sauce, salt, ginger root, green onion, sugar, cooking wine, and various plant-derived seasonings such as star anise (octagon). Marination time is usually 1–3 h. Some of the flavor ingredients are added during cooking to "fix" the taste. After cooking, additional sugar, MSG, and other condiments may be added to adjust or further enhance the color and flavor of the product.

Braised and seasoned meats are always cooked in marinades. Cooking is done on high heat first, which is then gradually reduced to medium or low heat. The brief high-temperature heating enables quick denaturation of proteins and destroys potential pathogens. Subsequent medium- or low-heat cooking lasts for several hours, during which meat is gradually tenderized, resulting from solubilization of collagen and disruption of myofibrils. Because muscle cells and the myofibril matrices are disrupted, flavor ingredients from the marinade can readily penetrate into the meat, thereby greatly enhancing the palatability and flavor of the cooked product.

Many cooked meats are prepared by briefly (10–30 min) boiling unmarinated raw meat first to get rid of undesirable odors, foaming agents, and debris. This pretreatment is followed by two to three stages of cooking of the cleaned meat in a marinade as described above. The disadvantage of the pretreatment is that some water-soluble nutrients (minerals, vitamins) may be lost. However, the dilemma can be resolved by saving the plain broth, referred to as "white soup," which, after skimming off the foam, is placed back in the pretreated meat during subsequent cooking.

4.2 FIVE-SPICE BRAISED PORK

Pork is cut into rectangular blocks measuring approximately 3 × 10 × 16 cm and weighing about 0.75 kg. Loin cuts are the most popular raw material although cuts from other parts of the pork carcass can be used as well. The skin should be kept on, but the bones can be removed. Three or four holes are poked on the meat surface to facilitate marinade penetration. A typical product formulation is shown in Table 7.4. The product is referred to as "five-spice"

TABLE 7.4. Product Formulations for Braised and Seasoned Meats.[a]

Ingredient	Five-Spice Braised Pork (kg)	Five-Spice Braised Beef (kg)	Lamb Braised in Soy Sauce (kg)	Pork Ribs Braised in Soy Sauce (kg)
Salt	2.5	1.5	1.5	1.5
Sugar	1.0	—	—	2.5
Soy sauce, brown	2.5	5.0	5.0	5.0
Sesame oil	—	—	0.75	—
Cooking wine	1.5	—	—	1.5
Green onion, shredded fresh	1.0	—	0.04	—
Ginger root, shredded fresh	0.1	—	0.04	0.25
Fennel	0.1	0.08	0.08	0.125
Cassia bark	0.1	0.08	—	0.125
Tangerine peel	0.1	—	—	—
Prickly ash	—	0.08	0.08	—
Clove	—	0.08	0.08	0.25
Round cardamom	—	—	0.08	—
Dahurian angelica root	—	—	0.08	—
Monosodium glutamate	—	—	—	0.1
Sodium nitrite	0.025	—	—	0.015
Fructus seu semen amomi	—	0.08	0.08	—

[a]Based on 50 kg of raw meat material.
Source: Adapted from Xia et al. (1987).

meat because its formulation calls for the following five spices, with each having a distinctive flavor: star anise, cassia bark (cinnamon), clove, prickly ash, and fennel seeds. Initially, meat cuts are rubbed with a mixture of sodium nitrite and salt at a rate of 25 g nitrite and 2.5 kg salt per 50 kg meat. The meat is allowed to cure in a tank for 24 h if the room temperature is greater than 10°C or for 36–48 h if the room temperature is less than 10°C. Subsequently, pork is placed in a large cooker (e.g., wok) and mixed with all other ingredients and pork broth (or a small amount of water). The mixture is quickly brought to a boil with high heat. At this time, the cooking temperature is lowered to a simmering temperature (70–80°C) at which pork will be slowly braised for 2 h. The finished product will have a reddish-brown and shiny appearance with a strong aroma and taste. The slow, moist heating also tenderizes meat by the disruption of myofibrils and conversion of cross-linked collagen to soluble gelatin. Thus, the final product also has a "melt-in-mouth" characteristic.

4.3 FIVE-SPICE BRAISED BEEF

To prepare this product, boneless beef cuts, each weighing approximately 1 kg with less than 4 cm in thickness, are used. The location of cuts is not restricted; it can be from the chuck (shoulder), the arm, the rib, the loin, or the round (leg). When rib cuts are used, they are not deboned. The product formulation slightly differs from that for five-spice pork (Table 7.4); sodium nitrite is not used because beef is a well pigmented muscle compared to pork. To process, beef cuts are stacked in a cooker, and water is added and brought to the level slightly above the surface of meat so that all cuts are submerged in water during cooking. When water starts boiling, all spices and ingredients are added, and the meat is cooked on high heat for 4 h. During cooking, meat cuts should be turned over once, and additional water or beef broth may be added to compensate for the water evaporated. The salt level should also be adjusted to obtain a desirable taste. With all meat submerged in the soup, the product will be braised at medium-high heat for 4 more hours, during which the meat will be turned over at 1-h intervals. After cooking, a layer of soy sauce will be poured onto the surface of the meat to enhance the product appearance.

4.4 LAMB BRAISED IN SOY SAUCE

The formulation for this product is shown in Table 7.4. The forequarter of the lamb carcass is cut into 0.5-kg pieces. The cuts should be pretreated by soaking in cold water for half an hour and then thoroughly rinsed. Raw lamb cuts placed in a cooker (wok) are covered with lamb broth or water. After cooking at medium-high heat for 30 min, spices (except salt, cooking oil, and sesame oil) contained in a cheesecloth sack are added (submerged in) to the broth. After cooking for another 30 min, salt is added, and the mixture is cooked for 30 more minutes. The final step of processing is done by frying the cooked lamb for 10 min in mixed cooking and sesame oils that have been heated to a smoking point. Fried lamb should have a strong aroma and taste and an appealing golden brown appearance.

4.5 PORK RIBS BRAISED IN SOY SAUCE

This product originated from the city Wuxi in the province of Jiangsu. It is a popular processed meat known for its salty-sweet taste and delicate aroma. The formulation of the product is not much different from other marinated and cooked meats, except that it requires a relatively large quantity of sugar and cooking wine (Table 7.4). A cure solution is prepared by dissolving approximately half of the required salt in the nitrite solution. Marination is done by blending pork ribs (usually spare ribs, 7–10 cm in length) with the cure solution and setting overnight to allow the cure to penetrate the meat. Marinated ribs are then boiled in water for a few minutes while stirring. Foams and floating particles are skimmed off, and the broth is saved for later use. The pretreated ribs are rinsed in cold water and replaced in the cooker on top of three cheesecloth sacks containing onion, ginger root, cassia bark, fennel, and star anise. Wine, brown soy sauce, the remaining salt, and broth (saved from the previous boiling) are added. Additional water may be needed to assure that the ribs are submerged with a minimum of 3 cm water cover. The mixture is rapidly brought to the boiling point. After boiling for 30 min, ribs should be cooked at medium heat for 2 h. Sugar is added at this time, and the ribs are cooked at high heat for 10 more minutes to obtain a thick, viscous coating. Fully cooked ribs are transferred to pans, and the concentrated soup is mixed with MSG before pouring over the ribs.

5. SMOKED AND BAKED MEATS

Smoke has long been utilized by the Chinese to enhance the flavor and to preserve the shelf life of cooked meats. In actuality, Chinese people in the Stone Age would cook the entire carcass of hunted animals over an open flame, which not only provided heat but also generated smoke. Smoking and baking are usually done simultaneously in an oven, and meats being processed this way are large pieces of cuts. Traditional smoked and baked meats include *cha shao,* smoked pork, baked meat, bacon, and smoked ham. The last two items are relatively new products processed using the technology and recipes adopted from Western countries.

The most popular smoked and cooked meat product is *cha shao* (meaning cooked in forks or sticks), so named because it is held in fork-like sticks and cooked in flames in a way similar to cooking of roasted chickens in the United States. The processing is done as follows. Pork, with skin, bones, and excess fat removed, is cut into 3-cm-wide strips. The strips are bundled and cut into smaller pieces (about 15 cm long). After rinsing, they are mixed (by massaging or brushing) with a special sauce comprised of 2.5 kg light-colored, premium-grade soy sauce, 3 kg cane sugar, 2.5 kg malt sugar, 1 kg salt, 1 kg liquor (100 proof), 150 g five-spice powder (prickly ash, star aniseed, cinnamon, clove, and fennel seeds), and 100 g rice wine, per 5 kg of meat. After a 40-min marination, meat strips are put on stainless steel hooks and hung on racks in an oven. Heat and smoke are generated from burning charcoal. Dripping oil should be collected to avoid excessive smoke and fire. After 15 min of baking, the rows of meat strips should be turned to receive uniform heat and smoke. At this time, more charcoal will be added, and the meat will be cooked for an additional 30 min. After cooling for a few minutes, cooked meat should be briefly soaked in malt syrup and reheated in the oven for 3 more minutes to partially dry up the malt syrup. The finished products have a characteristic roasted reddish-brown appearance, a strong nutty aroma, and a sweet taste.

6. SAUSAGES

6.1 PRINCIPLE OF PROCESSING

Chinese-style sausages are made from ground or comminuted meat through a series of processing steps, including marination, mixing, stuffing, cooking, and smoking. Today, over 20 varieties of sausages are manufactured and sold in the Chinese food market. The sausage processing technology originated from Europe, particularly from Italy, France, Germany, Poland, Holland, and Russia. However, the product recipes and formulations have been developed to uniquely suit Chinese consumers' taste preference. Today, several food companies in the United States are also making Chinese-style sausages, which are sold in supermarkets usually run by Chinese-Americans. However, the flavor of these products deviates considerably from the authentic Chinese sausages. For example, paprika and tripolyphosphate are commonly used in Chinese-style sausages made in the United States, but they are seldom used in those made in China. Pork remains the far predominant species used to make Chinese-style sausages, but other species, including beef, lamb, rabbit, chicken, and duck, are also used. Unlike their European-style counterparts, many Chinese-style sausages are prepared from larger meat and fat particles; are more heavily seasoned and spiced; and are drier, sweeter, and firmer in texture. Based on the product formulation and the specific processing procedure, Chinese sausages can be classified into three major types: (1) traditional sausages, (2) starch-type sausages, and (3) emulsion-type sausages. Several popular Chinese sausage products are described below.

6.2 TRADITIONAL SAUSAGES

Traditional Chinese sausages are made from pork and contain a small amount of grain alcohol. A large amount of sugar is also added, making the finished product taste quite sweet. Pork sausage and *Nan* sausage (named after the city Ji Nan) are two major products, and their processing formulations are shown in Table 7.5. Pork sausage is fully cooked and ready to consume, while *Nan* sausage, an uncooked product, is partially dehydrated and "cured" at room temperature for 30–40 days in a well-ventilated room. Because of dehydration, the product is fairly shelf-stable but must be cooked before consump-

TABLE 7.5. Product Formulations for Selected Chinese Sausages.

Ingredient	Pork Sausage (kg)	*Nan* Sausage (kg)	Ham Sausage (kg)	Starchy Sausage (kg)
Lean trim (>90% lean)	50.0	35.0	40.0	40.0
Fatty trim (>90% fat)	5.0	15.0	10.0	10.0
Water	5–10	—	—	40.0
Salt	1.75	1.5	1.5	2.0
Sugar	1.25	1.0	—	—
Soy sauce	—	2.0	—	5.0
Rice wine	0.25	—	—	—
Chinese whiskey (grain alcohol)	—	0.5	—	—
Sesame oil	—	—	—	1.0
Ginger root, shredded	—	—	—	0.50
Garlic, shredded	—	—	—	0.50
Fennel powder	0.10	—	—	—
Five-spice powder mixture	0.10	0.15	0.10	0.10
Black pepper	0.10	—	—	—
White pepper	—	—	0.10	—
Monosodium glutamate	0.10	—	—	0.05
Starch	2.5	—	2.5	18.5
Carmine	0.006	—	—	—
Nitrite	0.0025	—	0.0075	0.0075
Prickly ash	—	0.10	—	0.50[a]
Fructus seu semen amomi	—	0.05	—	—

[a]Prickly ash used in starchy sausage is soaked in 40 kg water before use.
Source: Adapted from CSTIB/CFRI (1985) and Xia et al. (1987).

tion. Taiwan sweet sausage is another typical Chinese-style sausage that is probably familiar to many U.S. consumers because it is available in most Asian food markets in the country.

The processing procedures for pork sausage are outlined in Figure 7.3. Pork trims (~90% lean) and fatty tissue (>85% fat) are cut into approximately 0.8-cm small cubes. Diced meat is presalted at 1–2°C for 12 h and then ground through a 0.45-cm plate. Ground salted meat is cured for another 12 h before it is mixed with the fat (also presalted), the red pigment carmine, and all other ingredients, including water by tumble blending. The meat batter is stuffed in a pork small intestine to make approximately 3.8-cm diameter links (12 cm in length). Before cooking, sausage links are poked with fine needles to release air pressure. Cooking is done (45 min) in a 65–80°C oven heated either by

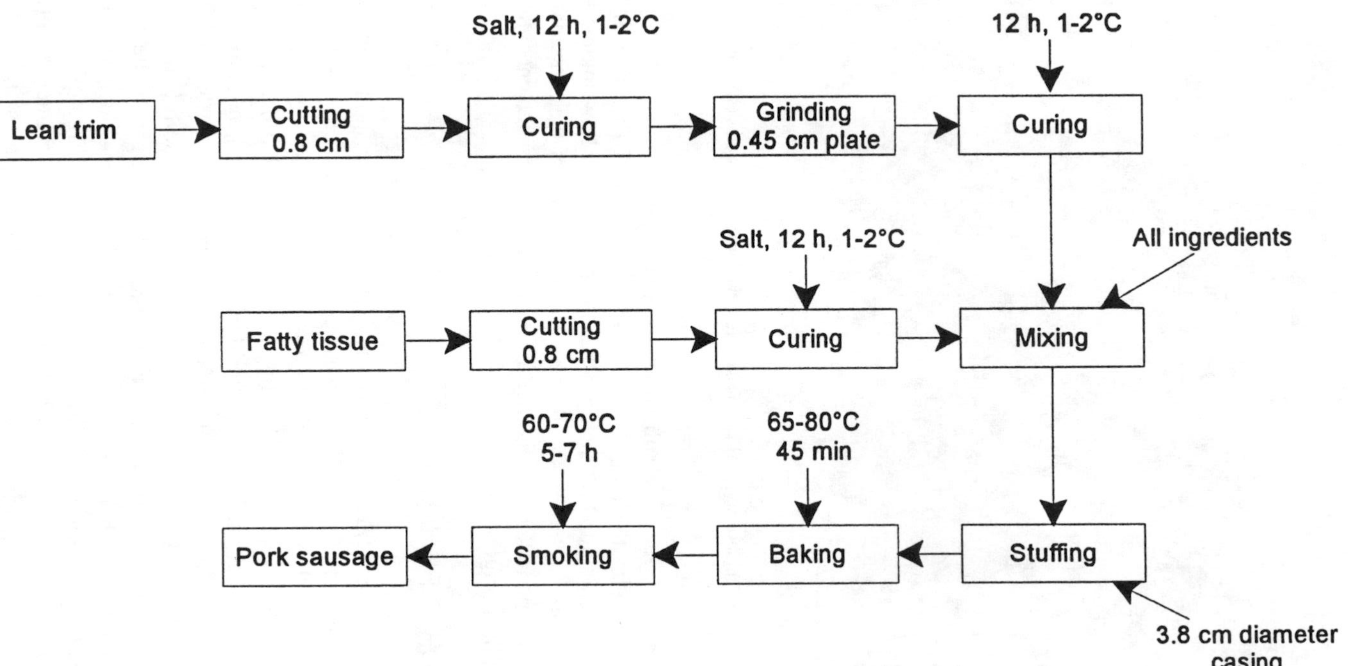

Figure 7.3 Typical processing procedure for Chinese-style pork sausage.

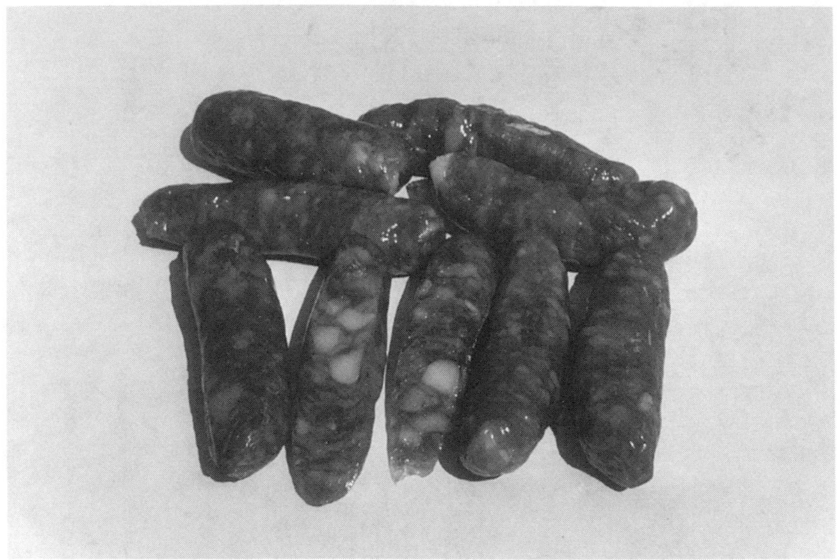

Figure 7.4 Chinese-style pork sausage. Note the large fat particles (white).

charcoal or natural gas. The subsequent smoking process takes 5–7 h in a smoke chamber with the temperature maintained at 60–70°C. The finished product appears shiny and reddish-pink and has considerable wrinkles on the surface; which are caused by shrinkage and denaturation of the proteinaceous casing. The fat particles are distinctly visible, appear translucent, and feel rubbery and firm (Figure 7.4).

6.3 STARCH-TYPE SAUSAGES

Starch is a common ingredient in Chinese-style sausages, and the amount of its addition is not regulated. When sausages contain a particularly high level (>10%) of starch, they are considered a special type ("starch-type") product. The added starch acts as a water-binding agent through gelatinization and, additionally, binds meat particles and modifies the texture and mouthfeel of the finished product. This group of products contains a large amount of added starch extracted from beans, yams, or grains (Table 7.5). Both lean and fatty trims are required. The fatty trims are cut into a 2-cm length, while the lean tissue is ground through 0.5 to 0.7-cm plates. The processing procedure is similar to that for traditional Chinese sausages (shown in Figure 7.2), except that the pork casing should be stuffed to only about 80% fullness to prevent rupture during cooking, which is caused by swelling of the starch granules. Cooking of starch-type sausages is done in a specific amount of boiled water (100°C) to facilitate water absorption by starch granules and subsequent gelatinization and is stopped when the water temperature drops down to 75°C (in about 20 min). Smoking is done as described above.

6.4 EMULSION-TYPE SAUSAGES

While most traditional Chinese-style sausages are made using distinctly visible coarse meat and fat particles, some new varieties of Chinese sausages are made from emulsified batters. These products are similar in texture and appearance to frankfurters and bologna produced in Europe and the United States. The processing technology is introduced largely from Western countries. However, the flavor characteristics of Chinese-style emulsion products differ greatly from those of their Western-style counterparts because of the different spice ingredients used. For example, the so-called ham sausages (which are emulsion-type products) typically contain five-spice powder, white pepper, and starch and do not contain paprika and several other seasonings commonly added to frankfurters in the United States (Table 7.5). Ham sausages are stuffed either in a beef small intestine or in cellulose casings. Different cooking conditions are utilized. Traditional ham sausages are cooked in hot water (80–85°C) for 70 min. When smoking is desired, cooked sausages will be smoked for up to 6 h. In recent years, many ham sausages, stuffed in plastic or cellulose casings as 12- to 15-cm links, are sterilized (>121°C). Because of sterility, the cooked products do not require refrigeration for storage and, hence, can be sold in various grocery stores or supermarkets that are not equipped with refrigeration systems.

Among the most widely consumed high-temperature cooked ham sausages today are *zheng da* ham sausage and *xi wang* (meaning hope) ham sausage. Both products appeal particularly to children, especially those in the elementary school-age group. They are also popular products among tourists since they are easy to carry and are shelf-stable at ambient temperature.

7. CANNED MEATS

Canned meats represent a small fraction of processed meats consumed in China. Canning has the advantages of extended shelf life, ease of transportation, and readiness to consume. There are three main groups of Chinese canned meats: (1) fresh processed meats, (2) flavored meats, and (3) further processed meats. Essentially, all the meat products described above can be processed in the can.

7.1 FRESH PROCESSED MEATS

This group of meat products is minimally processed to preserve the flavor of fresh meat; that is, only salt, white pepper, bulk onion, laurel, and a few other ingredients are added. Pork skin is often included in the product formulation to

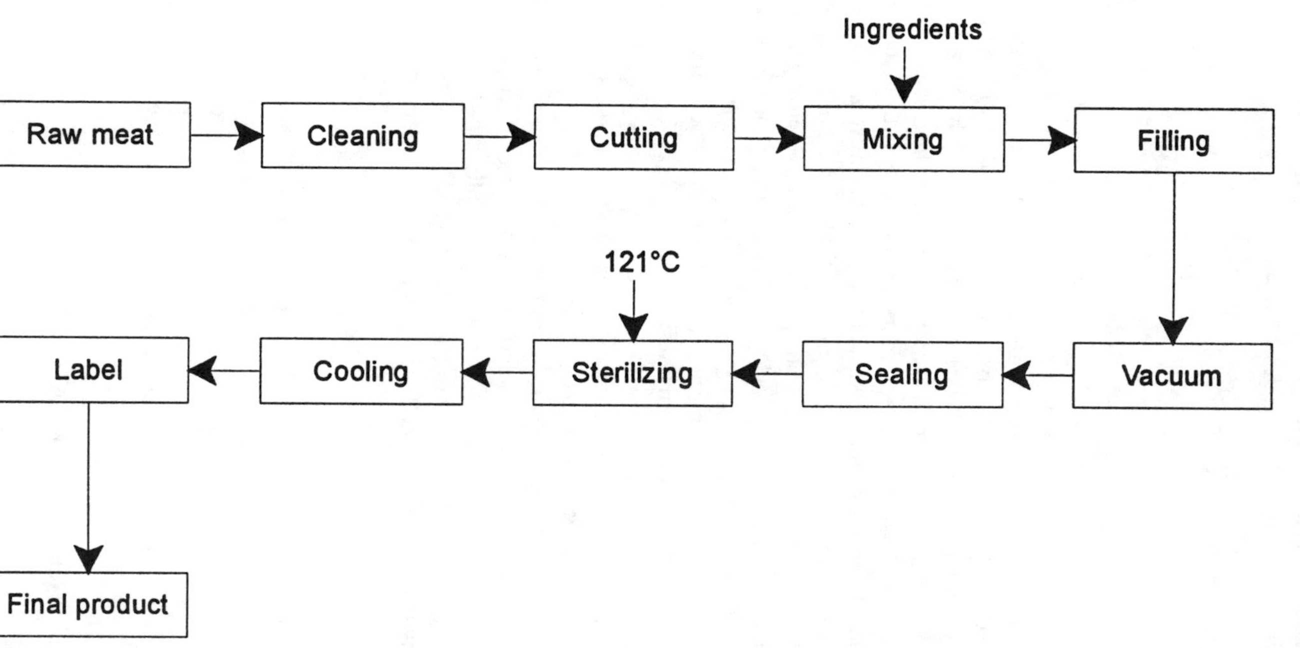

Figure 7.5 Typical processing procedure for fresh processed canned meats.

enhance the meaty flavor characteristics. Heavier flavoring agents such as soy sauce and wine are not used. Beef, pork, and lamb are all popular canned meats. The processing procedure is outlined in Figure 7.5.

Meat is typically cut into approximately 3-cm lengths, with each piece weighing 50–80 g. Before filling, meat should be mixed with pork skin, salt, and various seasonings by blending in a tumbler for about 5 min. Metal cans with a corrosion-resistant enamel and jars are common containers. Cooking is done in a 121°C retort with high-pressure steam.

7.2 FLAVORED MEATS

Most canned Chinese meats belong to the group referred to as "flavored meats" so named because various flavor-enhancing materials are formulated into the products. Though the specific seasonings used vary considerably, depending on the geographic locations, the following additives are common in flavored canned meats:

(1) Flavorings: bulk onion, laurel, white pepper, green onion, ginger root, garlic, prickly ash, fennel seeds, star anise, bark cassia, clove, nutmeg, chili pepper, curry, and other natural substances
(2) Taste-enhancing substances: salt, sugar, soy sauce, MSG, wine, grain alcohol, vinegar, fermented black bean, and tomato sauce
(3) Colorants: Synthetic colorants are prohibited from use in canned meats (and in any other Chinese meat products); however, natural plant pigments, such as red pepper powder, red tomato powder, caramel, and soy sauce are widely utilized

Pork braised in brown soy sauce and beef cooked in curry powder are two popular canned products. Processing of both products are essentially similar to that for fresh canned meats described above.

7.3 FURTHER PROCESSED MEATS

This group of canned meats encompasses a variety of salt- and nitrite/nitrate-cured and smoked meats that are packaged in conventional metal cans or glass jars. Because the curing process is involved, the products require a substantially longer time to prepare compared to fresh processed canned meats. The bright pinkish color, the high water-holding property, and the smoke flavor are some of the unique characteristics of these products. Salted beef, pork feet, and sausages are some of the popular further processed canned meats. However, the best known and most widely consumed further processed canned product is "luncheon meat."

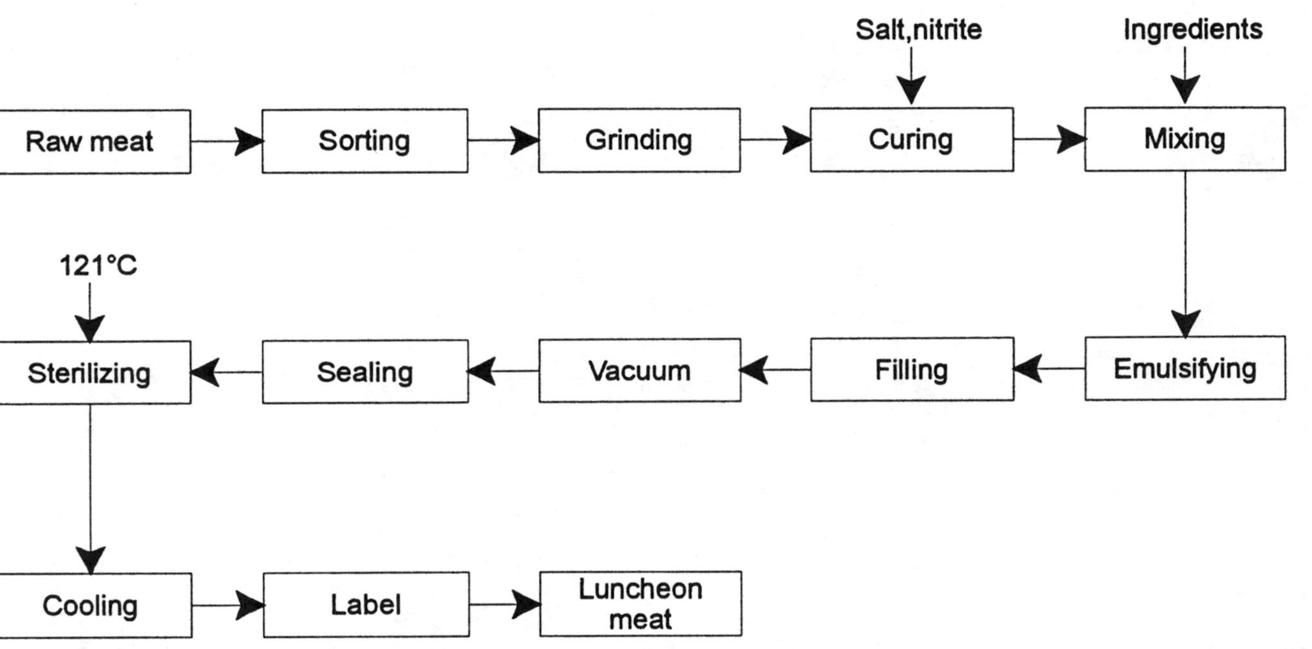

Figure 7.6 Schematic representation of Chinese luncheon meat canning process.

Unlike in the United States where the term *luncheon meat* is used indiscriminately to indicate any processed, ready-to-consume meat destined for a sandwich or deli, Chinese luncheon meat refers exclusively to one particular type of product with a typical formulation such as the following: 8 kg coarsely chopped pork lean (leg, >90% lean), 8.5 kg coarsely chopped pork fatty trims (rib/loin, ≤60% fat), 17.5 kg finely chopped pork lean (leg), 8 kg finely chopped pork fatty trims (rib/loin), 5 kg ice flakes, 3 kg starch, 0.93 kg salt, 20 g nutmeg apple powder, 50 g white pepper powder, 14 g brown sugar, 4.7 g sodium nitrite, and 2 g sodium tripolyphosphate (optional). The major processing steps are presented in Figure 7.6.

Raw meat must be kept at refrigerator temperature, and under no circumstances should the temperature exceed 15°C even during the summer. Meat is ground through 0.7- to 1.3-cm orifice plates, depending on the fineness desired. Curing is done in a 2–4°C cooler and lasts for 24–72 h. To ensure a uniform and smooth texture, the blade of the chopping knife in the bowl chopper should be kept as sharp as possible. Chopping (emulsification) time ranges from 2 min to 5 min, depending on the specific product specification. To prevent chopped meat from adhering to the interior wall of the can, the wall can be precoated with vegetable oil.

8. DRIED MEAT PRODUCTS

8.1 SHREDDED MEAT

Most shredded meat is made from pork. *Taicang* shredded pork is perhaps the most famous in this product category. Its processing is described below.

Meat from the front (picnic shoulder) and rear legs of the pig is used. As a pretreatment, skin, bones, and the adipose tissue are removed, and the meat is thoroughly cleaned. Well-trimmed lean meat is then fabricated into large pieces, with each weighing 1–1.5 kg. The product formulation (based on 50 kg meat) for *taicang* shredded pork is displayed in Table 7.6.

To cook, meat is boiled with ginger root for about 2.5 h in a steam-jacketed vessel. The exact boiling time will depend on the texture desired for the final product. It is imperative that meat is not overcooked, or else the shreds will become too short to provide desirable chewiness and mouthfeel. Foam and grease are skimmed off during cooking. At this time, meat is transferred to another cooking vessel where it will be mixed with 13 kg of meat broth and red and white soy sauces. The mixture will be boiled until approximately 4.5 kg of moisture (from broth) is evaporated. After addition of sorghum wine, the mixture is further cooked. When the broth is mostly dried and forms a thick sauce itself, the steam is turned low, and the loosely structured meat is stirred while cooking for 45 min to facilitate further dehydration. Sugar and MSG are then added and mixed with the meat for an additional 15 min. To obtain shreds,

TABLE 7.6. Product Formulations for Dried Meat Products.[a]

Ingredient	Shredded Pork (kg)	Curry Pork Jerky (kg)	Beef Jerky (kg)
Red soy sauce	3.5	2.0	1.5
White soy sauce	3.5	—	—
Sugar	5.6	7.0	11.0
Salt	0.83	1.5	2.0
Ginger root	0.14	—	—
Monosodium glutamate	0.15	0.30	0.30
Sorghum wine (100 proof)	0.14	1.0	—
Rice wine	—	—	—
Curry powder	—	0.25	—
Five-spice powder	—	—	0.25
Fennel powder	—	0.10	0.10
Nutmeg apple powder	—	—	0.10
Chilli pepper powder	—	—	0.25

[a]Based on 50 kg of meat (pork or beef).
Source: Adapted from Xia et al. (1987).

meat strings are heated for an additional period of time until their final moisture content is reduced to less than 18%. The dehydrated, disrupted meat is fed into a shredding machine where its fibers will be ripped into small, fine shreds.

Subsequently, meat shreds are separated from clumps and particles. This can be done by passing the shredded meat through a vibrating sieve where nonshred particles will drop into a container while the shreds will be retained on the screen. Shredded meat, after final sorting and inspection, is packaged in plastic bags, cups, bottles, or tin cans. They should be stored in a well-ventilated dry place. Since water activity of fully cooked meat shreds is extremely low, they can be stored unrefrigerated.

8.2 BEEF AND PORK JERKIES

Both beef and pork are used to make jerkies. Raw materials used must be low in connective tissue (collagen), and tender cuts (e.g., longissimus dorsi from loin; semitendinosus from leg or round) are preferred. The processing requires a relatively simple procedure that consists of two major steps: braising and lengthy baking. To prepare, meat is boiled briefly in water until proteins are fully denatured. It is then cut into pieces of specific shapes (e.g., 0.2 × 2.5 × 3.5 cm) and braised in a sauce containing various seasonings (Table 7.6). Occasional stirring is necessary to aid in dehydration. When the broth is almost completely dehydrated, stirring should become more frequent. Cooked pork or beef strips are further dehydrated and cooked by baking in a 60–70°C oven for 6–7 hours. For pork jerky, curry should be sprayed onto braised meat before the long baking process. Crushed dry chili pepper may also be added to make hot-spice pork and beef jerkies. Finished products have a firm and chewy texture.

9. OTHER MEAT PRODUCTS

9.1 PORK AND BEEF TENDON

An intermediate product, commercial tendon, is prepared from pig and cattle trotters through a series of washing and soaking steps to remove pigments, fat, and odors. Figure 7.7 summarizes the major steps involved in the production of dry tendon from the pig. The finished tendon, high in collagen, has a white color and minimal animal flavor. A processed beef tendon without drying is shown in Figure 7.8. The tendon is used to make various specialty products and delicacies via further processing. A large portion of processed pork tendon is exported.

9.2 MEATBALLS

Meatballs are made from virtually all species, including pork, beef, lamb, chicken, fish, and shellfish. The processing involves comminution, blending, forming, and cooking. Since meatballs are a special type of restructured product, adhesion or binding of meat particles is the most critical functional property. Traditional Chinese-style meatballs are bound by starch and egg white, both of which can form a cohesive gel upon heating. When a sufficient amount of salt ($>2\%$) is also added, myofibrillar proteins are extracted. The salt-soluble proteins will form a viscoelastic bind (gel) upon cooking, thereby also contributing to the firmness and integrity of cooked meatballs. With the development of various new binding technologies, more and more gelling substances are now utilized to serve as meat- and water-binding agents in Chinese meatballs. Among them are soy flour, soy protein concentrate, milk solids, and whey protein concentrate. Some meatballs (e.g., beef balls) also contain added collagen or tendon. Because collagen is converted to gelatin via hydrolysis during cooking, it can contribute to the smooth texture of meatballs by binding both meat particles and water.

To add flavor, soy sauce, sesame oil, garlic powder, clove, black mushroom, chopped green onion, white pepper, and sugar are commonly used to formulate Chinese meatballs. Chopped water chestnuts can also be used to modify the product texture. After comminuted meat is thoroughly mixed with salt and other ingredients, it is formed into balls either by hand or by a machine. The diameter of balls range from 2–3 cm. Freshly made meatballs should be frozen

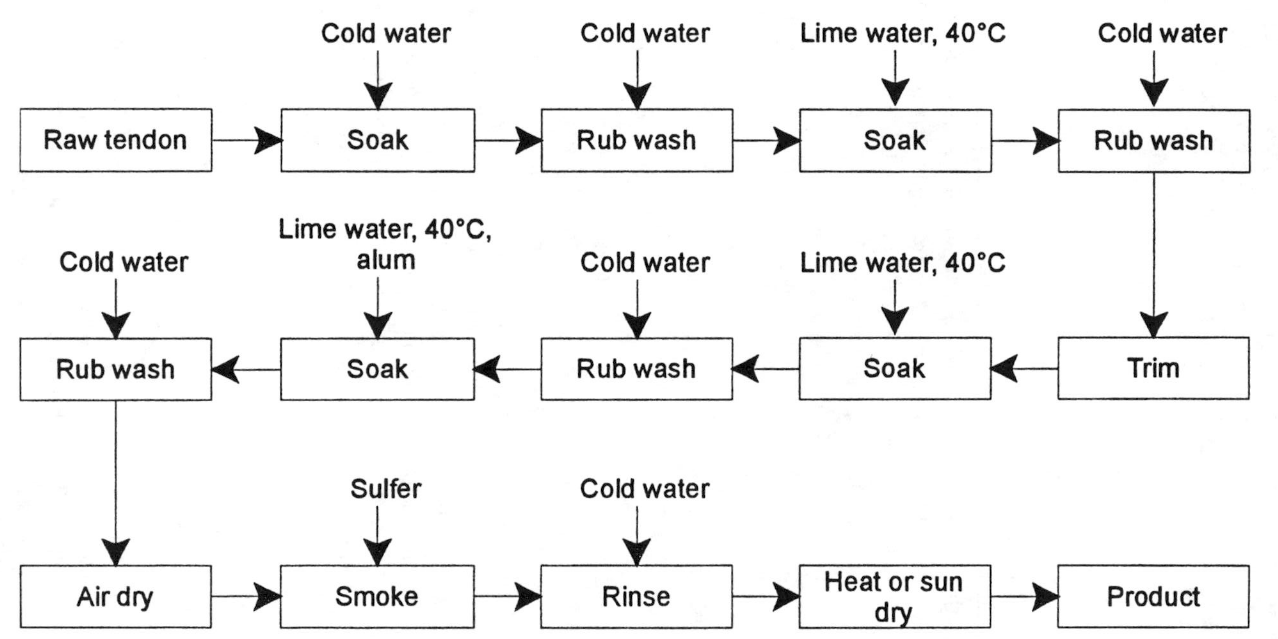

Figure 7.7 Schematic representation of pork tendon processing (adapted from CSTIB/CFRI, 1985).

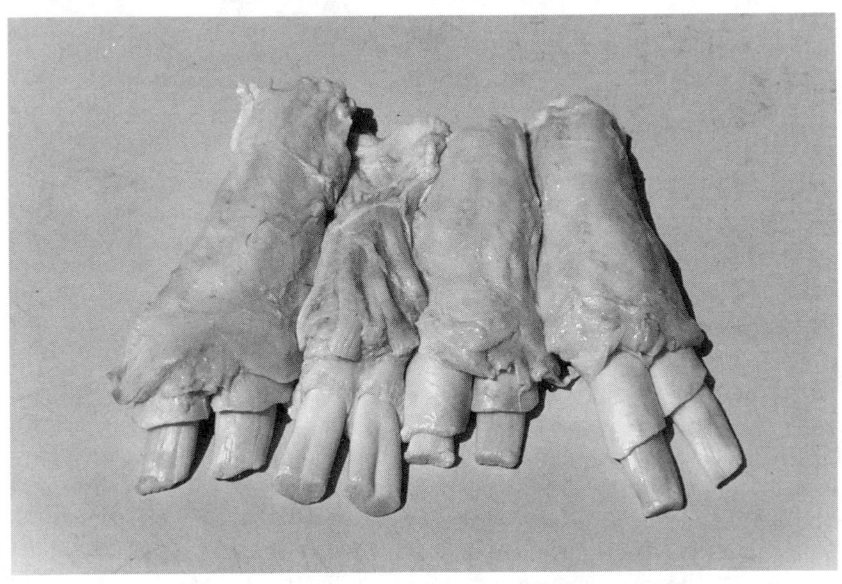

Figure 7.8 Processed beef tendon.

Figure 7.9 Chinese-style meatballs made from beef, pork, and fish.

immediately to preserve their spheric shape. However, if they are made precooked, they are cooked either by steaming (100°C) or by deep-frying. Steamed meatballs are tender and juicy, and the fresh meat flavor is maintained. On the other hand, fried meatballs are of a stronger aroma and are preferred by many consumers. To alleviate the strong flavor of lamb (or mutton), deep-frying is an excellent method of cooking.

Figure 7.9 shows photographs of meatballs made from beef (dark), pork (intermediately light), and fish (light). The color intensity reflects the content of the major muscle pigment myoglobin.

10. SUMMARY AND OUTLOOK

Meat processing in China has had several thousands of years of history. Currently, more than 100 kinds of processed meats are available, which range from whole cuts to finely comminuted products. Among the various processed products, pork remains as the most predominant meat species. Chinese meat products are distinguished from Western meats in many ways, especially in flavor, because of the complex product formulations and the processing procedures used. The transformation and scale-up of the traditional culinary arts to a large commercial production has resulted in a fairly rapid development of the Chinese meat industry in recent years. Modern meat-processing technologies from Western countries are gradually being incorporated into the traditional Chinese meat-processing line, and this certainly has facilitated the advancement of the country's meat industry. The main technological hurdle that hinders further development of the Chinese meat industry is the lack of a sufficient transportation and distribution system.

Chinese meat products have been known and enjoyed worldwide for hundreds of years, but this has been mainly through food services (e.g., restaurants) and trading at neighboring borders (in ancient times). However, as the entire Chinese food industry is becoming internationalized, more and more traditional and contemporary Chinese meats are expected to enter the global marketplace. There is also a clear trend that cultural foods are becoming more and more popular in Western countries. Hence, there seems to be great potential for Western countries, including the United States, to produce Chinese-style meat products based on the formulations and the traditional Chinese processing methods to meet the increased consumers' demand.

11. REFERENCES

AMI. 1997. *Meat and Poultry Facts.* American Meat Institute, Washington, D.C.

CASTP. 1993. *Chinese Agricultural Product Specialization and Regional Development.* Chinese Agricultural Science and Technology Press, Beijin, China.

CSTIB/CFRI. 1985. *Food Processing Technology and Formulations.* Changsha Science and Technology Information Bureau/Changsha Food Research Institute, Changsha, China.

Hedrick, H.B., Aberle, E.D., Forrest, J.C., Judge, M.D., and Merkel, R.A. 1994. *Principles of Meat Science,* 3rd ed., Kendall/Hunt Publishing Company, Dubuque, IA.

SSFI. 1961. *Food Technnology,* 1st ed., Shanghai School of Food Industry, China Ministry of Light Industry. China Finance and Economy Press, Beijin, China.

SSVM. 1990. *Meat Processing Technology.* Shanxi School of Veterinary Medicine. Chinese Agricultural Press, Beijin, China.

USDA. 1995. *Agricultural Statistics 1995–96.* National Agricultural Statistics Service, U.S. Government Printing Office, Washington, D.C.

Xia, G.J., Gao, Y.S., and Zhao, G.J. 1987. *Processing of Meat Products.* China Food Publishing Co., Inc, Beijing, China.

CHAPTER 8

Traditional Poultry and Egg Products

TSUN CHIEH CHEN

1. INTRODUCTION

AMONG the many traditional Oriental foods, poultry products are recognized not only as a delicacy, but also for nutritional values. Pork and chicken are the main choices of animal meat in oriental cooking, with chicken being the main ingredient, especially in Chinese cooking.

Like in many other areas, poultry has become one of the fastest growing meat sources in the Orient. For example, China ranked second in world poultry production, although the per capita consumption of poultry is still very low compared with those of further developed countries. Between 1985 and 1994, poultry meat production in China increased nearly fivefold, from 1.6 million tons to 7.6 million tons. During the same period, egg production almost tripled, from 5.4 million tons to 14.7 million tons (Anon., 1996). China has been the number one egg-producing country in the world for many years. In 1995, poultry meat and egg production reached 8 million tons and 15 million tons, respectively.

For purposes of trading, poultry is classified into chickens, turkeys, ducks, geese, guineas, and pigeons. Each species has its own unique characteristics that influence cooking methods and organoleptic properties. In the United States, poultry are further divided into many market classes according to their species, age, and sex. In some cases their age is limited to a span of several weeks; in other cases they may be classified simply as young or old.

As in most countries of the world, the combination of a more healthful image, competitive prices, and the recent development of further-processed products has been largely responsible for the rapid growth of poultry consumption. Fast-food stores in the Orient now offer poultry items such as chicken nuggets, fillets, and tenders like are offered in Western societies.

Eggs are a highly versatile food containing many essential nutrients. Nutritionists classify eggs in the protein food group with meat, poultry, and fish. A serving of two eggs is considered a dietary replacement for 2–3 ounces of lean meat. Eggs are also economical. They are quick, convenient, and easy to prepare and are appetizing and easily digestible. The mild flavor of eggs and some of the natural compounds in the white and yolk supply functional properties such as leavening, binding, thickening, retardation of sugar crystallization, emulsification, clarification, coating, coloring, and high nutritive value and make them an ideal ingredient to combine with other foods.

Poultry meat and eggs have been used for human food since the dawn of history. For centuries, ways to eat chicken have been developed in all corners of the world. Chicken is versatile; it can be broiled, roasted, baked, steamed, fried, boiled, fricasseed, barbecued, and more. Although many freshly prepared culinary types of poultry and egg products are available, traditional processed or preserved poultry and egg products are limited.

Drying is one of man's oldest and most widely used methods of food preservation. The sun-drying of foods yields highly concentrated materials of enduring quality. Fermentation and pickling have also been used in the preservation of foods for centuries. The traditional processed poultry and egg products also rely greatly on these basic preservation methods. The following sections deal with traditional preserved poultry and egg products.

2. TRADITIONAL POULTRY PRODUCTS

2.1 BEIJING ROASTED DUCK

Beijing roasted duck, or *Beijing kao ya* is a well known Chinese dish. The tender, crispy, and delicious flavor of this roasted duck has been enjoyed by many food lovers worldwide. Several methods for preparing this roasted duck have been reported.

The following ingredients are reported by Fu (1969):

Duck (specially fed—fat)	One (about 5 lb)
Sweet bean paste	3 Tbs
Honey or maltose, or syrup mixture	3 Tbs
Sesame oil	1 Tbs
Wine	2 Tbs
Sugar	2 Tbs
Vinegar	1 Tbs

Scallions (2 in. long)	20 pieces
Hot water	1 cup
Flour wheat tortilla or thin plain flour pancakes	20 pieces

Roast duck preparation is as follows (Fu, 1969):

(1) A specially fed duck should be "New York dressed" (a carcass with head, feet, and viscera intact).
(2) Pump air between the skin and meat at the neck opening until the whole duck becomes puffed.
(3) Eviscerate the carcass through a cut under a wing. The body cavity should be rinsed until the water becomes clean.
(4) In a large bowl, baste the duck thoroughly with the warm syrup mixture, using two small sticks to brace the wings away from the body.
(5) Tie a string around the neck and hang the duck in a drafty place. Let air dry for about 6 h.
(6) Roast the prepared carcass in a rotisserie for about 30–40 min. The duck should be hung vertically to allow fat to drip for an even roasting. Roast at high heat for about 20 min and then at low heat for another 20 min. The skin should turn brown and crisp.

In serving Beijing roast duck, all of the duck skin is sliced into thin slices, followed by the meat parts. The slices of skin and meat are wrapped with bean paste sauce and green onion strips in tortillas or thin plain flour pancakes.

Recently, a modified method for roast duck preparation was reported. Well-fed and well-finished ducks are dressed and eviscerated commercially with head and feet still attached. The surface of the carcass is rinsed with hot water and patted dry. The interior and exterior surfaces are then smeared twice with carbonated soft drink for color formation before air drying for about 24 h. The duck skin should be in a wrinkled condition before roasting in an oven at 190°C for 30 min.

2.2 PRESSED-CURED DUCK

Pressed-cured duck is a unique preserved duck meat product in China, with that from Nanking City being famous worldwide. It has been used as a tribute item for emperors for many years and is a popular gift item during the lunar New Year season. In the past, pressed-cured duck from Nanking has been shipped to Hong Kong and the Southeast Asia region.

As with other traditional foods, the processing methods and quality of pressed-cured duck varies greatly in China. According to a survey conducted by the Livestock Research Institute of Taiwan, the pressed-cured duck had the following quality attributes (Huang, 1994):

Moisture	50–57%
pH	5.69–6.88
A_w	0.79–0.91
TBA value	0.89–9.27 mg/kg
VBN value	37.8–57.2 mg%
Residual nitrite	$<$70 ppm
Total plate count	log 4.0–7.0/g

Apparently, as the data indicate, there is a wide variation in quality for the marketed pressed-cured duck. Some brands exhibited relatively high microbial counts, while others showed high rancidity levels. With this in mind, a modernized processing process and quality assurance program would be very beneficial for the marketing of pressed-cured duck.

Traditionally, pressed-cured duck was manufactured during the cold season. The low environmental temperatures assist in the production of a quality product. The general saying in China is, "The colder the season, the better quality of the pressed-cured duck." The following steps should be employed for processing pressed-cured duck:

(1) Selection of duck: The duck should come from a healthy flock, be well fleshed, and have fat covering under the skin, as well as the entire carcass. The fat not only contributes to the overall appearance, but contribute to the texture of the final product.

(2) Slaughtering: Prior to processing, a fasting time of 4–8 h is required. Fasting reduces the chance of carcass contamination from the undigested feed particles. It is important that the bird be fully rested before killing, which results in better bleeding. After defeathering, eviscerating, and final washing, the carcasses should be chilled rapidly to an internal temperature of 5°C or lower.

(3) Shaping: The carcasses with the neck and head intact should be free of bruises and broken bones. In most cases, the feet and wing tips, including winglets, are cut off for restaurant usage. Starting from the abdominal cavity, the breast of the carcass is split along the middle line until it reaches the neck section. For some processors, the breast cutting and eviscerating operations are combined.

After a thorough washing, the leg and wing joints are manually dislocated, and the carcass is pressed flat. To maintain the flat shape, the rib bones may have to be slit either manually or by using a knife. If a knife is used, care must be taken not to damage the muscle tissues.

(4) Curing: The curing process is for flavor and color development, as well as to provide a preservation effect for the product. Traditionally, pressed-cured duck is cured using a dry salt mixture technique. The salt mixture composition varies with each processor. Listed below are two examples (the amount of ingredients used depends on the weight of the prepared carcass):

- old formula:

Saltpeter	1%
Salt	12%
Pepper seed	0.3%

The pepper seeds are heated in a frying pan for flavor development and cooled before using. For curing, the prepared carcasses are rubbed evenly with salt and half of the saltpeter and pepper seed and then pressed with a heavy stone. The pressed carcasses are rotated twice daily, and the drippings are discarded. After 2 days, the rest of the saltpeter and pepper seeds are applied, and the carcasses are pressed for 3 more days before sun-drying.

- modified formula (Huang, 1994):

Salt	2.8–3.6%
Sodium nitrite	0.01%
Sugar	1.8%
MSG	0.50%
White pepper powder	0.05%
Fennel powder	0.12%
Pepper seed	0.10%

The pepper seeds are heated with a small amount of salt for flavor development. The mixed ingredients are rubbed evenly on the prepared carcasses, interior and exterior, and placed in a cooler for 4–5 days.

The cured duck is rinsed with 40°C water to remove salt particles and are further cleaned with 15–20°C water. The washed carcasses are allowed to drip dry.

(5) Drying: The old method for drying was performed by hanging the cured carcasses in the sun to dry for 4–5 days. The first day of drying should be under bright sunlight, and for the remaining days a less intense light is required. During the sun-drying, ventilation conditions are important to obtain a quality product. When the carcass color becomes reddish, the skin

turns yellow, and oozing fat is seen; the product is then ready for packaging and sale.

The modified method of drying involves the drying of cured carcasses on a screen tray with circulated 45–50°C hot air for 3 h, followed by 15°C air movement for another 5–8 h. These two stages of the drying operation are designed to prevent heat-induced surface discoloration. The cooled carcasses can be dried again with circulated 45–50°C hot air to the desired moisture content.

Regardless of the drying method, sliced bamboo sticks can be used on the cavity side of the carcass to offer a stretching effect.

(6) Smoking: Smoked pressed-cured duck provides a unique taste and also influences the color of the product. Preserving meat by smoking acts as an antioxidant, bactericidal, and bacteriostatic agents and provides a protective film.

The smoking of cured duck enhances its flavor, color, and extends the storage shelf life. In general, moderately dried cured ducks should be smoked at 35–40°C for 3–5 h.

(7) Aging: The freshly prepared pressed-cured duck is not very flavorful. For best results, the product should be aged at 10°C for 2 weeks at a relative humidity of 65%. Aging increases the free amino acid content of the product and enhances its flavor.

(8) Packaging: Traditionally, pressed-cured ducks are wrapped with a colored, plastic-type paper for marketing.

Pressed-cured duck is a delicious specialty product. Before preparation, the cured duck is soaked in water to soften the texture and reduce the salt content before cooking. After being chopped into smaller pieces, the duck can be cooked by simmering or steaming with other flavoring ingredients.

2.3 CAMPHOR AND TEA SMOKED DUCK *(CHANG CHA YA)*

The origin of this camphor and tea smoked duck was in Szechuan, China. Traditionally, ducks were raised in the rice field with insects being their main diet. Before slaughtering, they were grain fed. Therefore, the fat content of these ducks were lower and the meat texture more chewy than their farm-raised counterparts. In general, only male carcasses are used for this smoked duck processing.

The main ingredients are

Duck	About 3 kg in weight
Salt	3 Tbs
Peppercorn	2 Tbs
Saltpeter (optional)	2 tsp

Green onion, ginger, and wine	(optional)
Camphor woodchips	2 cups
Black tea leaves	1/2 cups
Fruit peels (orange or lemon)	Small amount

2.3.1 Preparation Method 1

Marinate the duck carcass in a mixture of salt, peppercorn, and wine for several hours before the addition of green onion and ginger, and then steam cook to doneness. After that, the carcass is removed and air dried. The dried cooked duck is smoked with a smoke generated from tea leaves and camphor woodchips for approximately 1 h. During smoking, the duck carcass is rotated for an even flavor penetration.

For serving, the camphor and tea smoked duck can be deep-fat fried to a golden color and cut into smaller pieces.

2.3.2 Preparation Method 2

Stir-fry peppercorns and salts in a dry pan over low heat for about 1 min. After cooling, mix with the saltpeter. Rub the carcass with the prepared curing mixture both on the surface and in the cavity and allow to cure for 6 h or overnight.

Hang the cured carcass by the neck to allow for wind-drying in a shady area before smoking. The smoke is generated from a mixture of tea leaves, camphor woodchips, and fruit peels until brown in color. Steam cook the smoked duck for 2 h before deep-fat frying. When the skin becomes crispy and dark colored, it is ready to cut into pieces and serve.

2.4 DUCK LIVER SAUSAGE

Duck liver sausage can be prepared by using the following ingredients:

Pork back fat	600 g
Lean pork	1,200 g
Duck liver	1,200 g
Wine	3 Tbs
Soy sauce	1 cup
Sugar	1.5 oz
Salt	2 Tbs
Ginger juice	1 tsp

Pork back fat, lean pork, and duck liver should be cut into slices about 1.5 cm × 1.5 cm in size. The remaining ingredients are added and mixed well until evenly distributed. The mixture is allowed to stand for 20 min before stuffing into hog casings and made into 5- to 6-in. links. Make sure there are no air pockets in the sausage.

Traditionally, the surface of the liver sausage is pasteurized by rinsing with hot water (180°F) before sun- or hot air-drying. One can also smoke this sausage after it has dried.

3. TRADITIONAL EGG PRODUCTS

3.1 SALT-CURED OR BRINE-PICKLED EGGS

Traditionally, salt-cured eggs are made from duck eggs; however, recently, chicken eggs have also been used. Salt-cured duck eggs, or *Hueidan,* are a popular processed egg product in the Oriental market. There are two methods used in the preparation of this product, namely, immersion and smearing methods.

For the immersion method, 100 eggs are immersed in a container containing a mixture of 1.2 kg salt, 1 L of wine, 2–3 kg of wood ash, and 2.5–3.0 L of red tea for 1 month.

For the smearing method, the liquid mixture as mentioned for the immersion method is mixed with red clay at a ratio of 1 : 3 (V/V) to form a paste. The egg surfaces are smeared with this paste and dusted with rice hulls before aging in a container for approximately 1 month.

Regardless of the preparation method, the eggs used should be fresh in quality and free from any cracks. Salt-cured egg contents are high in ash and free amino acid contents. The yolk particles aggregate and release yolk oil, which offers a unique texture and flavor for the product. For serving, the cured eggs are boiled in water. It is a popular side dish for the Oriental breakfast.

The yolks of salt-cured eggs are also used in manufacturing many traditional Oriental foods, such as mooncake.

There are several other modified salt-cured egg immersion formulas available. For example:

(1) For 25 eggs, a mixture of 30 ml alcohol, 300 g salt, 750 ml liquid tea, 220 ml water, and 1.3 g sodium carbonate
(2) For 300 eggs, a mixture of 4 kg wood ash, 4 kg salt, 0.05 kg starch, 0.05 kg vegetable oil, 0.05 kg wine, and 4 L of water was used to coat the eggshell and was stored for 20–30 days in a sealed jar.
(3) For 300 eggs, a mixture of 6 kg wood ash, 6 kg salt, and 5.55 L of water was used to coat the eggshell and was stored in a sealed jar for 20–30 days.

It is of interest to find the effect of ethnic background on the preferences of salt-cured eggs. Trongpanich and Dawson (1974) evaluated fresh duck eggs and brine-pickled eggs for flavor, texture, and color by a panel consisting of equal numbers of American and Thai ethnic origin. Cooked and raw duck eggs were held in saturated salt brine for periods up to 4 weeks. Their report indicated that American panel members preferred the flavor, texture, and color of the control eggs and rejected or disliked the flavor, texture, and color of the salt-pickled eggs. Thai panel members preferred the flavor, texture, and color of all salt-pickled eggs, especially those held in brine for 4 weeks at 21°C.

3.2 *PIDAN*

Pidan has been an alkaline-treated ethnic food for many generations in China. The unique color and flavor of this product has won its name of the "thousand-year-old egg" in Western society. Depending on the processing methods, several types of *pidan* are available, such as the pine-floral *pidan (Songhua dan),* soft-yolked *pidan (Tangsin dan),* and hard-yolked *pidan.*

In traditional *pidan* processing, many variations of methods can be found from different sources. Wang and Fung (1996) reviewed processing methods and chemical changes of *pidan* and indicated that the traditional processing of *pidan* is classified into three types: rolling powder method, coating method, and immersion method.

3.2.1 Rolling Powder Method

Fresh duck eggs are used. Eggs are coated with a thin layer of mud paste and rolled in the rolling powder in which all ingredients have been included before being packed and sealed in the jar. The powder-rolled eggs are allowed to ferment for 20–30 days at room temperature.

The ingredients used for the rolling powder method vary slightly according to the season. According to Liu and Zhang (1989), for 1000 duck eggs used for preparing the powder, the following ingredients are used:

Ingredient	Spring and Fall (kg)	Summer (kg)
Na_2CO_3	1.5	1.5
CaO	3.8–4.2	4.5–5.5
Salt	1.4	1.4
Tea	0.1	0.1
Water	3.0	3.0
Yellow earth	2.0	2.0

The rolling powder method produces hard-yolked *pidan.* The advantages of this method is a low cost and ease in handling.

3.2.2 Coating Method

A muddy paste containing the coating ingredients is prepared. The following ingredients as reported by Liu and Zhang (1989) will accommodate 1000 duck eggs used in preparing muddy paste:

Ingredient	Amount (kg)
Na_2CO_3	10
PbO	0.45
CaO	25
Salt	4
Tea	5
Dry mud	25
Wood ash	25
Water	50

Fresh duck eggs are completely coated with the prepared muddy paste and rolled in rice hulls. The rice hulls prevent the coated eggs from sticking together. They are then placed into jars, sealed with mud, and allowed to ferment for 40 days and 50 days in summer and winter, respectively.

The sealing step is very important for producing a high-quality product. A well-sealed jar prevents the mud coating from becoming dry. This coating method produces soft-yolked *pidan.*

Another coating method, as reported by Lin (1974), is described as follows: For 100 eggs, mix 125 g sodium carbonate, 625 g wood ash, 1000 g lime, 100 g salt, and 500 g of water to form a paste. The surface of the eggs is coated with about 1 cm thickness of paste and then rolled in rice hulls to prevent them from sticking together. The coated eggs are then packed in an earthen jar or wood barrel and sealed. After 5–6 months of pickling, the product is ready for sale.

3.2.3 Immersion Method

In this method, all of the ingredients are mixed into a pickling solution. The duck eggs are immersed in the solution for 45 days at 20–25°C. After the pickling process, the eggs are removed, washed with water, and allowed to air dry. The eggs are further coated with liquid paraffin or mud before packaging and marketing.

The following pickling formulation was reported by Zhang (1988) for 1000 duck eggs:

Ingredient	Amount (kg)
Na_2CO_3	3.6
PbO	0.37
CaO	14
Salt	2
Tea	1.5
Dry mud	0.5
Wood ash	1
Pine needle	0.25
Water	50

According to Wang and Fung (1996), the pickling ingredients are unevenly distributed in the brine, and there may be three different layers of pickling brine existing in a jar. Therefore, in an improved method, the brine is allowed to flow during the fermentation for an even development of desirable color and aroma.

Because of the toxicity of PbO, ZnO is used to replace PbO because it plays the same role without producing toxins. Liu and Zhang (1989) recommended the following ingredients for a modified immersion method for 1000 duck eggs:

Ingredient	Amount (kg)
Na_2CO_3	4
ZnO	0.1
CaO	8
Salt	2
Tea	2
Ginger	0.5
Cinnamon	0.3
Water	50

In the immersion method, ginger and cinnamon are added to the pickling solution to reduce the pungent taste of the product. For the five-spice flavored *pidan,* five spices (prickly ash, star aniseed, cinnamon, clove, and fennel) and tea are first boiled in water and then mixed with the other ingredients.

There are many other variations for the improved immersion methods. For example:

Ingredient	Amount
NaOH	40 g
Salt	100 g

Tea	5 g
Water	1 liter
Lemon essence	2 drops
$FeCl_2$ or FeO	trace

Water and tea are boiled and filtered. Other ingredients are added to the filtrate before immersing the eggs (20 pieces). The pickling time is 1 week for the summer and 2 weeks for the winter. In order to evenly distribute the pickle solution, an aeration method can be used. Air can be bubbled from the bottom of the jar through glass or plastic tubing.

Lin (1974) suggested immersing 1000 duck eggs for 2 week in a mixture of 5.63 kg sodium carbonate, 5.63 kg salt, 0.38 kg charcoal, 3.6 L tea solution, and 36 L of water at an ideal storage temperature of 30°C.

The formation of *pidan* is caused by the chemical reaction of sodium hydroxide and the egg components. In earlier formulas, sodium hydroxide is generated to form the action of sodium carbonate and calcium oxide with water in the pickling or mud coating. The alkali penetrates the eggshell and membrane and causes chemical changes of the egg components, which results in a gelation of albumen. The decomposition of proteins produce polypeptide and amino acids. Cysteine and cystine produced by protein hydrolysis can be continually decomposed into ammonia and hydrogen sulfide, which contributes to the unique flavor of *pidan.*

Milliard reactions between the glucose of the albumen and amino acids, combined with the pigment of the tea, contribute to the development of the brown-colored albumen gel. The albumen gel in the *pidan* solubilizies at 40°C, which contributes to its digestibility. Hydrogen sulfide from the decomposed protein reacts with the iron in the yolk, which gives *pidan* its typical dark-green–colored yolk.

The pine-floral–like structures between the yolk and albumen of *pidan* come from the degraded protein products such as alkalinized amino acids. Usually, high-quality *pidan* has more pine-floral crystals.

3.3 *ZAODAN*

Zaodan is one of the traditional Chinese duck egg products. It is characterized by its specific flavor. Eggs are immersed in a mixture of vinegar and salt to soften the shell before being pickled in a mixture of rice wine fermentation sediment (*zao,* in Chinese) and vinegar. After 5–6 months of pickling, a unique-flavored egg product can be obtained. For 100 eggs, the pickling solution contains a mixture of 10 kg of rice wine fermentation sediment, 2.7 kg of salt, and a small amount of vinegar.

Another method calls for the pickling of duck eggs in *lao-tsau* and 12.5% salt for 4 months at 30°C. *Lao-tsau* is a well known fermented rice product in

China with a soft-juicy texture, sour-sweet taste, and a slight alcohol flavor. A quick way to prepare *lao-tasu* is by inoculating steamed glutinous rice with a commercial starter containing organisms identified as *Rhizopus orizae, R. chinesis, Aspergillus oryzae, Mucor* sp. *Amylomyces rouxii,* and *Saccharomycopsis fibuligera* (Hesseltine, 1983) and fermented for 2–3 days at 25–30°C.

3.4 EGG SOLIDS

Dried eggs or egg solids have been produced in China for many years. They have been used in the baking industry and have the advantage of offering convenience and a long shelf life.

In the early 1900s, China was the main producer of egg solids. In 1925, they exported over 61 million pounds of egg solids, of which 36% were shipped to the United States (Termohlen et al., 1938).

Egg solids prepared from native liquid egg are brown in color and subjected to undesirable changes during storage; these include loss of solubility, decreased functionality, and the formation of brown color and an objectionable flavor. In the early 1930s, Westerners were unable to produced egg solids comparable to the Chinese product in storage stability and functionality because the Chinese were keeping their methods secret. The need for dried egg products during World War II resulted in numerous studies and elucidated the modern desugarization processes.

Traditionally, Chinese used spontaneous microbial fermentation to remove glucose prior to the drying of liquid eggs. Egg whites and sometimes yolk and the whole egg containing products are fermented before drying to prevent a Maillard reaction. The liquid eggs are fermented at 23.9–29.4°C, and the fermentation time depends on the amount of initial bacterial contamination. For liquid whites, the body is very stringy at first and becomes watery at the end of fermentation.

Stuart and Goresline (1942) tested eight samples of commercially fermenting egg white and found a predominance of *Aerobacter aerogenes* or *Escherichia freundii* with few other contaminants. Egg white fermented by either of these organisms yielded a bright, crystalline, granular product.

4. MEDICINAL EFFECTS OF POULTRY AND EGG PRODUCTS

The nutritional value of chicken meat was recognized early in history. The medicinal effects of chicken meat have been reported in many classical Chinese medicine books. Traditionally, the Chinese emphasis is more on the food remedy than in medical treatment. The unique nutritional value of chicken meat has made it one of the best foods for illnesses, as well as for postnatal nutrition.

Many medicinal effects involving the use of chicken meat have been recorded and passed down from generation to generation. These formulations have been tested through experience and common knowledge, and even though there has been no scientific testing, the formulations have been proven to work. For example:

(1) Chicken and wine cooked together is useful for treating a cough.
(2) Chicken and beans cooked together is beneficial for curing edema.
(3) The male chicken cooked with wine is beneficial for old-age weaknesses.
(4) The hen and some Chinese herbs cooked together will cure arthritis, dizziness, etc.
(5) Broth from the stewed spent hen and old ginger increases the strength of postnatal women.

Poultry eggs, such as chicken and duck, are also very popular in the Orient. Eggs are nutritious and possess many medicinal effects and have been widely accepted by the consumer since the prehistoric era. Eggs, alone or in combination with other ingredients, have been used for curing many diseases. For example:

(1) Stir-fried eggs with white wine cures flu and diarrhea.
(2) Partially cooked egg white cures a sore throat.
(3) A mixture of three eggs and white honey will cure a fever.
(4) A mixture of egg and tapioca starch cures tuberculosis.
(5) Fried eggs with pepper may cure stomach pain.
(6) The mixture of yolk, white wine, and honey has been used to cure skin burns.
(7) The application of egg whites on the hair not only cures a dendra problem, but also makes the hair more shiny.

In addition to the chicken meat, many internal organs, such as the gizzard, gizzard lining, and bile, have been widely used in traditional Chinese medicine.

5. RECENT TECHNICAL INNOVATIONS IN POULTRY AND EGG PROCESSING

Poultry and egg production is increasing in most countries of the world. The relatively low capital investment has allowed poultry production to expand rapidly in many Oriental countries. Modern poultry product technologies have been imported from Western societies. The following paragraphs briefly review the recent technical innovations in poultry and egg processing.

5.1 POULTRY PROCESSING

The technology of the poultry meat industry differs from red meat industries. Poultry processing refers to slaughter, feather removal, and evisceration. The conversion of poultry carcasses into a number of products is called further processing. As broiler meat dominates poultry meat, broiler processing and further processing steps will be considered as the normal procedure.

In recent years, automatic materials handling and automated eviscerating systems have been the worldwide technology in poultry processing. Following are the basic poultry processing procedures:

(1) Assembly of live birds
(2) Hanging of birds on the kill line
(3) Humane slaughter
(4) Scalding
(5) Defeathering
(6) Evisceration and inspection
(7) Washing and chilling
(8) Packaging

The majority of chilled poultry is packaged as a whole carcass, parts, or deboned or ground products. An individual package is weighed, priced, and printed with the store's label and bar code for automated checkout. The packaged poultry is placed in cardboard boxes and delivered to the warehouse for redistribution.

5.2 POULTRY FURTHER PROCESSING

In the 1960s, 80% of all broilers produced in the United States were sold in whole form. Today, only 20% are sold as whole birds. The poultry processing industry started to talk about "second processing" to cover all the operations between the chiller and shipping dock of a slaughter plant selling raw products. By the time the poultry markets switched away from ice-packed whole birds, the term *further processing* was introduced to plants that produce formed, formulated, breaded, and cooked products.

According to Baker and Bruce (1989), further processing of poultry can be defined as the conversion of raw carcasses into value-added, more convenient-to-use forms such as cut portions; pieces that have been battered, breaded, and precooked; cold cuts or nuggets; burger patties; hot dogs; and so forth. The concept of further processing incorporates convenience to the consumer and great utilization of the commodity, as well as economic advantages to the producer.

The term *further processing* encompasses such processes as portioning, deboning, size reduction, tumbling, massaging, reforming, emulsifying, breading, battering, deep-fat frying, marinading, cooking, smoking, canning, and dehydrating. Poultry carcasses that undergo any of these processes may be referred to as "further processed." Further processing leads to a better utilization of poultry meat; can upgrade off-cuts; offers portion control, convenience, variety, and relatively consistent product quality.

Further processing of poultry has grown for a number of reasons:

(1) The poultry producers taking advantage of overproduction through further processing for a better profit
(2) The resulting low price of poultry meat relative to other meat
(3) The development of deboning machines
(4) Improvements in processing, storing, and marketing further processed items to prevent decline in quality
(5) The demand of convenience products in retail and food service markets, probably the most important factor for the growth of further-processed poultry

Today, many further-processed poultry products are available worldwide, such as the following examples:

(1) Fried poultry products
(2) Ground poultry products
(3) Deboned poultry meats such as cutlets, fillets, etc.
(4) Poultry entrees such as kiev, cordon bleu, etc.
(5) Poultry roasts and rolls
(6) Poultry lunch meat, bologna, ham, salami, pastrami, etc.
(7) Poultry frankfurters
(8) Poultry nuggets
(9) Poultry burgers and sausages
(10) Marinated poultry products
(11) Canned poultry products
(12) Cured and smoked poultry products
(13) Barbecued poultry products
(14) Dehydrated poultry products
(15) Battered and breaded poultry products
(16) Sauce and gravy poultry products
(17) Others, such as chicken hot-wings, etc.

5.3 EGG PROCESSING

The term *egg products* refers to processed and convenience forms of eggs for commercial, foodservice, and home use. These products can be classified as refrigerated liquid, frozen, dried, and specialty products.

During the last century, several technical breakthroughs were made by the industry to improve the quality of egg products, including:

(1) Gelation control of yolk—The use of additives such as corn syrup, sugar, or salt prevents gelation of frozen yolk and whole egg.
(2) Improved equipment—High-speed machines automatically break and separate shells, yolks, and whites.
(3) Pasteurization—Many methods were proposed to pasteurize liquid eggs to assure that they are free of pathogens.
(4) Stabilization—Removal of glucose in egg whites or whole eggs prior to drying greatly improves storage stability of egg solids. There are four methods available for removing glucose from eggs: (a) spontaneous microbial fermentation, (b) controlled bacterial fermentation, (c) yeast fermentation, and (d) enzyme fermentation.
(5) Addition of ingredients—The use of various additives improves physical and functional performance of egg products. For example:
    - Carbohydrates help to preserve the whipping properties of dried whole egg and yolk products.
    - Gums and starches improve the quality of products that are precooked, frozen, thawed, and reheated.
    - Scrambled egg mixes contain nonfat dried milk solids and vegetable oil to improve texture and appearance.
    - Sodium silicoaluminate is added as a free flow agent for egg solids.

6. CONCLUSION

Poultry meat will continue to be one of the world's fastest growing meat sources; therefore, the Asian poultry industry will continue to grow. In addition, more and more poultry processors are moving into further processing. Convenience, as demanded by consumers, will continue to be the main thrust of the development of value-added poultry products.

7. REFERENCES

Anon., 1996. China's burgeoning poultry output. *Broiler Industry* (12):48.

Baker, R. C., and C. A. Bruce, 1989. Further processing of poultry. In: *Processing of Poultry,* G. C. Mead, ed., Elsevier Applied Science, New York, pp. 251–282.

Fu, P. M., 1969. *Pei Mei's Chinese CookBook.* Chinese Cooking Class Ltd., Taipei, Taiwan.

Hesseltine, C. W., 1983. Microbiology of oriental fermented foods. *Ann. Rev. Microbiol.* 37:575–601.

Huang, C. C., 1994. Manufacturing of Chinese pressed-cured duck. *Modern Meat Products* 23:43–44.

Lin, G. N., 1974. *Animal Product Technology.* Fu-Wen Press, Tainan, Taiwan.

Liu, Y. C., and S. A. Zhang, 1989. *Egg Processing and Technology.* Agriculture Press, Beijing.

Stuart, L. S., and H. E. Goresline, 1942. Bacteriological studies on the "natural" fermentation process of preparing egg white for drying. *J. Bacteriol.* 44:916–919.

Termohlen, W. D., E. L. Warren, and C. C. Warren, 1938. *The egg-drying industry in the United States.* U.S. Dept. Agr., Marketing Inform. Ser. PSM-1.

Trongpanich, K. and L. E. Dawson, 1974. Quality and acceptability of brine pickled duck eggs. *Poultry Sci.* 53:1129–1133.

Wang, J. and D. Y. C. Fung, 1996. Alkaline-fermented foods: A review with emphasis on pidan fermentation. *Crit. Reviews in Microbiol.* 22(2):101–138.

Zhang, B. F., 1988. *Chinese Traditional Egg Products.* Chinese Commercial Press, Beijing.

CHAPTER 9

Traditional Oriental Seafood Products

YAO-WEN HUANG
CHUNG-YI HUANG

1. INTRODUCTION

In Asia and the Far East, rice is the most important staple food, while fish and fishery products are used widely as ingredients for the daily diet. Fish, in one form or another, is eaten almost every day. The poor eat fish more often than the rich do (Kreuzer, 1974; Floyd, 1985). Fishery products have been an important high-quality protein food in Asian diets. Like most Western countries, more than half of the fish and shellfish are consumed fresh and chilled. However, many popular traditional fishery products in Asian countries are cured, dried, fermented, and minced (Anon., 1984). Although those products are processed in a similar manner for each method, the flavor and taste are different in different regions. This is primarily because of the raw materials and ingredients used, the method of preparation, and the culture differences from country to country. Typical Asian fishery products are discussed in detail.

2. DRIED SEAFOOD PRODUCTS

Drying is an ancient method for preservation of food, but many dried fish have been used as an intermediate product for value-added products in recent years. Asia has long been one of the world's leading production and consumption areas for dried fish products. Dried fishery products are important processed products in Asian countries. Hong Kong has been ranked as the largest Asian trading center for dried fishery products, followed by Singapore (Moen, 1983).

The dried products can be classified in the following categories according to their processing method: (1) plain dried; (2) boiled and dried; (3) salted and dried; and (4) boiled, roasted, and dried. The typical oriental dried fishery products are discussed below.

2.1 PLAIN DRIED

Fish and shellfish are dried under the sun or in a drier after being washed. These products include small fish such as sardines, large fish such as Alaska pollack and eel, shellfish such as shrimp and squid, and parts of fish such as herring roe and shark fins.

2.1.1 Sardines

Dried sardines are important fishery products in Japan, Korea, China, and other countries in Southeast Asia. It is an important animal protein source. In Japan, the raw fish used for drying include sardine *(Sardinia melanosticta),* round herring *(Etrumeus micropus),* and anchovy *(Anchovia indicus)* (Tanikawa, 1971). These fish, after being washed and exposed to the sun, are air-dried. It may take several days to dry the fish down to 20–30% of its original weight. In recent years, the mechanical drier has been gaining popularity to produce higher quality products.

2.1.2 Cod and Alaska Pollack

These products are popular in Korea and Japan. Cod *(Gadus macrocephalus)* and Alaska pollack *(Theragra chalcogramma)* are eviscerated, and the back bone is removed (Tanikawa, 1971). The fishes are then dried in the sun until the moisture is lower than 30%. The by-products such as the stomach and gills are also air-dried. The dried stomach, amber in color, and dried gill, red in color, are exported to China for the preparation of highly priced cuisine.

2.1.3 Pike Eel

The dried eel is a popular fishery product in China (Lo, 1988) and Taiwan (Wu, 1980). The Pike eel *(Muraenesox cinereus)* is abundant along the coast of China and Taiwan. The other species used for production is *M. tablabonoides.* Large eels with a length of 1 m are split from the back and then eviscerated and washed with 3% salt solution. The flattened eel is then air-dried without direct sunlight until the weight is reduced by 80%. The product is traditionally produced only during the winter. However, mechanical driers, used in recent years, have allowed eel to be dried year around. The split eel is dipped in a

0.02% BHT solution for 10 min in order to prevent rancidity of fish oil. The eel is then dried at 24°C for 2–4 days.

2.1.4 Shark Fins

Shark fins are mainly used in Chinese cuisine. The major producers for dried shark fins are China (Borgstrom and Paris, 1965; Lo, 1986), Japan (Tanikawa, 1971), Thailand (Borgstrom and Paris, 1965), and Taiwan (Wu, 1980). A set of shark fins includes one dorsal fin, two pectoral fins, and one caudal fin. The caudal fin has the highest quality of meat and is the most expensive.

There are two kinds of shark fins produced in Japan (Tanikawa, 1971): the white fin prepared from *Carcharhinus gangeticus* and *Cynias manazo,* and the black fin produced from *Isuropis glauca, Prionace glauca,* and *Heterodontus japonicus.* In Taiwan, shark fins are made from fins of two species: *Rhynchobatus djiddensis* and *Sphyrna zygaena* (Wu, 1980). The shark fin is also produced in Malaysia, India, and Pakistan (Moen, 1983). Singapore and Hong Kong play important roles in the import and export trade of shark fins. The main suppliers of shark fins to Singapore are India, Sri Lanka, Pakistan, Taiwan, and Hong Kong, while the major suppliers for the Hong Kong market are China, Singapore, the United States, Brazil, Mexico, and United Arab Emirates (Ferdouse, 1997).

The processing of shark fins is described as follows (Ka-keong, 1983; Limpus, 1991). The fins are cut off from the base of the shark body, avoiding the attachment of the body meat as much as possible. The cut fins are washed to remove blood and are sometimes bleached for 30–40 min using a 3% hydrogen peroxide solution. Processed fins are air-dried for 2–4 days until the moisture is reduced by 75%. Shark fins may be marketed in this form.

Dried shark fins are usually further processed into ready-to-cook fin-rays. The final processing is as follows: the dried fins are soaked in water for 8–12 hrs to make them soft. Fins are then boiled for 5 min until the fin needles expand and are exposed. The membrane on the surface of the fin needles is removed carefully. The fin-rays are then dried under the sun or in a mechanical drier.

2.1.5 Herring Roe

Herring roe is popular in Japan. The roe is removed from freshly caught herring and soaked in seawater for 4–5 days to remove the blood and increase firmness. After drainage, the roes are air-dried for a week. The yield of dried herring roe is 20% of the raw roe (Tanikawa, 1971). The quality of dried roe is judged by its color, shape, and size. High-quality dried roe should be yellow in color, approximately 6 cm in length, and weigh more than 5.6 g.

2.1.6 Cephalopods

Cephalopods include squid *(Loliginidae* and *Ommastrephidae),* cuttlefish *(Sepiidae* and *Sepiolidae),* and octopus *(Octopodidae).* Squid is the most important food cephalopods. The product form for international trade include fresh, frozen, canned, and dried (Anon., 1988). There are four main squid-consuming countries in the world, namely Japan, Korea, Spain, and Italy (Kreuzer, 1986). Among them, Japan is the world's largest squid-producing nation, followed by Korea. Other high squid-consuming countries in Asia include Taiwan, Thailand, India, Malaysia, and the Philippines (Anon., 1991).

Major cephalopods species caught by Japanese fleets are squid species, *Todarodes pacificus, Ommastrephes bartrami,* and *Illex argentinus;* cuttlefish species, *Sepia esculenta;* and octopus species, *Octopus dofleini* and *O. ocellatus.* The important species caught and used in Korea for dried squid include *Ommastrephes bartrami, Illex argentinus,* and *Todarodes pacificus* (Roh, 1992). In China, the most common species of squid is *Loligo chinensis* spp. (Lo, 1986), while that of the cuttlefish includes *Sepiella maindroni. Sepia esculenta, Sepiella maindroni,* and *Loligo* spp. are abundant along the China Sea, while *L. formosana* is found around the coast of Taiwan (Wu, 1990). The most abundant squid species in Japan is *Ommastrephes sloani pacificus* (Tanikawa, 1971). However, the harvesting ground has been expanded to the southern Pacific (especially near New Zealand), and most of them are harvested from the northern Pacific, the southern Pacific (especially near New Zealand), and the southwestern Atlantic (especially near Argentine). Dried products are mainly produced in Japan, Korea, Taiwan, and China, but some dried squid and cuttlefish are also produced in Thailand (Moen, 1983).

The processes of the dried cephalopod product consist of thawing, cutting, washing, drying, and shaping. Frozen squid and cuttlefish are thawed in a walk-in cooler. The belly side of the mantle is split, and viscera are removed. The head is also split at the central line to remove the jaw of the mouth. At this time, the transparent tendinous matter of squid or the hard shell of the cuttlefish is removed with a knife. The cut and split squids are then washed with water containing 2–3% salt to remove mucous substances. They are finally washed with fresh water. Stainless steel sticks are used to stretch the mantle section of the squid prior to drying. The drying process is carried out in the sun during the day and in a mechanical drier at 24°C during the night for several days. During the sun-drying, the squid or cuttlefish should be uniformly exposed to the sun. Uneven drying will cause a product red in color and may cause spoilage. After the drying process, the squid or cuttlefish should have a moisture level of 20%. The dried squid or cuttlefish is then flattened using a roller machine (Figure 9.1).

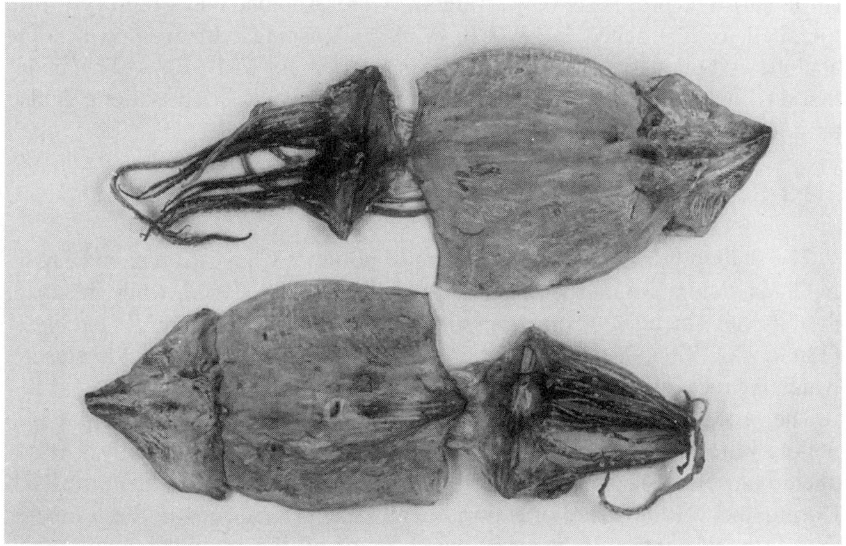

Figure 9.1 Dried squid is a popular seafood item in Asian markets.

2.2 BOILED AND DRIED

2.2.1 Sardines

Sardines, a group of more than 200 species of small pelagic fish, including *Clupeidae, Engraulidae* and *Chirocentridae,* are used for boiled, dried products (Tanikawa, 1971). The important species in Japan, Korea, and Taiwan is *Sardinops melanosticta,* while *Sardinella aurita* is the important species along the coast of southeastern China (Lo, 1988).

The fish is washed and placed in boiling water (0.5% salt is added) for 15 min until the fish floats. Cooked sardines are dried in the sun until the moisture is reduced to a level of 30–40%. Since herring is larger than the sardine, the boiling method is replaced by a steam cooker with a pressure of 2 psi for 15 min.

2.2.2 Abalone

The abalone is a delicate, high-priced seafood product. In China and Japan, abalones (*Haliotis* spp.) are also prepared in the boiled, dried form, as well as being a canned product (Tanikawa, 1971). The major species are *H. gigantea* (in China and Japan), *H. discus hannai* (in China and Japan), and *H. sieboldi* (in Japan). The process is very straightforward. Abalone meat, removed from the shell, is first salted for 3–7 days. After washing with fresh water, the abalone is boiled for 5 min and then sun-dried for 7–10 days. The major markets include Japan, China, Singapore, and Taiwan. Dried abalone is also produced in Malaysia for export (Moen, 1983).

2.2.3 Scallops

The scallop grows in the cold waters of northern China, Korea, and Japan. In China, the important species is *Chlamys farreri* (Lo, 1988), while in Japan, the species include *Pecten vessoensis, P. laqueatus,* and *P. pectinata* (Tanikawa, 1971). The edible portion of the scallop is the adductor muscle, which is processed into dried products.

The processing method is relatively simple. The fresh scallop is put into boiling water for 3 min to open the shell for easy removal of the body. The adductor muscle is separated from the mantle and viscera and then washed. During the washing step, the thin membrane that surrounds the adductor muscle must be intact. If the membrane is removed, the muscle fiber will break up. The washed muscle is boiled in a 7% salt solution for 20 min and then dried in the sun for 1–2 days.

2.2.4 Oyster, Clam, and Mussel

Oysters (*Ostrea gigas* in Japan; *O. rivularis* and *Crassostrea rivularis* in China; *C. gigas* in Korea), clams (*Meretrix meretrix* in China; *M. meretrix lusoria, Venerupis philippinarum,* and *Mactra sachalinensis* in Japan), and mussels (*Volsella* spp. in Japan; *Mytilus* spp. in China) are commonly used to produce boiled and dried products (Tanikawa, 1971; Lo, 1988). The major producing countries include China, Japan, Korea, and Malaysia (Moen, 1983).

In general, the shellfish is first boiled in a 3.5% salt solution so that the firmness of the muscle is increased. After the shell is opened, the meat is removed and washed prior to being dried in the sun for 2 days.

2.2.5 Sea Cucumber *(Beche-de-Mer)*

Sea cucumbers, also known as *beche-de-mer* or sea slugs, are a highly appreciated delicacy in most Asian countries and in countries where ethnic Chinese communities exist. The Asian market for the sea cucumber is estimated to be worth US$60 million (around 13,000 metric tons) annually (Ferdouse, 1997). China, Singapore, Taiwan, Hong Kong, Malaysia, Korea, and Japan account for almost 90% of the total trade of sea cucumbers, in which 95% of the sea cucumber product is in dried form. The international trade in dried sea cucumber is concentrated in three major markets, namely Hong

Kong, Singapore, and China (Krishnasamy, 1991). Hong Kong and Singapore are the main re-exporters in the trade.

Sea cucumbers belong to the group of sea animals called *Holothurians,* consisting of more than 650 species. Only about ten species have commercial value, including white teatfish *(Microthele fuscogilva),* black teatfish *(M. nobilis),* sandfish *(Metriatyla scabra),* blackfish (*Actinopyga* spp.), deep water redfish *(Actinopyga echinites),* prickly redfish *(Thelenota ananas),* and surf redfish *(A. mauritiana)* (van Eys, 1986). Major producing countries of sea cucumbers include Japan, Korea, Sri Lanka, Indonesia, India, Malaysia, and the Philippines (Meon, 1983; van Eys, 1986).

In Japan, the dried sea cucumbers are produced from *Stichopus japonicus* (a true sea cucumber having many large papillaes), *Holothuria argus, Pstsdyivhopud nigripunctatus,* and *Cucumaria japonica* (Tanikawa, 1971). There are many processing methods but all are quite straightforward (Bruce, 1983; Sachithananthan, 1986; Robertson et al., 1987). In general, the sea cucumber is first cut open to remove the intestines. The abdominal cavity is then cleaned with a thin brush and drained. The prepared sea cucumbers are mounted on a stick to keep their shape. They are put into a 3% salt solution to boil for 1–1.5 h and then dried in the shade for 5 days. The dried product is then kept in a room for 2 days to equalize the moisture content and is sun-dried again for 2–3 days (Figure 9.2). The yield is 5–10% of the weight of the raw sea cucumber.

Figure 9.2 Dried sea cucumber is a delicacy in addition to shark fin and abalone for Chinese consumers.

The best quality of dried products is light blackish brown in color, while the product having white powder coated on the surface of the body is of inferior quality.

In the Asian markets, the white teatfish appears to be the preferred species and one of the most expensive fishery products. It is followed by the sandfish (van Eys, 1986); however, the various markets have their own preferences. For example, people in both Japan and Taiwan consume mainly *S. japonicus,* while people in Taiwan also like two other species, *Actinopyga* spp. and *M. scabra.*

2.2.6 Shrimp

Shrimp caught by Asian countries contribute more than 60% of the world's supply (Infofish, 1991). China and India account for more than 50% of the Asian production, followed by Indonesia, Thailand, Bangladesh, Malaysia, Taiwan, and Vietnam. The major species include Indian white shrimp *(Penaeus indicus),* banana prawn *(P. merguiensis),* giant tiger prawn *(P. monodon),* kuruma prawn *(P. japonicus),* red-tail prawn *(P. penicillatus),* and small shrimp (*Metapenseus* spp. and *Acetes* spp.). In addition to the shrimp caught by fishing vessels, cultured shrimp are also very important in Asia. The major species cultured in Asia is the black tiger shrimp *(Penaeus monodon),* except in China where the white shrimp *(P. orientalis/chinensis)* predominates. Important producers of cultured shrimp are China, Indonesia, Thailand, Taiwan, and the Philippines. The traditional markets for all shrimp produced in Asia are Japan and Singapore. The product types are mainly in individual quick-frozen (IQF) and frozen block of shell-on or peeled shrimp. However, in recent years the United States and European markets have become consumers of Asian shrimp products.

Only small shrimps are used to produce dried products in either peeled or shell-on forms. White shrimp *(Exopalaeman carinicauda),* green shrimp *(Macrobrachium nipponensis),* and small shrimp (*Acetes* spp.) are used in China (Lo, 1986), while *Metapenaeus joyneri, Penaeuls japonicus, Palaemon nipponensis,* and *Sergestels luceus* are commercially important species in Japan (Tanikawa, 1971). The process involves washing, boiling, and drying. However, some products are directly dried without boiling (Tanikawa, 1973; Wu, 1980). In general, either peeled or whole shrimp are heated in water containing 2–5% salt and boiled for 5–10 min for inactivation of enzymatic activity. Boiled shrimp are then dried in the sun or in a mechanical drier until the moisture is reduced to levels of 10–40%, depending on the market need. The dried whole shrimp may be further peeled using a peel separator. The dried shell-on shrimp is commonly called "shrimp skin" in the marketplace. The quality of dried shrimp is judged by its light red color, good curved shape, and uniform size.

2.3 BOILED, SMOKED, AND DRIED *(KATSUOBUSHI)*

2.3.1 Skipjack

Katsuobushi is a general term that describes the process of boiling, smoking, drying, and molding of skipjack *(Katsuwonus pelamis)* flesh. Dried skipjack sticks are a special Japanese-style cured product. It is prepared by a combination of boiling, smoking, drying, and mold fermentation (Tanikawa, 1971; Wu, 1980; van Eys, 1983). Because of the appearance and texture of the product resembling a stick of hardwood, it is called dried skipjack stick. Skipjack with a low fat content is the most desirable fish for producing this product. Albacore, big-eye tuna, and mackerel are also used.

The basic process is as follows: (a) filleting the fish into two (2 kg or less) or four pieces, depending on the size of fish; (b) boiling the fillets at 80–90°C for 60 min; (c) cooling and removing small bones and debris; (d) the first smoking for 30 min at 85°C; (e) shaping and pasting cracks using the same flesh in the fish head; (f) continuing smoking up to ten times for 2 weeks until the moisture content is reduced to 20%; (g) the first drying for 1–4 days; (h) shaping with shavings; (i) boxing for 14 days to allow mold growth on the surface; and then (j) drying. The molding and sun-drying are repeatedly applied four times (Figure 9.3).

Figure 9.3 Traditional Japanese fish product, *katsuobushi,* is a high-value seafood item.

The entire process takes about 3 months. The combination of sun-drying and molding reduces the moisture of the finished product to 18% or less. The molds, *Aspergillus glaucus, A. ruber, A. repens,* and *Penicillum glaucum,* grow on the fillet to reduce moisture and fat content and increase inosinic acid level. The yield of dried sticks is 20% of the original weight. The product is shaved with a blade, and the thin flakes are added to the hot water to make a soup base.

Because of the intensive labor and time-consuming process, the intermediate product has been imported from the Philippines, Malaysia, Thailand, and Taiwan. The intermediate product is called *ara-busi,* which has passed through all of the processing stages except molding (van Eys, 1983).

2.4 SEASONED, DRIED PRODUCTS

2.4.1 Fish Flake

The fish flake, also called fish floss, is a traditional and popular fish product in China, Japan, Indo-China, southeastern Asia, and other areas with populations of Chinese origin. Fish flakes are a dried, seasoned fish muscle fiber product. Normally, the fish flake is eaten with rice, rice porridge, or by itself as a snack.

The best raw material is low in fat and low in connective tissue. The most commonly used species are amber fish *(Decapterus maruadsi),* big-eye snapper *(Priachanthus tayenus),* gold thread *(Nemipterus vigatus),* lizard fish *(Saurida tumbil),* mackerel scad *(Metalaspis cordyla),* and skate (*Raja* spp.). Usually fish flake production involves the following steps: brining the fish with 25% w/w salt for 2–5 days, boiling/steam cooking, collecting the flesh, stir-frying until the moisture is 25–40%, seasoning, and stir-frying again (Wu, 1980). The formula for the seasoning is 10–25% sugar, 3–6% salt, 10–25% vegetable oil, and a touch of soy sauce, sesame, seaweed, and the five spice. Bulk materials such as wheat flour, soybean flour, and bread crumbs are also added. The fish flake is a ready-to-eat product high in protein and rich in lysine and methionine.

2.4.2 Tuna

Seasoned tuna cubics are a unique product used to replace starchy snack foods (Wu, 1980). Tuna such as albacore, yellowfin tuna, big-eye tuna, bluefin tuna, and skipjack are commonly used as raw materials. The processes used to make the seasoned, diced tuna is described below. Fish is steam-cooked, cooled, and trimmed. The flesh is then cut into cube-size pieces (1–1.2 cm). The cubes are added to a kettle with sugar, salt, soy sauce, and five spice and steam-cooked at a pressure of 1 kg/sq cm for 1 h. The final step is drying for 1 h using a mechanical dryer.

2.4.3 Squid

The seasoned squid is a ready-to-eat, value-added product. The species used are the same as with dried squid. The process of preparing the squid is as follows (Wu, 1980). The mantle of the squid is cut open and skinned in warm water (50–60°C). The squid is then cooked at 65–80°C for 3–5 min. After cooling, the squid is cleaned and seasoned using a seasoning solution for a minimum of 4 h at 20°C. The formula for seasoning includes 6% sugar, 2% sorbitol, 2.5% salt, and 0.2% phosphate. It is then dried at 40–45°C for 12–20 h. The squid is finally roasted for 3–5 min at 110–120°C and then pressed prior to cutting into strips. A second seasoning may be applied as necessary. In the final step, the squid strips are dried to a moisture level of 25%.

2.5 SALTED AND DRIED PRODUCTS

Salted fish is an ancient preserved product. In Japan, the principal fish used for salted products include sardine, mackerel, saury, herring and its roe, cod and Alaska pollack and their roe, salmon roe, and squid. In China, however, the most important species is *Ilisha elongata.* In most cases, dried salt up to 25% is used to cure the fish for 2–20 days, depending on the size of the fish. Consumption of salted fishery products has been decreasing in recent years. However, the unique flavor of salted fish makes it a specialty product.

2.5.1 Sardines

The salted and dried sardines are a popular preserved fish product in tropical areas. In India, sardines are used to produce a slated and dried product (Hiremath et al., 1986). The oil sardine *(Sardinella logiceps)* is a substantial marine fishery resource on the west coast of India. The process method involves dressing the fish, brining with saturated salt solution for 6 days, and applying a pressure of 0.3 kg per square inch over a period of 15 h. The finished product has a 4-week shelf life at an ambient temperature of 28°C.

2.5.2 Jellyfish

Jellyfish is an important item in traditional Chinese cuisine. Some people believe that the Chinese pioneered the art of processing jellyfish for human consumption (Morikawa, 1984; Subasinghe, 1992). Jellyfish is not a fish. It is a close relative of sea anemones and corals. The jellyfish body consists of a hemispherical, saucer-shaped transparent umbrella. Sometimes, it has numerous fine marginal tentacles. The mouth is on the undersurface of the umbrella.

The fresh jellyfish has 96% moisture and 1% collagenous protein (Huang, 1988). After processing, the salted jellyfish has an average of 67% moisture and 5–6% protein. Chinese value the product for its unique texture, which is a combination of tenderness, crispiness, and elasticity.

There are at least five commercially important species of jellyfish in the eastern and southeastern Asian regions. The most popular edible jellyfish species in China, Japan, and Korea are *Rhopilema esculenta* and *R. hispidium.* Other species include *Aurelia aurita, Lobonema smithi, Lobonemoides gracilis, Hispidium* spp., *Dactylometra pacifica,* and *Stomolophus nomurai* (Morikawa, 1984).

Traditional processing methods require a long period of time, ranging from 19 days to 37 days (Wootton et al., 1982; Krishnan, 1984; Subasinghe, 1992). Huang (1988) developed an innovative processing method to produce salted American cannonball jellyfish *(Stomolophus meleagris)* in 1 week. The value of cannonball jellyfish as a food resource and its potential to support a U.S. fishery was first explored by Huang (Rudloe, 1992).

This time-saving method is as follows. Jellyfish is first washed to remove the mucous substances. After draining, the jellyfish is treated in three salting stages with different mixtures of salt and alum over a week's time. Seven and one-half percent salt and 2.55% alum are used to make a brine solution for the first salting, which lasts for 2 days. In the second salting, 15% slat and 1% alum are used for 2 days. For the last salting, the dried salt is applied (Figure 9.4).

The primary jellyfish-producing countries include China, Thailand, Malaysia, Indonesia, Korea, Burma, and the Philippines. Commercially, jelly-

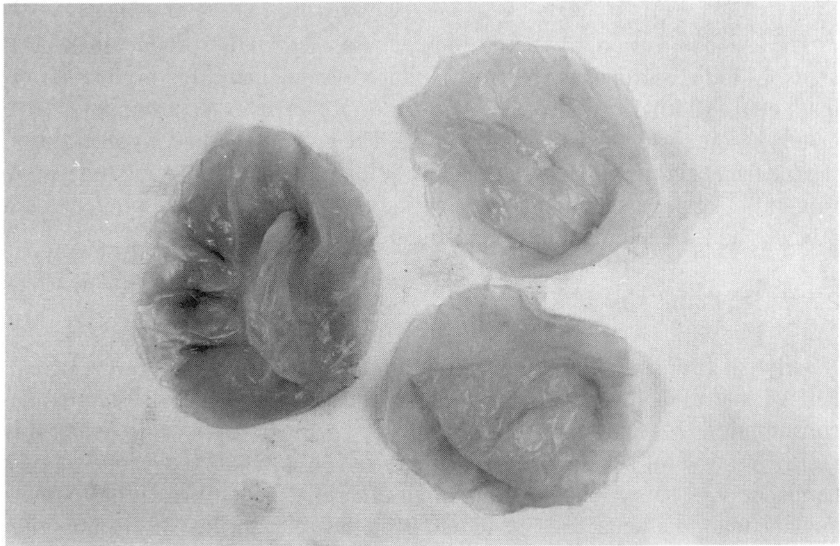

Figure 9.4 Salted jellyfish is a high value seafood product prized by its unique texture.

fish can be classified into different types, according to the species and processing method (Morikawa, 1984). The "white-type" jellyfish is produced in China at Gaundong, and the "red type" is produced in China at Dalian. The "Javanese-type" jellyfish is produced in Indonesia and the Philippines. The "Thai type" and "Malaysian type" are also popular products that are produced in Thailand and Malaysia, respectively. In recent years, the United States has been shipping salted American jellyfish to markets in Japan and Hong Kong.

2.5.3 Mullet Roe

The mullet *(Mugil cephalus)* is indigenous to tropical and subtropical waters. The roe of the mullet is matured during the winter season. Salted and dried mullet roe is a special and typical value-added seafood product typically consumed in Japan and Taiwan. Japanese call the product *karasumi.* Although there are salted fish roes of salmon, herring, and cod available in the market, the salted and dried mullet roe has a unique flavor. The processed roe contains 30.5% water, 35.5% protein, and 25.7% lipid. A high level of wax esters in the lipid also provides a unique chewy texture (Wu, 1980).

The processing method for mullet roe is as follows. A roe is carefully removed from the fish and washed thoroughly by using a 3% salt solution to clean the blood vessels on the surface. The roe is then covered with salt at a level of 10–15% of the total roe weight. After salting for 4–5 h, the roe is immersed in fresh water for desalting and then pressed with weight overnight. The roe is then covered with cloths and air-dried for 3–5 days until it is reduced to 70% of its original weight. The salted and dried mullet roe needs to be refrigerated in order to prevent oxidation of lipids.

In recent years, because of the high demand and good price in the Japanese market, the frozen raw mullet roe has been shipped from Brazil and the United States to Taiwan for further processing. However, the product made from the imported roe has a lower quality than those made from fish roe harvested from Asian waters.

3. FISH SAUCE AND FERMENTED PRODUCTS

3.1 FISH SAUCE PRODUCT

3.1.1 Fish Sauce

Fish sauce is a popular product in southeast Asia, especially in Thailand, Cambodia, and Vietnam. Fermented fish sauces and fish pastes are of greater importance than fish preserved by salting and drying in these regions. Fermentation results from the action of enzymes and microorganisms in the fish. Hydrolysis of the fish proteins occurs, resulting in free amino acids, peptides, and ammonia. In the presence of high concentrations of salt, pathogenic microorganism growth is controlled, and the salt results in a desirable taste and aroma. The liquid portion of fermented fish is called fish sauce, and the solid portion is called fish paste. Both products are used as condiments in the preparation of many seafood dishes and in dips and sauces.

Fish sauces and pastes are prepared from different kinds of small fish and shellfish. In Vietnam, species mainly include clupeids and carangids such as *Decapterus, Engraulis, Dorosoma, Clupeodes,* and *Stolephorus.* In the Philippines, sardines, anchovies, ambassids, and shrimp are commonly used, while in Japan, oysters, baby clams, and visceral masses of salmon are used, in addition to sand fish *(Arctoscopus japonicus)* and sand lance *(Ammodytes personatus).*

The preparation of fish sauce is simple and straightforward. Small ungutted marine fish are washed, drained, and kneaded. Salt ranging from 20–40% is added and mixed. The mixture is normally kept in earthenware pots, which are tightly sealed and sometimes buried in the ground for several months (van Veen, 1965; FAO, 1990). The fermented fish is filtered through a cloth bag using pressure. The filtered liquid is boiled and sold in glass containers as fish sauce.

In Japan, the fish sauce is sometimes further processed by adding a small amount of soy sauce or wheat *koji* to provide a soy sauce aroma (Tanikawa, 1971). Fish sauce is used as a condiment for cooking. The names of fish sauce are varied in different countries. Fish sauce is known as *nuoc-mam* in Vietnam and Cambodia, as *nam-pla* in Thailand, as *petis* in Indonesia, as *budu* in Malaysia, as *patis* in the Phillippines, as *yu-lu* in China, and as *shotturn* in Japan (Figure 9.5).

3.1.2 Fish Paste

Fish pastes are consumed in greater quantities than fish sauces and play a larger role in nutrition because of the increased consumption. Pastes may be prepared from fresh fish or the residue from fish sauce production. Like fish sauce, shrimp and fish pastes have different names in different countries. Shrimp paste is called *belachan* in Malaysia, *trassi* in Indonesia, and *ngapi* in Burma; however, fish paste is called *badging* and *belibel* in the Philippines, *kapi* in Thailand, *pra-hoc* in Cambodia, *mam-cho* in Laos and Vietnam, *yu-jiang* in China, and *gyomiso* in Japan. Some fish paste may be dehydrated into a form of powder (Figure 9.6).

The preparation of fish or shrimp pastes varies among countries. The basic process involves the addition of 15–25% salt to fish or shrimp and then mixing in other materials such as wheat flour, bran, rice, or soybean (FAO, 1990). In Japan, fish are deheaded, crushed, and then mixed with 10% wheat bran and 15% salt and inoculated with *Aspergillus oryzae.* The mixture is kept for 2 weeks at 25–30°C for the fermentation process to occur (Tanikawa, 1971).

Figure 9.5 Fermented fish sauce products, popular items in southeastern Asia, made from shrimp, fish, squid, and oyster.

Figure 9.6 Fish paste is also in a dried powder form in the market.

In the Philippines, the residue of fish sauce is used for processing *bagoong.* A red rice product called *angkak,* fermented by *Monascus ouroureus,* is added to the fish sauce residue, which makes the finished product pink in color. In Vietnam, Cambodia, and Lao, glutinous or roasted rice and molasses are added in the preparation of fish paste, *mam-cho.*

3.1.3 Oyster Sauce

In China, oyster sauce is very popular and commonly used in Chinese cuisines. The oyster sauce is not a fermented product. For the preparation of the sauce, ground oyster or concentrated juice left from the production of cooked dried oysters is mixed with various ingredients (Chen, 1992). Sugar is commonly added to enhance the taste and color. Starch is also added to increase the viscosity. Today, the commercial product contains 2–8% oyster juice only and no oyster meat. Oyster sauce is primarily produced by China, Hong Kong, Malaysia, and Thailand. Korea only produces concentrated oyster juice for further oyster sauce production.

3.2 FERMENTED PRODUCTS

3.2.1 Sea Urchin

The fermented sea urchin gonad is a unique Japanese delicacy. Japan annually harvests one-fifth of the world's harvest. However, in order to meet the demand for the sea urchin gonad in Japan, tons of fresh, chilled, frozen, and salted gonads are imported from the United States, Canada, Mexico, Chile, Korea, China, the Philippines, and Taiwan (Ramachandran and Terushige, 1991). Male and female gonads are difficult to differentiate, and both are consumed. The roe or milt of the sea urchin, fermented with salt, is consumed mainly in Japan. Twelve species of sea urchin are commercially harvested worldwide, but green sea urchins, *Strongylocentrotus pulcheriimus,* are the most expensive, followed by red sea urchins, *S. franciscanus* (Ramachandran and Terushige, 1991). Other species, *S. intermedius* (purple sea urchin), *Pseudocentratus depressus,* and *Heliocidaris crassispina,* are also commonly used in the preparation of fermented products (Tanikawa, 1971).

The processing method for the fermented sea urchin gonad is straightforward. The reproductive organs, removed from the shell, are washed. The unwanted material sticking to the gonads are carefully removed with saltwater rinsing. The gonads are then mixed with salt. The salting process varies with the type of products desired and the processors (Ramachandran and Terushige, 1991). For preparation of the watery, fermented sea urchin roe, *mizu uni,* 30–40% salt by weight of wet roe is added to each layer for fermentation. For processing of the fermented sea urchin, *doro uni,* the clean roe is first washed with diluted alcohol (6–9%) and drained and then mixed with 25–40% by weight of salt. For the most expensive product, pasty fermented sea urchin, *neri uni,* the roe is spread over a dressing table on which salt has been sprinkled. The roe is then covered with 20–30% of salt and are shelf stable. The yields of these products are 8–12% of the raw material. In recent years, the processors have been using lower levels of salt, between 5–10%, followed by alcohol treatment (5–7% w/v). The low salt product needs to be distributed and stored at a refrigerated temperature and has a 3- to 5-month shelf life.

3.2.2 Fermented Squid

Fermented squid is a typical Japanese product. Common squid, *Ommastrephes sloani pacificus,* is the primary source of raw material. Another squid, *Watasemia scintillans,* is also used (Tanikawa, 1971). The belly side of the mantle of the squid is split using a knife. The ink sac, liver, and cephalopodium are removed without damage to the mantle. The washed mantle is then cut into rectangular strips (3 × 0.5 cm). The head is split at the central line to remove the jaws and eyes. Salt, ranging from 15–30%, is added to the squid and mixed in a barrel. During the first week, the mixture is stirred two or three times a day. The barrel is then sealed and kept for 3 months. At the end of storage, the amino acid level of the fermented squid has increased.

4. MINCED FISH PRODUCTS

Minced fish, called *surimi* in Japan, is the deboned fish meat that is washed two to three times using fresh water. The washing cycle removes water-soluble proteins. The washed fish meat is primarily the salt-soluble protein that can form different shapes after ground with salt and heated. When salt is added to washed minced fish meat during grinding, myosin is dissolved out from the fish meat to form *sol,* which is very adhesive. The *sol* converts into gel, forming a network structure that provides elasticity. The quality of the finished product depends on the elasticity. The washed minced fish *(surimi)* is an intermediate material for further products. After mixing with cryoprotectant (4% sucrose, 4% sorbitol, and 0.5% sodium tripolyphosphate), the washed minced fish meat can be stored frozen up to 1 year for future use. In China, a popular minced fish product is the fish ball. It is also very popular in southeastern Asian countries, including Singapore, Malaysia, and Thailand. In Japan and Korea, however, fish cake *(kamaboko), chikuwa,* and fish sausage are commonly available in fish markets (Figure 9.7).

4.1 FISH BALL

The fish ball is a very important processed fishery product in China, Taiwan, Thailand, Indonesia, Malaysia, and Singapore. There are many kinds of products such as plain fish balls and fish balls stuffed with seasoned ground pork.

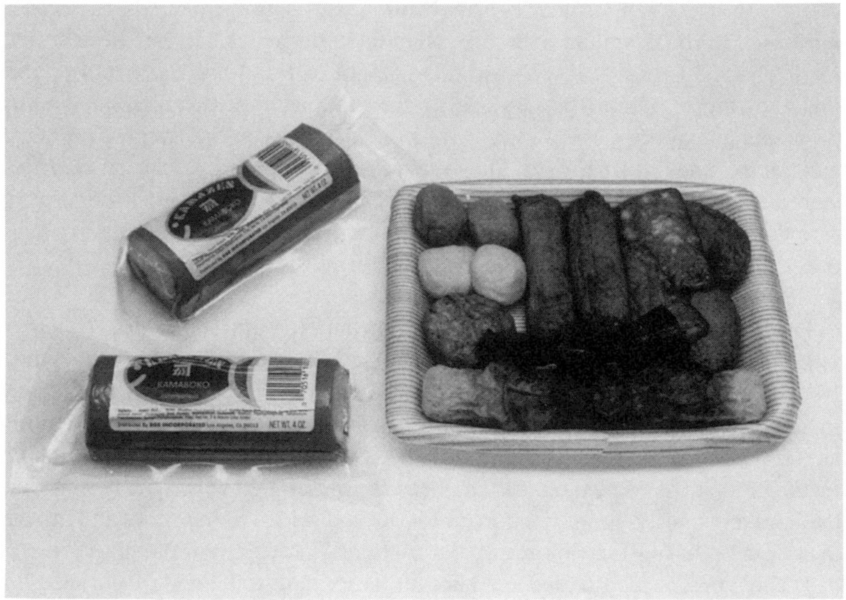

Figure 9.7 Typical *surimi*-based products include *kamamoko,* fish ball, and *chikuwa.*

Fish balls can be made from finfish, shrimp, or cuttlefish. One of the important quality attributes of the product is its elasticity. While this attribute is prized by Oriental people, it is not favored by Western consumers.

In Taiwan, the principle sources of fish used for raw material in manufacturing fish balls are sharks, lizard fish, pike eel, and marlin (Jeng, 1977). The processing method includes deboning, washing, dewatering, mixing, forming, and boiling. The starch and ingredients such as sugar, salt, and polyphosphate are used in the mixing steps. The spherical shape is formed by machine or by hand. The shaped minced fish meat, either plain or stuffed with ground pork, is then boiled. After the fish ball is cooled, it is sold as is in the market, or it may be packaged and sold as a frozen product.

The fish ball is a popular and important animal protein source for the daily diet for people. In addition to finfish, shellfish such as shrimp and cuttlefish are also commonly used for producing similar minced products.

4.2 *KAMABOKO*

Kamaboko is a Japanese term for the most popular minced fish product in Japan. It is also consumed in Taiwan and Korea. *Kamaboko* is a Japanese-style product made from washed minced fish meat followed by steaming. The premium product is made using the following fish: sand borer *(Pterothrissu gissu),* big-eye *(Scombrops boops),* sea bream (*Pagrosomus* spp.), eel *(Muracenesox cinereus),* lizard fish *(Saurida argyophanes),* and cuttlefish *(Sepia esculenta).* Other fish, including croaker *(Nibea mitsukuri)* and blue shark, are also used.

In preparation of *kamaboko,* seasonings and other materials are added to the washed minced fish meat. An example of ingredients used is as follows: 3.75% salt, 0.7–1.5% sugar, 1.7–5.5% starch, 1.1–5.5% rice wine, 1.0% sodium glutamate, and 0.2–0.3% sodium tripolyphosphate. The seasoned fish meat is formed into different shapes on a thin wooden plate and then steamed. Sometimes broiling is employed instead of steaming and is then called *yaki-kamaboko.*

4.3 *CHIKUWA*

Inexpensive fish such as shark, atka mackerel, Alaska pollack, and flat fish are commonly used as raw materials to prepare a minced fish meat. The minced fish meat is seasoned with the same ingredient and formula used for *kamaboko* and then molded around a brass pipe. The fish meat is then broiled for 4–5 min by passing it over a long furnace. When the broiling is completed, the brass sticks are removed, and the product is wrapped with parchment or paraffin paper.

5. SEAFOOD SNACKS

5.1 FISH CRACKERS

Fish crackers called *keropok* are a popular snack food in Southeast Asian. The crackers are the expanded snack product after deep-frying. The main ingredients are starch, water, and fish or shrimp. Although people still follow the traditional processing method to produce fish crackers, an improved method was developed by Yu (1986).

The new method of producing fish crackers is as follows: fish is first deboned and mixed with tapioca flour *(Manihot utilissima)* or sago flour *(Metroxylon sagu).* Salt (2%), water (25–35%), and sometimes sugar (1%) are added, mixed, and kneaded. The dough is stuffed into cylindrical casings (25 cm long and 4–6 cm in diameter). The rolls are then steamed or boiled for 90 min. After being cooled, the rolls are sliced (3–5 mm thick) and dried. For initial drying, a temperature range of 40–50°C is ideal. A final temperature range of 65–70°C should be used in order to maintain a moisture content of 10% (Figure 9.8).

An extrusion method for processing fish crackers has been reported by Suknark et al. (1997). The basic materials remain the same, while extrusion is

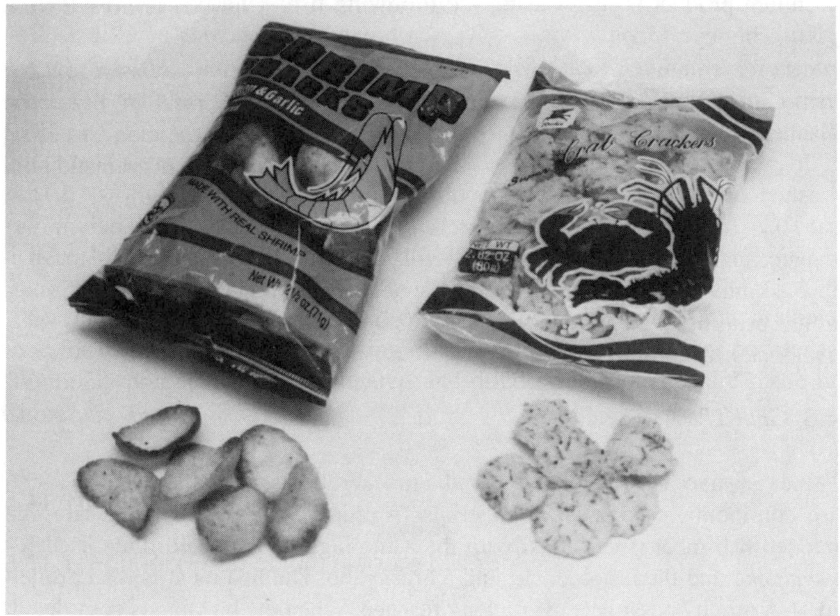

Figure 9.8 Fish snack made from shrimp and crab in addition to fish.

used instead of slicing and drying. In markets, the fish cracker is an intermediate product for ready-to-cook products. In addition, a packaged, deep-fried, ready-to-eat product is also available in the market.

6. MARINE PLANTS

The use of marine plants, both for food and medicinal purposes, was recorded in China thousands of years ago. Presently, seaweed remains widely used as food items in certain Asian countries. The world production of seaweed approximates 4 million metric tons annually. More than 80% of this production is from the Asian Pacific region. A large part of this is produced and consumed in China, Japan, and Korea (McHugh and Lanier, 1984; Anon., 1992). The seaweed industry is also established in other Asian countries, including the Philippines, Taiwan, Indonesia, Malaysia, Thailand, and India.

The important species for direct human consumption in Asia include (1) brown seaweeds: kelp *konbu (Laminaria japonica), wakame (Undaria pinnatifida* and *U. peterseniana),* and *Hizikia fusiforme;* (2) red seaweeds: purple laver, *nori (Porphyra tenera, P. yezoensis,* and *P. haitanensis);* and (3) green seaweeds: green lavers *ao-nori (Enteromorpha* spp., *Ulva,* and *Monostroma nitidum).* The largest brown seaweed-producing country is China, followed by Japan and Korea (McHugh and Lanier, 1984), while the largest green seaweed-producing country is Korea, followed by Japan and Fiji. Other Asian countries such as the Philippines, Malaysia, and Indonesia also consume red or green types of seaweed.

Although directly used in a regular Asian diet, the seaweeds are mostly used for extracting colloid substances in the developed Western countries. The red seaweeds, including *Chondrus ocellatus* and *Gigartina tenella* (Irish moss), are used for carrageenan extraction, and *Gelidium amansii* is used for agar extraction. Principal carrageenan producers are the Philippines, Indonesia, and China (Richards-Rajadurai, 1990). The seaweed colloids, including agar, carrageenan, and alginate, are used as functional ingredients in many food systems.

6.1 KELP

Kelp, a brown seaweed, is one of the most important food products consumed in China, Japan, and Korea. The primary processed products are dried kelp, salted kelp, and seasoned kelp. Once taken from the sea, kelp should be dried immediately. It takes 1–2 days to dry under the sun. Dried kelp is consumed directly or used for extracting iodine, algin, and manitol. Processing of salted kelp involves washing, boiling, salting, dehydrating, resalting, and packing (Suo and Wang, 1992).

For the preparation of seasoned kelp products, ingredients, including soy sauce, sugar, salt, and spice, are heated and dissolved. The kelp is then added and cooked. After cooking, the seasoned kelp is cut into strips and packaged. This product is consumed as snacks in Taiwan and Japan.

6.2 *WAKAME*

Wakame is the Japanese name of the raw material and the salted product of large brown seaweed. When reconstituted, it is one of the softest brown seaweeds. It is traditionally used in Japanese soybean paste *(miso)* soup.

Fresh *wakame,* either harvested from farms or collected from the wild, is blanched in fresh water at 80–90°C for 1 min (Lisac, 1984). The cooled *wakame* is then mixed with dry salt in a 60:40 ratio and kept overnight. After being drained, the salted *wakame* is packed and stored at −10°C. This is an intermediate product and has a shelf life of up to 12 months. The further processing involves removing the mid-rib (central thick, hard part). The final product contains 50% moisture and less than 40% salt. Before eating, the salted *wakame* needs to be desalted using fresh water.

6.3 LAVER

Laver *(nori)* is also a popular seaweed for food in China, Taiwan, Korea, and Japan. It can be used for making dried sheet or seasoned roasted products.

After harvesting, laver is washed, and impurities are removed. The fronds are cut into fine pieces and homogeneously mixed with fresh water. The suspended pieces are put into frames containing blinds and are sun-dried (Tanikawa, 1971). A typical Japanese dish is a rice roll wrapped with dried laver. Pickled vegetables, boiled eggs, and sliced raw fish flesh are commonly stuffed in the vinegar-seasoned rice roll.

Laver is also used for the preparation of seasoned and roasted sheets or seasoned paste. The seasoned paste is a popular product in which soy sauce and sugar are added to the crushed laver and boiled for several hours. The finished product is packaged in a glass container. In Taiwan, laver *(Monostroma nitidum)* is commonly used to prepare dried and seasoned paste.

7. CONCLUSION

Traditional seafood products have a long history in Asian countries. Today, the production of those products is not mainly for the purpose of preservation. Consumer demand is still focusing on the typical flavor and texture of the product. Because of health concerns, the salt level of cured products needs to be reduced. The advancement of processing machines would help produce a higher quality product, especially the mechanical drier replacing the sun-drying method. For the future of those products, quality and safety concerns will be the opportunity and challenge.

8. REFERENCES

Anon. 1984. *Fishery Statistical Bulletin for South China Sea Area 1982.* Southeast Asian Fisheries Development Center, Bangkok, Thailand.

Anon. 1988. Cephalopods. *Infofish International* (2188):44–47.

Anon. 1991. *ABD/Infofish Global Industry Update: Cephalopods.* Infofish, Kuala Lumpur, Malaysia.

Anon. 1992. Seaweed and seaweed products. *Infofish International.* No. 1/92, p. 43.

Chen, S-P. 1992. Traditional oyster products. *Infofish International.* No. 5/92, p. 27–32.

Borgstrom, G. and Paris, C. D. 1965. The regional development of fisheries and fish processing. In: *Fish as Food. Vol. 3. Processing: Part 1.* Academic Press, New York, pp. 301–409.

Bruce, C. 1983. Sea cucumbers—Extraordinary but edible all the same. *Infofish Marketing Digest.* No. 6/83. pp. 19–23.

FAO. 1990. Fish. In: *Utilization of Tropical Foods: Animal Products.* FAO Food and Nutrition Paper 47/8. Food and Agriculture Organization of the United Nations, Rome. pp. 21–40.

Ferdouse, F. 1997. Beche-de-mer and sharkfin—Markets and utilization. *Infofish International.* No. 6/97. pp. 23–29.

Floyd, J. M. 1985. The role of fish in Southeast Asian diets: Focus on Indonesia, Malaysia, the Philippines and Thailand. *Infofish Marketing Digest.* No. 4/85, pp. 31–34.

Hiremath, G. G., Sudhakara, N. S., and Serrao, A. 1986. Salted and pressed sardine. *Infofish Marketing Digest.* No. 2/86, pp. 21–22.

Huang, Y. W. 1988. Cannonball jellyfish *(Stomolophus meleagris)* as a food resource. *J. Food Sci.* 53:341–343.

Infofish, 1991. *ABD/Infofish Global Industry Update: Shrimp.* Infofish, Kuala Lumpur, Malaysia.

Jeng, S. S. 1977. V. Minced fish products. In: *Fishery Products of Taiwan,* Chung et al., eds., JCRR Fisheries Series 25B, Taipei, Taiwan Joint Commission on Rural on Commission. pp. 39–42. (In Chinese).

Ka-keong, E. L. 1983. Shark fins—Processing and marketing in Hong Kong. *Infofish Marketing Digest.* No. 5/83, pp. 35–39.

Krishnan, S. G. 1984. XV. Diversification of products and markets—Salted jellyfish—A potential diversified product for Japan/Hong Kong markets. *Seafood Exp. H. (India)* 16:23.

Kreuzer, R. 1974. Fish and its place in culture. In: *Fishery Products,* R. Kreuzer, ed., Fishing News (Books), Ltd. and Food and Agriculture Organization, United Nations. Surrey, England, pp. 22–47.

Kreuzer, R. 1986. Squid—Seafood extraordinaire. *Infofish Marketing Digest.* No. 6/86. Infofish, Kuala Lumpur, Malaysia, pp. 29–32.

Limpus, L. G. (Comp.) 1991. Draft code of practice for the full utilization of sharks. *FAO Fisheries Circular.* No. 844, pp. 62–65.

Lisac, H. 1984. Japan—Some traditional cured products. *Infofish Marketing Digest.* No. 1/84, pp. 40–42.

Lo, C. S. 1988. Agriculture Section, *Cihai* (Dictionary) 2nd ed., Shanghai Dictionary and Books Publisher, Shanghai, China, pp. 625–696. (In Chinese).

McHugh, D. G. and Lanier, B. V. 1984. Korea's edible seaweed trade. *Infofish Marketing Digest.* No. 3/84. pp. 17–19.

Moen, E. (Comp.) 1983. Cured fish: Market patterns and prospects. *FAO Fish. Tech. Pap.* 233:92–125.

Morikawa, T. 1984. Jellyfish. *Infofish Marketing Digest.* No. 1/84. pp. 37–39.

Ramachandran, A. and Terushige, M. 1991. Sea urchin for Japan. *Infofish International.* No. 5/91, pp. 20–23.

Richards-Rajadurai, N. 1990. Carrageenan—Multipurpose gum from the sea. *Infofish International.* No. 5/90, pp. 18–20.

Robertson, G. W., Hotton, C., and Merritt, J. H. 1987. Drying Atlantic sea cucumber. *Infofish Marketing Digest.* No. 3/87, pp. 36–38.

Rudloe, J. 1992. Jellyfish: A new fishery for the Florida panhandle. Report for the Apalachee Regional Planning Council, Blountstown, FL.

Roh, J. D. 1992. The squidmeal industry in Korea. *Infofish International.* No. 3/92, pp. 26–30.

Sachithananthan, K. 1986. Artisanal handling and processing of sea cucumbers (sand fish). *Infofish Marketing Digest.* No. 2/86. pp. 35–36.

Subasinghe, S. 1992. Jelly fish processing. *Infofish International.* No. 4/92, pp. 63–65.

Suknark, K., Philip, D. and Huang, Y. W. 1997. Processing of fish crackers using extrusion method. Abstract. Institute of Food Technologists Annual Meeting, Orlando, FL, June 20–24.

Tanikawa, E. 1971. Dried marine products. In: *Marine Products in Japan.* Koseisha-Koseikaku, Tokyo, Japan, pp. 212–294.

van Eys, S. 1983. Katsuobushi—A Japanese speciality. *Infofish Marketing Digest.* No. 2/83, pp. 23–27.

van Eys, S. 1986. The international market for sea cucumber. *Infofish Marketing Digest* (5186):41–44.

van Veen, A. G. 1965. Fermented and dried seafood products in Southeast Asia. In: *Fish as Food. Volume III, Processing: Part 1,* G. Borgstrom, ed., Academic Press, New York, pp. 227–250.

Wootton, M., Buckle, M. A., and Martin, D. 1982. Studies on the preservation of Australian jellyfish (*Catostylus* spp.). *Food Technol. Australia* 34:398.

Wu, H. S. 1990. Dried products. In: *Fishery Processing Industry in Taiwan,* H. S. Wu, ed., Taiwan Fisheries Bureau, Department of Agriculture and Forestry, Taiwan, pp. 100–130. (In Chinese).

Yu, S. Y. 1986. Better, crispier fish crackers. *Infofish Marketing Digest.* No. 6/86. pp. 33–35.

CHAPTER 10

Fruit Products

JOHN X. SHI
BOR S. LUH

1. INTRODUCTION

Most Asian countries are located in the tropical, subtropical, and temperate zones. In these regions, agriculture dates back several thousand years, with fruits representing a major source of food and nutrition. There are many varieties of fruits that grow abundantly in different regions. Many of them are eaten fresh or processed into homemade products. Since fruits are agricultural commodities, they have inherent variabilities caused by variety differences, and they are also subjected to the vagaries of climate and other environmental influences. Fruit preservation began in ancient times in China, India, and Persia. Since ancient times, several techniques have been used to preserve fruits and their products: drying, concentration, freezing, fermentation, and chemical preservation using vinegar, wine, sugars, honey, herbs, and spices. With the advent of modern food processing technologies, some of the traditional methods were developed for commercial production. There are many kinds of subtropical fruit products made from apples, apricots, citrus fruits, dates, grapes, plums, prunes, strawberries, and more in the world market. Tropical fruit processing is a rapidly expanding industry and has become an important source of export revenue in China, India, Malaysia, and Vietnam. Banana, mango, pineapple, lychee, and papaya products are particularly important tropical items with commercial significance in the international trade. The annual output of some fruit products in Asia is listed in Table 10.1.

It is impossible to discuss the full range of fruit products in the brief space available here, but the most important commercial items will be covered in this chapter.

TABLE 10.1. The Annual Output of Some Fruit Products in Asia (×1000 MT) (1989–1994).

	1989	1990	1991	1992	1993	1994
Dried fruits	159	168	185	154	178	222
Jam and jelly	79	77	173	154	177	160
Canned Fruits	1730	1333	1560	1817	1611	1595

Source: United Nations, 1996.

2. DRIED FRUIT PRODUCTS

The drying of fruits was practiced long before biblical times in China, India, and Persia. The ancient Chinese and Hindus dried fruits in the sun 5000 years ago. Dates, figs, apricots, and raisins, now important in world commerce, were dried by early inhabitants in the Middle and Near East. Dates and figs were especially essential food items in the Mediterranean region before history began to be written. Sun-drying is still used for fruits in many regions today. In fact, considerably more fruits are preserved by drying than by any other means. The quality of the final fruit products is improved by mechanical dehydration. Dried fruit products are available in many forms: whole, slices, pieces, bars, powders, flakes, or leathers. Techniques such as sun-drying, tunnel drying, cabinet drying, fluidized-bed drying, drum drying, spray drying, vacuum drying, foam-mat drying, osmotic dehydration, and freeze-drying have been established as commercial processes for some time. Dried fruits are more concentrated forms than products preserved in other ways. They are less costly to produce and require less storage space than canned or other preserved fruits. It can readily be seen that the cost of transportation will be much less for dried than for canned, frozen, or fresh fruits. For these reasons, dried fruits are usually considerably less costly to the consumer than the equivalent quantities of canned, frozen, or other preserved fruit products.

Artificial dehydration affords a means of producing dried fruits of new forms and of better quality than is possible by sun-drying. The dehydrated fruits are superior in flavor and cooking quality to the sun-dried fruits. Recently, the demand has been increasing for dried fruits with higher than usual levels of moisture, for example, intermediate moisture fruit (IMF) products. Such fruit products are softer in texture and are more appealing to consumers.

2.1 DRIED APPLE SLICES

The apple is a highly remunerative deciduous fruit that is grown in temperate regions. Some major apple-producing countries are Russia, Korea, Turkey,

Iran, Japan, India, and the northern part of China. After partial dehydration, apple chips are used as a snack food that has a crispy texture, nonsticky surface, and a fresh fruit flavor. Best varieties for drying are late autumn- and early winter-ripening varieties. There is an increasing demand for dehydrated apples of high quality, although the tendency in the past has been to produce quantity rather than quality. The clean, white fruit that is well trimmed and carefully dried and packed is most in demand. Dried apples are graded on texture and appearance rather than on flavor. Varieties that are firm and can yield a dried product of white color are preferred.

The peeled and cored apples are cut as rings, segments, chops, or cubes and treated with a citric acid solution and a bisulfite dip to hold color temporarily. The sulfured slices are dehydrated at 60–70°C for 6–8 h in a tunnel dryer. Among the different treatments, a 2500-ppm SO_2, 1-h dip of apple rings resulted in the best dehydrated product in hot-air dehydration. The finished product contains 20–22% moisture. The dehydrated products are packed in moisture-proof containers.

2.2 DRIED APRICOTS

The apricot is native to China and Siberia. Records suggested it was introduced to the Mediterranean about 100 B.C. Dried apricot products are produced from plump, fresh fruits that are fully ripe on the trees. Dried apricot products are usually sun-dried. The final dried apricots have a moisture content of 15–20%. The special flavor, texture, and nutritional value are winning over a new generation of consumers who use dried apricot as snacks (Figure 10.1). Iran and China are the major apricot-producing countries. The 'Royal' and 'Blenheim' cultivars are the principal ones used for drying because of their high flavor, density, and solid-to-acid ratio that gives a unique sweet-tart flavor. Predrying treatments include (a) selecting and sorting of fresh fruits, (b) washing, (c) cutting into halves and removing of stones, (d) steam-blanching, (e) spreading of fruits on drying trays, (f) sulfuring with burning sulfur or gaseous SO_2 for 3 h duration, and (g) placing trays in a dry yard in full sun. In artificial dehydration, good-quality dehydrated apricots are obtained at temperatures below 80°C. The dried fruits can be stored at 4°C and 75% relative humidity for least 6–9 months. Low-moisture apricot products are used in dry turnover pies, tart filling mixes, and fruit cocktail-type mixtures in combination with other dehydrated fruits to be consumed as stewed fruit or compote.

2.3 DRIED BANANA SLICES

Bananas are grown in humid tropical regions and constitute one of the largest fruit crops of the world. India is the leading banana producer in Asia. Other producers are Taiwan, Indonesia, and the Philippines. The dried banana

Figure 10.1 Dried apricots.

has a pleasant flavor and may appear in various forms. Halves, slices, or whole fruits may be sun-dried or artificially dehydrated in cabinet or tunnel dryers.

Diced banana products are used as raisin substitutes in food ingredients. The conventional technique for drying banana chips is to use firm, ripe fruit with a solid texture and total sugar content of about 2%. Thin banana slices (about 2 mm thick) are soaked in a solution containing NaCl, citric acid, and potassium metabisulfite for 30 min, or some alternative browning inhibitors such as SO_2 solution, aluminum chloride, citric acid, or ascorbic acid, and then blanched in hot water or steam. The final product has a unique color and flavor and 14–15% moisture content. The slices are wrapped individually in plastic pouches and then packed by the dozen in polyethylene bags and encased in cartons. The final products may be stored for a year at room temperature (24–30°C) and are commonly exported. They can be eaten as a snack food or used in making fruit cake and bakery products. Banana powder, flour, and drum-dehydrated flakes are products that have been produced commercially in other parts of the world, but each has problems with regard to color and flavor stability (Somogyi and Luh, 1986; Kotecha and Desai, 1995).

There is considerable interest in developing a market for banana chips as snack items made by deep-frying dried banana slices. To enhance product quality, the slices are treated with sugar solution, a gelatinous substance, or artificial flavors and coloring agents and then fried in hydrogenated fat or edible oil. This snack is considered a "natural" or "healthy" food. The Philippines is the main exporting country. Banana chips made in the Philippines are normally made of green banana with intermediate moisture content, dipped in or coated with cane sugar solution, and fried briefly a second time in hot oil to the final moisture level (Sole, 1996).

2.4 DRIED CHINESE DATES

Chinese dates are grown in the Hunan, Shandong, Zhejiang, and Shanxi provinces of China. Most fruits are dried in the sun or by dryers. Depending on the technique used for drying, the final dried date products have different names: red date and black date (Figure 10.2). The final product has a moisture content of 18–20%.

(1) Red dates: Full ripe Chinese dates are blanched and dried as whole fruits by the sun. The product has a dark-red color, golden-yellow meat, elastic texture, and sweet taste.
(2) Black dates: The product has a dark-violet color, wrinkled surface, sweet taste, and elastic texture. Fully ripe fruits are selected, blanched, and then dried and fumed at 60–70°C for 20–24 h (Chow, 1991).

Figure 10.2 Dried Chinese dates.

TABLE 10.2. Color of Chinese Red Date Related to Processing Temperature.

Sugar Content (%)	Temperature (°C)	Skin Color	Meat Color	Flavor
32.06	40	red	yellow	sweet
48.12	50	red	yellow	sweet
65.80	68	dark red	little red yellow	sweet
62.30	78	brown	brown	scorched smell

Source: Chow, 1991.

Some treatment results are shown in Table 10.2. Chinese date products have a special function of invigorating blood circulation according to Chinese traditional medicine.

2.5 DRIED FIGS

Dried figs are eaten as a snack or used as cake fillings. Figs are grown in mild-temperate climates and are now commercially produced in most of the countries bordering the Mediterranean Sea. Dried figs are considered to be the first fruit products preserved by drying. Figs are dried as whole fruits. They are usually sun-dried, but dehydration is also practiced, to a limited extent, to produce low-moisture figs. Turkey is one of the most important fig-producing countries. The 'Smyrna' fig, a white cultivar, is the principal cultivar grown in the Mediterranean countries for drying. Commercial drying of figs is an industry of great economic importance in these areas.

Commercially, figs are peeled by immersion for 1 min in boiling lye water or a boiling solution of sodium bicarbonate and then prepared for drying. Fruits are allowed to ripen fully and are partially dried on the tree, harvested, and then exposed to SO_2 fumes for about 1.5 h, placed under the sun, and turned daily to achieve uniform drying. The product is pressed flat after sun-drying for 5–6 days. Whole fruits can also be dehydrated in a drier at 85–90°C for 9–12 h to obtain good-quality products (Desai and Kotecha, 1995).

Dried figs are a good nutrient and energy source because of their carbohydrate content and the concentration effect caused by moisture removal. Figs are an especially good source of fiber, which aids in the anticonstipation process. Dried figs contain 5.6% fiber. In addition, the potassium salts of organic acids in figs help maintain the acid–alkaline balance in the body by neutralizing excess acids present. Dried figs exert a positive effect on the alkaline reserves in the body. Figs and fig extracts have been used for medicinal purposes, such as in the treatment of Ehrlich sarcoma. Dried figs have long been appreciated for their laxative action. The latex is widely used for treating warts, skin ulcers, and sores and is taken as a purgative and vermifuge. A sore throat can be relieved by gargling with a decoction of the fruits (Morton, 1987).

2.6 GRAPE RAISINS

Dried grapes, commonly known as raisins, represent the greatest quantity of dried fruits produced today. The drying of grapes to produce raisins is an ancient form of food preservation. It is not precisely known when and where the production of raisins first occurred, but it is likely that this practice dates back to biblical times in the Middle East. Raisin production spread from the Middle East to other parts of the world. Raisins form an important by-product industry in several grape-growing countries. Now raisins are produced primarily by sun-drying. Important raisin-producing countries are Turkey, Iran, Afghanistan, and Lebanon (Table 10.3). Many of the products from Afghanistan are exported to Russian and India (Table 10.4). Raisins are popular as dried fruit products and are commonly used in many traditional recipes of Asian countries in bakery and confectionery products. The quality of raisins depends on the size of the raisin berries; the uniformity and brilliance of the berry color; the condition of the berry surface; the texture of the skin and pulp in the berry; moisture content; chemical composition; and absence of decay, molds, yeasts, and foreign matter. Based on the method of preparation and grape variety used for raisin making, they are called various names.

Most of the raisins produced are from four varieties: 'Thompson Seedless,' 'Black Corinth,' 'Muscat of Alexandria,' and 'Fiesta.' The most important raisin grape is the 'Thompson Seedless.' Before drying, grapes are often dipped in alkali to remove the waxy layer and thus hasten drying. Raisins produced from the cultivars 'Thompson Seedless' are known as 'Sultana' in West Asian countries. In Middle East countries they are called 'Kismish.' Raisins

TABLE 10.3. The Raisin Production in Asia (×1000 MT) (1989–1995).

	1989/1991	1993	1994	1995	1996
Asia	472	501	506	506	507
Turkey	341	360	360	360	360
Syria	9	8	11	12	13
Oman	7	7	7	7	8
Iran	65	90	90	90	90
Afghanistan	44	28	28	28	28

Source: FAO, 1998.

TABLE 10.4. Exports of Raisin from Asia (MT) (1992–1994).

	1992	1993	1994
Asia	206,256	226,534	269,977
Afghanistan	28,000	25,000	23,000
China	1459	1767	2573
Cyprus	274	278	385
Hong Kong	3640	4220	6475
India	31	30	30
Indonesia	51	43	34
Iran	53,797	63,572	5000
Macao	390	301	300
Kuwait	6	35	49
Malaysia	7	41	42
Pakistan	1655	2819	1435
Singapore	2480	4317	4674
Syria	38	130	80
Thailand	4	30	26
Turkey	107,696	117,964	173,250
United Arab Emirates	6640	6000	7620

Source: FAO, 1997.

produced from 'Black Corinth' are also called 'Zante Currant,' 'Currant,' or 'Zante.' Raisins produced from 'Muscat of Alexandria' also have many synonyms. 'Monucca' is the common name in Russia and Afghanistan. The raisins may be called "naturals," "golden bleached," or "sulfur bleached," depending upon the making method. These products have a light color and glossy appearance (Somogyi and Luh, 1986; Chow, 1991; Patil et al., 1995).

Golden bleached raisins are also produced from the 'Thompson Seedless.' They are brilliant lemon yellow to golden yellow in color, tender in texture, and sometimes a little sticky. Development of the desired color is controlled with the help of sulfur dioxide during drying and storage. The dehydration process is done at 60–70°C.

Natural raisins are produced practically without any pretreatment from the 'Thompson Seedless' cultivar. They are dark or greyish black or brownish in color with a tough and dry skin, are meaty, and are of a characteristic oxidized flavor. The naturals are in high demand for eating by hand or for dessert purposes. They are soft, amber to brown in color, and shiny because of the use of a mineral or olive oil dip before drying. They are dried to about 15% moisture (Somogyi and Luh, 1986).

2.7 DRIED LONGANS

Longan fruits are produced in southern China, Taiwan, India, Thailand, Cambodia, Laos, Vietnam, Malaysia, and the Philippines. The dried products are black, leathery, and smoky in flavor. They are mainly used in making infusion beverages. The United States imports longan products mainly from Taiwan. Before dehydration, the fruits are first heated to shrink the flesh and to facilitate the hulling. Then the seeds are removed and the flesh (mesocarp) dried with low-temperature hot air or over a slow fire. The fruit products can be administered as a stomachic, febrifuge, and vermifuge and are regarded as an antidote for poison. A decoction of the dried flesh is traditionally taken as a tonic and treatment for insomnia and neurasthenic neurosis (Morton, 1987).

2.8 DRIED LYCHEES

Lychees are considered to have originated in the Guangdong Province of China and have been grown in China for more than 4000 years. The Guangdong and Fujian Provinces in southern China remain the largest producers of lychee, followed by Vietnam, Thailand, India, Burma, Japan, the Philippines, Taiwan, Pakistan, and Bangladesh. Lychees are becoming popular fruits in Asian markets. A large proportion of lychees is traditionally dried in China, and often lychees are offered as presents. Dried lychee fruits, frequently referred to as lychee nuts, or lichi nut, offer interesting opportunities in domestic and foreign markets. During drying, the pericarp or outer skin gradually loses its original color and becomes cinnamon-brown and brittle, while retaining its shape. The pulp turns dark-brown to nearly black as it shrivels around the seed and becomes very pleasant in flavor and raisin-like in texture.

In China, lychees are preferably dried in the sun by hanging wire trays and brought inside at night or during showers. They are dried by means of brick stoves during humid weather. Now, electric oven-drying and hot air-drying at about 65°C are employed in major producing areas. Drying at higher temperatures gives a bitter flavor. Dried final products can be stored in tin cans at room temperature for about a year without substantial change in texture and flavor (Chow, 1991).

Ingested in moderate amounts, lychees are traditionally taken to relieve coughing and to have beneficial effects on gastralgia, tumors, and enlargements of glands. Fermented lychees are also used in Chinese medicine.

2.9 DRIED MANGOES

Mangoes are originally from the Indo-Malaysian region. The earliest growing area is from northeastern India and Burma eastward to Indo-China.

Now the production of mango has extended to many Asian countries and regions such as India, the Philippines, Indonesia, China, Thailand, Burma, Malaysia, Sri Lanka, and Israel. India, the Philippines, Pakistan, and Thailand are the leading exporters of processed mango products. The dried mangoes are utilized commercially as a substitute for the mangoes used in chutney manufacture.

Dried slices are prepared from ripe fruits. The peeled fruits are cut in halves lengthwise, spread on wooden trays, and sulfured for 1 h by exposure to burning sulfur. The slices are then dehydrated at 55–60°C in a cabinet drier until the product is pliable, soft, and nonsticky. The fruits dehydrated at a relatively low temperature such as 43°C are superior in quality. The peeled or unpeeled slices of raw mango are dried in the sun or in a cabinet dryer and then turned into powder as a souring agent in Indian cuisine. SO_2- preserved, solar-dried fruits tend to have better flavor than hot air dehydrated products (Somogyi and Luh, 1986). Dried products are packed in cardboard cartons lined with a polyethylene film. The major problem of this type of product is progressive browning during storage. Mango products have medicinal properties (laxative, diuretic, and fattening) according to Chinese traditional medicine.

2.10 DRIED PALM DATES

Long before recorded history, dates were an important crop in the desert regions of Middle Eastern countries and formed the basis for the survival of many ancient desert nomads. Palm date is believed to be indigenous to countries around the Persian Gulf (Table 10.5). Dates, a very rich source of energy, constitute a staple food in many developing countries from western Iraq to Arabia. Date palms are now grown commercially in hot, dry desert regions,

TABLE 10.5. Date Production in Asia (×1000 MT) (1989–1995).

	1989/1991	1993	1994	1995	1996
Asia	2272	2646	2972	2986	2969
Bahrain	16	19	19	20	20
China	18	30	30	35	38
Iraq	533	613	576	600	550
Iran	563	716	774	780	795
Israel	12	12	13	12	12
Oman	122	133	133	133	133
Pakistan	288	280	579	532	533
Saudi Arabia	526	563	568	586	597
United Arab Emirates	152	236	236	237	240
Yemen	22	22	21	23	24

Source: FAO, 1998.

such as in the oases of the Sahara and the Arabian deserts. Libya and Saudi Arabia are the leading producers of date products, with an annual production of about 400,000 tons in Saudi Arabia. Dried palm date products are processed, shipped, and consumed throughout the world.

There are several varieties of palm date: 'Zahidi,' 'Sayer,' 'Halwy,' and 'Khadrawy' are commonly grown in Iraq; 'Saidy' in the Libyan desert; and 'Halawy,' 'Barhee,' 'Chichap,' 'Shanker,' 'Burein,' and 'Shahaani' in India. Date fruits pass through five growth and developmental stages: *hababauk, kimri, khalal, rutab,* and *tamar.* Dates are boiled in water for a certain time and then allowed to dry until they are hard and wrinkled (Morton, 1987; Al-Hooti et al., 1997).

Dates are almost exclusively sun-dried under the natural climatic conditions of the desert areas. In the Middle East, dates are often allowed to ripen on the tree and dried directly in the sun after harvest to a moisture content of less than 20%. Dates are also dehydrated by hot air flow in a cabinet at about 76°C for several hours. Too high a temperature damages quality. Dry dates can be held at room temperature for years without substantial changes in color or overall quality.

The date products, because of their relatively high tannin content, are used medicinally as a detersive and astringent in people having intestinal troubles. In the form of an infusion, decoction, syrup, or paste, they may be administered as a treatment for sore throat, colds, and bronchial catarrh. Traditionally, they are taken to relieve fever, cystitis, gonorrhea, edema, liver, and abdominal troubles. It is also said that fruit products can counteract alcohol intoxication (Morton, 1987).

2.11 DRIED PEACHES

The peach originated in China. In China, it was mentioned in the literature of the sixth century B.C. In addition, it was cultivated in Iran for a long time. A peach suitable for drying should be large in size, with a high sugar content. It should be pulpy rather than juicy, with flesh of a rich golden-yellow color and pleasing flavor. Well-dried peaches should be golden-yellow, not gray-green or brown, and firm and pliable, but not syrupy or sticky. Low-moisture slices or diced peaches are utilized as pie, tart, and turnover fillings. Only a small proportion of peaches is dried. Peaches are cut in halves, stoned, and dried in a dehydrator or by sun-drying. Yellow-fleshed peach varieties are the most suitable for drying. The steps involved in common predrying treatments of peaches are (a) selection and sorting for size, maturity, and soundness; (b) washing; (c) peeling by lye solution or abrasion; (d) cutting into halves, slices, cubes, or segments; and (e) sulfuring for 4–5 h and then drying by sun-drying or hot air-drying at 55–65°C for 14 h. The dried products are translucent, of good flavor, and of quick-cooking quality.

The above dehydration procedure is also applicable to pear and prune products. Dried pears have a light yellow color, semitransparent and tender texture, fresh fruit flavor, and very attractive appearance. Dried prunes are used in fillings and other bakery spreads. Powder is used as a sweetening and flavoring ingredient in whole wheat or rye breads.

2.12 DRIED PERSIMMONS

Persimmons originated in China. They now have spread all over Asia. Japan is the largest producer and *kaki* is their popular name in Japan. The Japanese dry large quantities of persimmons, which are used as confection or food. Other persimmon-producing countries are China, Israel, the Philippines, Indonesia, India, Burma, Vietnam, and Korea.

Dried persimmon products have white "persimmon sugar" on the surface, a soft texture, and a sweet taste. Large quantities of persimmons are preserved by sun-drying. The fruits are picked when mature and still firm. The fruits are first treated with ethylene gas to speed up the metabolism and to reduce the tannin content. Then fruits are peeled and hung up by their stems and left to dry in the sun for 30–50 days. Kneading every 4–5 days is necessary to produce a uniform texture and improved flavor. They are taken down and sweated for 10 days in heaps. The dried final products are shaped into a flat form by pressing. Finally, sugar crystals appear on the surface. That contributes to the acceptability of the final products. Persimmons are dehydrated to 20% moisture in a shelf oven with hot air circulation at 40–45°C (Somogyi and Luh, 1986; Chow, 1991). In Indonesia, ripe fruits are stewed until soft and then pressed flat and dried in the sun. Intestinal compaction from consumption of persimmons in Israel has been eliminated by drying the fruits before marketing, and some dried fruits are now being exported to Europe.

A decoction of the calyx and fruit products is traditionally taken to relieve hiccups, coughs, and labored respiration in China and Japan.

2.13 SOME OTHER DRIED FRUITS

Some dried fruits such as dried guavas, papayas, and pineapples are available on the market. The dehydration procedures are similar to the procedure of mango dehydration.

(1) Dried guavas: Guava products are of commercial importance in India. Dried guavas are used for making ice cream, confections, fruit juices, and jelly. Because of its relatively high pectin content, guava can be used to make jelly from low-pectin fruits.
(2) Dried papayas: Dried papayas have a light yellow color, sweet flavor, and better sensory qualities.

(3) Dried pineapples: Dried pineapples have an excellent appearance, with a particularly appealing pineapple flavor and odor.

2.14 FRUIT POWDERS AND CAKES

Tropical fruit powders have a large export potential because of their inherent advantages of weight reduction and product stability. Owing to recent advances in flexible film packaging, food powders have become more common in the market. They are especially attractive for military rations, to backpackers and campers, and for use as infant foods. The unripe starchy fruits are cut into slices, dried either in the sun or in a cross-flow air dryer, powdered, and packed. Foam-mat dehydration is a relatively innovative method of preparing dehydrated food powders. Fruit powders are often used in making cakes.

2.14.1 Mango Powder

A dehydrated mango powder has been developed for use in baby foods or reconstituted for beverages in India. The dried mango powder is also used for fruit cakes. It is a common method to peel and slice mangoes and then dry them in the sun. This indigenous practice is prevalent in India.

Mango baby foods are prepared from sulfated mango pulp using several techniques such as drum drying, puff drying, foam-mat drying, and spray drying. Mango slices are blanched in boiling water and then dehydrated in dryers where the temperature and humidity are controlled. The product obtained by the drum-drying process not only maintains all the qualities of a good product, but also retains high percentages of ascorbic acid ($>87\%$) and β-carotene ($>97\%$). Ascorbic acid is retained better in the freeze-dried powder than in the drum-dried powder. SO_2 in the pulp increases the stability of the ascorbic acid. The drum-dried powder has some nonenzymatic browning during storage. The powder produced by the foam-mat drying of mango pulp is inferior in color, flavor, and acceptability to that produced by spray drying (Kalra et al., 1995). A reduction in viscosity of mango pulp by pectin enzymatic treatment and then spray drying will result in a good mango powder that is relatively inexpensive. The products have good rehydration capacity and show reasonable retention of flavor and nutrient constituents (Somogyi and Luh, 1986).

2.14.2 Some Other Fruit Powder

Apples, hawthorns, and oranges are made into powder for several special uses:

(1) Apple powder: Apple powder is a special product mainly in the treatment of infant diarrhea.

Figure 10.3 Hawthorn cakes.

(2) Banana powder: Banana powder has a light yellow color, but low flavor.

(3) Coconut milk powder: The coconut milk is dehydrated into powder after removal of cream. Coconut milk powder contains 24% protein, 6% oil, 5% moisture, and 25% starch ingredient. This stable product can be reconstituted with water.

(4) Guava powder: Guava powder has an attractive color, aroma, and taste.

(5) Hawthorn powder: Hawthorn fruits are only produced in China. Ripe fruits are selected, washed, sliced, pitted, cooked to a soft texture, mashed into puree, mixed with sugar at a ratio of fruit to sugar of 10:8 (w/w), and then spread on a tray and dehydrated by hot air at 65°C for 4 h. The powder is made into cakes and then cut into thin circular pieces and packed (Figure 10.3). It has a sweet-sour taste and a fresh flavor. The products have some medicinal properties, especially in reducing blood pressure.

(6) Orange powder: Orange powder has a yellow color and fresh flavor and is easy to make into a beverage with water.

2.15 DRIED FRUIT LEATHERS AND BARS

Fruit leathers, known commercially as fruit rolls, are manufactured with fruit purees into leathery sheets. The leathers are eaten as a confection or used as a sauce. Fruit leathers are well established products overseas, particularly in the North American market. The fruit puree is mixed with sugar or other ingredients as desired, spread in a thin layer, dried, and then wrapped in plastic film and stored at room temperature. The dried products have a bright translucent appearance, chewy texture, and distinct fruit flavor. They can be prepared from a wide variety of fruits, including apple, apricot, banana, blackberry, cherry, grape, guava, hawthorn, papaya, peach, pear, pineapple, plum, raspberry, strawberry, and so on. Overripe fruits with high sugar and flavor but low fiber content are suitable for making fruit leathers.

2.15.1 Mango Leathers

Mango leathers in India, also known as *amawat, ampapad,* or *tundra,* are thick, dried products of mango puree. In India, mango leather processing is an important rural industry, with an annual production over 500 tons. The product is prepared as follows: the puree is squeezed manually and passed through a bamboo sieve to remove the fiber; the puree is then spread on date palm leaf mats to dry in the sun; when one layer dries up, another layer of puree is laid over it, and this process is repeated until a product thickness of 3 cm is obtained, which usually takes about 40 days; the slabs of leathers are cut into approximately 500-g portions, which are packaged with cellophane, polyethylene film, or waxed paper.

A refined industrial method for the preparation of mango leathers has been developed (Rao and Roy, 1980). When the sugar content reaches as far as 35°Brix, potassium metabisulfite (1,000–2,000 ppm) is added. The mixture is spread over trays at an initial thickness of 6 mm, dehydrated to 2 mm in a forced-air drier at 80°C for about 2–5 h. Three or four additional layers are added until a total thickness of 6–8 mm is obtained, and then the mixture is dried below 20% moisture by air drying at 60–70°C for 20–22 h. The dried final product is then packaged in flexible film.

2.15.2 Other Products

Some other commercial products include kiwifruit leather, hawthorn leather, banana bars, and mango bars.

(1) Hawthorn leather: Hawthorn leather is usually made from hawthorn puree mixed with apple, date, or peach puree.

(2) Kiwifruit leather: Kiwifruit extract is mixed with apple pulp. Kiwifruit leather has a light green color and high kiwifruit flavor. Hawthorn leather has a light red or dark red color and a sour-sweet taste. The total sugar content is 60%, and total acidity is 0.8–1.0%.

(3) Mango bar: Mango bars are soft in texture, with an attractive color and a characteristic mango flavor. They can keep well in a 50–60% relative humidity range.

3. CURED FRUITS

One of the oldest methods of curing fruits is by adding sugar, honey, salt, spices, and herb ingredients. These products are commonly called preserved fruits or candied fruits. Cured fruits can be prepared from many kinds of fruits, such as peaches, pears, grapes, plums, crabapples, currants, Chinese olives, and more. Fruits for curing should be in the firm-ripe stage. Cane sugar, beet sugar, corn syrup, honey, salt, and some herb or spice flavoring ingredients are commonly used. In general, most curing fruits consist of 65–70% sugar, with water activity values of 0.6–0.8 to control microbial growth. Cured fruits are the most popular products among Asian people. In recent years, consumer awareness of the adverse effect of high sugar and preservatives in cured fruits has resulted in a new approach to develop some low-sugar cured fruits. Cured fruits are excellently served with a variety of entrees, confections, and snacks. In fact, the demand is far more than the supply. Some products are shown in Figure 10.4.

Figure 10.4 Cured kumquats, mangoes, prunes, plums, and strawberries.

3.1 CURED CHINESE OLIVES

There are several kinds of cured Chinese olives with different tastes and flavors. Processing methods include selecting ripe fruits, soaking in NaCl solution for 4 days, washing to remove salt from fruits, drying, soaking in a sugar solution (sugar concentration, 20%) for 24 h and cooking with the same sugar solution, repeating this process three times with a high concentration of sugar solution (50–60°Brix), dehydrating at 55–60°C by hot air, and finally mixing the products with herb flavoring ingredients, spices, or pigments (Zhang, 1994; Fang et al., 1995).

(1) Aroma cured olive: product with black color, crispy texture, and sweet taste, cured by 35% sugar and 35% salt
(2) Multi-taste olive: light green color, translucent and crispy texture, salty-sweet taste, cured by 35% sugar and 35% salt
(3) Sweet-cured olive: dry product, cured by 38% sugar and 50% salt, with a dark brown or dark red color and translucent texture
(4) Sweet-cured olive *(soo larm):* semidry product, cured by 38% sugar and 35% salt, with lime juice added
(5) Sweet-cured olive *(wo sang larm):* semidry product, cured by 38% sugar and 35% salt, with some sweet herb added
(6) Sweet-cured olive *(lar chow larm):* semidry product, cured by 38% sugar and 35% salt, with some spices added
(7) Sweet-cured olive *(wong cho larm):* semidry product, cured by 38% sugar and 35% salt, with some sweet herb and yellow pigment added
(8) Salted dry olive: dry product, cured by 35% salt, wrinkled surface, high fresh flavor

3.2 CURED CHINESE DATES

Cured Chinese dates, also called Honey Dates in China, have a dark red color, translucent and soft texture, and nonsticky surface. After a surface scratching treatment, ripe fruits are sulfured for 2–3 h, cooked in a sugar solution (65°Brix), soaked in a sugar solution (70°Brix) for 30–45 h, and dehydrated at 60–70°C by hot air until the moisture content of the product is 20–25%. The products are pressed flat and then dehydrated again. The final products have a moisture content of 16–18% and a total sugar content of about 65%.

3.3 CURED CITRUS PEELS

Cured sweet orange or lemon peels are well known products of commerce. Processing procedures include soaking in NaCl solution, washing, soaking in a sugar solution, drying, soaking in a sweet herb solution, drying, soaking in a sugar solution, drying, and mixing with a sweet herb powder. The following is a recipe of cured citrus peels: orange peel 10 kg, prune juice 6 kg, NaCl 2 kg, sweet herb 300 g, sugar 600 g, sweetener 20 g, some citric acid. The product has a multi-taste, yellow-brown color, and high flavor.

3.4 CURED KUMQUATS

Kumquats, also called Golden Orange in China, originated in northern China. Kumquats are produced primarily in China and the Philippines, with limited production in other areas of the world. Kumquats are candied or cured whole and are unique in that the entire fruit, including the peel, is generally eaten. The cured products have a golden color, translucent texture, dry surface without sugar particles on the surface, and strong fresh flavor. Fresh fruits of proper ripeness are washed after harvest and sorted. Overripe fruits with an undesirable soft texture and unripe material that lacks flavor are not used. The peel of each kumquat fruit is scored with slight cuts to speed up the soaking process. The fruits are first soaked in 30°Brix sugar solution and cooked for 5 min. The cooked kumquats are cooled immediately and allowed to stand overnight. The next day kumquats are removed, and the sugar solution is raised to 40°Brix. The fruits are cooked in 40°Brix solution for 2 min, cooled rapidly, and allowed to stand overnight. This procedure is repeated with a 10°Brix increase in the sugar solution until the sugar concentration reaches 70°Brix. Then the kumquats are placed on trays and dried in an oven at 60°C for 8 h. After cooling, they are packed in bags and stored in boxes (Chen, 1989). Chinese-cured kumquats are often exported and served as dessert in Chinese restaurants.

3.5 CURED LIMES

Limes, lemons, and citrons are usually referred to as acid citrus fruit group. Limes originated in Southeast Asia. To cure limes, fully ripe, juicy limes are selected and washed with water. The limes are cut into halves. The juice is extracted from the limes, and salt is added (250 g salt for 1 kg limes). To the salt–lime mixture, a 4–5% citric acid solution is added to cover the limes. The mixture is then exposed to the sun for about a week so that the limes become soft. During this period, the original yellow-green color changes to brown. The final product is then put into sterilized bottles and stored in a dry place.

3.6 CURED PRUNES AND PLUMS

Cured prune and plum products are produced mainly in southern China and Malaysia. The product has a bit of a sour and sweet taste. It can stimulate the appetite. Processing methods include selecting ripe fruit, soaking in NaCl solution for 20 days (plum/NaCl ratio 100:25), washing to remove NaCl, precooking in water at 85–90°C, sulfuring for 2 h, soaking in a sugar solution with some herb ingredients (fruit to sugar ratio 1:1) for 7 days, boiling the mixture until the sugar concentration reaches 75%, and then a drying. These products have a light brown or dark brown color, translucent texture, and can maintain its original shape. There are several kinds of cured prune and plum products with distinctive taste and flavor (Fang et al., 1995).

(1) Sweet-cured prune: dry surface, cured by 55% sugar and 50% salt, with some sweet herb
(2) *Chen-pee (mei prune, chen-pee mei):* semidry surface, cured by 55% sugar and 35% salt, with some orange peel
(3) Seasoned prune: semidry surface, cured by 38% sugar and 35% salt, with some sweet herb
(4) Dried prune: dry surface, cured by 35% salt, moisture content 5–7%, stored for 1 year to keep good quality
(5) Salted cured prune: dry surface, cured by 35% salt, with some sweet herb, wrinkled surface, sour-sweet taste
(6) Half-dried prune: semidry surface, cured by 30% salt
(7) Preserved prune: wet surface, cured by 55% sugar and 35% salt, dark brown color
(8) Brine-cured prune: wet surface, cured by 25% salt, remains as fresh fruit, has green color and crisp texture
(9) Sweet prune cake: semidry surface, cured by 55% sugar and 35% salt, with some aromatic herb, sweet flavor, sour-sweet taste
(10) Sweet-cured plum (dried plum): dry surface, cured by 50% sugar and 50% salt, with some sweet herb
(11) Sweet-cured plum *(poo tow lee):* semidry surface, cured by 55% sugar and 35% salt, pitted, retains original fruit shape, has translucent and crisp texture, light brown color, and high fresh flavor
(12) Seedless preserved plum: semidry surface, cured by 55% sugar and 50% salt, with some sweet herb, tart-sweet taste, high sweet flavor, soft but firm texture, bright yellow color, sugar content 58–62%
(13) Salted-cured plum: dry surface product, cured by 55% sugar and 35% salt
(14) Salted dried plum: dry surface product, cured by 35% salt, firm texture, light yellow color, moisture content 12–18%
(15) Half-dried plums: semidry surface product, cured by 30% salt, firm texture, retains original fruit shape, moisture content 25–28%
(16) Brine-cured plum: wet product, cured by 25% salt, retains original fruit shape, fresh flavor

3.7 CURED RED BAYBERRIES

Red bayberries, also called *Yangmei,* are produced in the southern part of China. Cured red bayberries have a dark red color and round shape. The processing method includes selecting ripe fruits; soaking in NaCl solution (2%) for 5–7 days; washing to remove NaCl; drying; and mixing with sweet herb powder, water, and sugar. The mixture is cooked and finally dried again.

3.8 OTHER CURED FRUIT PRODUCTS

Apples, apricots, hawthorns, kiwifruits, mangoes, and peaches are cured to make snack foods with similar processing technology and procedures. Ripe fruits are selected, sliced, pitted, predried, cooked in a sugar solution (55–60°Brix) for 20 min, soaked in a sugar solution (65–70°Brix) for 12 h, and finally dehydrated by hot air. The final product contains 16–18% moisture and 68–70% sugar.

(1) Cured apples: Cured apples have a soft, translucent texture, a milk-yellow color, and sweet flavor. The surface is dry and nonsticky
(2) Cured apricots: Cured apricots have a round shape, red-yellow or light yellow color, translucent and soft texture, sweet-sour taste, and fresh flavor.
(3) Cured hawthorns: Cured hawthorns have a dark red or bright red color, translucent texture, fresh flavor, and sour-sweet taste.
(4) Cured kiwifruits: Cured kiwifruits have a fresh green color and fresh natural flavor, soft but not sticky texture, and sweet taste.
(5) Cured mangoes: Cured mangoes have a golden color, translucent texture, sweet taste, and fresh mango flavor.
(6) Cured pears: Cured pears have a yellow or light yellow color, soft and translucent texture, and strong fresh pear flavor.

4. FRUIT PICKLES AND CHUTNEY

Pickles as flavorful products represent a palatable and safe way to preserve foods. Pickles may be classified as relishes, fresh-pack pickles, and brined pickles. The skillful blending of spices, sugar, and vinegar with fruits gives a crisp, firm texture and a pungent, sweet-sour flavor. Pickled fruits contribute some nutritive value, contain little or no fat, and except for the sweet type, are low in calories. Fruit pickles are usually prepared from whole fruits and simmered in a spicy, sweet-sour syrup. They should be bright in color, of uniform size, and tender and firm without being watery. Pickle products are classified on the basis of ingredients used and the method of preparation. Brined pickles, also called fermented pickles, go through a curing process of about 3 weeks. Fresh-pack pickles, also called quick-process pickles, are brined for several hours or overnight and then drained and combined with boiling hot vinegar, spices, and other seasonings. These are quick and easy to prepare. The products have a tart, pungent flavor. Pickled fruits include sweet pickled fruits and salt pickled fruits. Some popular pickled fruits are apple, mango, papaya, peach, pear, grape, and plum (sweet pickles), as well as olive (salt pickle).

Chutney is a mixture of fruits, sugar, salt, herb, and spices. The product varies from hot and spicy to mild and tangy. Though chutney is usually associated with Middle Eastern or Indian food, it is served as a condiment complementing a wide variety of foods. It is important to select firm fruits and to use pure granulated salt, white sugar, fresh spices for best flavor, and a high-grade cider or white distilled vinegar of 4–6% acidity. Fruits may be slightly underripe. Uniodized table salt can be used, but the materials added to the salt to prevent caking may make the brine cloudy. Iodized table salt may darken pickles. Cider vinegar, with its mellow acid taste, gives a nice blending of flavors but may darken fruits. White distilled vinegar has a sharp, pungent, acetic acid taste and is desirable when a light color is important, as with pickled pears.

4.1 MANGO PICKLES

Mango pickles and chutneys are the traditional export items of India, Pakistan, and Bangladesh. The export of mango pickles and chutneys is directed mainly to Canada, the United States, Denmark, Germany, and the U.K. Mango pickles are prepared from unripe fruits and chutneys from half-ripe fruits.

There are numerous recipe combinations of mango pickles because of regional variations in taste. Peeled or unpeeled raw mango slices are mixed with 10–20% powdered salt to extract some moisture from the slices for 2 months. A partially ground mixture of spices, including garlic and onions, is added to the drained slices. The spices are coriander, fenugreek seeds, nigella, fennel, cumin seeds, powdered tumeric, and red chilies. The whole mixture is poured into a clean jar and covered with mustard oil. The addition of a small amount of sugar along with the spices produces a tasty blend. If the variety is not sour enough, a small amount of acetic acid may be added.

4.1.1 Salted Mango Pickles

Pickling in salt brines is essential for the proper development of flavor and texture of the mango slices. The brining process requires a minimum of 3–6 months. The methods of salting mango slices vary slightly in different parts of India but may be classified as intermittent-salting method or single-salting method. In the intermittent-salting method, the slices are mixed with 5% salt and stored in wooden tubs for 1–2 days. As water is gradually extracted from the slices, the resulting saline solution is drained off, and the slices are recharged with 5% fresh salt. The process is repeated until a total of 15% salt has been added. The salted slices are then packed tightly in wooden casks. In the single-salting method, the slices are interlayered with 15–20% salt in wooden barrels, tightly packed to the top, and pressed. The supernatant layer of brine, containing the leached principles from the fruits, undergoes spontaneous fermentation by a halophilic yeast, which ultimately results in the formation of mycoderma at the surface. After a 7- to 8-day storage, the saline solution is drained off, and the space made available is replaced with cured mango slices. The barrels are topped off with about 2 kg of salt and then tightly sealed.

4.1.2 Sweet Mango Pickles

Sweet mango pickles can be prepared from ripe mangoes according to the following recipe: mango slices, 3.6 kg; sugar, 1.8 kg; vinegar, 150 ml; water, 1 cup; whole cloves, 2 tablespoonsful; and cinnamon, 2 sticks. The syrup is first prepared with sugar, water, and vinegar, and then slightly pounded spices are added to the syrup. The fruit slices are then added, and the whole mass is boiled until the fruit becomes clear.

4.2 OLIVE PICKLES

Most olive trees are grown in Mediterranean countries. Turkey is a leading producer and processor. The pickled olives have a bright yellow color, crisp texture, salty taste, and pleasing flavor after fermentation.

For pickling, the harvest time would be from April until August. The olives are placed in a large container, covered with 1–1.5% lye solution, and kept at 25–28°C for 4–8 h. They are subsequently washed with water for 24–36 h and again covered with brine in a fermentation tank. Upon attaining equilibrium, the product contains 6–8% NaCl. The brine should be fortified with sufficient salt to make an 8% salt concentration and acidified with 0.5% lactic acid or 0.25% acetic acid (Luh and Mahecha, 1971; Luh and Woodroof, 1988). Spices are added to the fermented olives. The fruits are graded according to shape, size, and color. A desirable flavor is developed during curing without being excessively sour, salty, or spicy. The skin of the pickled olive is tender and firm but not hard, rubbery, or shriveled. The inside is tender and firm, not soft or mushy. The well-pickled olives are packed by weight into No. 1 or No. 10 cans. A pH range of 7.0–7.5 is most desirable for color retention during canning and subsequent storage (Luh and Martin, 1996).

4.3 DATE CHUTNEY

Date fruits are washed in water and left to air-dry. The fruits are separated from the stalks. For the preparation of chutney, the cleaned date fruits are peeled, pitted, and cut into slices. The mustard oil is heated in a frying pan. The spices are mixed with the sliced date fruits and fried lightly in mustard oil with constant stirring to make the fruit soft (25–30 min). After adding vinegar, heating is continued on a medium flame for another 10 min with constant stirring. Sugar is mixed into the contents thoroughly, and the chutney is cooked further to a finishing temperature of 105–106°C. The product is ready for use (Yousif et al., 1985; Al-Hooti et al., 1997). A typical date chutney recipe is as follows: date fruits, 1000 g; sugar, 600 g; salt, 100 g; vinegar, 200 ml; mustard oil, 50 ml; black pepper, 4 g; cloves, 1 g; and red chili powder, 9 g (Al-Hooti et al., 1997).

4.4 MANGO CHUTNEY

Unripe or half-ripe green mangoes are used to make chutney, which is a valuable mango product in the world market. The manufacture of mango chutney is essentially a two-stage process: the first stage is the pickling of green mango slices, and the second is the production of the spiced and flavored final product known as mango chutney. Most mango chutney is manufactured in India, Pakistan, and Bangladesh. The preparation comes in a multitude of forms and varieties, which are influenced by consumers' preferences and regional customs. Six principal types of chutney are as follows:

(1) Sweet chutney: This chutney contains higher amounts of sugar than the others.
(2) Hot chutney: This chutney has less sugar and more chili peppers.
(3) Major gray chutney: This chutney is sweet, with small amounts of chili peppers and more garlic.
(4) Colonel skinner chutney: This sweet chutney contains raisins, with chili peppers added in the form of chips or rings.
(5) Kashmir chutney: In this sweet chutney, the mango slices are pulped, resulting in a thick paste.
(6) Bengal club chutney: This sweet chutney contains mustard, and the mango slices are cut into cubes.

A typical sweet mango chutney recipe and processing method are as follows: mango slices, 4.54 kg; sugar, 4.54 kg; salt, 280 g; cardamom, 47 g; cinnamon, 47 g; cumin, 47 g; garlic, 28 g; onion, 140 g; vinegar, 600 ml; red chilies, 70 g; and green ginger, 560 g. Under-ripe, but fully developed, fruits are peeled, sliced or graded, and mixed with salt. The slices are taken out of the saline after a few hours and cooked with an equal quantity of sugar to a thick consistency. Green mangoes are peeled and simmered whole in a boiling sugar solution until they are almost clear. The sugar solution is prepared by mixing sugar, water, and lime juice in a ratio of 1 : 1 : 0.5. Spices are added to this solution according to taste. Raisins, almonds, dates, and vinegar are also added to this mass and then cooked to acquire a thicker consistency. The product thus obtained is known as mango chutney.

5. CANNED FRUITS

Canning simply means sealing the fruits in airtight tins so that they may be preserved for a long time without spoilage. Fruits are canned whole, in halves, or in quarters, depending on the fruit size. Some products are shown in Figure 10.5.

5.1 CANNED CITRUS FRUITS

Citrus fruits are believed to be native to Southeast Asia. Their use and cultivation in China is recorded in 2200 B.C., and the first Chinese book on citrus appeared in A.D. 1178. Mandarin oranges are the main sweet oranges con-

Figure 10.5 Canned bananas, jackfruits, longans, lychees, and pineapples.

sumed in Asia. Mandarin oranges are canned in the form of segments in China, India, and Japan on a large commercial scale. The raw material is a popular fruit known as 'Mikan' or 'Tunkan.' The products have a golden yellow color and sweet flavor and can be packed in No. 2 or No. 10 cans with light syrup (14–18° Brix), heavy syrup (18–22°Brix), or extra heavy syrup (23–35°Brix).

5.2 CANNED DURIANS

Durians are the large fruits covered with hard, hexagonal, stubby spines. It is a heavy fruit reaching the size of a honey melon. The ripened fruits are light yellow. Durian is a delicious tropical fruit and well known throughout Southeast Asia. Thailand, Malaysia, South Vietnam, and the southern Philippines are important producers of durians. Durian flesh is canned in syrup for export in Thailand.

Durian is a good source of iron, B vitamins, and ascorbic acid. The thick, pudding-like texture of the aril is caused by gums, pectin, and hemicellulose (Martin, 1980). The flesh of durian is said to serve as a vermifuge. It is also widely believed that durian flesh can act as an aphrodisiac in Thailand (Morton, 1987).

5.3 CANNED GUAVAS

Guavas are native to Central America and were adopted as a crop in Asian countries. The major producers of guavas in Asia are Malaysia, Indonesia, and India. The canning of guava was initiated in India in the early 1940s. In terms of the quantity of production, canned guavas are probably more important than any other processed commodities of guava, except the juice products. They are frequently seen on the markets in India, Pakistan, and Indonesia. Whole guava or slices are packed in No. 2 cans with 45°Brix syrup containing 0.25% citric acid at a pH of 3.8. The products made from immature fruits are commonly used to halt gastroenteritis, diarrhea, and dysentery throughout the tropical area.

5.4 CANNED LOQUATS

The loquats, also called Japanese plums, probably originated in China and are adapted to a subtropical or mild–temperate climate. Today, China, India, Israel, and Japan are leading producers of loquats. Japan, Taiwan, and Israel have exported canned loquats in syrup to the world market. Canned loquats are consumed largely as dessert fruits. Canned products retain a gold color and fresh flavor. The fruits are also used in gelatin desserts and as pie filling or are chopped and cooked as a sauce. The final pH of the canned product is less than 4.0. The loquat products are traditionally considered to act as a sedative and are taken to halt vomiting and quench thirst.

5.5 CANNED LYCHEES

The excellent flavor and aroma of this fruit is largely retained after canning. There are well developed substantial canning industries in China, Taiwan, and smaller industries in Thailand and India. Peeled, seeded lychees canned in sugar syrup have been exported from China, Taiwan, and India for many years. Canned lychees are the most important processed product of lychees. Firm, ripe lychees are selected, washed, peeled, and pitted by machine or by hand, filled into No. 2 cans, processed for a maximum of 10 min in boiling water, and then cooled immediately (Luh et al., 1986; Yu and Gao, 1994). Excessive heat processing of lychees may cause phenols, leucoanthocyanins, and flavanols to leach into the syrup, causing pink discoloration in the canned products. Browning or pink discoloration of the fresh fruit is prevented by the addition of 4% tartaric acid or by using 30°Brix syrup containing 0.1–0.15% citric acid to achieve a pH of 4.5 (Chakraborty et al., 1974; Luh et al., 1986; Hwang and Cheng, 1986). The canned fruits can be made into jelly, sherbet, or ice cream. A highly flavored squash can also be prepared from lychee fruits.

5.6 CANNED MANGO SLICES

Mangoes are commercially canned in India, Burma, China, Taiwan, Pakistan, the Philippines, and Thailand. Canned mango slices are exported by Thailand, the Philippines, and India. 'Dusehri' and 'Bauganapalli' are the major varieties exported as canned slices from India to Europe.

Mangoes are usually canned in the form of slices. Only ripe, but firm, mangoes of the appropriate variety are used for this purpose. Some Indian varieties are good for canning. Two well-known methods for canning mangoes are the open-kettle method and the cold-pack method. With the open-kettle method, mangoes are washed and peeled, cut, and packed in standard cans. Then a boiling sugar solution (25°Brix) is added to fill the containers, which are then sealed, sterilized for about 20 min, and cooled immediately afterwards.

In the cold-pack method, the slices are packed uncooked and the boiling sugar solution is added afterwards to fill the cans. The cans are then partially sealed and processed in a pressure cooker for 15 min at 2.2 kg of pressure. After processing, they are finally sealed and stored after cooling. The addition of calcium salts to the cover syrup inhibit the degradation of protopectin to a considerable extent during the prolonged storage of canned mango slices (Luh et al., 1986).

5.7 CANNED PEACHES

As an important industry, a larger quantity of peaches is canned in China. Its delicate flavor, which persists after canning; firm texture; attractive appearance; and moderate price have combined to give the peach its present popularity. Canned peaches are of a uniform large size; of symmetrical shape; of a sharp yellow color; of close tender fiber, not coarse or ragged; and of good cooking quality. Peaches are usually classified either as clingstone or freestone types. Freestone fruits can be easily separated from the pit. But clingstone peaches are firmer in texture than freestone fruits and hold their shapes well after canning. Yellow-fleshed cultivars are most common and preferred for canning. After being peeled, peaches are cut in halves or quarters, washed in NaCl solution (2%), precooked for 5–10 min, filled into No. 10 cans, sealed, sterilized, and cooled. Peaches are canned in a sugar syrup of 40–42°Brix with 0.1% citric acid (Luh et al., 1986; Chow, 1991; Zhang, 1994).

5.8 CANNED PINEAPPLES

Most of the Asian pineapples are canned. Canned pineapples are important value-added products in the world market. The highest grade is the skinned, cored fruit sliced crosswise and packed in syrup. In the 16th century, India, Malaysia, and the Philippines began to produce pineapples. Pineapple canning was started in Singapore in about 1892. The chief sources of the world's canned pineapple and pineapple juice are Bangladesh, India, Malaysia, Taiwan, Thailand, and the Philippines (Table 10.6). Thailand is the leading producer and exporter in the world canned pineapple product market. 'Cayenne,' the major canning cultivar, is grown exclusively in China, Thailand, and the Philippines. It is increasing in popularity in other tropical areas. The 'Singapore' cultivar is grown in Malaysia.

In general, the fruits are sorted by size and shape. They are free from major blemishes. The ends, shell, and core are removed, yielding a cylinder of flesh, which is cut into the usual slices, chunks, and smaller pieces. Flesh remaining

TABLE 10.6. Export of Canned Pineapples from Asia (MT) (1992–1994).

	1992	1993	1994
Asia	875220	906482	930256
China	15052	10413	11997
Hong Kong	3803	2143	1779
Indonesia	71530	99773	99121
Malaysia	50191	52218	51000
Philippines	197542	193582	215227
Singapore	41820	40635	38663
Thailand	495239	507563	512266

Source: FAO, 1997.

on the trimmed shell and ends, as well as fragments from cutting operations, are used for crushed pineapple and juice. Processing steps are as follows: blanch for 4 min, remove peel about 3 cm thick, remove eyes, cut into rings, wash thoroughly in cold water, pack in cans, add syrup, seal, sterilize, and cool (Luh et al., 1986; Chow, 1991; Zhang, 1994).

5.9 OTHER CANNED FRUIT PRODUCTS

Other canned subtropical fruits such as apricots, cherries, kiwifruits, pears, and plums, as well as some canned tropical fruits such as canned jackfruit, mango, and papaya, are very popular in Asian areas.

(1) Canned apples: Firm flesh and high quality of fresh apples are desired for canning. Apple slices are packed in No. 10 cans with a light or heavy syrup or apple juice. Canned apples have a light yellow color and sweet flavor.
(2) Canned apricots: Canned apricot has a bright yellow color and the characteristic tart-sweet flavor of fresh apricots used for baby food formulations. Apricot slices are packed in No. 10 cans with light or heavy syrup or apricot juice.
(3) Canned cherries: Sour cherries are canned for pie fillings. Cherries whole or in pieces, are packed in No. 10 cans with heavy (up to 45°Brix) to light (less than 16°Brix) syrup.
(4) Canned jackfruits: Jackfruits are produced in the Philippines, Malaysia, Thailand, Cambodia, Laos, and Vietnam. Crisp types of jackfruits are preferred for canning. Products are more attractive than the fresh pulp and are called "vegetable meat." The Chinese consider jackfruit pulp a nutritious tonic.
(5) Canned kiwifruits: Kiwifruits are native to the province of Hubei, in the Yangtze Valley of northern China. Kiwifruit slices are packed in No. 10 cans with light (18°Brix) syrup. The product has a fresh green color.
(6) Canned mangoes: Mango pieces are packed in No. 2 or No. 10 cans with 18–22° Brix syrup containing 0.25% citric acid to adjust the pH of the product to less than 4.1. The product has a yellow color and a high sweet flavor.
(7) Canned papayas: Local names for papayas are *kapaya, papias,* or *tapayas.* Canned papaya chunks or slices are packed in No. 2 cans with 40°Brix syrup containing 0.75% citric acid at a pH of 3.8. It is a popular ingredient in fruit salad preparation.
(8) Canned pears: Canned pears have a uniform shape, fine texture, and excellent flavor. Pear slices are packed in No. 10 cans with light or heavy syrup or pear juice.

(9) Canned plums: Plum canning is an important industry in several countries such as Russia. The large, sweet varieties of white plum, such as 'Green Gage' and 'Yellow Egg,' are in the greatest demand for canning purposes—'Green Gage' and 'Yellow Egg' cultivars for light color canned products and 'Lombard' cultivar for the dark color product. Plum slices are packed in No. 10 or No. 2 cans with 25°Brix syrup.

6. FRUIT PUREES

Fruit puree is generally macerated and mechanically screened in a pulper, which converts the pulp into a finer and smoother puree. Puree is an intermediate product that can be further processed into different products. Purees of apples, apricots, bananas, grapes, mangoes, oranges, peaches, pears, and plums are the common products in the world market. The pulp is then concentrated, deaerated, heated to achieve microbiological and enzymatic stabilization, and cooled prior to storage in sterilized tanks.

6.1 BANANA PUREE

Banana puree is by far the most important processed product from the pulp of ripe fruits. The puree has a creamy white to golden yellow color, free from musty or off-flavors. Banana puree is an important infant food. Puree canned in drums by an aseptic canning process is a new product for the baking and ice-cream industry. The puree can be successfully canned by the addition of ascorbic acid to prevent discoloration and used as an ingredient in dairy desserts, bakery items, drinks, and processed foods and sauces and as a part of special diets in hospitals and nursing homes. In canning banana puree it is necessary to lower the pH to approximately 4.2 with citric acid. Sugar is added to balance the sugar/acid ratio.

6.2 MANGO PUREE

Mango puree is the most common semiproduct of this fruit. Mango puree, prepared from ripe fruits, is particularly relished for its succulence and exotic flavor. It can be used for making jams, jellies, beverages, and various dairy and bakery products that contain mango as an ingredient. As a commercial product, mango puree in a fresh, processed, or chemically preserved form appears promising because of the ease of handling and the low cost of production. The puree requires a heat treatment sufficient for the inactivation of enzymes. A plate heat exchanger is usually used to heat the product to 108–112°C for 2 min, and then the puree is cooled rapidly. Sucrose and high-conversion corn syrup are acceptable sweeteners in the preparation of mango puree (Wu and Sheu, 1996). The puree can be stored under suitable conditions (by freezing or chemically preserving) for 6–8 months.

6.3 PEACH PUREE

Fruits are washed, trimmed, and cooked, and the pulp is passed through a continuous rotary unit with perforation, followed by the addition of ascorbic acid (0.14%). Clingstone peaches are preferable for the preparation of baby food because of the yellow color of the puree, better storage ability, nonmelting flesh, and thicker consistency. The peach pulp is blended with sugar syrup, sterilized at 110°C, and deaerated. It can either be canned or preserved in jars. The final peach baby food has soluble solids 21–22%, a pH of 3.0–4.1, and 0.35–0.45% acidity (as citric acid) (Yu and Gao, 1994; Zhang, 1994).

6.4 SOME OTHER FRUIT PUREE

Some subtropical fruit puree such as kiwifruit puree, plum puree, and some tropical fruit purees such as guava and papaya puree are available on the market.

(1) Guava puree: Guava puree, also known as guava pulp, is a relatively liquid product. Guava puree is an excellent source of ascorbic acid, niacin, thiamin, riboflavin, calcium, iron, phosphorus, and dietary fiber. The ascorbic acid contents in guavas range from 100–1000 mg/100 g fruits. It is most commonly used for the preparation of nectars, various juice drink blends, ice-cream toppings, and jams and jellies.
(2) Kiwifruit puree: Kiwifruit puree is used as a fruit topping and in yoghurts. The product has a fresh green color and high natural flavor.
(3) Papaya puree: Papaya puree is a liquid product with a light yellow color and sweet flavor, which is prepared by the maceration of fresh papaya flesh into a semifluid-like product. Many papaya products, such as juices, nectars, jams, jellies, syrups, toppings, and dried fruit rolls or leathers can be made from papaya puree.
(4) Plum puree: Plum puree is used in ice-cream mixes, confectionery products, and meat sauces. The product has a light yellow color and tart-sweet flavor.

7. FRUIT JAMS

Jam is a combination of crushed, ground, or chopped fruits and sugar that is cooked to a smooth consistency, thick enough to be spread. Fresh, frozen, or canned fruits may be used in preparing jam. The fruits are cut into small, thin slices and cooked with a little water to soften the slices. The cooked slices are then mashed and mixed in a ratio of 60% sugar to 55% mashed fruit slices. The exact levels depend on the pectin content. Citric acid at the rate of 5 g/kg of mashed fruit slices is added to improve the sugar/acid blend of the product, to effect proper inversion of the sugar, and to avoid undesirable crystallization in the final product during storage. The mixture is then cooked with continuous stirring until it attains a thick consistency corresponding to 65–68°Brix. The jam is filled hot into clean, sterilized dry containers, sealed airtight, cooled, and stored.

Many subtropical fruits such as apples, apricots, plums, and strawberries, as well as some tropical fruits such as bananas, guavas, mangoes, papayas, and santol fruits, are commercially processed into jams.

(1) Apple jam: Apple jam has a sweet flavor and light yellow color. Apple jam is very popular in Russia, Korea, and the northern part of China.

(2) Banana jam: Banana is an important component of mixed-fruit jam in the Indian fruit-processing industry. The product has a light yellow color. *Panchamrutham,* as well-known as *Prasadam* in South Indian temples, is made from Virupakshi banana and has a property similar to jam.

(3) Guava jam: Guava jam was once a by-product of the guava jelly industry. A standard jam may be made by combining 45 parts puree with 55 parts sugar, but a better-flavored, fancy-quality jam will result when 50 parts puree are mixed with 50 parts sugar.

(4) Mango jam: Mango jam has a light yellow color and high sweet flavor. A typical formulation is mango puree, 45 kg; sugar, 55 kg; pectin, 0.4 kg; and citric acid, 0.4–0.5 kg (sufficient to adjust the pH to 3.2).

(5) Papaya jam: Papaya jam is very popular in the Philippines, most of which is a mixed-fruit product. To prepare papaya jam, the fruit/sugar ratio is 1:1, For papaya-orange jam, the papaya/orange/sugar ratio is 4:4:3. For papaya-guanabano jam, the papaya/guanabano/sugar ratio is 1:1:1.5. For papaya-pineapple jam, the papaya/pineapple/sugar ratio is 1:1:1.5. For papaya-tamarind jam, the papaya/tamarind/sugar ratio is 1 : 0.5 : 1.5.

(6) Santol fruit jam: Santol fruits are produced in Cambodia, India, Malaysia, South Laos, and the Philippines. In India and the Philippines, with the seeds removed, santol fruits are made into jam or jelly. In the Philippines, santol fruits are peeled chemically by dipping in hot water for 2 min or more and then into a lye solution at about 90°C for about 4 min and washed in cool water to remove the outer skin. The Philippines exported santol jam to Europe and North America. The preserved pulp is used medicinally as an astringent, as is the quince in Europe (Morton, 1987).

(7) Strawberry jam: Strawberry jam is very popular among consumers. Strawberry jam has a bright red natural color and a garden-fresh, sweet, and tart flavor.

8. FRUIT JELLIES

Fruit jelly is one of the oldest and most popular fruit by-products. The jelly is beautifully bright and transparent. A good jelly gelatinizes on cooling and is firm enough to hold the shape of the container when turned out of the container. It must be soft enough to quiver upon shaking but must not flow. It must be clear and transparent and should retain the flavor of the fruits.

Three substances are essential to the preparation of a normal fruit jelly: pectin, acid, and sugar. Fruit gives jelly products its characteristic flavor and furnishes at least part of the pectin and acid required for successful gels. Flavorful varieties of fruits are often used for jelly products because the fruit flavor is diluted by the large proportion of sugar necessary for proper consistency and good keeping quality.

Jellies are made from a mixture of not less than 45 parts by weight of fruit juice ingredients and 55 parts by weight of sugar ingredients. Fruit jelly differs from fruit jam in that fruit jelly is made from fruit juice and contains little or no insoluble solids. Therefore, the product is clear and transparent. The mixture is concentrated by heating until the soluble solids content is not less than 65%. To achieve the proper pectin/sugar/acid ratio, the juices of underripe or overripe fruits may be blended, and the juice of different fruits may be added. Optional sugar ingredients may be added, which are sucrose, inverted sugar syrup, corn syrup, and honey. Spices, sodium benzoate, benzoic acid, mint flavoring, and food grade green coloring may be used, provided their presence is suitably declared on the label. The procedure of making jellies is very similar to that used for making jams.

9. FRUIT JUICES AND CONCENTRATES

Large quantities of citrus juices, apple juice, and grape juice are processed throughout the world. The remarkable rise of the fruit juice industry makes fruit juices quickly become popular. Juices appeal to young and old alike because of their fine fresh flavor and attractive appearance. Most of the juices are inexpensive, often costing less than equivalent quantities of fresh squeezed juices. Physicians generally recommend feeding fruit juices to infants and children and also to adults. Changing occupations and food habits have had much to do with the rapid adoption of fruit juices in the diet. Fruit juices are believed to be nonfattening and have been in great demand by persons trying to reduce weight and those desiring to keep youthful figures.

Fruit juices must be prepared from sound fruit only. Fruit juices are most palatable when first expressed from the fresh fruit, and any treatment applied to preserve or clarify them results in some injury to quality. Preservation must be accomplished with as little injury as possible to the fresh flavor and other desirable qualities of the products. Juices are extracted, pasteurized, and then

Figure 10.6 Coconut milk and guava, mango, and orange juice.

packaged in barrels, glass containers, or plastic containers. Formerly, it was customary to clarify fruit juices either by filtration or by fining before bottling. At present, most of the preserved juices on the market are cloudy or pulpy or clarified with a pectin enzyme. The citrus, apricot, peach, pineapple, and pear juices are most popular in the cloudy condition. Some commercial products of coconut milk and guava, mango, and orange juices are shown in Figure 10.6.

Juice concentrates, nectars, pulp juice, and fruit juice blends are of great commercial importance. Many fruit juices are either too acidic or too strongly flavored to be pleasant beverages without diluting or blending or both. Often these strong, tart juices are delicious after dilution with syrup or bland juice (Luh, 1980). Some juices do not have enough flavor, and they are greatly improved if the entire fruit, with the exception of the skin and seeds, is converted into a smooth pulpy beverage (Luh, 1991). The term *fruit nectars* is used by the industry to designate pulpy fruit juices blended with sugar syrup and citric acid to produce a ready-to-drink beverage. These beverages, although they resemble fruit juices in flavor, can't be called fruit juices because of the presence of added water, sugar, and acid (Luh and El-Tinay, 1993). The concentration of fruit juices (a thick viscous liquid or powder) for subsequent reconstitution into fruit juices, preparation of carbonated beverages, or for jellies has been produced on an industrial scale.

9.1 APPLE JUICE

Apple juice is a popular drink and one of the important breakfast items available to consumers. It ranks second to orange juice in fruit juice consumption in the world. Most of this juice is consumed fresh, directly after pressing, or from barrels in which it is preserved with benzoate of soda. Apple juice contains a considerable proportion of the soluble components of the original apples, such as sugars, acids, and other carbohydrates. Apple juice possesses a rich apple flavor and should be tart. Malic acid is the predominant acid in apple juice.

Apples used for juice processing should be fresh and firm. Apples should not be allowed to become overripe and mealy in texture before crushing. The preparation of clarified apple juice involves grading and pressing the apples, clarification with pectin enzymes, filtration, and packaging. Traditional packaging involves pasteurization at 80–88°C, filling, and sealing the juice in glass containers or metal cans. Several distinct forms of apple juice available on the market include clarified apple juice, natural apple juice, pulpy apple juice, and apple juice blends in combination with other juices/extracts. The characteristics of apple juice are considered to be close to the juice that comes directly from the fruits. Commercially, this is accomplished through the addition of ascorbic acid or through heating the pressed juice to flocculate unstable compounds. Ascorbic acid helps preserve the light color of the juice by reversing the oxidation of juice constituents. The juice is then immediately pasteurized and bottled (Chow, 1991; Yu and Gao, 1994). Apple cider is considered naturally fermented apple juice. Some new products are developed such as shelf-stable juice or sweet apple cider.

9.2 CARAMBOLA JUICE

Carambola originated in Sri Lanka. The major production areas are in East Asia, including Indonesia, Malaysia, Sri Lanka, Taiwan, the southern part of China (Fujian, Guangdong, and Hainan), and Vietnam. In Sri Lanka and India it is called *kamaranga* or *kamruk.* In Vietnam, it is called *khe* or *khe ta.* In China, it is called *yang tao.* There are two distinct cultivars: sweet and sour cultivars. The sour cultivar is rich in flavor, with more oxalic acid. The sweet cultivar has a mild flavor, is rather bland, and has less oxalic acid. The sweet cultivar is for fresh consumption and juice processing; the sour cultivar is processed into jam, jelly, canned fruits, sweetened nectar, or other preserves. Juice products are by far the most important processed commodities of carambola fruit. Carambola juice is served as a cooling beverage. The total consumption of carambola juice in Taiwan in 1992 was over 17,000 metric tons and rated second among all domestic fruit juice products, second only to guava nectar (Wu and Sheu, 1996). Yellow, but not fully ripe, fruits are selected for juice processing. Juice extraction is by crushing and pressing fruits.

Fermented carambola juice is a traditional health drink in China and India. Carambola juice is good for soothing some uncomfortable body conditions, such as to quench thirst, to increase salivary secretion, and to allay fever. Carambola is recommended as a diuretic in kidney and bladder complaints and is believed to have a beneficial effect in the treatment of eczema (Wu et al., 1993).

9.3 COCONUT MILK

Coconut is the most suitable for use with the widest variety of foods. About 84% of the world coconut production is concentrated in Asia. Coconut palms are mainly distributed over the tropical coasts of the Philippines, Indonesia, Malaysia, Sri Lanka, and India. The Philippines contribute about 60% of the world's coconut production (Woodroof, 1979).

Seventy-two percent of the coconut is edible. Fresh coconut meal (kernel) is rich in fat and carbohydrates and contains a moderate amount of proteins. Coconut milk is very finely ground coconut meat. The process of coconut milk production is essentially a wet-milling process. The kernel is washed, sterilized, and scraped into a fine meal. The meal is squeezed in a special press to obtain milk. It is a white emulsion and has poor stability. Coconut milk is adjusted to pH 4–6 to get a good stable condition. Bottled or canned coconut milk is manufactured on a large scale in the Philippines, Thailand, and Malaysia.

9.4 GRAPE JUICE

Grape juice is an important commercial product. In many developed countries, colored, as well as white, grapes are used for making juice. The composition of grape juice is similar to that of whole grapes, except that crude fiber and oil, which are primarily present in the seed, are removed. The maturity of the fresh grapes is an important factor affecting composition and quality of the grape juice. Usually, soluble solids at 16°Brix or higher in grapes are considered ripe for juice processing. Grapes should be picked when slightly underripe in order that the juice may not be too low in acidity or too rich in sugar. In the case of colored varieties, it is necessary to heat the crushed berries for 10–15 min at 60–63°C to extract the coloring matter. White grapes are not heated. Juice is extracted from the crushed grapes by means of a basket press. The extracted juice is filtered through cloth and is bottled. The bottled juice can be preserved by pasteurization or by the addition of sodium benzoate. Some color will be extracted from many varieties of grapes. Even the

'Concord' variety will yield a pink-colored juice. The color of grape juice is largely caused by anthocyanin pigments. The soluble solid content is 15–18°Brix, and acidity (marc acid) is 0.4–0.6%. The principal acids of grape juice are tartaric and malic acids, although small amounts of citric, succinic, and lactic acids have been observed (Chow, 1991; Zhang, 1994; Patil et al., 1995).

9.5 GUAVA JUICE

Guava juice and nectar are among the numerous popular canned or bottled fruit beverages in tropical areas. A clarified guava juice can be used in the manufacture of clear guava nectar, clear guava jelly, or in various juice blend drinks. It has a light amber, yellow, or light pink color, depending on the cultivar, since most of the pigments in the guava remain with the solid material. In India, clear guava juice is prepared from the white-fleshed 'Allahabad' variety according to the following steps: washed and trimmed fruits are blanched for 3–5 min to inactivate the oxidative enzymes prior to pulping; after blanching, the fruit is passed once through a screw press fitted with a coarse sieve of 20–30 mesh; and the residue is passed through again for maximum recovery of the pulp.

Two types of cloudy guava juice are prepared in India: one product contains 55–60% pulp, and the other consists of 85% pulp. The soluble solid content of the guava pulp is around 13–14°Brix, and the acidity is 0.20–0.25%. The pulp, citric acid, and 70°Brix syrup are heated and blended, yielding a final 15°Brix and 0.20–0.30% acidity. Ascorbic acid is added at the rate of 0.01%. The prepared product is poured hot into bottles, capped, and processed in boiling water for 25 min, followed by cooling to ambient temperature. Guava juice is further processed and utilized in the form of concentrates, beverages, jellies, powders, and other products (Adsule and Kadam, 1995; Wu and Sheu, 1996).

Guava juice can be concentrated to five times its original content of total soluble solids. But the color of the juice changes to brown because of browning. Guava juice concentrate has been found to be suitable for drying into guava juice powder and ready-to-serve beverages.

9.6 KIWIFRUIT JUICE AND CONCENTRATE

Kiwifruit juice of optimal flavor is produced from ripe fruits of sound quality. After ripening, whole fruits are washed with water and peeled and then coarsely milled in a hammermill fitted with a 125-mm screen. Pressing is commonly undertaken using a rack-and-cloth press. The pulp of the puree is one of the processed kiwifruit products finding use in the international market. Although heat can be used to inactivate microorganisms in the production of kiwifruit pulp, undesirable changes also take place in the color and flavor of the product. The kiwifruit pulp is frozen immediately after manufacture and stored at −18°C. Frozen concentrated kiwifruit juice is a relatively new commercial product. Because of the high acidity of kiwifruit juice, it is desirable to dilute the juice with a sugar solution or with a low-acid juice (e.g., apple juice) to obtain a product with consumer appeal. The blending of kiwifruit juice with a bland clear apple juice is the one means of reducing the high acidity of kiwifruit juice while still maintaining a 100% juice product. Consumers have remarked that the lack of a distinctive flavor in the juice may be a factor limiting its universal acceptability. An acceptable kiwifruit drink is obtained by diluting the juice with an equal volume of sucrose solution (17°Brix) such that a final solution solids concentration of 15°Brix is obtained. The titratable acidity (as citric acid) of the ameliorated juice should be 0.75% w/v. As with other fruit juice, a sparkling kiwifruit juice can be made by dissolving carbon dioxide gas in the juice under pressure (Chow, 1991; Zhang, 1994). In China, the kiwi fruit juice is regarded as valuable for expelling stones, such as kidney stones, from the body.

9.7 ORANGE JUICE AND CONCENTRATE

Orange is one of the most widely favored fruits. Today, major producers in Asia are China, India, and Japan. Orange juice has a deep orange color. Sound and mature oranges of the highest possible juice quality are desired for the production of orange juice. The color, flavor, juice yield, and soluble solid content increases with the maturity of the orange. The best-quality orange juice is produced when the Brix/acid ratio is around 15 in the fruits.

The juice is extracted, screened, deaerated, deoiled, and pasteurized. The pH of orange juice varies from 3.3 to 3.8. 'Coorg' and 'Nagpur' orange juice is orange-yellow in color with good flavor, but it has a tendency to develop bitterness. The rate of deterioration in flavor increases with rising temperature. Orange juice should always be kept as cool as possible (Kale and Adsule, 1995).

Orange juice concentrate is prepared from either freshly extracted and pasteurized single-strength juice or from a storage and pasteurized single-strength juice. Dehydrated orange juice has been produced for several years. The product is obtained by adding a stabilizer such as alginate, sucrose, or corn syrup to the juice to assist the drying. The main technologies are spraying and drum drying. The final powder has less than 0.6% moisture and maintains its quality when stored at room temperature.

Orange products are traditionally taken to allay fever and catarrh. The roasted pulp is prepared as a poultice for skin diseases.

9.8 PASSION FRUIT JUICE

Passion fruit juice is a newcomer to the market. India, Sri Lanka, Indonesia, Thailand, Malaysia, Taiwan, and the Philippines are important sources of passion fruit products in the world market. Passion fruit juice, because of its unique intense flavor, high acidity, and yellow/orange pulp, has been described as a natural concentrate. Passion fruit juice is a highly palatable beverage when sweetened and diluted. It blends very well with other fruit juices.

Compressing roller extractors are the most popular for the extraction of passion fruit juice. Juice is preserved by heat processing or by freezing. Passion fruit concentrates are prepared by using a falling-film evaporator and a centrifugal evaporator. Concentrates are preserved through freezing or through the addition of chemical preservatives such as sodium benzoate and potassium sorbate. Dehydrated passion fruit powders can be prepared by freeze-drying or vacuum-puff freeze-drying.

The juice can be sweetened and then diluted with water or other juices (especially orange or pineapple juices) to make cold drinks. Passion fruit juice can be concentrated to a syrup, which is used in making sauce, gelatin desserts, candy, ice cream, sherbet, cake icing, cake filling, meringue or chiffon pie, cold fruit soup, and cocktails. The frozen juice can be kept for 1 year and is a very appealing product. The juice can also be dehydrated in a freeze-dryer or vacuum-dryer.

The juice is taken as a digestive stimulant and used in treatment for gastric cancer. There is currently a revival of interest in the pharmaceutical industry in the use of glycosides as sedatives or tranquilizers (Morton, 1987).

9.9 POMEGRANATE JUICE

The pomegranate is a subtropical fruit native to the Middle East. It has long been cultivated in the Middle East, the Mediterranean region, and other areas in Asia. The most important pomegranate growing regions are China, Afghanistan, Pakistan, Bangladesh, Iran, Iraq, India, Burma, and Saudi Arabia. There are some commercial orchards in Israel on the coastal plain and in the Jordan Valley. In some countries, such as Iran, pomegranate juice is a very popular beverage. The juice sacs are removed from the fruits and put through a basket press. Otherwise, the fruits are quartered and crushed, or the whole fruits may be pressed and the juice strained out. The juice has a pleasing flavor and attractive purplish-red color. The juice makes a delicious drink. But, because the rind and carpellary membranes contain tannin, it is difficult to extract the juice without getting more tannin into the juice. To reduce the tannin content in the juice, the fruits are shriveled before crushing, and gelatin is used to absorb the tannin in the juice. The total solids vary from

17.3% to 18.3%, and acidity (as citric acid) varies from 0.81% to 1.23%. The tannin content is about 0.17% in the juice. An attractive colored juice (purplish-red), large juicy grains, mild acid-sweet taste, and tannin content of not more than 0.25% are the qualities desired in the fruits used for the juice processing. Pomegranate cultivars such as 'Ahmar,' 'Asward,' 'Khandhar,' 'Mangulati,' 'Paper Shell,' and 'Wonderful' give juice of the desired qualities and are largely used for processing (Morton, 1987; Adsule and Patil, 1995).

The juice may be preserved by adding sodium benzoate, or it may be pasteurized for 30 min, settled for 2 days, strained, and bottled. For beverage purposes, the juice is usually sweetened. In Saudi Arabia, the juice sacs may be frozen intact, or the extracted juice may be concentrated and frozen for future use. Pomegranate juice is widely made into grenadine syrup for use in mixed drinks. In some Asian countries, it is made into a thick syrup for use as a sauce.

The juice is rich in citric acid and sodium citrate, which can be used for pharmaceutical purposes. Pomegranate juice enters into preparations for treating dyspepsia and is considered beneficial in leprosy.

9.10 OTHER TROPICAL FRUIT JUICES

Other tropical fruit juices, such as mango juice, papaya juice, and pineapple juice, are popular on the market.

(1) Mango juice: Mango juice has a red-yellow color and fresh-like flavor. Mango juice has a cooling effect and is used during hot weather in the North Indian region. Mango fruit has medicinal properties (laxative, diuretic, and fattening). It is also alleged to help cure cholera and plague.
(2) Papaya juice: Papaya juice is extracted and then pressed into nectar, a ready-to-drink beverage. Papaya juice has a deep, rich orange color and contains papain. It is also high in vitamins A and C and is considered a health food. Papaya juice concentrate is commonly sold to hospitals and health food stores in the Philippines.
(3) Pineapple juice: There is a growing demand for pineapple juice. Crushed pineapple juice, nectar, and concentrate are commercially prepared from the flesh remaining attached to the skin after the cutting and trimming of the core. All residual parts (core, skin, and fruit ends; undersize or overripe fruits) are cut into pieces, crushed, and pressed for juice. Pineapple juice as syrup is used in confections and beverages or is made into powder. Pineapple juice is traditionally taken as a diuretic and to expedite labor and also can be used to gargle in cases of sore throat and as an antidote for seasickness.

10. SUMMARY

Fruits constitute an important part of the human diet. It is one of the main food resources that humans need to ingest daily. Most fruits are consumed fresh with little preparation. Approximately half are graded, prepared, processed, packaged, and transported for year-round consumption. Types of fruit products can vary from country to country or even place to place in the same country because of environmental differences. Some fruit products are consumed directly as foods, while some are used as ingredients in confectionery, bakery, and diet foods. Some are also used in pharmaceutical products.

The main fruit products on the world market are canned fruits, cured fruits, pickled fruits, jams, jellies, fruit juices, and a variety of dried fruit products. The main processing technologies are canning, dehydration, fermentation, pickling, and preserving with sugar or salt. A combination of the above methods are often used in fruit processing. Some additives such as calcium salt, ascorbic acid, sugar, spices, and salt are used to enhance the color, flavor, texture, and nutritive value. These properties are generally recognized as the four quality factors in fruit products.

With the advance of modern industrial techniques, most traditional processing methods have shifted continuously from natural sun-drying, sugar preserving, and chemical preserving to thermal processing, concentration, and a combination of new techniques. Industrial production of fruit products must meet world market demands.

Today, consumer trends are toward convenience and high quality. This trend is best described as the "healthy-natured" food preference. High-quality fruit products usually imply fresh-like quality characteristics of flavor, texture, and appearance. The combined demands for fresh-like quality and convenience has brought about a new fruit product called minimally processed fruit products. A significant increase in demand for fruit juice is caused by the overall increase in "natural fruit" juice consumption as an alternative to the traditional caffeine-containing beverages. Some new techniques have emerged to improve fruit product quality, such as osmotic treatment, radiation, freezing, water activity control, aseptic packaging and control/modified atmosphere packaging. Fruit products with high quality in flavor, color, texture, appearance, nutritive value, and microbiological quality will provide consumers with a healthy product containing natural sources of color, flavor, vitamins, and minerals.

11. ACKNOWLEDGMENT

The authors wish to express their appreciation to Professor Robert L. Shewfelt (Department of Food Science and Technology, The University of Georgia) for reviewing the manuscript and for his valuable suggestions.

12. REFERENCES

Adsule, R. N. and Kadam, S. S. 1995. Guava. In: *Handbook of Fruit Science and Technology,* D. K. Salunkhe and S. S. Kadam, eds., Marcel Dekker, Inc., New York.

Adsule, R. N. and Patil, N. B. 1995. Pomegranate. In: *Handbook of Fruit Science and Technology,* D. K. Salunkhe and S. S. Kadam, eds., Marcel Dekker, Inc., New York.

Al-Hooti, S., Sidhu, J. S., Al-Otaibi, J. and Al-Ameeri, H. 1997. Processing of some important date cultivars grown in United Arab Emirates into chutney and date relish. *J. of Food Process. and Preservation* 21:55–68.

Chakraborty, S., Rodriguez, R., Sampathu, S. R. and Saha, N. K. 1974. Prevention of pink discoloration in canned litchi *(L. chinensis). J. Food Sci. Technol. (India),* 11:266–268.

Chen, A. O. 1989. Quality improvement of candied fruits. In: *Quality Factors of Fruits and Vegetables—Chemistry and Technology,* Joseph J. Jen, ed., American Chemical Society, Washington, D.C.

Chow, S. T. 1991. *Fruit Processing and Storage,* 2nd ed., Beijing Scientific Press, China.

Desai, U. T. and Kotecha, P. M. 1995. Fig. In: *Handbook of Fruit Science and Technology,* D. K. Salunkhe and S. S. Kadam, eds., Marcel Dekker, Inc., New York.

Fang, T. T., Shy, H. T. and Chen, H. E. 1995. Agricultural material handling and processing, II. Fruit curing technology. *Taiwan Farmer Handbook,* Taipei Press, Taipei, Taiwan, pp. 490–500.

FAO, 1997. *FAO Publication Yearbook Vol. 49–1995.* Basic Data Branch, Statistics Division, FAO. FAO Publications, Rome, Italy.

FAO, 1998. *FAO Publication Yearbook Vol. 50–1996.* Basic Data Branch, Statistics Division, FAO. FAO Publications, Rome, Italy.

Hwang, L. S. and Cheng, Y. C. 1986. Pink discoloration in canned lychees. In: *Role of Chemistry in the Quality of Processed Food,* O. R. Fennema, W. H. Chang and C. Y. Li, eds., Food and Nutrition Press, Inc., Westport, CT.

Kale, P. N. and Adsule, P. G. 1995. Citrus. In: *Handbook of Fruit Science and Technology,* D. K. Salunkhe and S. S. Kadam, eds., Marcel Dekker, Inc., New York.

Kalra, S. K., Tandon, D. K. and Sing, B. P. 1995. Mango. In: *Handbook of Fruit Science and Technology,* D. K. Salunkhe and S. S. Kadam, eds., Marcel Dekker, Inc., New York.

Kotecha, P. M. and Desai, B. B. 1995. Banana. In: *Handbook of Fruit Science and Technology,* D. K. Salunkhe and S. S. Kadam, eds., Marcel Dekker, Inc., New York.

Luh, B. S. 1980. Nectar, pulpy juices and fruit juice blends. In: *Fruit and Vegetable Juice Processing Technology,* 3rd ed. P. E. Nelson and D. K. Tressler, eds., AVI Publishing Co., Westport, CT.

Luh, B. S. 1991. The preparation of nectars, fruit juices and vegetable juice. *Fluss. Obst* 58:290–294.

Luh, B. S. and El-Tinay, A. H. 1993. Nectar, pulpy juices and fruit juice blends. In: *Fruit Processing Technology,* Steven Nagy, Chin Shu Chen, and Philip E. Shaw, eds., Agscience, Inc., Auburndale, FL.

Luh, B. S., Kean, C. E. and Woodroof, J. G. 1986. Canning of fruit. In: *Commercial Fruit Processing,* 2nd ed., J. G. Woodroof and B. S. Luh, eds., AVI Publishing Co., Westport, CT.

Luh, B. S. and Mahecha, G. 1971. Anthocyanins in Manzanillo olives. *J. Chinese Agric. Chem. Soc.,* Special issue, 1–24.

Luh, B. S and Martin, M. H. 1996. Olive. In: *Processing Fruits: Science and Technology, Vol. 2.* L. P. Somogyi, D. M. Barrett, and Y. H. Hui, eds., Technomic Publishing Co., Inc., Lancaster, PA.

Luh, B. S. and Woodroof, J. G. 1988. *Commercial Vegetable Processing,* 2nd ed. Van Nostrand Reinhold, New York.

Martin, P. W. 1980. Durian and mangosteen. In: *Tropical and Subtropical Fruits,* S. Nagy and P. E. Shaw, eds., AVI Publishing Co., Westport, CT.

Morton, J. F. 1987. *Fruits of Warm Climates.* Media, Incorporated, Greensboro, NC.

Patil, V. K., Chakrawar, V. R., Narwadkar, P. R. and Shinde, G. S. 1995. Grape. In: *Handbook of Fruit Science and Technology,* D. K. Salunkhe and S. S. Kadam, eds., Marcel Dekker, Inc., New York.

Rao, V. S. and Roy, S. K. 1980. Studies on dehydration of mango pulp. (i) Standardization for making sheet/leather. *Indian Food Packer* 34:64–71.

Sole, P. 1996. Banana. In: *Processing Fruits: Science and Technology, Vol. 2,* L. P. Somogyi, D. M. Barrett, and Y. H. Hui, eds., Technomic Publishing Co., Inc., Lancaster, PA.

Somogyi, L. P. and Luh, B. S. 1986. Dehydration of fruits. In: *Commercial Fruit Processing,* 2nd ed., J. G. Woodroof and B. S. Luh, eds., AVI Publishing Co., Westport, CT.

United Nations, 1996. *Industrial Commodity Statistics Yearbook-1994. Production and Consumption Statistics, 1996.* Department for Economic and Social Information and Policy Analysis, Statistics Division, United Nations. UN Publications, New York, USA.

Woodroof, J. G. 1979. *Coconut: Production, Processing, Products,* 2nd ed., AVI Publishing Co., Westport, CT.

Wu, J. S., Sheu, M. and Fang, T. 1993. Oriental fruit juices: Carambola, Japanese apricot (Mei), lychee. In: *Fruit Processing Technology.* Steven Nagy, Chin Shu Chen, and Philip E. Shaw, eds., Agscience, Inc., Auburndale, FL.

Wu, J. S. and Sheu, M. J. 1996. Tropical fruits. In: *Processing Fruits: Science and Technology, Vol. 2,* L. P. Somogyi, D. M. Barrett, and Y. H. Hui, eds., Technomic Publishing Co., Inc., Lancaster, PA.

Yousif, A. K., Hamad, A. M. and Mirandilla, W. A. 1985. Pickling of dates at the early khalal stage. *J. of Food Technol.* 20:697–702.

Yu, Z. H. and Gao, X. Y. 1994. *Fruit Processing.* Jin Dun Press, Beijing, China.

Zhang, D. B. 1994. *Fruit and Vegetable Processing Technology.* China Light Industry Press, Beijing, China.

Zheng, Y. J., Jiang, Y. and He, M. 1993. *Cured Fruits and Technology.* Jin Dun Press, Beijing, China.

CHAPTER 11

Vegetable Products

SAMUEL L. WANG

1. INTRODUCTION

THERE are a large number of species of vegetables grown and consumed in Asia. Over 3000 years ago, the Chinese began to domesticate and propagate wild vegetables in garden plots in populated areas and also learned to preserve vegetables by simple salting and drying. Over the years, hundreds of different techniques, recipes, and color and flavor variations have evolved throughout different regions in Asia. The Chinese-populated areas such as Mainland China and Taiwan, with a rich resource of vegetables, have the most abundant supplies of processed vegetables (Fang, 1982; Chiou, 1990; Xiao, 1990). Korea and Japan also rank high in the consumption of processed vegetables. *Kimchi* has achieved a predominant status in recent years in Korea, with more than 150 varieties, which boost the per capita consumption of vegetables in Korea to 150 kg per year (Cheigh, 1997). *Zukemono* (Japanese pickles) has also proliferated into a great variety of recipes and tastes in many regions in Japan (Inden et al., 1997).

For centuries, salting, drying, and fermenting were the three primary processing techniques used to preserve vegetables, but salting of vegetables is generally used in combination with the other two preservation methods. Technological developments in canning, sugaring, freezing, plastic packaging, and pasteurization over the past 200 years provided new means for packaging and distributing preserved vegetables. Value-added processing technologies are only adopted slowly by processors of vegetable products in recent years. Most of the newer processing techniques such as canning, freezing, refrigeration, and pasteurization are adopted to the processed vegetables in Asia without many alterations. Despite the more modern packaging appearance, preserved vegetables using traditional preservation techniques developed over a long period in each region in Asia tend to have a regional accent, which distinguishes Asian vegetable products clearly from vegetable products in other parts of the world. These distinctions, along with unique vegetables rarely used in other parts of the world, will be presented in this chapter.

Descriptions of processed vegetable products sometimes can be confusing if a vegetable product is only identified by a common name. For instance, *dongcai* made in Sichuan, China, is from *jiecai* (leaf mustard), but those made in Hebei, China, are from Chinese cabbage (Deng, 1989). Hence, their differences in color, flavor, and nutritional values are significant and will be listed separately under leaf mustard and Chinese cabbage.

Genuine Asian vegetable products produced in the greatest amount for the Asian population are mostly salted and fermented vegetables. However, salting may be the only common denominator as a vegetable pretreatment. After salting, variations in processing techniques began to evolve in different regions or countries. Vegetable stocks may be desalted and seasoned to a regional taste preference using different spices and seasonings. Subsequent processing methods determine the style, flavor profile, and shelf life of the fermented vegetables.

For example, Korean *kimchi* and Chinese *paocai* are similarly processed vegetables using similar raw materials. However, both Chinese *paocai* and Korean *kimchi* have hundreds of variations (Xiao, 1990; Cheigh, 1997). Traditional *kimchi* includes four categories: (1) vegetables cured in concentrated salt; (2) vegetables fermented with grains; (3) vegetables cured with soy sauce and vinegar; and (4) vegetables fermented with salt, spices, and other seasonings (Cheigh and Park, 1994), but today's popular *kimchi* products generally are from the fourth category, which is identical to Chinese *paocai.* Various fresh vegetables, including Chinese cabbage, cabbage, radish, cucumber, green onion, and baby oriental radish are used as starting materials, and fermentation can be carried out with or without added water. Korean *kimchi* and Chinese *paocai* generally have a short shelf life unless pasteurization and refrigeration are employed.

In Japan, traditional *zukemono* includes seven distinct categories: (1) pickles in salt, (2) pickles in rice-bran paste, (3) pickles in soy sauce, (4) pickles in soybean paste, (5) pickles in *sake*-lees, (6) pickles in vinegar, and (7) pickles in light seasoning brine. The first six categories are started mainly with desalted vegetable stocks and developed into different styles and flavors of finished pickles using fermented flavoring, sauces, or pastes as seasoning stock solution. Pickles in the seventh category are made of fresh vegetables, and are close to Western pickles or Chinese *paocai.* These types of pickled vegetables have a short shelf life and must be pasteurized in sealed containers or preserved with chemical preservatives and refrigeration to keep for more than a few weeks. Many packaged Japanese *zukemono* declare a shelf life of longer than 1 year.

Certain preserved vegetables, which are seasoned with spices and mixed with vegetable oil after salting, dewatering, and aging, have a longer shelf life without the aid of chemical preservatives. Aging gives rise to a finished product with flavor and aroma from degraded protein constituents and a brown appearance associated with the Maillard reaction in properly aged food products. For example, Chinese *datoucai, zhacai, luobogan,* or *dongcai* can be kept in sealed barrels for up to 9 months and have a very tasty flavor and long shelf life (Deng, 1989; Chen, 1993). A Chinese study of the aging effect of *zhacai* indicated that degradation of protein into peptides and amino acids is mostly caused by viable proteinases during the aging period of *zhacai;* the effect of fermentation on dry-salted vegetables is insignificant (Deng, 1992).

Vegetable products may be made from different parts of a vegetable plant such as roots, bulbs, stems, leaves, flowers, fruits, and seeds, depending on the vegetable type. While some vegetables may have more than one edible part, other vegetables may be limited to only one edible part. In addition, a vegetable may be processed with several different processing techniques, or several vegetables may be mixed and processed into one vegetable product. To have a clearer illustration, processing vegetables will be divided into 16 categories, according to their horticultural classification given in the *Encyclopedia of Chinese Agriculture—Vegetables* (Li, 1990): (1) root vegetables, (2) Chinese cabbages, (3) cole crops, (4) mustards, (5) solanaceous fruits, (6) vegetable legumes, (7) gourd vegetables, (8) bulb vegetables, (9) green vegetables, (10) tuber vegetables, (11) herbs, (12) sprouting vegetables, (13) aquatic vegetables, (14) perennial vegetables, (15) wild vegetables, and (16) edible fungi. These vegetables are listed alphabetically under each category, with the edible part and nutritional values given if available (Li, 1990). A total of 60 vegetables, which are available in processed form, are discussed in Section 3 of this chapter.

A large number of commercially processed products are found in the Oriental markets in North America. Except for those vegetable products with low acid, which must be sterilized, most vegetable products with either high acidity (pH less than 4.5) or low water activity are now packaged and pasteurized in cans, glass bottles, or plastic bags. Mild thermal treatment in a container is a technological development adopted for preserved vegetables. This packaging technique prolongs the shelf life and expands the distribution perimeter of this type of vegetable products. A number of the commercial vegetable products, most of which are made of mixed vegetables from each region, are cited as examples under Section 4.

2. VEGETABLE PROCESSING TECHNIQUES

2.1 DRYING

2.1.1 Applications of Drying Techniques

Solar energy has always been a natural and economical source for food drying, ever since the early humans learned to preserve vegetables by drying. Even though modern dehydration technology is available, sun-drying is still quite common in many Asian regions, especially those sunny areas. Several vegetables were preserved and sold primarily in dry form, for example, *jinzencai, meigancai,* edible fungi (white mushrooms, Shiitake mushrooms, Jew's ears), seaweeds (kelp, laver), *facai* (hair-like algae), chili pepper, and most herbs and spices. These products must be dried to a shelf stable moisture level lower than 8%. A substantial amount of cabbages, radishes, bamboo, and root mustards are partially dried outdoors in China prior to further processing (Gao and Hu, 1994).

Dehydrated vegetables commonly used in North America, including dehydrated cabbages, carrots, onion, and garlic, are processed in the same manner in Asia. A brief procedure of sun-drying and mechanical drying follows.

2.1.2 Procedure of Sun-Drying and Mechanical Drying

(1) Grade and sort vegetables.
(2) Peel and cut into desired pieces and shape.
(3) Blanch prepared vegetables with hot water, if needed, to inactivate enzymes. Blanching time lengths vary with temperature, size of vegetable, and load. Sulfiting by burning sulfur is used as an alternate technique to inactivate enzymes if the vegetable parts are too tender for blanching.
(4) Dry treated vegetables in different types of dryers or sun-dry to shelf-stable moisture level (3–8%, depending on commodity).

2.2 SALTING

Salting is a very important step for vegetable preservation in Asia. In the early days, salt was only known as a preservative that kept foodstuffs from getting spoiled. The effect of salting on vegetables is better understood now (Deng, 1989; Chen, 1993).

2.2.1 Benefits of Salting

(1) Salt at high concentration (10% or higher) creates a high osmotic pressure, which causes impairment of microbial cells.

(2) Salt reduces the water activity of vegetables and makes free water less available for spoilage microorganisms.
(3) Salt reduces the soluble oxygen in water and prohibits the growth of aerobic microorganisms.
(4) Salt solution creates high concentrations of Na cation, which disrupt normal metabolism of microorganisms.
(5) Salt solution inhibits the enzymic activities of exogenous enzymes of microorganisms.

2.2.2 Procedure of Salting

2.2.2.1 Dry Salting

Leafy vegetables are graded, sorted, and air-dried for a couple of days before salting. Dry salt is sprinkled onto the surface of vegetables at a ratio of one part of salt to ten parts of vegetables. Kneading, mixing, and pressing (with stone slabs or screw press) are performed to facilitate the exudation of juice. When bulk quantities of vegetables are salted, salt is placed between layers of vegetables. Root vegetables require cleaning, trimming, and cutting into smaller sections prior to air-drying. Dry salting of root crops, because of their bulky and dense texture, must be carried out in successive steps using a fraction of the required salt each time. Exuded juice, which helps keep the vegetables clean, is discarded each time after salting. Once salted and covered tightly in a barrel or an underground cement storage vat, the vegetables can be kept for 3–9 months, depending on temperature conditions.

2.2.2.2 Brine Salting

Concentrated brine with as high as 20% salt is used to preserve minimally processed vegetables. In this case, partially processed fresh vegetables, including bamboo shoots, lotus roots, gingers, egg plants, scallions, and cucumbers, are packed with salt brine and exported to Japan and North America annually for further processing. Heavily salt-brined vegetables are desalted and flavored to make pickled products in other parts of the world.

2.3 FERMENTING

Traditional methods of vegetable fermentation achieve inoculation from the natural environment. After vegetables a partially air-dried (with or without hot water blanching) and exposed to ambient temperature for a while, microorganisms begin to flourish on the vegetables. The vegetables are then put into a container and deprived of all sources of oxygen with a one-way pressure relief device using a water lock. The oxygen inside the container is exhausted, and the fermentation process commences. A short-term fermentation and/or a long-term ripening process are employed for making pickled vegetables.

2.3.1 Short-Term Fermentation

Short-term fermentation is carried out at low salt concentrations of 2–6% salt to generate lactic acid, which, along with different spices and seasonings, imparts unique flavors to pickled vegetables. Commonly used spices and seasonings include wine, sugar, chili peppers, cloves, paprika, star anise, licorice, cinnamon, orange peel, black or white pepper, ginger, and garlic. Commonly used vegetables include cabbages, cucumbers, leaf mustard, root mustard, radishes, napa, carrots, tomatoes, ginger, scallions, peppers, garlic, asparagus beans, and perilla.

Short-term fermentation can be divided into three stages:

(1) Initially, with a pH value of around 5.5, a number of microorganisms and lactic acid-producing bacteria are active in the vegetable mix at room temperature. Lactic acid, alcohol, and acetic acid, as well as other metabolites, are produced by different microorganisms. In 4–5 days, the total acidity will slowly increase to 0.2–0.3%, with a lowering pH value to force the less tolerant microorganisms to cease growth.

(2) As the pH value drops under 4.5, typical lactic acid fermentation begins to dominate the growth medium and accelerate lactic acid production to elevate the total acidity to 0.4–0.8%. Most of the aerobic and anaerobic microorganisms, except lactic acid bacteria and yeasts, are dormant. Pickled vegetables are ready for consumption at this stage, but the potential for spoilage still exists, and this type of pickled vegetables needs tightly sealed packaging and refrigeration to extend the shelf life.

(3) If fermentation by the lactic acid-producing microorganisms continues, the total acidity eventually rises to above 1.2%, leading to a halt of fermentation. The fermented vegetables are called *suangcai* (sour vegetable, sauerkraut). Commonly used sour vegetables are made from cabbages, mustard greens, or Chinese cabbages (napa). This type of pickled vegetable may be fermented for up to 2 months and have a much longer shelf life under refrigeration if sealed and pasteurized.

2.3.2 Long-Term Fermentation

Long-term fermentation is in fact a process of aging (ripening) of partially dewatered vegetable stock with a dry-salting technique to produce intermediate moisture pickles that have a longer shelf life. The vegetable stock, with excess juice removed, is mixed with spices and seasonings such as chili pepper, cinnamon, clove, fennel, orange peel, and more packed tightly into large barrels for prolonged ripening. Aging of dewatered vegetable stock in a sealed barrel allows minor fermentation and promotes the breakdown of proteinaceous substances in the vegetables into peptides. The end products have a combination of fermented flavor, peptides, and added seasonings. Peptides from degraded proteins tend to form pigmentation after long-term storage. The end products are usually darker in color and quite aromatic in flavor. If oxygen and water are present, lactic acid or other spoilage microorganisms may cause poor flavor. This type of long-term fermented vegetables is more common in China than in other countries. Vegetables such as stem mustard, root mustard, leaf mustard, radish, cabbage, and fermented *zhacai, datoucai, xuelihong, dongcai, meigancai* are well known and consumed throughout China.

As higher quality fermented vegetable products are demanded by consumers in today's marketplace, the processing industry begins to inoculate salted vegetable stocks with prepared bacterial culture for standardized quality in appearance, texture, and flavor. *B. coagulans* culture has been used to improve the rate of fermentation and flavor quality of pickled cucumber and cabbage (Fang and Lin, 1985). On the other hand, aged types of pickles are generally high in salt content (>6%). Lowering of salt content is an objective for new process and product development research.

2.4 PICKLING

Pickling of vegetables was begun in Asia in the 7th century A.D. according to ancient Chinese literature (Li, 1990) or possibly as far back as the 4th century in Korea (Cheigh and Park, 1994). Pickled vegetables are divided into two types: fermented and nonfermented (Chen, 1993).

2.4.1 Fermented Type

The fermented type is based on lactic acid fermentation at a 2–6% salt level. The fermentation process is usually carried on until the total acidity reaches 0.4–0.8%, as described in Section 2.3.1.2. The fermented vegetables must be consumed within a short time period. Most *paocai* in China and *kimchi* in Korea fall into this category. In one Chinese publication (Xiao, 1990), 51 vegetables were listed as a raw material source, and six regional styles were cited, including Sichuan *paocai,* Korean *paocai* (a sizable group of Korean descendants living in Manchuria), Wuhan *paocai,* Dushan *paocai,* Henan *paocai,* and Guangdong *paocai.* In Korea, 30 vegetables were given as raw materials, plus additional materials from fruit, nuts, cereals, seafood, and mushrooms to make more than 100 styles of *kimchi* (Cheigh and Park, 1994). The Japanese depend on Mainland China and Taiwan for a large quantity of raw materials, such as radishes, bamboo shoots, ginger, and lotus, but add locally grown *sansai* (wild vegetable) and seaweeds (kelp, laver, etc.) for their pickled vegetables. Since brine-salted vegetables are used as raw materials, the fermented type of pickling is limited to a few items. Combining the vegetable variety and seasonings used, there are hundreds of variations in finished product flavors in these countries.

These pickled vegetables are flavorful and nutritious, owing to the presence of the vegetable nutrients and viable lactic acid bacteria, which are benevolent for the human digestive system. Nutritional aspects of *kimchi* have been studied in Korea (Cheigh and Park, 1994).

2.4.2 Nonfermented Type

The nonfermented type usually is made of desalted vegetable stocks, which have lost their original flavor because of heavy salting. In the past, they were flavored by adding fermented sauce preparation. Today, pickling with desalted vegetable stocks is mostly done by adding flavorings, coloring, and preservatives without fermentation. Flavors may be adjusted with sugar, acid, dill, chili pepper, cloves, soy sauce, garlic, ginger, and other seasonings. Both natural and artificial colorings are used to modify the product appearance. Chemical preservatives may be added to increase their shelf life. This type of pickled product is very popular in the oriental food markets because it can stimulate good appetite on plain rice meal when the homemaker does not have time to prepare other dishes and is usually more stable than the fermented type because of the use of higher levels of sugar and vinegar (sweet and sour) or chemical preservatives.

Vegetables such as garlic, scallions, gherkins, cucumbers, carrots, and ginger are favorite choices for making heavily seasoned, sweet and sour pickles because of their taste and shelf stability. Sweet and sour pickled vegetables are popular in most countries around the world. Nonfermented-type pickles are also made with salt, soy sauce *(jiangyou, shoyu),* or fermented sauces from cereal powder *(huangjiang, miso, mianjiang,* or *doubanjiang).* A number of pickled products, including most Japanese *zukemono,* are made in this manner. Since Japan relies heavily on imported salted vegetables for its processing industry, a large number of nonfermented *zukemono* products are made in Japan. *Nukazuke* (pickles in rice bran paste), *shiozuke* (pickles in salt), *shinzuke* (pickles in new and quick style), *shoyuzuke* (pickles in soy sauce), *kasuzuke* (pickles in *sake*-lees), *suzuke* (pickles in vinegar) and *misozuke* (pickles in fermented soybean paste) are listed as major categories (Anon., 1996; Inden et al., 1997).

There is little lactic acid fermentation involved in the nonfermenting type of vegetable pickles. The taste and appearance of these products are modified by seasoning brine, and their shelf life is extended with chemical preservatives.

The nutritional value of this type of pickles is not as good as the fermented type of pickles, since these vegetable products are salted heavily with exudation of cell contents and then desalted with excessive water leaching. However, many seasonings and flavorings come from fermented products such as wine, vinegar, *miso,* or soy sauce.

2.5 SUGARING

2.5.1 Principle and Applications of Sugar Preservation

Sugar preservation of vegetables has been used for a long time in Asia. Honey was used initially and replaced later by sugar. When the sugar content reaches 60–65%, the sugar-infused vegetables become shelf stable because microorganisms cannot survive under the high osmotic pressure and low water activity created by the high sugar content. Stability of sugared vegetables also improves because of removal of dissolved O_2 in the tissue. Many vegetables, including *donggua,* lotus root, lotus seeds, carrot, turnip, eggplant, chili pepper, tomato, ginger, and bitter melon, have been preserved with sugar.

Desirable quality aspects of sugar-preserved vegetables include wholesome appearance, uniform sugar distribution, and clear or translucent vegetable tissue free of particles.

2.5.2 Typical Procedure of Sugar Preservation

(1) Selection of raw material (mature and firm vegetables preferred)
(2) Preparation and grading (peeling, cutting, pitting, slashing, or puncturing of vegetables)
(3) Firming of tissue with calcium solution and blanching if needed
(4) Incorporation of sugar by increments of sugar solution from 15% up to 65% and soaking
(5) Brief heating and/or vacuum used to accelerate penetration of sugar molecules into the plant tissue. Use of vacuum infusion is a relatively new technique that shortens the processing time and reduces possible burnt flavor.

2.6 CANNING

Many vegetables that are indigenous to Asia have been canned mainly for export. Canned products, including bamboo shoots, water chestnuts, asparagus, and mushrooms, which are favorites of Westerners, were the backbone of the food industry for Taiwan in the 1960s and 1970s, and for China in the 1980s and 1990s (Song and Tan, 1995; Ye and Xi, 1996). Canning of oriental vegetables also shifted to Thailand and other Southeast Asia countries in the

1990s because of lower production costs. In Thailand, 60% of canned oyster mushrooms, salted fermented mustard greens, sweet potatoes, and bamboo shoot strips, halves, or tips are for export. Major importing countries are Japan, Mid-Eastern, and Western European countries (Anon., 1996). In India, Pakistan, Sri Lanka, and neighboring countries, bitter gourd, drum stick, *parwal, kudo, tinda,* and bread fruit are canned for export.

In recent years, vegetables commonly consumed in other areas of the world are grown and canned by Asian countries for exportation, for example, canned green beans, tomatoes, baby corns, and mushrooms, which are used domestically as well.

For the oriental food markets in North America and Western Europe, many salted, fermented, pickled, or sugared vegetable products receive mild heat processing. Metal cans and glass bottles are equally important for canned value-added vegetables. Glass jars appear to have a better image than the metal cans and are preferred for pickled products. Plastic packages (soft cans) are used for high-acidity vegetables or lightly salted vegetables with preservatives. Better barrier plastic bags, as well as laminated polymeric film bags, have been developed by packaging companies and used by manufacturers in Japan and Taiwan. Japanese *zukemono* are mostly packaged in plastic bags, but pasty vegetable products usually are placed into glass jars.

2.7 FREEZING

Frozen vegetables, especially individually quick-frozen (IQF) vegetables, are relatively new to the Asian countries. Production of this type of vegetable product is primarily for export (Song and Tan, 1995; Ye and Xi, 1996). Many popular vegetables consumed by people in Western countries, such as peas, green beans, corn kernels, and potatoes are processed by local processors under contract with foreign companies. In the last 15 years, more local companies began processing authentic Asian vegetables such as lotus roots, snow peas, vegetable soya beans, and asparagus and export to oriental markets around the world.

2.8 VEGETABLE JUICE PROCESSING

Processing of vegetable juice in Asia is relatively new and small. Some developmental effort has been made to simulate market demand for vegetable juice products such as V-8® or carrot juice. The processing technique is no different from that used in the West. Demand for western-style tomato juice and carrot juice is big in Japan. However, few popular Asian vegetables are suitable for juice making. The better-known vegetable juices include asparagus, water chestnut, and winter melon juice. For extending the shelf life, as well as improving the juice flavor, lactic acid fermentation has been applied to low-acid vegetable juice products (Wu, 1990). Fruit juice has also been added to control the pH value. Impact of this type of product in the marketplace is insignificant.

2.9 VEGETABLE SOUP PROCESSING

Vegetable soups are highly valued adjuncts on the dinner table of many Asian families. Soups are always made with long simmering time and seldom made without the use of meats, poultry, or seafoods. Development of commercial processing of vegetable soups in Asia began over the past 15 years, after the western-style products of U.S. companies won acceptance in Asia.

Vegetable soups are either canned in concentrated forms or packaged in dry mix form. Packaged vegetable soups are mostly made by multinational companies and consumed by overseas Asian ethnic groups. A couple of leading brands of ethnic vegetable soup mix products are available in food markets. Major vegetables used in soups include cabbage, carrots, tomatoes, onions, celery, potatoes, seaweeds, mushrooms, garlic, ginger, winter melons, sweet peas, water chestnuts, lotus roots, luffa melon, bitter melon, day lily, *baihe,* and bamboo shoots. *Miso* and *tofu* are also popular ingredients for vegetable soups.

For flavor development, meat, chicken, or bone stocks are used as the soup base, to which vegetables, spices, and seasonings are added. At present, Taiwan, Hong Kong, and Japan are principal manufacturers and buyers of processed vegetable soups. Canning is the main processing method for vegetable soups.

3. PROCESSING VEGETABLES: CLASSIFICATION AND UTILIZATION

Since one vegetable may have several common names used by different people in different countries, Latin nomenclature is listed with each vegetable to avoid confusion.

3.1 ROOT VEGETABLES

3.1.1 Carrot (*Daucus carota* L. var. Sativa DC.)

Carrot *(Hongluobo, Huluobo, Ninjin)* is rich in carotenoids and very nutritious. Carrots have been made into many different processed products, such as canned, frozen, or dehydrated baby carrots, carrot slices and dices, pickled vegetables using sliced carrots as a component, and canned carrot juice.

3.1.2 Edible Burdock (*Arctium lappa* L.)

Edible burdock *(niubang, gobo)* is a nutritious root vegetable and is very popular among Japanese. Edible roots are preserved in pickled or dried form.

3.1.3 Radish (*Raphanus sativa* L.)

Radish (turnip, *lobok, luobo, daikon*) is a succulent root vegetable rich in fiber, minerals, vitamin C, and sulfur compounds. It is the favorite vegetable in Asia for processing. Radish is used in literally hundreds of processed vegetable products in Asia.

Processing of salted dry radish *(luobogan)* in China started in 200 B.C. with a salting and sun-drying technique. To improve the taste, oil and seasoning were added to semimoist *luobogan* and packaged in jars or plastic bags. Products are also aged after salting, fermenting, and seasoning. Fermented-type pickling with low salt and seasoning to different acid levels is practiced throughout China. Mixed vegetables are usually used.

Pickled radish in Japan is called *takuan zuke.* Radish is salted; desalted; seasoned with acid, sugar, and monosodium glutamate; and colored with yellow pigment or spice without further fermentation.

Radish is also heavily used in Korean *kimchi.* Cut-up radishes are seasoned and fermented with other vegetable ingredients. The flavor can be modified by animal product, which is not used in Chinese or Japanese products.

3.1.4 Rutabaga (*Brassica napobrassica* Mill.)

Rutabaga *(yanggeda, yangdatoucai, kabu)* is grown in a limited area in Asia but is pickled in the same manner as root mustard to make *datoucai* in China.

3.2 CHINESE CABBAGES

3.2.1 Chinese Cabbage [*Brassica campetris* L. ssp. *Pekinensis* (L.) Olssen]

Chinese cabbage *(napa, dabaicai, hakusai)* is a very versatile vegetable with excellent flavor. It is used widely in China and Korea for fermented vegetables at different acid levels. It can also be used to make *dongcai,* a semidried and ripened vegetable product originating in North China and *xuecai* in South China.

Korean *kimchi* depends on Chinese cabbage as a base material. Several *kimchi* products, including *baechu, tongbaechu, bossam, yanbaechu,* and *nabak,* are based on cabbage as the main ingredient.

3.3 COLE CROPS (*BRASSICA OLERACEA* L.)

3.3.1 Cabbage (*Brassica oleracea* L. var. Capitata L.)

Cabbage (*ganlan, baoxincai, gaolicai, kyabetsu*) contains 2.7–3.4% carbohydrate, 1.1–1.6% protein, 0.5–1.1% fiber, and 40 mg per 100 g vitamin C. This vegetable is used widely in processing in Asia, as well as the rest of the world. Because of the presence of glucosinolate compounds, tasty substances are formed after fermentation and aging. Cabbage can be made into fermented-type pickles with other vegetables or can be dehydrated for soup mixes.

3.4 MUSTARDS (*Jiecai, Gai Choy, Karashina*)

All mustards contain glucosinolates, a class of sulfur-containing compounds, which, after hydrolysis, produce thiocyanates with a unique aroma. They are ideal raw vegetable materials for processing. This group of vegetables contains rich amounts of proteins, vitamins, carbohydrates, and minerals.

3.4.1 Root Mustard (*Brassica juncea* Coss. var. Megarrhiza Tsen et Lee)

Among different varieties of root mustards *(datoucai, gedacai),* there are four shapes of enlarged roots in various lengths ranging from 10 mm to 20 mm and widths ranging from 7 mm to 11 mm. Root mustards contain 1.2% protein, 6.1% carbohydrate, 2.1% fiber, and minerals and vitamins. All roots are suitable for processing *datoucai.*

3.4.2 Stem Mustard (*Brassica juncea* Coss. var. Tsatsai Mao)

Stem mustard *(Jingjiecai)* is primarily for making *zhacai* (*zhatsai,* preserved mustard). It contains a rich amount of protein, which supplies flavorful peptides after fermentation and aging.

3.4.3 Leaf Mustard (*Brassica juncea* Coss. var. Foliosa Bailey)

Leaf mustard is rich in thiocyanates and isothiocyanates, which break down into compounds with unique flavor after aging and fermentation. Several well-known fermented leaf mustards, including *dongcai* and *meigancai,* are made in China.

3.4.4 Seed Mustard (*Brassica juncea* Coss. var. Gracilis Tsen et Lee)

Seed mustard *(jiecaizi)* is used to produce mustard seeds, which can be either in paste or powder form.

3.5 SOLANACEOUS FRUITS

3.5.1 Chili Pepper (*Capsicum frutescens* L.)

Chili pepper *(lajiao, kara togarashi)* is used to provide a hot taste to foods. It is made in dehydrated form or paste form. Chili sauce is also salted and mixed with oil.

3.5.2 Eggplant (*Solanum melongena* L.)

Eggplant *(qiezi, nasu)* is a very popular vegetable in the Mideast. Stuffed eggplant in cans is made for export to North America. Eggplant is also preserved with sugar and is used in pickled products in Japan.

3.5.3 Tomato (*Lycopersicon esculentum* Mill.)

Processing of tomatoes *(fanqie, xihongshi)* is relatively new in Asia. Most of the processed products such as catsup, tomato juice, or tomato soup are introduced from the West. Pickling of green tomatoes is practiced in Japan.

3.5.4 Sweet Pepper (*Capsicum annuum* L. var. Grossum)

Sweet pepper *(tianjiao, pemen)* is seldom processed, except by dicing and freezing for export.

3.6 VEGETABLE LEGUMES (*Leguminosae*)

Seeds and tender pods are the edible parts of this type of vegetable. They contain relatively high total solids, protein, carbohydrates, and minerals.

3.6.1 Green Beans (*Phaseolus vulgaris* L.)

Green bean *(yundou, sijidou)* is a popular vegetable in North America. A large portion of green beans are processed in frozen or canned form for export.

3.6.2 Asparagus Bean (*Vigna unguiculata*)

Fresh asparagus bean *(douquat, chang doujiao, chang jiangdou)* contains 11–15% total solids, 2.9–3.5% protein, 5–9% carbohydrates, and vitamins and minerals. It is only used as an ingredient in pickling products.

3.6.3 Broad Bean (*Vicia faba* L.)

Fresh broad bean (faba bean, *candou*) contains 22–36% total solids, 9–13% protein, 12–15% carbohydrates, and several vitamins. It is mostly dried or canned. Dry bean contains up to 30% protein and can be fried and salted as a snack.

3.6.4 Sword Bean (*Canavalia Adans*)

Sword beans *(jackbean, daodou)* can be pickled or dehydrated. Dry sword bean contains 25–27% protein and can be ground into powder form.

3.6.5 Snow Peas (*Pisum sativum* L.)

Snow peas *(wandoujia, helandou, sayaendo)* are mostly consumed in stir-fry dishes. Frozen snow peas are processed in Asia for exportation.

3.6.6 Vegetable Soybeans (*Glycine max* L. Meir.)

Vegetable soybean *(maodou)* is shelled and used in stir-fry dishes. Shelled beans are frozen by local processors for export.

3.7 GOURD VEGETABLES

3.7.1 Cucumber (*Cucumis sativus* L.)

Cucumber *(huanggua, hugua, kyuri)* is one of the most important pickling vegetables. It can be pickled whole or sliced with dry-salting or salt-brining.

3.7.2 Bitter Gourd (*Momordica charantia* L.)

Bitter gourd (bitter melon, *kugua, nigauri*) is a very popular vegetable among the Asians despite its bitter taste. However, it is only canned in water for the export market.

3.7.3 Tinda (*Colocynthis citrulla*)

Tinda is one of the most popular and favorite vegetables in the Indian Peninsula. It contains 1.7% protein, 5.3% carbohydrate, 0.1% fat, 0.6% minerals, and 28 IU/100 g carotene. Tinda has a texture similar to cucumber and must be cooked with spices for dishes.

3.7.4 Wax Gourd (*Benincasa hispida* Cogged.)

Wax gourd *(winter melon, donggua)* is usually consumed as fresh produce. Candied *donggua* is processed for desserts using a sugar preservation technique.

3.8 BULB VEGETABLES

3.8.1 Garlic (*Allium sativum* L.)

Garlic *(suantou, dasuan, ninniku)* has a high total solids (close to 30%), 4.4% protein, 23.6% carbohydrate, 0.2% phosphorus, and a rich quantity of allicin, which produces the unique pungency.

Garlic can be processed into garlic puree, garlic oil, dehydrated garlic flakes or powder, garlic with hot chili sauce, and canned garlic pickles. Salt-brining of garlic is done to remove the pungent odor prior to seasoning and aging. Pickled garlic is usually seasoned with a sweet and sour taste. Finished products are mostly packaged in glass jars.

3.8.2 Onion (*Allium cepa* L.)

Onion *(yangcong, yuancong, congtou, tamanegi)* has 12% total solids, of which 1.8% is protein, 8.0% is carbohydrate, and a small amount is vitamin C and phosphorus, iron, and calcium. Cooked onion is diced for freezing or chopped for dehydration. Miniature-sized onions are pickled with vinegar and sugar, and large onions are made into onion rings.

3.8.3 Scallion (*Allium chinensis* G. Don)

Pickled sweet and sour scallions *(jiaotou, xie, rakkyo)* make very tasty appetizers. Salted scallions with acetic acid are exported for further processing.

3.9 GREEN VEGETABLES

3.9.1 Asparagus Lettuce (*Lactuca sativa* L.)

Asparagus lettuce (stem lettuce, *wojushun*) is commonly made into a non-fermented pickle product, either singly or in mixed formula.

3.9.2 Celery (*Apium graveolens* L.)

Celery (Chinese celery, *qincai, xiqin*) is a common vegetable in both the East and West. The large size *xiqin* is mainly used as a vegetable and has been processed by dehydration and freezing of chopped celery pieces. Small-sized celery has a stronger flavor and is used in Asia in fresh form. Dried celery seed is an important spice with a pleasant characteristic odor containing 2.5% volatile oil, 15% nonvolatile oil, and 12% protein (Lewis, 1984).

3.9.3 Coriander (*Coriandrum sativum* L.)

Coriander *(yansui, xiangcai)* is an aromatic herb and is used as a garnish for Chinese dishes. Drying of coriander leaves is rarely, if ever, done. Coriander seed, however, is an important spice for processing and is used in a lot of Western dishes. India is a major producing country, which produces and uses a huge amount of coriander seeds, as well as essential oil. Coriander seed contains 11% moisture, 14% protein, 18% fat, 30% fiber, and 0.5% volatile oil (Lewis, 1984).

3.10 TUBER VEGETABLES

3.10.1 Ginger (*Zingiber officinale* Rosc.)

Fresh ginger *(jiang, huangjiang, shoga)* contains 13–15% total solids, 0.6–1.4% protein, 8.5% carbohydrates, and special aromatic compounds such as zingiberene, curcumene, and sesquiterpene. It is a very popular condiment in oriental foods. Ginger can be made into ginger powder, sugared preserve, ginger juice, and pickled ginger.

3.10.2 Parwal (*Trichosanthes dioica*)

Parwal is a popular vegetable in India, which contains 8.0% total solids, 2.0% protein, 2.2% carbohydrates, 0.3% fat, 0.5% minerals, and vitamins A and C. Parwal is mostly consumed fresh, and a small portion is canned or dehydrated for export or military use.

3.10.3 Potato (*Solanum tuberosum* L.)

Like tomato, potato *(malingshu, tudou, yangyu, gagaimo)* is rarely processed in Asia. In recent years, potato chips, french fries, and other processed potatoes are gaining popularity through the introduction of the fast-food business in Asia.

3.10.4 Sweet Potato (*Ipomoea batatas* Lam.)

Sweet potato *(ganshu, fanshu, digua, hongshao, satsumaimo)* is a staple food for rural people in Asia. Drying of sweet potato into sticks or powder is commonly practiced in Asia; sugared and dried sweet potatoes are made in China.

3.11 HERBS

3.11.1 Fennel (*Foeniculum vulgare* Mill.)

Fennel *(huixiang, xiaohuixiang)* is rich in volatile oils such as anethole, linalool, and anisic aldehyde and is a popular condiment. Fennel seeds are ground or extracted to produce spice (Lewis, 1984).

3.11.2 Perilla (*Perilla frutecens* L.)

Perilla *(zisu, shiso)* is purple in color and rich in volatile oils. It is a common ingredient used in *zukemono* in Japan.

3.12 SPROUTING VEGETABLES

3.12.1 Bean Sprout (*Phaseolus radiatus* L.)

Bean sprout *(lyudouya, ah choy, moyashi)* is germinated mung bean with the shell removed. Bean sprouts are very nutritious; they contain, on a dry weight basis, 4.4% aspartic acid, 1.9% tyrosine, 1.7% valine, 1.6% leucine, 1.5% phenylalanine, 1.2% isoleucine, 1.1% lysine, 1.0% threonine, 2.6% proline, 2.0% glutamic acid, 1.7% arginine, 1.2% alanine, 1.1% serineare, and minor amounts of glycine, histidine, and cystine. Processing of bean sprouts usually occur in canned chow mein products as an ingredient in North America. A small quantity is used as an ingredient in pickled products.

3.13 AQUATIC VEGETABLES

3.13.1 Chinese Arrowhead (*Sagittaria sagittifolia* L.)

Tuber of Chinese arrowhead *(cigu)* contains 66% starch, 5.6% protein, 25.7% carbohydrates, and 0.26% phosphorus. *Cigu* can be dehydrated into powder form for preservation.

3.13.2 Kelp (*Laminaria japonica* Aresch)

Kelp *(haidai, kunbu, kumbu)* is a popular vegetable in the oriental diet. Dried kelp contains 8.2–9.0% protein, 55% carbohydrates, 1.2% calcium, 0.2% phosphorus, 0.12–0.15% iron, iodine, and vitamins. Kelp is mainly dried in strips and is also used in pickled products in Japan and Korea.

3.13.3 Laver (*Porphyra C.* Ag.)

Laver *(zicai)* is a popular aquatic algae used extensively in Japan and coastal China since the 6th century. It is usually dried in sheets for use. Laver contains, on a dry weight basis, 24–28% protein, 31–50% carbohydrates, 3.4% alanine, 3.2% glutamic acid, 2.4% glycine, 2.6% leucine, 1.4% isoleucine, 0.5–0.8% phosphorus, and 1200 individual units of carotene.

3.13.4 Lotus (*Nelumbo nucifera* Gaertn.)

Lotus *(lianou)* is an important and unique vegetable in Asia. The root portion contains 22% total solids, 20% starch, 0.03–0.05% vitamin C, and various sugars. The seed contains 17% protein, 62–66% carbohydrates, and minerals. Both roots and seeds are edible, but different varieties are grown for harvesting roots or seeds. Lotus roots are processed by canning peeled and sliced roots, drying lotus powder, and salt-brining of peeled and sliced lotus roots for further processing. Seeds are dried after skin and embryo removal, sugared as candied lotus seeds, or canned as sweetened paste *(lianrong)*.

3.13.5 Water Chestnut (*Eleocharis tuberosa* Roem. et Schult)

The edible part of water chestnut *(mati, biqi, kuwai)* is the slightly flattened tuber in the soil under water. It is high in total solids with 0.8–1.5% protein, 13–21% carbohydrate, and minerals. It has a very crisp texture, even after cooking and canning. Water chestnut is an excellent garnish for stir-fry dishes. Processing includes canning of peeled water chestnut, drying of water chestnut starch, and packaging of water chestnut juice.

3.14 PERENNIAL VEGETABLES

3.14.1 Bamboo Shoot (*Bambusoideae*)

Bamboo *(zhushun, takenoko)* originated in China and spread to most of the world, except Europe and Africa. Four of the 30 species are edible: *gangzhu (Phyllostachys), cizhu (Sinocalamus), cizhu (Bambusa),* and *kuzhu (Leioblastus).*

Bamboo shoot has 10% total solids; 3.3% protein; and abundant fiber, calcium, and other minerals. It is a popular garnish in stir-fry dishes. Bamboo shoots are mostly canned in water or in seasoned oil. Dried bamboo is also available.

3.14.2 Day Lily (*Hemerocallis lilio-asphodelus* L., *Hemerocallis minor* Mill., *Hemerocallis fulva* L.)

Fresh day lily flower *(huanghuacai, jinzencai)* has 2.9% protein, 11.6% carbohydrate, 0.033% vitamin C, 0.07% phosphorus, and 0.12% carotene. To process dried *jinzencai,* it is treated with sulfur dioxide and blanched before drying.

3.14.3 Lily (*Lilium* L.)

The edible part of lily *(baihe, yuri)* is scale-like stem, which is rich in protein (3.3%), sucrose (10%), reducing sugar (3%), pectin (5.6%), starch (11%), calcium, phosphorus, and vitamins B_1 and B_2. Lily is mostly dried for therapeutical food use.

3.14.4 Drum Stick (*Moringa oleifera*)

Pulp and seeds, as well as young leaves of drum stick *(moringa),* are consumed in India. Drum stick contains 2.5% protein, 3.7% carbohydrate, 4.8% fiber, 2.0% minerals, and 184 IU carotene. Canned and dried *moringa* are processed primarily for export and military ration.

3.14.5 Okra (*Hibiscus esculentus* L.)

Okra *(huangqiukui, okura)* has a wide appeal in many parts of the world. In addition to the tender pods, leaves, buds, and flowers are also edible. It contains 2.2% protein, 7.7% carbohydrate, 1.2% fiber, and 0.7% minerals and is high in calcium, phosphorus, magnesium, and potassium. Processing methods include canning, freezing, pickling, and drying. Acid is often used to reduce the processing for better texture.

3.15 WILD VEGETABLES

3.15.1 *Facai* (*Nostoc commune* var. flagelliforme Born. et Flah)

Facai is a type of wild algae only found in the northwestern region of China and is so-named in Chinese because it resembles human hair after drying. Dried *facai* contains 20.3% protein, 56.4% carbohydrates, 2.6% calcium, and 0.2% iron. Because of its scarcity, *facai* is highly valued by the Chinese and is used often to symbolize prosperity, since it sounds like the Chinese New Year's greeting *"Gong Hee Fat Choy."*

3.15.2 Wild Brake (*Pterikium aguilinum* L. Kuhn)

The edible young leaves of wild brake *(juecai)* are harvested for stir-fry or salad dishes. Brine-salted wild brake is in large demand for exportation. Dried wild brake is processed by cleaning, blanching, and sun-drying.

3.16 EDIBLE FUNGI

3.16.1 Enoki Mushroom [*Flammulinavelutipes* (Curt. Ex Fr.) Sing.]

Enoki mushrooms (lily sprout, *jinzengu, enoki dake*) have a complete profile of amino acids, especially lysine, arginine, and leucine. It is also rich in polysaccharides. Canned enoki mushrooms are mainly for export.

3.16.2 Jew's Ear [*Auricularia auricula* (L. Ex. Hook.) Under.]

Jew's ear *(heimuer, yun er, chuan er)* is generally preserved by sun-drying and some canning. The dried *heimuer.* contains 9.4–10.6% protein, 0.2–1.2% lipids, 65.5% carbohydrates, and 4.2–7% fiber. It is regarded as a functional food by Asian people.

3.16.3 Oyster Mushroom [*Pleurotus ostreatus* (Jacq.ex Fr.) Quel.]

Oyster mushroom *(baoyugu, caopi ceer, pinggu)* is one of the major cultivated edible mushrooms. On a dry weight basis, it contains 19.5% protein; 3.8% fat; 50.2% carbohydrates; 6.2% fiber; and calcium, phosphorus, and vitamins B_1, B_2, and C. Drying is the only processing technique.

3.16.4 Shiitake Mushroom [*Lentinus edodes* (Berk.) Sing.]

Cultivation of shiitake mushroom *(donggu, xianggu)* started in the 14th century in China, and its production is the second largest of the major cultivated edible mushrooms in Asia. A large part of the production is dehydrated for year-round consumption. Drying improved the flavoring effect of shiitake mushroom. Dry shiitake mushrooms contain 13% protein, 1.8% fat, 54% carbohydrates, and several vitamins.

3.16.5 Straw Mushroom [*Volvariella volvacea* (Bull. ex Fr.) Sing.]

Straw mushroom *(caogu)* is mostly canned for distribution. Fresh straw mushroom contains 2.6% protein, 2.6% sugars, 2.2% fat, and 0.2% vitamin C.

3.16.6 White Fungus (*Tremella fuciformis* Berk.)

White fungus *(tremella, yinger)* is generally dried for distribution. Some canning in syrup is done. Dried white fungus contains 5.0–6.6% protein, 0.6–3.1% fat, 68–78% carbohydrates, and 1.0–2.6% fiber. China is the largest producing country, and the Chinese regard *yinger* as a health supplement for respiratory functions of humans.

3.16.7 White Mushroom [*Agaricus bisporus* (Lange) Imbach]

Cultivation of white mushroom *(yanggu, mogu, baimogu)* was started in France in the 17th century and introduced to China in 1935 and to Taiwan in 1950. Production of *yanggu* flourished in these areas for export markets. A substantial amount of canned and dried *yanggu* was shipped to consuming countries in North America and Europe. Fresh white mushroom contains 3.7% protein; 0.2% fat; 3.0% sugars; 0.8% fiber; 0.1% phosphorus; and some calcium, iron, vitamins B_1, B_2, and C. Processing of white mushrooms includes canning, drying of slices, marinading in vinegar and oil, and salt-brining.

4. COMMERCIALLY PROCESSED VEGETABLES FROM ASIA

4.1 ASSORTED CHINESE PRODUCTS (INCLUDING PRODUCTS FROM CHINA, TAIWAN, HONG KONG, AND THAILAND)

4.1.1 Canned Products in Metal Cans

Product	Ingredients
Bamboo shoots	Bamboo shoots, water
Bitter melon	Bitter melon, water, salt
Lily sprout mushroom	Lily sprout, water, salt
Lotus root	Lotus, water
Mushroom	Mushroom, water
Pickled cabbage	Cabbage, salt
Pickled cucumber	Flower cucumber, water, soy sauce (soy bean, wheat, water, salt), sugar, MSG, citric acid
Pickled scallion	Scallion, vinegar, sugar, salt
Szechuan *zhacai* (Preserved vegetable)	Chinese radish, chili powder, salt, spices
Tianjin *Dongcai* (Preserved cabbage)	Tianjin cabbage, garlic, salt
Water chestnut *(Qingshui mati)*	Water chestnut, water
Young corn	Corn, water, salt

Figure 11.1 Canned Chinese vegetable products imported from China.

Figure 11.2 Chinese *dongcai* packaged in a traditional ceramic jar.

4.1.2 Packaged Products in Glass Jars

Product	Ingredients
Chili garlic sauce	Water, chili , salt, sugar, vinegar, garlic
Chili radish in soy sauce	Radish, red chili, sesame oil, salt, lemon acid
Fermented cabbage with braised gluten	Cabbage, gluten, water, soy sauce, chili, salt
Ginger in water	Ginger, vinegar, sugar, salt
Ground chili paste	Chili, distilled vinegar, salt, potassium sorbate, sodium bisulfite
Pickled cucumber	Cucumber, sugar, spice, soy sauce
Pickled scallion	Scallion, sugar, salt, acid, chili
Preserved bamboo shoots in chili oil	Bamboo shoot, chili oil, salt, citric acid, lactic acid
Preserved radish with chili	Radish, chili, sesame oil, soy bean sauce
Salted cucumber	Salted cucumber, water, soy sauce, MSG, red pepper, citric acid, lactic acid
Salted lettuce in soy sauce	Salted lettuce, water, soy sauce, MSG, citric acid, lactic acid
Salted radish strips in dressing sauce acid	Radish, soy sauce, sugar, water, red pepper, sesame oil, citric acid, lactic
Seasoned cabbage strips in chili paste	Seasoned cabbage, chili paste, water, soy bean oil, salt, sugar, acetic acid, sorbic acid

4.1.3 Packaged in Plastic Bags

Product	Ingredients
Chili radishes	Radishes, sugar, chili, potassium sorbate, sesame oil, salt, soy sauce
Preserved mustard greens *(Meigancai)*	Mustard greens, sugar, salt, water
Sour mustard	Mustard, water, salt, sugar, spices, food coloring yellow No. 5, sodium benzoate

Figure 11.3 Pickled Chinese vegetables packed in glass jars imported from Taiwan.

Figure 11.4 A variety of canned and glass-packed vegetables from Hong Kong and Thailand.

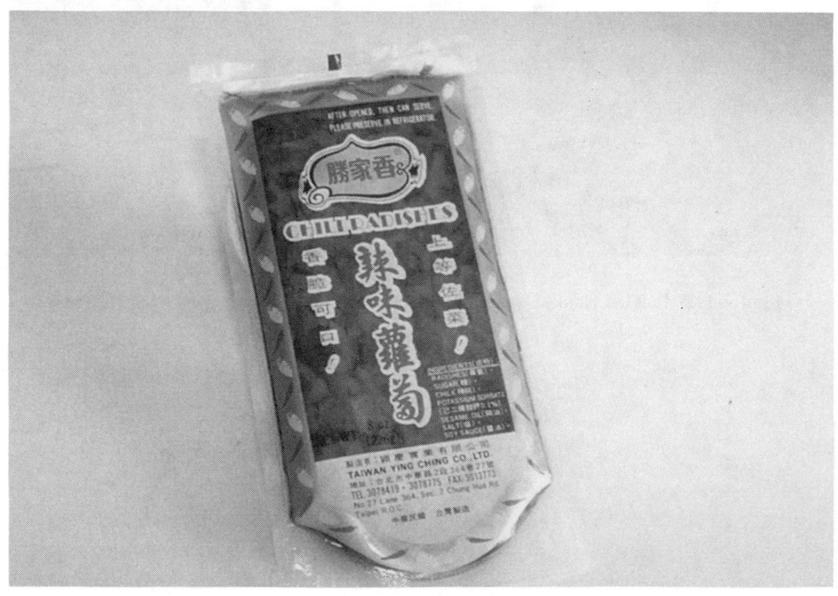

Figure 11.5 Chili radish packaged in a new barrier film package from Taiwan.

4.2 ASSORTED JAPANESE PICKLES PACKAGED EITHER IN PLASTIC BAGS OR GLASS JARS

4.2.1 *Shio* and *Nuka Zuke*

Product	Ingredients
Bibinba sansai	Flowering fern, soy bean, bracken, parsley, sesame, fungus, soy sauce, sugar, sesame oil, MSG, sodium benzoate
Ikiiki ipponzuke (pickled whole radish)	Radish, water, salt, sugar, MSG, sodium metaphosphate, aspartame, citric acid, potassium sorbate, malic acid, glucose, acetic acid, yellow No. 5
Kimuchi takuan (pickled radish)	Radish, garlic, onion, red pepper, MSG, FD&C yellow No. 5
Pacikko cucumber	Cucumber, perilla seeds, ginger, sesame seeds, amino acid, glucose, corn syrup, MSG, citric acid, chili pepper, sorbic acid, coloring
Takuan 500-katsuo (pickled radish)	Radish, rice bran, salt, high-fructose corn syrup, soy sauce (water, soybean, wheat, salt), dried bonito extract, dried bonito, distilled vinegar, spice, MSG, citric acid, sorbic acid, saccharin, FD&C 5 and FD&C 6
Takuan zuke nuka (pickled icicle radish in rice bran)	Radish, rice, glucose, sugar, MSG, salt, citric acid, red pepper, turmeric

4.2.2 *Shin* and *Zhacai* (Chinese Style) *Zuke*

Product	Ingredients
Aoshibazuke	Eggplant, cucumber, *myoga* plant, ginger, beef steak leaves, vinegar, water
Kyo murasaki (amasu sliced *shoga)*	Pickled sliced ginger
Mixed kelp	Kelp, agar agar, radish, sesame, vinegar, salt, sugar, MSG
Preserved *zhacai*	Preserved turnip, soy sauce, sugar, sesame oil, salt
Shichimi menmu (seven tastes)	Bamboo shoot, *kikurage* mushroom, sugar, corn syrup, MSG, sesame oil with chili, citric acid, chili pepper, sorbic acid, coloring
Shichimi zhacai	Turnip, bracken, bamboo shoot, sugar, MSG, citric acid, sesame oil, sorbic acid
Shinshinzuke (7 wild vegetables)	Radish, egg plant, sword bean, lotus root, perilla leaves, ginger, corn syrup, sugar, soy sauce, MSG, citric acid, coloring

4.2.3 *Shoyu Zuke*

Product	Ingredients
Okutone sansai	Mushrooms *(shimeji, enoki, kikurage),* bamboo shoot, bracken, soy sauce (water, soybean, wheat, salt), sugar, rice, vinegar, MSG, sorbic acid, artificial coloring
Shiso kombu	Kelp, perilla seed, soy sauce, sugar, MSG, sorbic acid
Tamari zuke	Garlic, soy sauce, sugar, MSG, sorbic and coloring
Takana zuke (pickled vegetable)	*Takana* vegetable, soy sauce (water, soybean, wheat, salt), MSG, salt, vinegar, FD&C yellow 5
Yamagobozuke (mountain burdock)	Burdock, soy sauce (water, soybean, wheat, salt), FD&C yellow 5

4.2.4 *Su Zuke*

Product	Ingredients
Pickled ginger	Ginger, vinegar, salt, FD&C red 40
Pickled ginger	Sliced ginger, vinegar, salt, water, citric acid, potassium, FD&C red 40
Pickled cucumber	Cucumber, salt, vinegar, water, rice, MSG, sorbic acid
Rakkyo zuke (pickled scallions)	Pickled Japanese scallions, scallions, sugar, vinegar, salt, MSG

Figure 11.6 Assorted vegetable *zukemono* packaged in film packages imported from Japan.

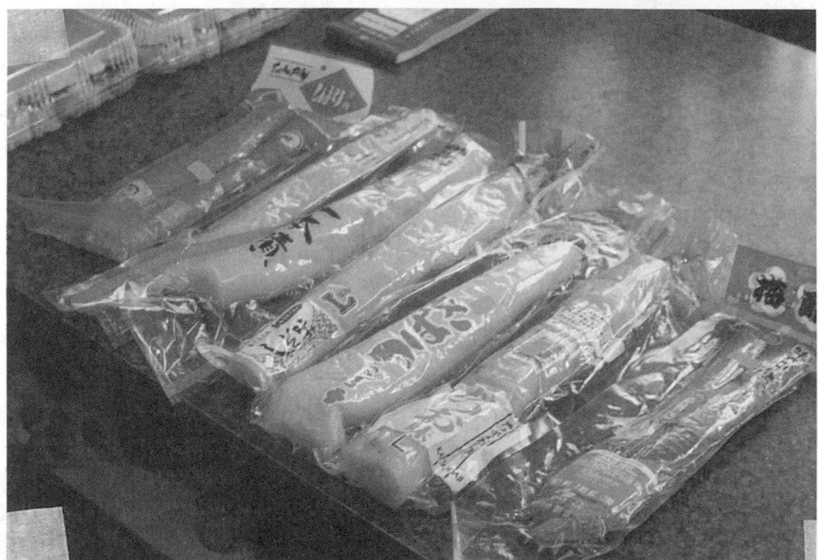

Figure 11.7 Assorted Takuan *zukemono* vacuum-packaged—imported from Japan and Korea.

4.3 ASSORTED KOREAN VEGETABLE PRODUCTS

Product	Ingredients
Baechu kimchi	Korean cabbage, water, hot pepper, salt, ginger, sugar, garlic, spices
Bak kimchi	Chinese cabbage, radish, garlic, pear fruit, onion, welsh onion, ginger
Donchimi	Whole radish, salt, spices
Green pepper *kimchi*	Green pepper, ginger, hot pepper, anchovy, sesame, soy sauce, malt, green onion
Hot pepper sauce	Ground hot pepper, vinegar, salt, potassium sorbate
Indian mustard *kimchi*	Mustard, hot pepper, anchovy sauce, shrimp sauce, garlic, onion, shallot, carrot, rice
Kaktugi kimchi	Diced radish, water, hot pepper, salt, ginger, sugar, garlic, spices
Kanpyo (dried gourd slice)	Gourd
Radish *kimchi*	Chinese cabbage, radish, pepper powder, garlic, ginger, fermented fish sauce, onion
White *kimchi*	Chinese cabbage, radish, red pepper, green mustard, pear, garlic, onion, shallot, water, dropwort, rice

Figure 11.8 Radish and Indian mustard *kimchi* on display in the Kimchi Museum in Seoul, Korea.

4.4 CANNED VEGETABLE PRODUCTS FROM INDIA AND SRI LANKA

Product	Ingredients
Garlic pickle	Garlic, ginger, chili, mustard, salt, sugar
Garlic pickle	Fresh garlic, sesame oil, chili powder, mustard, sugar, saffron, cumin seed, dill seed, acetic acid, salt

4.5 CANNED VEGETABLE PRODUCTS FROM MIDEAST COUNTRIES

Product	Ingredients
Fried eggplant	Eggplant, vinegar, garlic, refined sunflower oil, salt, tomato sauce
Okra	Okra, water, salt, citric acid
Pickled cucumber	Cucumber, water, salt, vinegar, garlic, dill, spices, sodium benzoate
Pickled hot pepper	Hot pepper, water, salt, vinegar
Stuffed cabbage	Cabbage leaves, rice, onion, tomato, sunflower oil, mint, parsley, spices, salt, water
Stuffed eggplant	Eggplant, rice, onion, tomato, sunflower oil, mint, parsley, salt, water
Stuffed green pepper	Green pepper, rice, onion, tomato, sunflower oil, mint, parsley, spices, salt, water

5. MARKET TREND OF ASIAN VEGETABLES

China is the biggest producer of processed vegetables, followed by Taiwan and Thailand. The annual output and exports from China and Thailand in the past few years have been increasing (Ye and Xi, 1996; Song and Tan, 1995), while output and export in Taiwan decreased (Wu, 1996). The Asian population of Chinese descent is the largest group of consumers of processed and preserved vegetables. In addition to the Chinese, the Japanese and the Koreans are also heavy users of processed and preserved vegetables. In addition, other Asian countries such as Malaysia, Thailand, Singapore, and Vietnam, which have a large population of Chinese immigrants, also adopted processed vegetables brought in with the Chinese.

Traditional processed vegetables have been enjoyed by Asians for more than 1000 years without any noticeable changes. The market was mostly local, and their quality was never challenged. This type of vegetable became an indispensable part of the Asian diet, especially in the winter. Although the knowledge in processing was limited to salting, drying, and fermentation, very tasty and often salty products were developed for the local population and became known in other areas. As recently as the 1940s, processed vegetable products were not distributed beyond a few neighboring cities. People in other areas would have to ask friends who visit the locale to bring back a few packages. Because of the lack of transportation and interprovincial commerce, many local communities developed their own tastes over the years, and these recipes became the unique renditions of some very well known products today that are available internationally.

While traditional processing techniques have not changed much up to the 1970s, newer packaging techniques were adopted by many leading manufacturers to replace traditional packaging, such as clothes and kraft papers, with ceramic jars and metal cans. This was because of the need to transport salted, aged, or pickled vegetables to distant markets. Not only do the vegetables need protection, but the flavor quality must also be retained for a longer period. Since mild heat treatment is the only way to keep the crisp texture of many pickled vegetables, the packaged pickled vegetable products still require relatively high salt, high acidity, and high sugar as primary deterrents of microbial growth inside the containers.

Glass bottles and plastic packaging were adopted in the 1980s. Although both packaging materials have been used for preserved vegetable products, consumer attitude in each country dictates which packaging material is more acceptable. Plastic film bags of *zukemono* are commonplace in the Japanese market, while processors in Taiwan, Hong Kong, Korea, and Thailand use mostly glass jars for pickled vegetables. Plastic film bags are only used for dried vegetables.

Concerns about high salt content in pickled vegetables have been expressed by consumers in more nutrition-conscious societies in Japan, Taiwan, and North America. Pickles with lower salt are gaining more market share, while the overall sales of traditional pickled vegetables lose consumer appeal steadily, especially among the younger generation. The salt content of most pickled vegetables in Japan, Korea, and Taiwan has been lowered in recent years. Developmental efforts are required by the research community to lower the salt usage of pickling further, while keeping the original flavor appeal of the pickled vegetables.

6. ACKNOWLEDGMENTS

The author wishes to express his sincere thanks to the following professionals from various institutions in the United States, Canada, China, Taiwan, and India, who provided valuable information: Prof. Daniel Y. C. Fung , Dr. John X. Q. Shi, Dr. Gopi Paliyath, Dr. Mikio Chiba, Dr. Tibor Fuleki, Prof. Weitang Liu, Ms. Judy Wanner, Prof. Tzuu-Tar Fang, Dr. Min-Sai Liu, Prof. Jun Hu, Prof. Tao Han, and Dr. V. Prakash. In addition, assistance provided by friends of various ethnical backgrounds is appreciated.

7. REFERENCES

Anon. 1996. *Annual Agricultural, Forestry and Fishery Report.* Ministry of Agriculture, Forestry and Fishery, Kyoto, Japan, pp. 403–414.

Cheigh, H. S. 1997. Production, characteristics and health functions of *kimchi.* In: *Quality of Fresh and Fermented Vegetables, Proceedings of VII International Symposium on Vegetable Quality,* October 27–30, 1997, Seoul, Korea, pp. 16–28.

Cheigh H. S. and K. Y. Park. 1994. Biochemical, microbiological and nutritional aspects of *kimchi* (Korean fermented vegetable products). *Critical Reviews in Food Science and Nutrition* 34(2):175–203.

Chen, X. P. 1993. *Processing Technology of Fruit and Vegetable Products.* Agriculture Publishing Co., Beijing, China, pp. 158–175.

Chiou, Z. R. 1990. *Famous Quality Fermented, Sauced and Pickled Vegetables—300 Home-style Recipes.* Jin Dun Publishing Co., Beijing, China.

Deng, G. S. 1989. *Storage and Processing Technology of Vegetables.* Agriculture Publishing Co., Beijing, China, pp. 218–271.

Deng, Y. 1992. A study of the transformation of constituents during aging of Sichuan *zhacai. J. of Food Science* 10:8–12.

Fang, T. T. 1982. Indigenous pickles in Taiwan. *Food Industries* 14(10):1–13.

Fang, T. T. and L. F. Lin. 1985. Application of pure culture of *Bacillus coagulans* to sauerkraut fermentation and cucumber pickling. *Agri. Res. Report.* National Taiwan Univ. 24(2):1–15.

Gao, X. Y. and J. Hu. 1994. *Practical New Vegetable Processing Techniques.* Shanghai Science and Technology Publishing Co., Shanghai, China, pp. 39–160.

Inden, H., Y. Kawano, Y. Kadama and K. Nakamura. 1997. Present status of vegetable pickling in Japan. *Proceedings of VII International Symposium on Vegetable Quality,* October 27–30, 1997, Seoul, Korea, pp. 29–35.

Lewis, Y. S. 1984. *Spices and Herbs for the Food Industry.* Food Trade Press, Orpington, England.

Li, S. X. 1990. *Encyclopedia of Chinese Agriculture—Vegetables,* 2nd ed. Agriculture Publishing Co., Beijing, China.

Song, Y. L. and Tan, S. C. 1995. *Handbook of High-Yield Cultivation and Processing of Vegetables Destined for Export Market.* Agricultural Technology Publishing Co., Beijing, China, pp. 1–6, 53–255.

Wu, S. L. 1996. Dehydration and pickled products. In: *Annual Report on Agricultural Industry.* FIRDI, Taiwan, pp. 86–100.

Wu, Y. X. 1990. *Fruit Products and Vegetable Processing.* Chongqing Publishing Co., Chongqing, China, pp. 198–206.

Xiao, Y. C. 1990. *Home-style* Paocai—*100 Recipes.* Jin Dun Publishing Co., Beijing, China.

Ye, X. C. and Z. F. Xi. 1996. *Processed Vegetables for Exportation.* Agriculture Publishing Co., Beijing, China, pp.1–159.

CHAPTER 12

Fats and Oils in the Asian Diet

PETER J. WAN

1. INTRODUCTION

FATS and oils are important components of our daily diet. They are nutritionally important as a rich energy source, with about 9 kcal per gram and as a carrier for the oil-soluble vitamins. They are frequently used to preserve foods and to make foods more palatable with the desired aroma, appearance, and texture. Fats and oils can be derived from either animal fatty tissues or plant sources. Fatty acid esters of glycerol or triglyceride, as shown in Figure 12.1, are the primary components of fats and oils.

Fats and oils are conveniently termed because those triglycerides stayed as solid and liquid at ambient temperature, respectively. While the supply and demand of edible fats and oils are primarily determined by the economics of each type of oilseed in industrial nations, they are still largely dictated by the local dietetic culture, regional supply, and processing technology available in the developing countries. For instance, in recent years, the urban population in China is quickly catching up with the Western diet pattern and have access to refined-bleached-deodorized (RBD) oils. But this is only 20% of the nation's 1.2 billion consumers. The majority of people in China still rely on relatively authentic means to obtain their daily cooking oils. This includes freshly rendered lard from pork fatty tissue or other animal fats and vegetable oils from mechanically pressed oilseeds with minimum refining and purification. This type of oil consumption pattern will likely stay with the developing nations in the foreseeable future. As the economy continues expanding, the consumers in the large cities of developing Asian countries will use more RBD vegetable

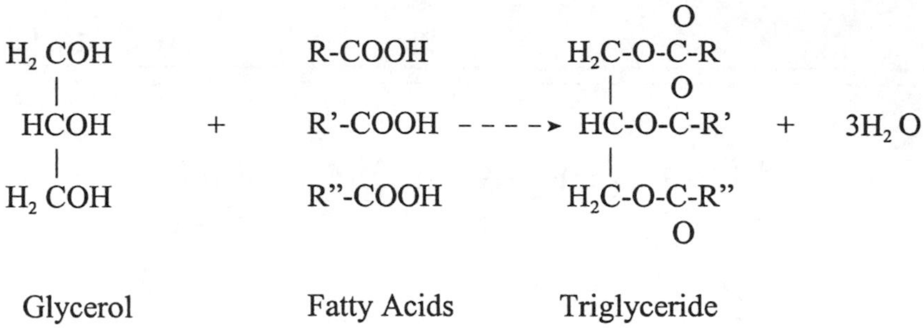

$$
\begin{array}{lclclcl}
H_2COH & & \text{R-COOH} & & \overset{O}{\overset{\|}{}}H_2C\text{-O-C-R} & & \\
| & & & & |\quad O & & \\
HCOH & + & \text{R'-COOH} & \dashrightarrow & HC\text{-O-C-R'} & + & 3H_2O \\
| & & & & | & & \\
H_2COH & & \text{R''-COOH} & & H_2C\text{-O-C-R''} & & \\
 & & & & \quad O & & \\
\text{Glycerol} & & \text{Fatty Acids} & & \text{Triglyceride} & &
\end{array}
$$

Figure 12.1 Schematic description of triglycerides.

oils and a variety of specialty oils at a premium price, like those in the industrialized nations. Even though the large population in the rural area will likely retain their ancient and unique dietetic culture, the supply and demand of oils and fats in Asia will certainly experience a dramatic increase and rapid change.

2. SOURCES OF FATS AND OILS

Worldwide, more than 70% of the edible oils and fats are obtained from plant sources (Salunkhe, et al., 1992). The world production of major oilseeds from 1980 to 1995 are shown in Table 12.1. The annual total oilseed production and each of the top five major oilseeds increased steadily throughout this period, and soybean amounted to more than half of the world oilseed output.

TABLE 12.1. World Production of Major Oilseeds.

	Production (MMT)			
Oilseed	1980	1985	1991–92	1994–96
Soybean	83.48	101.14	107.38	137.77
Cottonseed	27.16	32.3	36.62	33.02
Peanut	18.90	20.99	22.18	26.28
Rapeseed	10.57	19.04	28.27	30.28
Sunflower	13.17	18.87	21.82	23.37
Coconut	4.55	4.96	4.73	5.47
Palm kernel	1.81	2.6	3.41	4.54
Sesame	1.93	2.3		
Safflower	1.02	0.83		
Total	162.59	203.03	224.41	260.73

MMT = million metric tons.
Source: FAO (1980, 1985) and USDA (1996).

TABLE 12.2. Production of Ten Major Oilseeds (1000 metric tons) in the World and Asian Countries or regions during 1996–1997 Season.

Countries	Soybean	Cottonseed	Peanut	Sunflower Seed	Rapeseed	Sesame Seed	Palm Kernels	Coconut	Linseed	Castorseed	Grand Total
Bangladesh								15	46		
China	13100	7481	7097	1420	9201	575	3		480	220	39577
India	4000	5900	5550	1380	6150	800	4	680	319	770	25553
Indonesia	1517	22	686				1223	1355			4803
Iran	140	365		46		19					
Iraq				67		14					
Israel		97	16	18							
Japan	148		21		1						
N. Korea	360										
S. Korea	160		13		2	32					
Malaysia							2635	60			2695
Pakistan	3	3230	80	190	410	41			5	8	3966
Philipines	5	17	25				19	2130		7	2202
Sri Lanka						5		100			
Syria		471		10		9					
Taiwan	10		65								
Thailand	359	49	108			32	87	100		15	750
Turkey	50	1259	56	670	2	29					2067
Vietnam	190		250			30		215			
World	131566	34411	20240	24651	31159	2708	4998	5537	2240	1121	258631

Source: Mielke, 1998.

Not all the oilseeds harvested each year were used for oil production. Part of peanuts and some of the soybeans are used for human consumption or as animal feed. Palm plantation around the world increased most dramatically during recent decades and will continue its upward trend for the foreseeable future. Oilseed production in Asian Countries during the 1996–97 season is summarized in Table 12.2. Global production of edible oils in 1995–96 was 71.6 million metric tons (MMT) compared with 53.3 MMT in the 1987–88 crop season (Table 12.3). Production and disappearance of 17 major fats and oils in Asian countries are given in Tables 12.4 and 12.5, respectively. Most of the major oil types increased impressively during the past decade. Soybean oil accounts for 28.1% of the total in 1995–96. During the same period, the annual production of palm oil almost doubled and increased from 8.5 MMT to 15.6 MMT, or 16% to 22 %, of the world vegetable oil production. The major and minor oil sources for the Asian diet and around the world are briefly described in the following sections. The fatty acid composition and some selected properties of the major vegetable oils are summarized in Table 12.6.

2.1 MAJOR VEGETABLE OIL SOURCES

2.1.1 Soybeans

Soybean *(Glycine max)* is native to China and has been grown as a food crop for thousands of years. On an average, whole soybean contains 40% protein, 20% oil, 34% carbohydrates, and 5% ash. Soybean has been the largest oilseed crop in the world for the last few decades. The United States, Brazil, China, and Argentina are the major soybean-producing countries, with the United States providing nearly half of the world production (*Soya Bluebook,* 1998). Most of the Asian countries are cultivating soybeans for their domestic consumption. In the region, soybeans are used to produce numerous types of traditional soyfoods, such as *tofu,* soymilk, soy sauce, *miso, tempeh,* and more (for details, see Chapter 6). In the industrial nations, soybeans are primarily planted for oil and protein feed. Direct use of soybean as food in Western society is very limited because of the beany or grassy flavor and, to a lesser degree, because of the presence of flatulent sugars such as stachyose (3.8%) and raffinose (1.1%).

In most Asian countries, soybean oil is primarily used for cooking and frying. In China, it provides 30% of total edible oil consumption. In Western nations, soybean oil is widely used for cooking, salad dressing, hydrogenated base stocks for the formulation of margarine, and baking and frying shortenings.

2.1.2 Cottonseed

Cotton *(Gossypium)* has been grown primarily for fiber since the beginning of civilization. Its by-product, cottonseed, is the second major oilseed and fifth

TABLE 12.3. World Production of Major Vegetable and Marine Oils

Oil Type	Production (MMT) 1987–88	Percent of Total Production	Production (MMT) 1995–96	Percent of Total Production
Soybean	15.50	29.1	20.10	28.1
Palm	8.52	16.0	15.58	21.7
Rapeseed	7.03	13.2	8.74	12.2
Sunflower	7.00	13.1	11.14	15.5
Cottonseed	3.31	6.2	3.97	5.5
Peanut	2.69	5.0	3.96	5.5
Coconut	2.71	5.1	3.17	4.4
Olive	1.63	3.1	1.46	2.0
Fish	1.41	2.6	1.44	2.0
Palm kernel	1.14	2.1	2.08	2.9
Corn	1.13	2.1		
Linseed	0.65	1.2		
Sesame	0.58	1.1		
Total	53.30		71.64	

MMT = million metric tons.
Source: USDA (1988) and USDA (1997).

largest vegetable oil source worldwide. Its meal is mostly used as feed protein for ruminant livestock. The major cotton-producing countries include China, the United States, USSR, India, Pakistan, Brazil, Turkey, and Egypt. Since the 1970s, white undelinted cottonseed after the ginning process has been a preferred feed for dairy cows in the United States. As of now, nearly 50% of the cottonseed produced in the United States is fed directly to livestock. In the developing countries, a large portion of cottonseed is either mechanically pressed or prepress solvent extruted for oil. Sometimes it is used as animal feed or fertilizer because of a lack of transportation and available processing facility.

Cottonseeds contain 5–12.8% moisture, 15.2–22% oil, and 17.1–21.3% protein. For each 100 pounds of cottonseed, there is about 16% crude oil, 45.3% meal, 25.4% hulls, and 8% linters. Refined cottonseed oil has been used for cooking and frying in the United States for many decades. Its unique fatty acid profile (Table 12.6) and slightly nutty flavor allow cottonseed oil to offer a quality flavor with good stability for snackfoods. Therefore, it is still the preferred oil by potato chip makers. Its use as a food oil in the Orient is somewhat limited, partially because the screw-pressed cottonseed oil contains a reddish natural pigment, gossypol. With the increasing oil-refining capability in the region, most of the cottonseed will likely be processed into various specialty oils for human consumption and protein meal as animal feed.

TABLE 12.4. Production of Seventeen Major Oils and Fats (1000 metric tons) in the World and Asian Countries or Regions during 1996–97 Season.

Countries	Soybean	Cotton Oil	Peanut	Sunflower Oil	Rapeseed	Sesame Oil	Corn Oil	Olive Oil
Bangladesh					112.8			
China	1260	915.4	1582.4	304.7	3010	163.4	42.8	
India	603.7	551.8	1573.2	436.7	2138.8	210		
Indonesia		2	11.3	2.8				
Iran	27.5	51.5				19		
Iraq				27.1		5		
Israel	89.3	14.6		9.4	3.3			6
Japan	686.5	6.8			845.3	39.9	99.2	
Jordan								16.5
N. Korea	30.1							
S. Korea	241.5	5.1			0.4	22.2	43.2	
Lebanon								6.5
Malaysia	64.9							
Pakistan	6.3	367		71	134.7			
Philipines	22.4	17						
Singapore	3.9		1.7			1.6		
Sri Lanka								
Syria	2.5	62		3.4				138
Taiwan	457.9		6.8			5.9		
Thailand	105.4	6.4						
Turkey	26.6	186		491.9		16.7	30.9	217
Vietnam	17.5		22.2			4		
Word Total	20812.4	3966.1	4281.5	9317.6	11484.3	762.7	1881.7	2761.4

Source: Mielke, 1998.

TABLE 12.4. *(continued)*

Countries	Palm Oil	Palm Kernels	Coconut	Butter	Lard	Fish	Tallow	Linseed	Castor Seed	National Total
Bangladesh			29.3	14				12.5		169
China	15			68	2528.7	12	630.5	120	79.6	10734
India	17		414.5	1254.9			139.1	88.8	290.3	7718
Indonesia	5150	507.3	806.2			3.3				6206
Iran				83						162
Iraq				7.5						40
Israel										123
Japan			25.9	70.4	74.1	51.2	41	31		1971
Jordan										17
N. Korea										30
S. Korea				44	51.7	2.1	14.9	1.6		427
Lebanon										
Malaysia	9057	1157.1	35.1							10258
Pakistan			4.7	409.8						993
Philipines	70	8.2	1218.1							1318
Singapore			5.6							14
Sri Lanka			52.4							52
Syria				15.5						221
Taiwan					68.4					539
Thailand	390	36.4	53						12.1	600
Turkey				102		4.8	23.2			1099
Vietnam			110.3		65.5					219
World Total	17617	2134.2	3262.8	5700.9	6117.5	1259.7	7406.3	650	458.8	99535

TABLE 12.5. Disappearance of Seventeen Major Oils and Fats (1000 metric tons) in the World and Asian Countries or Regions during 1996–97 Season

Countries	Soybean	Cotton Oil	Peanut	Sunflower Oil	Rapeseed	Sesame Oil	Corn Oil	Olive Oil
Bangladesh	274.7				114.8			
China	2816.4	928.3	1609.8	315.4	3226.6	159.4	44	2.1
Hong Kong	24.9		18.6	5.9	106.5	2.5	15.9	
India	677.4	574.2	1533.2	845	2091.3	210		
Indonesia	23	1	11.4	1			1.5	
Iran	374.6	51.5		250.3		19	0.6	
Iraq	31.6			57.8		5	20.9	
Israel	96.1	23.2		14.8	3.3		7	6.1
Japan	689.8	19.8		13.5	850.5	38.7	103.6	24.2
Jordan	14						8.5	17.4
N. Korea	67.4							
S. Korea	277.7	20.8			11.4	22.8	75.7	2.2
Kuwait	3.9			3.7			22.2	1.5
Lebanon	24			12.2			6.9	7.7
Malaysia	20.1		2	10.7	2.8		2.3	
Pakistan	227	378.1		114	143.4			
Philipines	58.2						2.1	
Saudi Arabia	5.1						51.9	6.5
Singapore	24.5		3	2.9	6.5		4.5	
Sri Lanka								
Syria	18.5	63.5		18.9				96.5
Taiwan	519.7		6.8	28	3.7	3.3	4.5	
Thailand	125.6	6.4		9.8				
Turkey	173.4	183.8	0.1	609.8	4.4	16.7	112.5	82.4
Vietnam	56.9		22.1		2	3		
World Total	21134	4043.6	4288.5	9511.1	11513.2	763.7	1877.9	2207.7

TABLE 12.3. *(continued).*

Countries	Palm Oil	Palm Kernels	Coconut	Butter	Lard	Fish	Tallow	Linseed	Castor Seed	National Total
Bangladesh	132.7		34.3	14.5			22.1	13		607
China	1653.3	10.7	40.9	75.6	2553.3	18.6	797.2	142.9	102.2	14497
Hong Kong	7.1			3.7	24.7		2.4	6.7		222
India	1279.7	2.9	420.6	1256.4			141	90.8	91.3	9214
Indonesia	2758.2	92.1	261.5	7.7		6.6	1.2	1.9		3167
Iran	80.4	1.8	3.5	95.3			11			870
Iraq	175.7			10.5			2			304
Israel	9	2.4	0.5				1.6			164
Japan	379	54.3	52	75	74.6	121.9	150.8	31	21.6	2701
Jordan	50.4	2.5	0.5	14.3			2.1			111
N. Korea										90
S. Korea	192	7.4	43.9	45.6	53.1	14.1	77.1	6.1	2.6	853
Kuwait	8	5		3.3						48
Lebanon	2.2	1.1	2.9	3.5						62
Malaysia	1217.5	649.3	21.9	7.5			1.5	1.5	2.6	1939
Pakistan	1050.4	4.7	8.2	411.2			56.7			2394
Philipines	94.7	8.5	288.4	11			16.3	2.2		482
Saudi Arabia	178.5	5.6	1.4	29.1			5.7			287
Singapore	167.4	13.3	24	10.6				0.9		260
Sri Lanka	60.2	9.5	46.5				4.8			121
Syria	21	4.9		19.3						243
Taiwan	74.9	1.9	9.1	10.6	74.6	9.2	79.4	3.2	2.1	831
Thailand	400.5	40.1	54.3	20.1		4	3.7		23.6	690
Turkey	203.5	44.3	14.7	104.5			134.3	3.1		1692
Vietnam	80.1		109.2	6.5	65.5					345
World Total	17135.2	2120.3	3200.4	5747	6130.9	1315.5	7378.6	666.1	486.9	99520

TABLE 12.6. Fatty Acid Profile and Physical Properties of Some Common Oils.

Fatty Acid and Properties	Soybean	Cottonseed	Peanut	Rapeseed	Canola	Sunflower	High-Oleic Sunflower	Palm
Caproic C6:0								
Caprylic C8:0								
Capric C10:0								
Lauric C12:0								0.1
Myristic C14:0	0.1	0.7	0.1		0.1	0.1		1.0
Palmitic C16:0	10.6	21.6	11.1	3.5	4.1	7	3.7	44.3
Palmitoleic C16:1	0.1	0.6	0.2		0.3	0.1	0.1	0.1
Stearic C18:0	4	2.6	2.4	1.2	1.8	4.5	5.4	4.6
Oleic C18:1	23.3	18.6	46.7	14.2	60.9	18.7	81.3	38.7
Linoleic C18:2	53.7	54.4	32	13.8	21	67.5	9	10.5
Linolenic C18:3	7.6	0.7		9.1	8.8	0.8		0.3
Arachidic C20:0	0.3		1.3		0.7	0.4	0.4	0.3
Gadoleic C20:1			1.6	10.9	1	0.1		
Behenic C22:0	0.3		2.9		0.3	0.7	0.1	0.1
Erucic C22:1				46.9	0.7			
Lignoceric C24:0			1.5		0.2			
Total saturated gravity %	15.3	24.9	19.3	4.7	7.2	12.7	9.6	50.3
Specific gravity	0.917–0.921	0.916–0.918	0.910–0.915		0.914–0.920	0.915		0.892–0.893
Refractive index	1.470–1.476	1.468–1.472	1.467–1.470		1.470–1.474	1.472–1.474	1.456–1.458	1.456–1.458
Iodine value	123.0–139.0	99.0–113.0	84.0–100.0		110.0–126.0	125.0–136.0	18.0–88.0	46.0–56.0
Saponification value	189–195	189–198	188–195		182–193	188–194	188–194	196–202
Melting point, °C	−23.0–−20.0	10–16	−2			−18.0–−16.0	7.2	36.0–45.0

Fatty Acid and Properties	Palm Kernel	Coconut	Safflower	High-Oleic Safflower	Sesame	Corn	Rice Bran	Olive
Caproic C6:0	0.2	0.4						
Caprylic C8:0	3.3	7.6						
Capric C10:0	3.4	7.3						
Lauric C12:0	48.2	48.2		0.1				
Myristic C14:0	16.2	16.6	0.1			0.1	0.2–0.4	
Palmitic C16:0	8.4	8	6.8	3.6	7.8–9.1	10.9	15.4–20.6	9
Palmitoleic C16:1		1	0.1	0.1		0.2	0.2	0.60
Stearic C18:0	2.5	3.8	2.3	5.2	3.6–4.7	2	1.6–2.2	2.7
Oleic C18:1	15.3	5	12	81.5	45.3–49.4	25.4	41.1–46.7	80.3
Linoleic C18:2	2.3	2.5	77.7	7.2	37.7–41.2	59.6	29.6–34.9	6.3
Linolenic C18:3			0.4	0.1		1.2	1.0–1.5	0.7
Arachidic C20:0	0.1		0.3	0.4	0.4–1.1	0.4	0.7–0.8	0.4
Gadoleic C20:1	0.1		0.1				0.5–0.8	
Behenic C22:0			0.2	0.2		0.1	0.3–0.4	
Erucic C22:1				1.2				
Lignoceric C24:0							0.6–0.9	
Total saturated gravity%	82.3	91.9	9.7	0.3	0	13.5	0	12.1
Specific gravity	0.860–0.873	0.917–0.919	1.448–1.450	10.8	0.916–0.921	0.915–0.920	0.908–0.917	0.909–0.915
Refractive index	1.448–1.452		7.5–10.5	1.467–1.469	1.463–1.474	1.470–1.474		1.468–1.471
Iodine value	13.0–23.0	0.910–0.920	23.0–26.0	85.0–95.0	103–150	103–128	100–105	80–88
Saponification value	230–254	250–264	23.0–2.0	186–199	186–193	187–193	190	188–196
Melting point, °C	24.0–26.0	23.0–26.0	−18–−16			−12–−10		

2.1.3 Peanut

Peanut *(Arachis hypogaea),* also called groundnut, is the third largest oilseed crop of the world. It is native to South America and was first found in Brazil or Peru as early as 950 B.C. (Higgins, 1951). Nearly a third of the total world peanut production is concentrated in India (32%), which is naturally the predominant source of edible oil for India (Salunkhe et al. 1992). China (26%) and the United States (8%) are the other two major producing nations. These three nations collectively contributed over 66% of the world's peanut production of 22.8 MMT in 1988 (FAO, 1988).

The kernels of peanuts, on average, contain 5% moisture, 28.5% protein, 47.5% lipids, 13.3% carbohydrates, 2.8% crude fiber, and 2.9% ash. It is a rich source of protein and oil. Because of its pleasant nutty flavor and texture, a large portion of the peanut is used for confectionery, peanut butter, and snacks. The lesser grade of peanut is usually processed for peanut oil and meal.

2.1.4 Rapeseed

Rapeseed *(Brassica napus* and *Brassica campestris)* is a major oilseed crop in Far East and in northern Europe and America. It refers to more than one plant species and is often used to denote the seeds derived from oil-yielding members of the Brassica family, including some mustard seeds (Bengtsson et al., 1972). It is grown for edible or industrial uses. As a source of edible oil, it can be tracked back at least 1300 years ago in China (Hung, 1994). It is now the principal source of cooking oil in China and provides nearly 50% of China's domestic consumption of food oil.

Rapeseed normally contain 6–9% moisture, 38–50% oil, 36–44% protein, 11–16% fiber, and 7–8% ash. Oil of rapeseed contains a substantial amount of long-chain fatty acids and a relatively low total saturated fatty acid (Table 12.3). Selective breeding has resulted in the drastic reduction of erucic acid (C22:1) from 20–50% to 0.2–2.0% and increases of oleic acid from 12–15% to 60%. Canola® is a registered trademark of the Canola Council of Canada for the seed, oil, and meal derived from rapeseed, which is low in erucic fatty acid and glucosinolates. In the United States, low erucic acid rapeseed (LEAR) oil has been approved as a food substance generally recognized as safe (GRAS) in 1985, and in 1988, the U.S. FDA agreed that rapeseed oil containing less than 2% erucic acid can be identified as canola oil. With the desirable fatty acid profile the double low (low in erucic acid and glucosinolates) varieties of rapeseed have been quickly adopted by most of the growing regions in recent years.

2.1.5 Sunflower

Sunflower *(Helianthus annuus)* is primarily grown for its oil. It probably originated in the Southwest United States to Mexico area (Heiser, 1976) and was taken to Europe in the 16th century. It was considered as an ornamental flower until the 19th century and is now a major oilseed crop in Russia. Russian breeders were the first to successfully develop a high-oil variety sunflower seed, which has thin black hulls as opposed to the traditional confectionery type (Robertson and Russell, 1972). The confectionery sunflower seeds are larger than the high-oil variety and are consumed in the shell or hulled. Whole sunflower seeds contain 34–52% oil. The kernels have an oil content ranging from 46.7% to 64.7%. Oleic (C18:1) and linoleic (C18:2) acids make up more than 80% of the total fatty acids (Table 12.6). The proportion of oleic and linoleic acid is highly influenced by the climate and temperature of the growing region. Sunflower grown in the northern region (above 39° latitude) of the United States will have a higher percentage of linoleic and a lower percentage of oleic acid than oil from seed grown in southern regions. The high degree of unsaturation has given sunflower oil a healthy image among consumers.

2.1.6 Oil Palm

Oil palm (*Elaeis guineensis* Jacq.) is an important source of edible oil in the tropical region of Africa, Asia, and Central and South America. Two functionally different types of oil can be derived from palm fruit: palm oil from fruit mesocarp and palm kernel oil. Palm oil is commonly used for cooking, while palm kernel oil is used as a confectionery fat because of its high lauric acid (C12:0) content and high melting point. Asia accounts for about two-thirds of the world production of palm oil and palm kernel oil, with Malaysia as the leading producing country, followed by Indonesia (FAO, 1987). Oil palm is also grown in China and Thailand but to a lesser extent. The oil palm fruits are oval-shaped sessile drupes. The mesocarp or pulp is orange or reddish-brown in color and is oily and fibrous. Oil palm seed is the nut that remains after the removal of soft oily mesocarp during palm oil extraction. It consists of shell or endocarp and one or more kernels. Palm oil, palm kernel oil, and palm kernel cake or meal are three commercial products derived from oil palm fruit. Palm oil represents 20–24% of the weight of palm fruit. The mesocarp or pulp of palm fruit has an average oil content of about 56% on wet basis or 70–75% on dry-weight basis.

2.1.7 Coconut Palm

Coconut palm *(Cocos nucifera)* is one of the most useful trees in the world and is a source of food, fuel, and shelter for people in humid, tropical regions.

About 84% of the world coconut production is concentrated in Asia, with Indonesia, the Philippines, and India as major producing countries (FAO, 1988). Oil is the principal constituent in coconut endosperm or copra. The oil content ranges from 34% to 45% in the ripe endosperm and from 60% to 75% in copra. The oil has a characteristic odor that is largely caused by the presence of a small amount of lactones. Coconut oil contains about 84–93% triglycerides, 2–12% diglycerides, and 1–7% monoglycerides. Among the triglycerides, 84% are trisaturates, 12% are disaturated monounsaturated triglycerides, and 4% are monosaturated diunsaturated triglycerides. Therefore, it stays solid at room temperature. Traditionally, matured coconuts are dehusked and broken, and the copra is dried. It is either utilized for extraction of oil in an expeller press or processed into a variety of confectionery and bakery items as a supplement to give a characteristic flavor to such products. The oil is used either in crude form or after refining for both edible and industrial purposes, while the press cake containing a significant level of proteins is utilized as cattle feed and manure (Salunkhe et al., 1992).

2.1.8 Safflower

Safflower *(Carthamus tinctorius)* is mainly grown for its edible oil in the region of arid and semiarid tropics in India, Mexico, the United States, Ethiopia, and Australia (Salunkhe et al., 1992). It is a minor oilseed crop and accounts for 0.5% of the worldwide oilseed crops (FAO, 1985). But safflower is considered a premium cooking oil because of its high content of linoleic acid, low saturated fat, and a characteristic pleasant flavor. Safflower is the world's oldest crop, grown for centuries in India and other parts of Asia, Africa, and Europe as a source of dye until recently. About 25 species of the genus *Carthamus,* are known and many of them are indigenous to the Mediterranean region.

2.1.9 Sesame

Sesame *(Sesamum indicum)* is one of the world's oldest oilseeds known, and there is no record about its origin. It has been cultivated for centuries in Asia and Africa. Almost all sesame in the world is grown and produced in developing countries (FAO, 1987). It is the major source of edible oil in Burma. The oil is principally used for cooking and flavoring. A small percentage of oil is also used in pharmaceuticals, cosmetics, and perfumery industries and for the manufacture of soaps, paints, and insecticides. Sesame seeds are also used for the preparation of sweets, as a condiment for culinary purposes, and for confectionery and bakery products. Sesame cake obtained from mechanical press is rich in protein, methionine, calcium, phosphorus, and niacin. The defatted sesame meal is usually fed to cattle (Salunkhe et al., 1992).

2.1.10 Corn

Corn *(Zea mays),* or maize, is a major grain after wheat and rice and is native to South America (Beadle, 1978). Asia produces about one-fifth of the world's production of corn (FAO, 1985). More than 90% of the corn is processed and fed to animals in the Western world, whereas in Asia and Africa, majority corn production is utilized for human consumption. Whole kernels of corn consist of 82.3% endosperm, 11.5% germ, and 5.3% pericarp. The oil content of corn ranges from 1.2% to 5.7% and up to 19.5% for the high-oil variety. The majority of the corn oil is in the germ. The germ constitutes 10–14% of the seed weight of regular corn and up to 24.7% for the high-oil corn. Regular corn germ contains 34% oil while that of the high-oil variety contains over 50% oil . Therefore, corn is considered an important source of edible oil. Currently, only about 4.3% of the vegetable oil produced in the United States comes from corn although it has a potential of 25% of the soybean oil production (Weber and Alexander, 1975). The corn oil utilization for edible purposes in the United States is about 7.2% of total oil consumption (USDA, 1986). Corn can be processed by dry milling or dry degerming to produce low-fat cornmeal or wet degerming, which is the most frequently used process to obtain starch, sweeteners, and corn oil.

2.1.11 Rice

Rice *(Oryza sativa)* is the second largest cereal grain produced in the world, after wheat. It is the primary food source for nearly half of the world's population. Most of the rice production is concentrated in the developing countries of Asia (FAO, 1985). Southeast Asia or Africa is considered to be the origin of rice (Chatterjee, 1951). About 4000 of the 7000 known rice species have been identified in India. Only *Oryza sativa,* with three subspecies, is broadly cultivated (Houstan and Kohler, 1970). The cultivars of *Oryza glaberrima* are cultivated in Africa. Food products from rice are treated in Chapter 2.

Rice bran contains 17–23% oil, which is also rich in tocopherols and tocotrienols. The rough rice can produce about 67% milled rice and about 10% bran in a modern rice milling system. But in the developing countries of Asia, rice is commonly processed by traditional methods such as huller mill, shelter-huller, or shelter mill equipped with a pearling and polishing device. In huller mills, all the bran is lost with the husk, while in shelter mills, the bran is not efficiently separated. The oil in raw rice bran is quickly hydrolyzed by lipase. To make the rice oil process economical, the raw rice bran has to be immediately stabilized after milling. The interest in rice bran oil in industrialized countries has escalated recently due to the health-promoting effect associated with the oil consumption (Orthoeter, 1996). However, until the transportation and processing facility in the rice-producing regions of Asia are adequately improved, rice bran oil will likely be underutilized.

2.1.12 Olive

Olive *(Olea europaea)* is a perennial evergreen tree and is believed to have originated in the Syro-Iranian region of Asia (Brousse, 1989). Its fruit is used for edible purposes and for the production of olive oil. Spain and Italy combined contribute about 63% and 65% of the world's olive and olive oil, respectively. Greece, Turkey, and Tunisia also produce significant amounts of olive oil. The world production of olive was 10,267,000 metric tons while that of olive oil was 1,965,000 metric tons during 1987 (FAO, 1987). The flesh of olive comprises 70–80% of the fruit (Fernandez Diez, 1971). The pulp is rich in oil and contains up to 75% oil on a dry basis.

2.2 MINOR VEGETABLE OIL SOURCES

There are several other oil crops that contain appreciable amounts of edible oil in their seeds or kernels. These crops are considered as minor oilseeds because of their limited cultivation or intended uses. The horticultural crops such as mango, grape, watermelon, avocado, and apricot are grown mainly as fruit crops. The seeds or kernels are the by-products of the fruit processing industry. These oilseeds are important or even sometimes major sources of oil for the people where these crops are grown (Salunkhe et al., 1992). Some of these minor oil sources are summarized in the following. For food products of fruits and vegetables, refer to Chapters 10 and 11 respectively.

2.2.1 Apricot

Apricot *(Prumus armeniaca)* is grown mainly in Turkey, Spain, the Untied States, Russia, Morocco, and some European countries as flesh fruit. It is consumed fresh or in processed forms. The fruits contain edible mesocarp and stone. The kernel in the stone contains 40–50% oil (Vaughan, 1970). The oil is composed mainly of oleic and linoleic acids, which together amount to about 90–93% of the total fatty acids, with small quantities of palmitic acid. The fruit crop can serve as an important source of edible oil if their stones are collected efficiently and used for oil extraction on a large scale.

2.2.2 Chinese Vegetable Tallow

Chinese vegetable tallow *(Sapium sebiferum)* is a deciduous tree endogenous to China. Chinese vegetable tallow is derived from the mesocarp and stillingia oil or lumbang oil from the kernel. The fruits yield 20–23% tallow and 10–17% stillingia oil. The tallow is rich in palmitic (62%) and oleic (27%) acids, while the stillingia oil is rich in linoleic (53%) and linolenic (30%) acids (Godin and Spensley, 1971). The tallow is mostly used for the manufacture of soaps and candles. It is also used for edible purposes in China. Stillingia oil is characterized as a drying oil used in paints and varnishes.

2.2.3 Cocoa

Cocoa *(Theobroma cacao)* is an important tropical crop and is considered to have originated in the Amazon and Orinoco areas of South America (Chatt, 1953). It has also flourished in West Africa. World production of cocoa beans was 2 million metric tons during 1987, about 50% of which was contributed by African countries (FAO, 1987). Other important producing countries are Cote d'Ivoire, Brazil, Ghana, and Malaysia. Cocoa beans contain about 40–59% fat (Godin and Spensley, 1971). They are processed mainly to obtain butter fat and powder by roasting, crushing, and pressing. Cocoa butter fat melts around 33.5°C and is composed of 24–29% palmitic, 34–36% stearic, 30–40% oleic, and 2.4% linoleic acids (Godin and Spensley, 1971). Cocoa butter is mainly used in making chocolate. It also has several uses in the pharmaceutical and cosmetic industries (Eckey, 1954).

2.2.4 Grape

Grape *(Vitis vinifera)* is a fruit crop and is grown mostly (85% of the world's crop) for the production of wine. About 60% of the world production of grapes (64.8 million metric tons during 1987) was contributed by European countries (FAO, 1987). Italy, France, Spain, Russia, and the United States are the major grape-producing countries in the world. Grape seeds, a by-product of wine manufacture, contain about 6–20% oil. The grape seeds are composed of about 40% kernels and 60% shells. Grape seed oil contains 7–9.5% palmitic, 3.5–5.5% stearic, 14–44% oleic, and 46–74% linoleic acids (Godin and Spensley, 1971). The oil is used as a substitute for linseed oil in the manufacture of paints, varnishes, and liquid soap. Refined oil is used for edible purposes (Godin and Spensley, 1971).

2.2.5 Mango

Mango *(Mangifera indica)* is one of the oldest tropical fruit crops and has been cultivated by man for over 4000 years. The mango tree is considered to have originated in the Indo-Burma region. About 77% of the 14.6 million metric tons of the 1987 mango crop was produced in Asian countries. India alone contributed about 63%, or 9.2 million metric tons, of the 1987 world mango production (FAO, 1987). Mango fruits are eaten for their juicy pulp. Its kernel is usually thrown away or dried for burning. It has only been realized recently that mango kernels are a good source of starch and fat. Mango kernels contain about 11.0% fat, 73.0% starch, 9.5% protein, and 3.6% ash (Bhattacharyya, 1987). The fat contains 35–50% oleic, 33–48% stearic, 5–8% palmitic, 4–8% linoleic, and 1–7% arachidic acids (Shah et al., 1983; Augustin and Ling, 1987). It melts between 34°C and 43°C.

2.2.6 Niger

Niger *(Guizotia abyssinica)* is widely grown as an oilseed crop in India and Ethiopia. India is the major niger-producing country in the world, with an annual production of over 120,000 tons, followed by about 80,000 tons in Ethiopia. On average, niger seeds contain about 40% oil, 23% protein, 10% crude fiber, 22% carbohydrates, and 4% ash. The oil content of the seed varies from 30–50%.

2.2.7 Perilla

Perilla *(Perilla frutescens)* is a coarse aromatic herb cultivated in northern India, China, Japan, Nepal, Burma, Bangladesh, and Korea (Mukherjee, 1940). China is considered to be the origin of perilla (Zeven and Zhukovsky, 1975). The perilla seeds are small but rich in oil, 30–40%, and protein, 16–24% (Godin and Spensley, 1971; Sharma et al., 1989). Perilla oil contains 63–70 linolenic, 14–23% oleic, 16% linoleic, and 6–12% saturated fatty acids (Godin and Spensley, 1971). The oil is locally used for edible purposes. In the United States, it is considered as a substitute for linseed oil in the manufacture of paints, varnishes, linoleum, oilcloths, printing inks, and so on (Godin and Spensley, 1971). Perilla seeds are relished by hill people in India in the form of chutney (Godin and Spensley, 1971).

2.2.8 Poppy

Poppy *(Papaver somniferum)* has been known since ancient times for its oil-rich seeds and alkaloid opium. It is grown mostly in eastern and southeastern Asia and southeastern Europe for the production of opium. It is believed to have originated in the Mediterranean region (Krzymanski and Johnson, 1989). Poppy seeds contain about 40–55% oil (Sangwan et al., 1985): 9–10% palmitic, 1.5–2.5% stearic, 10–30% oleic, and 62–63% linoleic (Godin and Spensley, 1971).

2.2.9 Watermelon

Watermelon *(Citrullus vulgaris)* is an important fruit crop and is grown mainly in Asia, Russia, and Europe. World production of watermelon in 1987 was 28.1 million metric tons, and more than half was produced in China, Turkey, and Russia (FAO, 1987). The seeds contain 27% oil and 20% protein.

The kernels contain 52–57% oil and 34% protein (Lakshminarayana et al., 1968; Oyenuga and Fetuga, 1975). Watermelon seed oil is composed of 11–19% palmitic, 13–17% stearic, 14–22% oleic, and 52–58% linoleic acids (Oyenuga and Fetuga, 1975). The oil is used for cooking and illumination.

2.3 ANIMAL AND MARINE FATS

Worldwide, the proportion of fat and oil production from animal tissues has declined steadily during recent decades (Love, 1996). It was projected that tallow and lard will only be 8.4 and 7.3%, respectively, of the world production of the 12 major fats and oils in 1995 (USDA Foreign Agricultural Service, 1994). Because of the limited supply, marine fat has never been a major source of cooking fat in the Orient, except for Japan.

2.3.1 Lard

For many generations, lard was the fat of choice for cooking and frying and for preparing dough and batters. This is still the case in the farming community of developing countries. In the United States, the periodic supply shortages of lard in the past have prompted the development of all-vegetable substitutes. Now in industrial nations, vegetable shortenings are largely replacing lard in bakery products. The common characteristics and composition of lard are give in Table 12.7. Ordinary lard is characterized by a translucence and a poor plastic range. This may be attributed to the highly symmetrical arrangement of the triglycerides, disaturated fatty acids at the 1,3 positions, and monounsaturated acid in the middle position of glycerol. These symmetrical triglycerides tend to form stable beta crystals upon cooling. The beta crystals of fats are largely responsible for the grainy mouthfeel or poor eating quality. However, this problem can be eliminated by interesterification of the lard with the presence of sodium methoxide as catalyst (O'Brien, 1998).

2.3.2 Tallow

Tallow is fat derived from the fatty tissue of ruminants. It is a by-product of the meat packing industry. It is normally water washed and vacuum dried and filtered with diatomaceous earth to remove proteinaceous material and residual moisture before use. Sometimes caustic refining or a neutralization step is added to remove excess residual free fatty acid. The neutralized tallow was the preferred frying fat for French fry potatoes by fast-food restaurants for many years. Since tallow contains a high percentage of saturated fatty acids (Table 12.7) and cholesterols, its uses for cooking and frying have been diminished drastically in the industrial nations over the last 20 years. Structurally, tallow contains highly asymmetric triglycerides that tend to solidify into fine beta-

TABLE 12.7. Physical Properties, Fatty Acid Composition, and Triglyceride Distribution of Lard, Tallow, and Milk Fat.

Properties		Lard	Tallow	Milk Fat
Specific gravity at 99/15.5°C		0.858–0.864	0.680–0.870	0.907–0.912
Iodine value		46.0–70.0	38.0–48.0	25.0–42.0
Saponification value		195–202	193–202	210–250
Unsaponifiable matter, %		1.0 max.	0.8 max.	0.4 max.
Titer, °C		36.0–42.0	40.0–46.0	34.0
Solids fat index, %				
at 10.0°C		29.0	32.6	33.0
at 21.1°C		21.6	22.1	14.0
at 26.7°C		15.3	19.6	10.0
at 33.3°C		4.5	13.8	3.0
at 37.8°C		2.8	8.6	
at 40.0°C		2.2	6.1	0.0
Mettler dropping point, °C		32.5	42.6	35.0
Fatty acid composition, %				
Butyric	C-4:0			3.6
Caproic	C-6:0			2.2
Caprylic	C-8:0			1.2
Capric	C-10:0	0.1		2.5
Decenoic	C-10:1			
Lauric	C-12:0	0.1	0.1	2.9
Decenoic	C-12:1			
Myristic	C-14:0	1.5	3.2	10.8
Myristoleic	C-14:1		0.9	0.8
Pentadecanoic	C-15:0	0.1	0.5	2.1
Palmitic	C-16:0	26.0	24.3	26.9
Palmitoleic	C-16:1	3.3	3.7	2.0
Margaric	C-17:0	0.4	1.5	0.7
Margaroleic	C-17:1	0.2	0.8	
Stearic	C-18:0	13.5	18.6	12.1
Oleic	C-18:1	43.9	42.6	28.5
Linoleic	C-18:2	9.5	2.6	3.2
Linolenic	C-18:3	0.4	0.7	0.4
Arachidic	C-20:0	0.2	0.2	
Gadoleic	C-20:1	0.7	0.3	0.1
Eicosadienoic	C-20:2	0.1		
Triglyceride composition, mole %				
Trisaturated (GS_3)		2.0–5.0	15.0–28.0	
Disaturated (GS_2U)		25.0–35.0	46.0–52.0	
Monosaturated (GSU_2)		50.0–60.0	0–64.0	
Triunsaturated (GU_3)		10.0–30.0	0–2.0	

Source: O'Brien (1998).

prime crystals. This beta-prime tendency gives tallow the desired functionality to provide foods with smooth texture, small air bubbles, good lubricity, and so on. But tallow generally has an undesirable flavor and odor and develops reverted flavor shortly after deodorization; therefore, it needs to be protected with added antioxidants such as butylated hydroxyanisole (BHA), butylated hydroxytoluene (BHT), or *terit*-butylhydroquinone (TBHQ) with a metal chelating agent like citric acid.

2.3.3 Milk Fat

Milk has long been considered a nutritional food for young and old. Cow's milk contains an average of 3.7% fat, 4.9% lactose, 3.5% protein, 0.7% minerals, and 87.2% water. To preserve the milk fat in the form of butter is an ancient technology. Butter was used in worship, medicinal, and cosmetic purposes (in addition to foods) long before the Christian era. Milk fat has therefore always had the highest economic value of any of the milk constituents. Because of the desired functionality and flavor quality, milk fat has always been the standard of many baking products. Milk fat is processed and used in one of the following three forms: (1) butter—a water in oil emulsion that contains no less than 80% milk fat with or without salt and with or without color; (2) anhydrous butterfat—prepared from milk or cream; and (3) butter oil—separated from butter. Milk fat contains a wide range of fatty acids (Table 12.7) and melts sharply, which gives a cooling sensation in the mouth and, in a broad temperature range from −30°C to 37°C, functions well in pastry applications (O'Brien, 1998).

2.3.4 Marine Oils

Approximately 28–30% of the 100 million metric tons (MMT) of annual world production of fish is processed to fish meal and oil, resulting in about 1.44 MMT of fish oil (Table 12.3). The principal sources for fish oil are the so-called industrial fishes, herring, menhaden, anchovy, sardine, and mackerel. Most of these species are bony, oily, and not popular as human foods (Pigott, 1996). The high contents of omega-3, double bond beginning at the third carbon, and fatty acids such as eicosapentaenoic (C20:5) and docosahexaenoic acid (C22:6) in the fish oils are considered responsible in increasing the high-density lipoprotein and decreasing the low-density lipoprotein in blood and other health benefits. The average and ranges of oil content of some species of fishes and fatty acid profiles of commercial menhaden oil are given in Tables 12.8 and 12.9.

TABLE 12.8. Average and Ranges of Oil Content of 17 Common Species of Fish.[a]

Species	Range Oil Content (%)	Average Oil Content (%)
Carp	1.0–12.0	5.0
Chub, lake	4.0–13.0	8.5
Cod, Atlantic	0.2–0.9	0.4
Flounder	0.3–3.4	1.4
Haddock	0.2–0.6	0.35
Mackerel, Atlantic	2.7–25	14.0
Ocean perch, Pacific	3.0–6.0	4.2
Perch, yellow	0.8–1.2	0.9
Pike, yellow	0.8–3.0	1.3
Rockfish (Sebastes)	1.2–4.3	2.5
Salmon, chum	2.2–7.3	4.0
Salmon, coho	3.3–11.2	7.0
Salmon, pink	3.2–11.6	6.5
Salmon, sockeye	7.8–13.7	11.3
Smelt, marine	4.6–8.8	6.3
Smelt, lake	1.5–3.3	2.4
Whitefish, lake	4.7–18.8	9.6

[a]Average value based upon fish taken during the main industrial harvest season, although even then a fairly large range of fat content usually occurs.
Source: Pigott (1996).

3. PROCESSING TECHNOLOGY OF FATS AND OILS

Processing of fats and oils involves first separating oil from the oil-bearing materials and then refining the oils to an acceptable quality for consumption or final applications. Because of the divergent nature of the various oil-bearing materials, there are many deviations in the preparation steps of the oil-bearing materials prior to the extraction of oils. Once the material is properly prepared, the oil may be obtained by heat rendering, such as fish and animal fats; by mechanical pressing for high oil-containing materials such as sunflower seed, rapeseed, and olive; and by direct solvent extraction such as soybean. To further improve the productivity and yield of oil, a combination of partially pressing out the bulk of the oil and solvent extraction has been used by most of the present operations of rapeseed and sunflowerseed mills. These combined processing steps are called prepress solvent extraction. Oilseeds containing less than 30% oil, such as soybean and cottonseed, are processed largely by low shear extruder or expander and then extracted by solvent. These seeds can

TABLE 12.9. Fatty Acid Composition in Large Batches of Commercial Menhaden Oil from 1977–1988.

Fatty Acid			
Nomenclature	Common Name	Symbol	Range (% of Total FA)
Tetracecanoic	Myristic	C14:0	7.2–12.1
Pentadecanoic		C15:0	0.4–2.3
Hexadecanoic	Palmitic	C16:0	15.3–25.6
Hexadeconic	Palmitoleic	C16:1	9.3–15.8
Hexadecatrienoic		C16:2	0.3–2.8
Hexadecatrienoic		C16:3	0.9–3.5
Hexadecatraenoic		C16:4	0.5–2.8
Heptadecanoic	Margaric	C17:0	0.2–3.0
Octadecanoic	Stearic	C18:0	2.5–4.1
Octadecenoic	Oleic	C18:1	8.3–13.8
Octadecatrienoic	Linoleic	C18:2	0.7–2.8
Octadecatrienoic	Linolenic	C18:3	0.8–2.3
Octadecatetraenoic		C18:4	1.7–4.0
Eicosanoic	Arachidic	C20:0	0.1–0.6
Eicosatetraenoic		C20:4	1.5–2.7
Eicosapentaenoic		C20:5	11.1–16.3
Docosenoic		C22:1	0.1–1.4
Docosapentaenoic		C22:5	1.3–3.8
Docosahexaenoic		C22:6	4.6–13.8

Source: Pigott (1996).

also be flaked first and then extracted with solvent. This approach is called direct solvent extraction. The oils obtained by the above-described procedures are normally called crude. Except animal fat, most crude vegetable oils will be further processed, according to crude composition and targeted final applications. These processes are summarized in the following sections. For a detailed description of the processing conditions, readers are referred to other references (Wan, 1991; Hui, 1996; Wan and Wakelyn, 1997; O'Brien, 1998).

3.1 PREPARATION OF OILSEEDS

Most of the oilseeds are prepared through the steps described in Figure 12.2. The conditions and steps used in seed preparation can vary, depending on the types of oilseeds and oil extraction technology applied. The primary goal of seed preparation is to remove foreign matter and to achieve maximum oil recovery with minimum loss of quality.

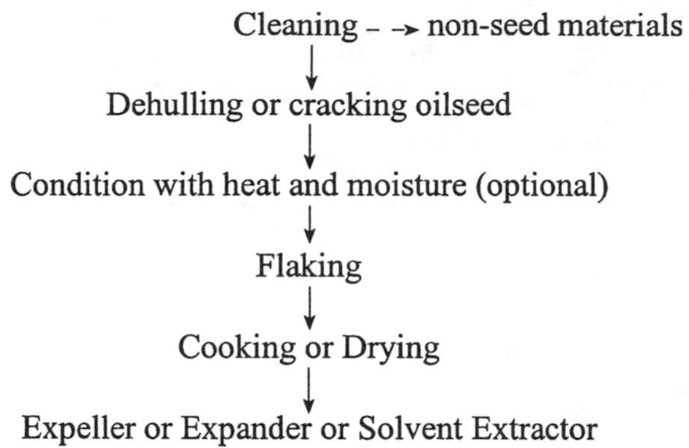

Figure 12.2 Preparation seed for extraction.

3.2 EXTRACTION FOR OIL

Processes to obtain oils or fats from the oil-bearing materials vary drastically because of the nature of the starting material. Animal fats mainly produced by rendering processes will be briefly described later. Once the oilseed is properly prepared, the vegetable oils can be obtained by one of the following processes: (1) mechanical means, such as the ancient edge stone, stone mill, press with lever, hydraulic press in the 19th century, and continuous-screw press at the beginning of the 20th century; (2) prepress solvent extraction; (3) direct solvent extraction; or (4) expander solvent extraction. Oilseeds containing a higher amount of oils than 25% can be processed by mechanical means such as screw press or can be prepressed to remove part of the oil, followed by solvent extraction. Olive oil is still mainly produced by pressing. Rapeseed, sunflower, and linseed are prepressed and then extracted by solvent. Seeds containing less than 30% oil may also be processed by direct solvent extraction without prepressing. To obtain food-grade soy flour, the direct solvent extraction is normally used. Oilseeds containing less than 30% oil are mostly processed by an expander, which is a low-shear extruder, and followed by solvent extraction. Now both soybean and cottonseed are primarily processed by expander-solvent extraction because of its high efficiency. Direct solvent extraction is only used for the production of food-grade soybean flour or to provide the desolventized flakes for the production of protein concentrate and to isolate the food ingredients. High oil content oilseeds such as rapeseed and sunflower seed, after cleaning, dehulling, or just cracking, will need to be prepressed to remove a major portion of the oil and then extract the pressed cake with paraffin solvent to obtain the remaining portion of the oil. Most of the solvent in the extracted oil was removed and recycled by a desolventizerstripper. The oil is handled and traded in its crude form. Most of the cottonseed oil is now miscella refined; that is, the oil is caustic refined with the presence of extraction solvent. This once refined oil is called partially bleached summer yellow (PBSY) by the trade and is more consistent in its color and overall quality.

3.3 RENDERING AND REFINING OF MEAT FATS

Tallow or lard can be processed by either a dry- or wet-rendering process from the fatty tissues. Dry rendering implies that no water or steam is added to the fatty tissues during the process. The fat released from the disintegrated fat cells by heat and agitation is strained or filtered to remove the cooked proteinaceous residues (the cracklings). This operation is similar to the traditional household practice in many families in the Orient. In wet-rendering processes, the fatty tissues are heated in the presence of water or by live steam under pressure to speed up the fat-releasing process. Fats rendered by the wet process generally have a lighter color and milder flavor than those that are dry rendered because of the low processing temperature. Lard produced by this process is called prime steam lard. If fatty tissues are handled and rendered properly, the resulting lard is suitable for cooking or food use without further treatment. The rendered fat may be caustically or chemically refined if its free fatty acid is higher than 0.3%. Edible animal fats may also be subjected to bleaching, hydrogenation, deodorization, interesterification, or fractional crystallization to improve their functional characteristics or to produce fats for specialized applications.

3.4 PROCESSING OF OILS

After being extracted, most vegetable oils are desolventized and traded as crude oils. These crude oils are then processed by the oil refiners into various final products: salad oil, cooking oil, frying and baking shortening, margarine, and more. The commonly used processing steps are outlined in the flowchart shown in Figure 12.3.

Soybean crude oil contains valuable phospholipids, and sometimes a hydration or water washing step is conducted to separate this lecithin fraction prior to caustic refining. This is also termed the degumming step. If the oil is relatively low in free fatty acid and phospholipids at this stage, it may go through the bleaching step, followed by steam deodorization without using caustic to refine. This is called physical refining. Companies that have reliable sources of good-quality soybean crude are adopting physical refining. This will eliminate the caustic refining and water washing after caustic refining and therefore eliminate the production of soapstock and minimize the amount of wastewater generated.

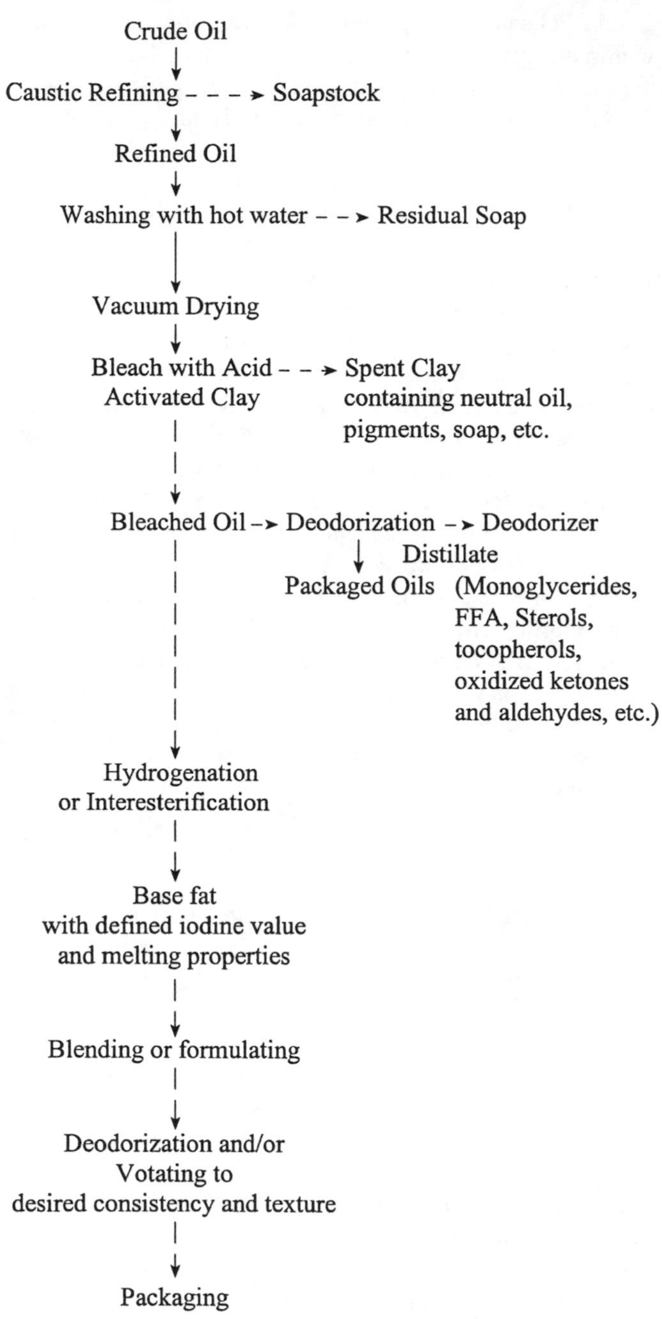

Figure 12.3 Flow diagram for the common practices of oil refining.

Most caustic (chemically) refined oils will normally be followed by water wash, drying under vacuum, bleaching with acid-activated clay to remove color bodies, residual soap, etc., and deodorizing for final packaging or going through hydrogenation or interesterification to create a specialty base fat for formulation or blending prior to deodorizing and packaging. For some vegetable oils that contain higher amounts of saturated fatty acid or wax may also go through a winterization step prior to deodorization and the packaging step. Winterization is a cooling or refrigeration process used to separate the solid fraction from a given oil such as palm and cottonseed oils. Sometimes, it is also used to separate wax from sunflower oil. It is an energy-demanding process.

3.5 UNIQUE SESAME PROCESSING TECHNIQUE

Sesame is believed to have been introduced to China during the East Han Dynasty and Warrior Era. The stone mill process was recorded as early as 1571. The procedure is even used now in the villages and small townships in China. This simple and novel procedure is described in Figure 12.4. The cooking step serves three purposes: (1) to crate the desired aroma, (2) to inactivate the hydrolytic lipases, and (3) to rupture the oil cells for quick and more complete oil release. This procedure apparently has been broadly applied by the Chinese for many centuries. Most of the mechanically or hydraulically

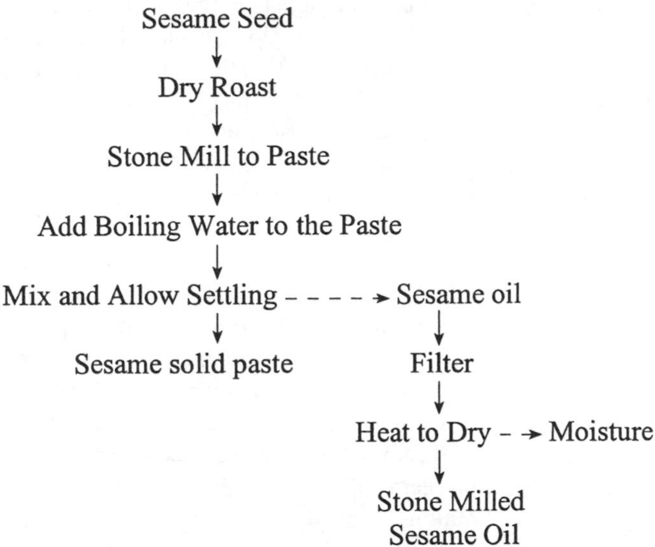

Figure 12.4 Flow diagram for sesame oil of a traditional little stone mill process practiced in China.

pressed vegetable oils were evolved with a similar concept. The pressed oil is then used directly for cooking without refining (Hung, 1994).

4. UTILIZATION OF OILS AND FATS

Traditionally, oils and fats have been used to cook and fry foods and to prepare baked products. These are still the major uses for the present day. Rural populations in developing countries still use oils with minimum refining or largely unrefined processes. Since Napoléon's time, margarine has gained acceptance in the Western diet as a butter substitute. For decades, its production relied on the hydrogenation and blending of vegetable oils to achieve a melting characteristic similar to that of butter. Similar processing steps were used to produce deep-frying and baking shortenings. Rapid growing fast-food services and food processors demand large volumes of vegetable oil-based shortenings to imitate the frying stability of beef tallow and the performance characteristics of pastry fat like tallow and butter. This pattern of oil and fat utilization plus the applications of fats and oils in dressings and mayonnaise are considered mature in developed nations. Developing countries apparently closely follow the footsteps of industrial nations.

5. FUTURE TRENDS

Fats and oils in the Oriental diet will continue dramatic changes both in quantity and quality. This will be accomplished by self-developed and by imported new processing facilities and technology from industrial nations. The rapid economic growth in this region will also stimulate more variety of edible oils to be available to consumers. Within the flux of changes, old technology will coexist with the new, and local dietetic culture will likely guide the formulation of oils to achieve the desired flavor quality and functionality. For example, one of the most popular cooking oils in China now is a blend of rapeseed, peanut, and sesame oils.

While developing countries in Asia are diligently catching up with industrial nations to expand their processing capacity and to adopt more efficient technology for oilseed and oil refinery, there are several factors influencing the supply and demand of edible oils in industrial nations. These factors will surely impact the oils and fats used in the Asian diet. One of the most powerful factors is the growing awareness of health effects of oils and fats. The abundant varieties of reduced calorie foods and fat-free products are the results of fear of obesity, heart problems, and other health-related issues associated with excess fat intake (Watkins et al., 1996). This trend has pushed the fat mimetic technology to the extreme. Numerous carbohydrate- and protein-based, structure-based, reduced-calorie fat replacers are available (Stanton, 1996). More recently, olestra (sucrose polyesters), a noncaloric heat-stable fat replacer is being marketed by Procter & Gamble under the trademark of Olean® (Haumann, 1997). Apparently, all these fat substitutes will play a role in fats and oils–containing foods. Their success and acceptance in the marketplace will be governed by their functional and health benefits at a reasonable cost.

Recent studies on the nutritional role of *trans* fatty acids in hydrogenated vegetable oils used in shortenings, margarine, and salad dressings (Mensink et al., 1991; Hunter, 1992; Mensink et al., 1992; Wahle and James, 1993; Weinberg, 1993; Mann, 1994; Chisholm et al., 1995; Samman, 1995; Ip and Marshall, 1996; Khosla and Hayes, 1996) have prompted serious discussions among professionals about the risk benefit of the hydrogenation process. Another major factor is the impact of regulatory pressure on the fats and oil industry. Nutritional labeling and clean air regulation will all add to the cost of the producers and users of fats and oils.

As the oil industry is striving hard to adjust to what the consumers want and to the numerous new rules from the government, some new hope seems to come from the rapidly progressing genetic technology (Wilkinson, 1997). It appears possible that minimally processed oils and fats will be used in the formulation of shortenings, margarine, dressings, and baking fats in the foreseeable future. But all these will come with a cost, and their successes rely heavily on the acceptance of consumers. While much progress has been made in the genetically modified oilseed area, the final product still relies on the existing processing capabilities to bring it to the marketplace. To continue to satisfy the consumer's need of quality and functional fats and oils, the physical refining, blending, and interesterification by chemical and enzymatic means will be the focus of future processing technologies. Genetically improved oilseeds will become the source of specialty oils or identity preserved oils. Their success will likely be dictated by the benefit they deliver at an affordable cost.

6. REFERENCES

Augustin, M. A., and Ling, E. T. 1987. Composition of mango seed kernel. *Pertanika* 10:53–59.

Beadle, G. W. 1978. Teosinte and the origin of corn. In: *Maize Breeding and Genetics,* E. B. Walden, ed., Wiley, New York, pp. 113–128.

Bengtsson, L., Hofsten, A. V., and Loof, B. 1972. Botany of rapeseed. In: *Rapeseed: Cultivation, Composition, Processing and Utilization,* L. A. Appelquist and R. Ohlson, eds. Elsevier, Amsterdam, pp. 36–44.

Bhattacharyya, S. K. 1987. Mango *(Mangifera indica)* kernel fat. In: *Non-traditional Oilseeds and Oils of India,* N. F. Bringi, ed., Oxford and IBH, New Delhi, pp. 73–96.

Brousse, G. 1989. Olive. In: *Oil Crops of the World,* G. Robbelen, R. K. Downey, and A. Ashri, eds., McGraw-Hill, New York, pp. 462–74.

Chatt, E. M. 1953. *Cocoa: Cultivation, Processing and Analysis.* Interscience, New York.

Chatterjee, D. 1951. Note on the origin and distribution of wild and cultivated rices. *Indian J. Genet. Plant Breed.* 11:18–24.

Chisholm, A., Mann, J., and Skeaff, M. 1995. Trans fatty acids: A cause for concern. *Int. J. Food Sci. Nutr.* 46(2):171–176.

Eckey, E. W. 1954. *Vegetable Fats and Oils.* Reinhold, New York.

FAO. 1977, 1980, 1985, 1987, 1988. *Production Yearbook.* Food and Agriculture Organization of United Nations, Rome.

Fernandez Diez, M. J. 1971. The olive. In: *The Biochemistry of Fruits and Their Products. Vol. 2,* A. C. Hulme, ed., Academic Press, London, pp. 255–79.

Godin, V. J., and Spensley, P. C. 1971. *Oils and Oilseeds.* Tropical Products Institute, London.

Haumann, B. F. 1997. Fat replacers. *INFORM* 8(12):1206–1211.

Heiser, C. B. 1976. Sunflowers. In: *Evolution of Crop Plants,* N. W. Simmonds, ed., Longman Green, London, pp. 36–38.

Higgins, B. B. 1951. *Origin and Early History of the Peanut. The Peanut—The Unpredictable Legume.* Washington: National Fertilizer Association.

Houstan, D. F., and Kohler, G. O. 1970. *Nutritional Properties of Rice.* National Academy of Sciences, Washington D.C.

Hui, Y. H. ed., 1996. *Bailey's Industrial Oil & Fat Products, Volume 4,* 5th ed., Wiley-Interscience, New York, pp. 1–391.

Hung, G. Z. 1994. A partial report on historical processing technology for vegetable oils. In: *Chinese Diet Culture, Vol. 1,* XiChing Li, ed., Economic Science Publishing Co., Beijing, China, pp. 144–152.

Hunter, J. E. 1992. Safety and health effects of isomeric fatty acids. In: *Fatty Acids in Foods and Their Health Implications,* C. K. Chow, ed., Marcel Dekker, Inc., New York, pp. 857–868.

Ip, C., and Marshall, J. R. 1996. *Trans* fatty acids and cancer. *Nutr. Rev.* 54(5):138–145.

Khosla, P., and Hayes, K. C. 1996. Dietary trans-monounsaturated fatty acids negatively impact plasma lipids in humans: Critical review of the evidence. *J. Am. Coll. Nutr.* 15(4):325–339.

Krzymanski, J., and Johnson, R. 1989. Poppy. In: *Oil Crops of the World,* G. Robbelen, R. K. Downey, and A. Ashri, eds., McGraw-Hill, New York, pp. 388–393.

Lakshminarayana, T., Surendranath, M. R., Kristappa, G., Vishwanatham, R. K., and Thirumala Rao, S. A. 1968. Processing of Indian water melon seeds. *Indian Oil Soap J.* 33:323–325.

Love, J. A. 1996. Animal fats. In: *Bailey's Industrial Oil & Fat Products, Volume 1,* 5th ed., Y. H. Hui, ed., Wiley-Interscience, New York, pp. 1–18.

Mann, G. V. 1994. Metabolic consequences of dietary *trans* fatty acids. *Lancet* 343:126–271.

Mensink, R. P., Louw, M. H. J., and Katan, M. B. 1991. Effects of dietary *trans* fatty acids on blood pressure in normotensive subjects. *Eur. J. Clin. Nutr.* 45(8):375–382.

Mensink, R. P., Zock, P. L., Katan, M. B., and Hornstra, G. 1992. Effect of dietary *cis* and *trans* fatty acids on serum lipoprotein levels in humans. *J. Lipid Res.* 33(10):1293–1501.

Mielke, T. 1998. *Oil World Annual.* ISTA Mielke Gmblt, Hamburg, Germany.

Mukherjee, S. K. 1940. A revision of Labiateae of the Indian empire. *Records Bot. Survey India* 14:186–91.

O'Brien, R. D. 1998. Fats and Oils—Formulating and Processing for Applications. Technomic Publishing Co., Inc., Lancaster, PA, pp. 33–36.

Orthoeter, F. T. 1996. Rice brand oil. In: *Bailey's Industrial Oil & Fat Products, Volume 2,* th ed., Y. H. Hui, ed., Wiley-Interscience, New York, pp. 393–409.

Oyenuga, V. A., and Fetuga, B. L. 1975. Biochemistry and nutritive value of water melon seeds. *J. Sci. Food Agric.* 26:843–54.

Pigott, G. M. 1996. Marine oils. In: *Bailey's Industrial Oil & Fat Products, Volume 1,* 5th ed., Y. H. Hui, ed., Wiley-Interscience, New York, pp. 225–254.

Robertson, J. A., and Russell, R. B. 1972. Sunflower: American neglected crop. *J. Amer. Oil Chem. Soc.* 49:239–244.

Salunkhe, D. K., Chavan, J. K., Adsule, R. N., and Kadam, S. S. 1992. *World Oilseeds—Chemistry, Technology, and Utilization.* AVI, New York, pp. 1–23, 140–216, 280–327, 371–379, and 449–529.

Samman, S. 1995. Dietary *trans* fatty acids and coronary heart disease. *Food Aust.* 47(3):S10–S13.

Sangwan, N. K., Dhindsa, K. S., and Gupta, R. 1985. Effect of variety and growing location on the proximate and fatty acid composition of opium poppy (*Papaver somniferum* Linn.). *Int. J. Trop. Agric.* 3:1–8.

Shah, S. G., Subrahmanynam, V. V. R. and Rege, D. V. 1983. Composition and characteristics of four non-traditional indigenous hard fats. *J. Oil Technol. Assoc. India* 16:20–22.

Sharma, B. D., Hove, D. K., and Mandal, S. 1989. Perilla: An oil and protein rich underexploited crop of North-eastern Hills. *J. Oilseeds Res.* 6:386–389.

Soya Bluebook, 1998. Soyatech, Inc. Bar Harbor, ME.

Stanton, J. 1996. Fat substitutes. In: *Bailey's Industrial Oil & Fat Products. Vol. 1,* 5th ed., Y. H. Hui, ed., Wiley-Interscience, New York, pp. 281–310.

USDA. 1986. *Oil Crop Situation Outlook Yearbook.* OCS-11 (July). USDA-Economic Research Service, Washington, D.C.

USDA. 1988. *Oil Crops Yearbook.* OCS-1988 (October). USDA-Economic Research Service, Washington, D.C.

USDA. 1996. *Oil Crops Yearbook.* OCS-1996 (October). USDA-Economic Research Service, Washington, D.C.

USDA. 1997. *Oil Crops Yearbook.* OCS-1997 (October). USDA-Economic Research Service, Washington, D.C.

USDA Foreign Agricultural Service, Feb. 1994. *Oilseeds and Products,* Counselor and Attache Reports.

Vaughan, J. G. 1970. *The Structure and Utilization of Oil Seeds.* Chapman and Hall, London.

Wahle, K. W. J., and James, W. P. T. 1993. Isomeric fatty acids and human health. *Eur. J. Clin. Nutr.* 47(12):828–893.

Wan, P. J., ed. 1991. *Introduction to Fats and Oils Technology.* Amer. Oil Chem. Soc. Press, Champaign, IL, pp. 1–330.

Wan, P. J., and Wakelyn, P. J., eds. 1997. *Technology and Solvents for Extracting Oilseeds and Nonpetroleum Oils.* Amer. Oil Chem. Soc. Press, Champaign, IL, pp. 1–353.

Watkins, B. A., Henning, B., and Toborek, M. 1996. Dietary fat and health. In: *Bailey's Industrial Oil & Fat Products, Volume 4,* 5th ed., Y. H. Hui, ed., Wiley-Interscience, New York, pp. 159–214.

Weber, E. J., and Alexander, D. E. 1975. Breeding for lipid composition in corn. *J. Amer. Oil Chem. Soc.* 52:370–373.

Weinberg, L. 1993. *Trans* fatty acids hidden in foods may raise cholesterol levels. *Environ. Nutr.* 16(7):1, 6.

Wilkinson, J. Q. 1997. Biotech plants: from lab bench to supermarket shelf. *Food Technology* 51(12):37–42.

Zeven, A. C., and Zhukovsky, P. M. 1975. *Dictionary of Cultivated Plants and Their Centres of Diversity.* Centre for Agricultural Publishing and Documentation, Wageningen.

CHAPTER 13

Perspectives on Alcoholic Beverages in China

TIEN CHI CHEN
MICHAEL TAO
GUANGSENG CHENG

1. INTRODUCTION

THIS chapter discusses Chinese alcoholic beverages and sketches their history, manufacturing methods, and role in Chinese culture.

China has one of the oldest civilizations, the largest population on earth, and one of the most influential cultures in East Asia. Her tradition in alcoholic beverage making has now been traced to neolithic times. Relying mainly on grains as the raw material, China developed the unique *qu* technology to convert starch into sugar and sugar into alcohol. China's alcoholic beverage culture is little known in the West though, which is steeped in grape-oriented tradition and malt-based saccharification practices.

In Section 2 we will retrace the history of alcoholic beverage making and consumption. Section 3 gives the classification and representative examples of alcoholic beverages, and Section 4 describes the *qu* technology and traditional manufacturing processes for yellow wine and white liquor. Secton 5 briefly surveys the place of wine and liquor in Chinese culture, including their role in medicine, in food making, and in artistic creation.

Before proceeding further, we must first address the semantic mismatches between Chinese and English terms. To the Chinese, every alcoholic beverage, regardless of the raw ingredients used, whether distilled or not, is *jiu* (sometimes spelled *chiu* or *chiew*); this word is commonly translated as wine. Most *jiu* is based on staples such as millets, sorghum, barley, wheat, rice, glutinous rice, or maize and also sometimes on starchy root vegetables such as sweet potatoes, potatoes, manioc, and legumes. Rather than using the more precise term *starchy plant based,* we shall call these "grain-based," because grains are by far the most respected and the most common ingredients. It may be worth noting that, in the classical Chinese literature, the soybean was considered to be one of the five grains.

In Chinese, different alcoholic beverages are distinguished, if need be, by prefixing; for example, those fermented from grapes are called grape wines, and distilled spirits from grapes are brandy wines. Grain-based, undistilled alcoholic beverages are called yellow wines, and their distilled counterpart are white wines because their most well-known members are, respectively, yellow and white (i.e., transparent), though a given white wine may be yellow, and a given yellow wine may actually be transparent, milky white, or red. To dispel some of the ambiguity, we designate high-proof, starch-based alcoholic beverages as white liquor, but sometimes the ambiguous translation *wine* is unavoidable because the alcoholic content is unspecified, as is often the case with synthetic and medicinal alcoholic beverages (see Table 13.1).

TABLE 13.1. Translating Names of Alcoholic Beverages.

Chinese Pinyin	Literal Meaning	Translation	Western Equivalent
Jiu	Wine	Wine	Alcoholic beverage
Bei jiu	White wine	White liquor	Vodka, whiskey
Huang jiu	Yellow wine	Yellow wine	Grain-based, undistilled wine
Putao jiu	Grape wine	Grape wine or brandy	Wine or brandy
Guo jiu	Fruit wine	Fruit wine or fruit brandy	Fruit wine or fruit brandy
Yao jiu	Medicinal wine	Medicinal wine	Medicinal wine, medicinal liquor
Peihe jiu	Synthetic wine	Flavored wine	Liqueur, cordial, vermouth
Beilandi jiu	Brandy wine	Brandy	Brandy
Xiangbin jiu	Champagne wine	Champagne	Champagne
Shao jiu	Burned wine	Burned wine	Liquor or spirit

2. HISTORY OF CHINESE WINES AND SPIRITS

Simoons (1991), Chang (1977), Anderson (1988), Yao (1989), and Wang (1989) present the history of Chinese wines and spirits (see Table 13.2).

2.1 MURKY LEGENDS, SOLID FACTS (5000 B.C. TO 18TH CENTURY B.C.)

The origin of alcoholic beverages in China, like that in other civilizations, is clouded in mystery. It was said that Yi Di, an official in the court of King Yu, the founder of the Xia Dynasty (about the 21st century B.C.), invented wine and offered it to the king, who became drunk, and said afterwards, "In future generations there must be those who lose their country because of this!" He then immediately issued the first prohibition decree ever.

As it turned out, fact was way ahead of legends. In the neolithic age, about 10,000 years ago, with the dawn of agriculture, drinks based on fermented grain products probably became available, and its consumption quite possibly was an integral aspect of religious rituals and the practice of medicine. The earliest grain under cultivation were millets in the north (about 6000 B.C.) and rice south of the Yangtze River (about 5000 B.C.). In North China, archeologists have found 7000 year-old pottery vessels with a small hole at the bottom, which they judge to be containers for making wine from grains. The hole was plugged during fermenting and then was unplugged to allow the clear wine to flow out below, leaving the lees behind (Wang, 1989, pp. 38–39).

2.2 SHANG INDULGENCE AND ZHOU SPECIALIZATION (18TH CENTURY B.C. TO 221 B.C.)

The court of Shang (18th–12th century B.C.) was fond of wines and used them extensively in rituals. Shang produced magnificent bronzeware for ceremonial eating and drinking. Shang oracle bones frequently used the the word *jiu* (wine). Two kinds of wine were in use: *li* was low in alcohol, a kind of beer; *chang*, flavored with aromatic herbs, was stronger and was often awarded to high officials. In 1987 a Shang bronze container was excavated; it contained ethyl formate, evidently a leftover from *chang* after 3000 years (Yao, 1989, pp. 119–120).

The last Shang king was said (by his enemies) to be a dissolute despot who built "a forest of meats and a pond of wine," such that 3000 revelers would drink like cattle at the same time. Judging by the increased lead content in Late Shang bronzeware, archeologists suspect that his irrational deeds had resulted from poisoning by lead dissolved in wine (Wang, 1989, p. 55), the same reason some historians gave for the decline and fall of the Roman Empire.

TABLE 13.2. Chronology.

About 3000 B.C.	Primitive wine making using special pottery with drain hole.
21st–18th century B.C.	Xia Dynasty in North China. Wine was said to have been invented and offered to its first king, who got inebriated and issued a prohibition decree.
18th–12th century B.C.	Shang Dynasty, known for use of wine in rituals. First mention in oracle bones of flavoring wine with herbs.
12th century B.C. –221 B.C.	Zhou Dynasty. Appointed court officials in charge of wines.
221–207 B.C.	Qin Dynasty. The great wall was built to ward off the Huns.
206 B.C.–220 A.D.	Han Dynasty. Beginning of Silk Road traffic, importing grapes and grape wine into China proper.
265–420	Chin Dynasty. Nomad invasions began, bringing with them kumiss.
420–589	Age of chaos. North (nomad) and South (Chinese) Dynasties. The first known book with a chapter on the making of wines.
581–618	Sui Dynasty. China reunified.
618–907	Tang Dynasty. Extensive traffic with Central Asia. Poetry in praise of grape wine. "Burned wine" mentioned in Sichuan.
960–1279	Song Dynasty. Burned wine used for snake bites. First complete book devoted to wine. Extensive ocean traffic with Southeast Asia and Arabia.
1271–1368	Yuan (Mongol) Dynasty. Importation of arrak from Central Asia.
1368–1644	Ming Dynasty. Importation of New World vegetables, including maize, potatoes, and sweet potatoes; these became ingredients for white liquor. Solid-phase fermentation was believed to have been invented during this period.
1644–1911	Qing (Manchu) Dynasty. Exposure to Western alcoholic beverage technology.
After 1911	Revolution toppled the Ching Dynasty and established the Chinese Republic. The Japanese invasion led to the World War II; then the civil war led to the founding of the Chinese People's Republic and the retreat of the Nationalist government into Taiwan. China has now embarked on scientific understanding and production of traditional and Western-style alcohol beverages.

The succeeding Zhou Dynasty (12th century B.C. to 221 B.C.) was far more ascetic, and early Zhou literature recorded an admonition to the vanquished Shang folk about the harm of their excessive drinking. The main ingredient of Chinese wines and liquors were (and are still today) grains and starchy vegetables; their conversion into wine requires first converting starch into sugars by saccharification, and then sugars into alcohol. Even before about the 12th century B.C., the Chinese had discovered the saccharification effects of sprouting grain and had invented *qu,* a grain-based solid embedded with dormant microorganisms, for more efficient conversion. The same *qu* also converts sugar into alcohol, but inefficiently; for this purpose, yeast, from the air or cultured (in *xiao qu* "small *qu*"), was used. Different Zhou court officials were engaged in the making, storing, and serving of wine. After 722 B.C. Zhou power fell into the hands of feudal lords. The king of Yue (about 500 B.C.) was said to have poured wine into a river before a battle, and all citizen-warriors who drank it became "a hundred times more willing to fight." The modern name of his capital is Shaoxing, now the most famous producer of yellow wine. A sealed bronze jar of wine from the kingdom of Zhongshan (4th century B.C.), near Beijing, was excavated recently; the wine was a brilliant green through interaction with the bronze vessel. It still has traces of alcohol and esters, and electronic microscopy revealed shells of yeast at the bottom. Wine also increased in variety: Qu Yuan (d. 278 B.C.), the great poet of the kingdom of Chu in Central China, wrote about the use of Sichuan peppercorns *(fagara)* and cinnamon for flavoring.

2.3 THE SILK ROAD AND THE FIRST BOOK ON WINE MAKING (221 B.C. TO 618 A.D.)

The First Emperor of Qin built the Great Wall to ward off the Hun nomads from the North, yet the Huns continued their raids, and the founder of the Han Dynasty and his armed expedition was sieged by them instead, starved for seven days, and escaped only through intrigue. Partly to avenge this indignity, his great-grandson WuTi sent Zhang Xian westward to find allies for a giant pincer attack on the enemy. Zhang was captured by the Huns on the way out and again on the way back, but succeeded in opening the famed Silk Road to the Western Lands (including the present Xinjiang Uigur Autonomous Region and parts of Central Asia) and beyond, through which soon came grapes and grape wine. Good grape wine was known then to remain drinkable even after several decades; most scholars attribute this to a high alcohol content, but it could also be because of an unusually high residual sugar. The art of grape wine making stayed in the hands of Turkic peoples along the Silk Road; even the venders of grape wine were usually Turkic women. In subsequent centuries, the Silk Road was blocked for long periods, and as a result, grape and grape-based wines and spirits never really took root in China Proper until the 20th century.

Nomads, invited to settle south of the Great Wall, soon overran North China. Their favorite drink was *kumiss,* fermented naturally from mare's milk. Even so, the first known concrete account of Chinese wine-making technology using *qu* was published in North China in *Qi Min Yao Shu* (Important Arts for Peoples' Welfare, by Jia Si-xie about 540 A.D.). This technique has since spread to Japan, Korea, Vietnam, and India. The book listed 33 different wines and mentioned the making of medicinal wine, notably with the skin *(pi)* of the plant *wujia (Acanthopanax gracilistylus). Wujiapi* liquor today remains the most popular medicinal wine.

2.4 THE BURNED WINE OF TANG, SONG, AND THE MONGOLS (618–1368)

The Tang Dynasty was marked by a great outpouring of poetry, much of it in praise of wine and drinkers. The most well-known imbiber-poet, called the Immortal of Wine, was Li Po (701–762), who wrote,

Don't you see the water of the Yellow River, coming from Heaven,
Charging into the sea, never to return?
Don't you see in mirrors in lofty halls, white hair which cause lamenting,
In the morning like black silk, but at twilight, snow?
In life, when in the mood, one should enjoy to the fullest measure,
Let no golden flask face the moon alone!

Most probably he was drinking yellow wine with about 16% alcohol or undistilled grape wine sold by Turkic hostesses. While recent excavations unearthed 2000-year-old stills, they were probably used only for practices in alchemy. Explicit mentioning of distilled liquor (*shao jiu,* meaning "burned wine," anticipating by at least three centuries the Dutch word *brandewijn,* which became the English "brandy") occurred near the end of the Tang Dynasty (about 900) (Temple, 1986) and seemed to be localized to the Sichuan Basin, surrounded by mountains in Central China. As the poet Yong Tao wrote,

Since reaching Chengdu, the burned wine matured
I wish to enter Changan nevermore.

This shows, not only that burned wine was available in Chendu, the queen city in Sichuan, but that it was probably not available in Changan, then the greatest cosmopolitan city in the world. That burned wine really was distilled was later seen in a text in the Song Dynasty (960–1279), which prescribed burned wine for external application on snake bites. Only distilled liquors can neutralize the effects of venom.

Song was economically prosperous but militarily weak, first losing northern China to the Jurchen nomads, and then was extinguished by the Mongols (1271–1368). A Jurchen still was excavated and was dated at 1161 or earlier. Sorghum, called *gaoliang,* apparently was first cultivated in late Song, and it became an important ingredient for distilled spirits. Zhu Gong's *Beishan Jiu Jing* (The Wine Classic of North Mountain) was the first book extant completely devoted to wine and wine making; it mentioned using heat to stop fermentation, anticipating Louis Pasteur by six centuries. *Wulin Jiushe* (Bygone Happenings in Hangzhou), a book written on the verge of the Mongol occupation of Hangzhou, the South Song Capital (called Kinsai by Marco Polo), listed 54 wines, three of them called "dew," which could well be distilled white liquor. In 1330, Hu Sihui, royal physician of the Mongol court, wrote *Yinshi Zhengyao* (Outlines on the Right Ways for Food and Drink), a book on healthful food and drinks; he listed three kinds of mare's milk, perhaps in different degrees of fermentation. He also listed 13 wines; most were medicinal wines with dissolved herbs, but also included was grape wine and arrack from Central Asia. Arrack became so popular that *Bencao Gangmu,* the great herbal encyclopedia by Li Shizhen (1518–1593), stated that the technique for distilling spirits was imported.

2.5 EXCHANGES WITH THE WEST (1368–PRESENT)

During the Ming Dynasty (1368–1644), white liquor became widely appreciated. Some of the famed white liquor factories today had their origin in Ming, most notably *fen* liquor from the north and *moutai* from the southwest. New World starchy vegetables (corn, potatoes, sweet potatoes) became available, probably through the Philippines, and they became standard ingredients for wine and liquor.

Foreign powers occupied enclaves in China during the (Manchu) Ching Dynasty, through which Western technology was brought in. The Russians in the northeast introduced vodka. The German influence in part of the Shandong Peninsula led to the first beer factory. Zhang Bishi, a Chinese merchant from Indonesia, established the Zhangyu Grape Wine Company in Yantai and in Shandong, using French and Italian grapes and hiring the Austrian consul to Yantai as winemaker; it produced the first brandy in time to win a gold medal in the Panama Exhibition in San Francisco in 1915.

In 1911 the Ching Dynasty was overthrown; then for more than 30 years, China was in almost constant turmoil. After the devastations of the war with the Japanese, followed by the civil war, the Chinese People's Republic was proclaimed in 1949 on the mainland, and intermittent internal strifes sapped the economic strength of the nation until the start of modernization and the free market economy in the 1980s. Now the traditional grain-based wines, grain, and starch-based liquors are produced under strict quality control; brandy, vodka, rum, Western-style whiskeys, and even grape wines with vintage labels and varietal names have also made their appearance, stimulated partly by cooperation with Western producers and the prospect of foreign markets. Taiwan, meanwhile, developed into one of the four "economic lesser dragons" of Asia. Since mainland alcoholic beverages were not easily available, Taiwan tried to determine and then reproduce their unique flavors scientifically using local resources.

With steadily increasing prosperity on both sides of the Taiwan Strait, beer is becoming indispensible in the average household. Good Chinese white liquors and imported alcoholic beverages are a common sight at banquets.

3. MAJOR ALCOHOLIC BEVERAGE TYPES

3.1 CLASSIFICATION

Heilongjiang Business Institute and Beijing Municipal Sugar, Tobacco and Wine Company (1980) and Zheng (1980) are good sources for reviewing classifications of alcoholic beverages (see Table 13.3).

In the People's Republic of China, alcoholic beverages are usually subdivided by processing technique into three categories:

- fermented wines *(niangzao jiu),* i.e., not distilled; up to 20% alcohol
- distilled liquors *(zhengliu jiu);* 32–66%
- flavored or synthetic wines (*peihe jiu,* "pieced-together wine"); 16–66%

Commercially, alcoholic beverages are classified into six types:

- white liquors *(bei jiu):* grain-based and distilled, somewhat like vodka or whiskey

TABLE 13.3. Standard Classification of Chinese Alcoholic Beverages.

Name	Base	Distilled or Fermented	Examples and Remarks
White liquor	Grain	Distilled	Moutai liquor
Yellow wine	Grain	Fermented	Shaoxing wine
Beer	Malt	Fermented	Flavored with hops
Grape	Grape	Either	Gold medal brandy
Fruit	Fruit	Usually fermented	Litchi wine
Flavored or synthetic	Grain or fruit	Either	With added flavoring/ medicinal agents, e.g., *Wujiapi*

- yellow wines *(huang jiu):* grain-based and fermented but not distilled
- beers *(pi jiu):* grain-based and hop-flavored
- grape wines *(putao jiu):* all grape-based alcoholic beverages, including brandy
- fruit wines *(guo jiu):* fruit-based, sometimes fortified with distilled spirits
- flavored or synthetic wines: usually based on yellow wine or white liquor

This classification scheme may appear arbitrary to the outsider; nevertheless it reflects the grain-centered Chinese imbibing tradition. But the sugarcane-based rum, now made in South China, is hard to place into these categories, as is kumiss fermented from mare's milk, popular for millenia in Inner Mongolia. Incidentally, a famous wine grape brought in from the Silk Road, still being used to make wine in the Northwest, is called mare's teat (perhaps because of its shape); because in Chinese the same word is used for both "teat" and "milk," it is sometimes incorrectly rendered in English as "kumiss grape."

3.2 YELLOW WINE (UP TO 20%)

In the People's Republic, all grain-based, undistilled wines are called yellow wines, named after the color of the most distinguished member of this category, and to distinguish from (distilled) white liquors. It is not always yellow in color.

Nowadays, if one mentions yellow wine, one usually means Shaoxing wine *(Shaoxing jiu)* or a wine made in the same style. Shaoxing is a city 100 miles southwest of Shanghai, with a wine-making tradition of at least 2000 years. Its standard offering (fermented from glutinous rice, about 15% alcohol) is called "prime red" *(yuanhong)* because of its red-painted jar. Higher grades include "rice added" *(jiafan)* to reach 16.5% alcohol; "flower engraved" *(huadiao)* is *jiafan* wine aged in sealed, decorated earthen jars. They are often likened to sherry, but the taste and aftertaste are quite different. Aging in sealed earthen jars improves its flavor and complexity apparently without limit, and a common practice for more than a millenium is to put jars of *huadiao* wine in storage upon the birth of a girl, to be opened on her wedding day; this wine is called "daughter's red" *(nuer hong),* referring to the prevailing decor color in a wedding.

Some wines classified as yellow deserve specific mention: Glutinous rice wine *(numi jiu)* (often with only 3% alcohol), also called sweet wine *(tian jiu)* is usually made by adding yeast cake powder to steamed glutinous rice for about a day. This sweet liquid, being too low in alcohol to transport, is usually produced locally or even in the household kitchen. A specialty in Sian is a milk-colored thick wine *(chou jiu),* essentially glutinous rice wine mixed with its own beaten wine lees. Rice wine *(mi jiu)* or clear wine *(qing jiu)* (8–18%) is a transparent wine fermented from rice, similar to Japanese *saki.*

3.3 WHITE LIQUOR, CALLED "WHITE WINE" *(BEI JIU)* (30–70%)

Distilled from fermented grains, usually sorghum, white liquor owes its special bouquet to the unique use of *qu,* the agent with stabilized mold and bacteria used to convert starch into sugar (see Table 13.4). The most well-known white liquor is Moutai (also spelled Maotai), used for official banquets in China. The bouquet is classified into five types: clear bouquet *(qing xiang),* rich bouquet *(nong xiang),* fermented-bean-sauce bouquet *(jiang xiang),* rice bouquet *(mi xiang),* and compound bouquet *(fu xiang).* The list of bouquets is being expanded slowly, and several others are awaiting certification (Zhu, 1991; Heilongjiang Business Institute and Beijing Municipal Sugar, Tobacco and Wine Company, 1980, pp. 56–58).

3.4 GRAPE AND FRUIT-BASED WINES

All wines and spirits based on fermented grape juice are classified as grape wine by Chinese tradition. These include flavored grape wines (such as ver-

TABLE 13.4. Famed White Liquors.

Name	*Qu* Type	Bouquet	Proof	Province
Moutai	Large *qu*	Beanpaste bouquet	106	Guizhou Xifeng
"Western Phenix"	Large *qu*	Clear bouquet	106 or 130	Shaanxi
Fen liquor	Large *qu*	Clear bouquet	120	Shanxi
Wuliangye (Five-grain nectar)	Large *qu*	Rich bouquet	104 or 120	Sichuan
Luzhou *laogau tequ* ("Special *qu* from the old cellar of Luzhou")	Large *qu*	Rich bouquet	110 or 120	Sichuan
Dong jiu	Medicinal *qu*	Other bouquet	116–120	Guizhou
Jiannan chun (Spring in Jiannan)	Large *qu*	Rich bouquet	120	Sichuan
Yanghe daqu	Large *qu*	Rich bouquet	110, 118 or 120	Jiangsu
Gujing gongjiu (Ancient well tribute liquor)	Large *qu*	Rich bouquet	120–124	Anhui
Chuanxing daqu	Large *qu*	Rich bouquet	116–118	Sichuan
Dah-chyu	Large *qu*	(not officially classified)	132.5	Taiwan

mouth), medicinal grape wines (such as ginseng grape wine), fortified grape wines (such as port), sparkling grape wines, and brandy–all classified as grape wines. The popular osmanthus flower aged wine *(guihua chenjiu)* is put in this category because of its grape base.

Fruit wines are fermented from fruit juice, usually with the addition of sugar. The fruit used includes apples, pears, peaches, plums, crabapples, cherries, oranges, tangerines, *lychees,* pineapples, mulberries, even tomatoes, and lesser known ones such as kiwifruit (the Chinese gooseberry that became popular worldwide after transplantation in New Zealand).

3.5 FLAVORED WINES AND MEDICINAL WINES

The term *dew wine (lu jiu)* is often used to cover the entire category, though the narrow sense of this term refers only to flower- or fruit-flavored, sweet alcoholic beverages.

Flavored or synthetic wines are made by adding extra ingredients to yellow wine, white liquor, or ethyl alcohol to alter their taste and bouquet. Flavoring agents include rose petals in rose dew liqueur *(meigui lu)* and bamboo leaves in bamboo-leaf green *(zhuye qing).*

Medicinal wines are flavored wines, using as flavoring agents herbs (such as ginseng in ginseng wine) or animal parts with medicinal properties (tiger bones, seal sex organs) for the health, rather than the enjoyment, of the drinker. Most medicinal wines are tonics or reinforcing wines *(bu jiu),* valued for their ability to reinforce the body against the elements. The line separating common flavored wines and medicinal wines is not easily drawn, because medicinal ingredients could possess perfectly agreeable flavors. *Wujiapi* wine (also spelled *ng ka py*) is a prime example; this medicinal wine is said to expel rheumatism, improve circulation, relax muscles and tendons, and nourish both the spleen and the stomach, yet to the imbiber, its unique taste probably is enough reward.

4. THE MANUFACTURING OF ALCOHOLIC BEVERAGES IN CHINA

Heilongjiang Business Institute and Beijing Municipal Sugar, Tobacco and Wine Company (1980) and Hsu (1975) discuss manufacturing of alcoholic beverages in China.

4.1 THE TRADITIONAL CHINESE ALCOHOLIC BEVERAGE INDUSTRY

China developed its alcoholic beverage industry in relative isolation, using its own resources, with little contact beyond its East Asia neighbors. China's unique practices are seen even today, particularly in the making of yellow wines and white liquors.

(1) An industry based on grains: The traditional Chinese alcoholic beverage culture is based on grains and starchy vegetables rather than sugar-rich grapes. Although grape wine was 2000 years ago and has been in continuous production in the Northwest, yellow wine has an even longer tradition and reached a much wider public. The development of white liquor, though hard to trace in detail, was a natural next step.
(2) Combining saccharification and fermentation: Agents are needed to convert starch into sugar in saccharification, before sugar can be fermented into alcohol. While the West also uses grains to produce liquors and beer, its traditional saccharification agent is barley malt. The Chinese knew 3000 years ago, but later forgot, how to use malt for wine making. This was because of China's unique *qu* technology, combining saccharification and fermentation that produces beverages of higher alcoholic content and greater complexity.
(3) A preoccupation with medicine: The Chinese believe that everything has medicinal value, and wine and spirits are no exception. The latter can be reinforced by medicinal herbs and animal parts to heal the sick and to safeguard the body. The extension to the use of herbal medicine in alcoholic beverage manufacture is only natural, and some of these herbs can favor the growth of beneficial microorganisms in *qu.*
(4) Conservatism: The making of wine and spirits is respected as an art. The time-consuming fermentation steps, primitive stills, and labor-intensive procedures are maintained, lest the quality of the product suffers. Mechanization tends to be adopted only in conveying ingredients and packaging and transport of the final products. Taiwan appears to be more innovative as it tries to reproduce mainland products using locally abundant ingredients.
(5) Adaptation to local conditions: China's grain orientation is partially explained by noting that grains are plentiful along the densely populated areas along the China coast and in central China, with thriving commerce and convenient transportation channels. Nevertheless, because of the regional economy, cultural traditions, and limited transportation until recently, grape wine remains a tradition in the Northwest; mare's milk is fermented and even distilled in Inner Mongolia; *qingke (Fagopyrum tartaricum),* a cold-resistent buckwheat, is used to make wine in Tibet.

While Western-style wines and spirits are made in China in much the same way as in the West, local conditions do make a difference. Part of the barley in many Chinese beers is replaced by the more abundant rice and some with maize and glutinous rice. Many wines are made from Chinese fruits little known in the West, and many modern Chinese grape wines and vermouths contain added Chinese medicinal herbs.

4.2 *QU* TECHNOLOGY

We shall concentrate on yellow wines and white liquors, employing *qu,* a fermentation containing stabilized microorganisms, in their traditional manufacture.

4.2.1 History

The Chinese have been using *qu* for more than 3000 years. In the *Classic of Books,* written more than 3000 years ago, there is the passage, "To make wine *(jiu)* and *li,* there are only *qu* and malt." *Li* is a sweet beverage of low alcohol content using malt as starter, perhaps close to beer. It disappeared from the Chinese beverage repertoire about 15 centuries ago, along with its method of manufacture from malt, but wine *(jiu)* made from *qu* continues to this day. The book *Qi Min Yao Shu* (~540 A.D.) gives no less than 10 methods for producing eight different kinds of *qu. Qu* is largely responsible for the production of yellow wines, white liquors, vinegar, soy sauce, bean sauces *(jiang),* and fermented beancurds.

4.2.2 *Qu* Making

The main ingredients for most *qu* are barley, wheat, rice, peas, and beans, with fillers such as rice chaff, corncobs, peanut shells, and various herbs and wild plants to favor the production of certain growths while suppressing others. Perhaps the champion of medicinal *qu* use is Dong liquor; it employs wheat-based large *qu* with 40 added herbs, as well as rice-based small *qu* incorporating 95 herbs.

The ingredients are crushed and pressed into shape and then kept in dark chambers with careful control of temperature and humidity and are turned periodically. It had been thought that cooking was required to prepare for the growth of microorganisms, but about 800 years ago, this was found unnecessary. Usually the microorganisms in the air within the chamber suffice to seed the *qu,* but contact with fully ripened *qu* gives better quality control.

4.2.3 Saccharification with Fermentation

The microorganisms in *qu* include the yellow mold *Aspergillus oryzae,* the black mold *Aspergillus usami,* and members in the Saccharomyces (yeast), Rhizopus, Mucor, and Monascus families. Their enzymes convert starch into sugar, and *qu* is sometimes called a starter for making alcohol from starch. It is interesting that in 1892, Albert Calmette studied *qu,* isolated *Amylomyces rouxii,* and developed the Amylo process for malt-free saccharification, which is extensively used in the industrial production of ethyl alcohol.

Actually, *qu* is much more than a starter. The enzymes also ferment sugar into alcohol, though not as efficiently as yeasts, which are often added to enhance fermentation. *Qu* also generates a large number of by-products that give the resultant wine and spirits their complex bouquet and taste.

4.2.4 Common Types of *Qu*

Large *qu* is in the form of rectangular bricks, which takes some 20 days under expert supervision to reach the desired microorganism growth. It keeps for months afterwards without noticeable deterioration. Relatively slow-acting, it is nevertheless responsible for the complex flowery bouquet of some of the most famous white liquors in China. Small *qu* is made in different shapes and often with liberal addition of Chinese herbs and is also called "medicinal *qu*" *(yao qu)* or "wine medicine" *(jiu yao).* It favors the growth of fewer varieties of microorganisms; hence, it is more efficient in making alcohol but gives a simpler bouquet. Other *qu* types for wine and spirits production include bran *qu* (with pure strain molds artificially introduced into bran), wine lees *qu* (wine lees with bran added), liquid *qu* (molds introduced into liquid culture), cellulose *qu* (with molds capable of digesting cellulose), and red *qu* with mainly Monascus molds attached to cooked, soured rice.

4.3 YELLOW WINE MANUFACTURE

Yellow wine will be represented by its most illustrious member, namely Shaoxing wine (see Table 13.5).

TABLE 13.5. Common Shaoxing Wines.

Prime red *(yuanhong):* Spread-rice type, limited aging (1–3 years).
Rice-added *(jia fan):* Spread-rice type with 10% more rice (or 10% less water) and longer saccharification and fermentation period. Limited aging.
Flower-engraved *(hua diao):* Spread-rice type. Rice-added wine aged several years.
Daughter's red *(nuer hong):* Spread-rice type, with longer aging than *hua diao.*
Fragrant snow *(xiang shue):* A fortified wine based on sprayed-rice type wine and lees, with added *qu* and white liquor distilled from wine lees for refermentation. Limited aging. Unlike other Shaoxing wines, this one is to be sipped at room temperature, not warmed.

4.3.1 Ingredients

The chief ingredient of Shaoxing wine is glutinous rice, and the excellent water near Shaoxing is also used. The *qu* used is made of wheat in the form of bricks. Pure yeast produces a sweet, uncomplicated wine, an example being the sprayed-rice type of wine *(linfan jiu)*. The best Xiaoshing wines are of the spread-rice type of wine *(tanfan jiu)*, using sprayed-rice wine as a natural source of yeast.

4.3.2 Sprayed-Rice Method

In the sprayed-rice method, the glutinous rice is sieved, washed to remove impurities, soaked, steamed, and then sprinkled with cold water to cool. An earthen container is dusted with yeast powder, and the cooled, cooked rice is made into the concave form of a bird's nest and put inside. Yeast powder is dusted on the "nest," and the jar is covered. After about 36 h, water and powdered wheat *qu* are introduced. When the temperature rises because of saccharification and fermentation, the mash rises to the top and is raked periodically with a special wooden rake to cool, to add oxygen, and to expel carbon dioxide. The raking continues until further raking meets little resistance. The lees sink gradually to the bottom. When the sinking is complete, the raw wine can be skimmed from the lees. The lees are used as a yeast source in the manufacturing of spread-rice wine, as will be shown presently. If only wine is needed, the lees are pressed to extract the wine.

4.3.3 Spread-Rice Method

The glutinous rice is sieved, washed, soaked for 2–3 weeks, and then steamed. The cooked rice is spread on a bamboo mat to cool. The cooled rice is mixed in a large jar with powdered wheat *qu,* wine lees from the sprayed-rice method, the water used in soaking rice (without the precipitates), and fresh water and is covered for saccharification and fermentation. During the next week, raking is conducted like in the sprayed-rice method, using highly skilled workers. When the residues sink to the vat bottom, the temperature of the mash approaches ambient conditions, the primary fermentation is complete, and no more raking is needed. For the next several days, secondary fermentation occurs, either in the same vat or in smaller jars to which the mash is transferred. In either case, the jars are sealed and kept at low temperatures for about 2 months. Then the raw wine can be extracted from the lees.

4.3.4 Pasteurizing and Aging

The wine produced by either method is extracted by pressing, colored with brown sugar, pasteurized, and then put in germ-free jars for aging. These operations can all be mechanized as an assembly line. The aging takes from several months to several years. Though the jars appear to be waterproof, they actually contain desirable microscopic pores to allow limited exchange with the environment. The longer the aging, the higher the production cost is, but, at least up to several decades, the better the wine.

4.4 THE PRODUCTION OF WHITE LIQUOR

Traditionally, white liquor uses *qu* as a saccharification and fermentation agent, relying a great deal on solid-phase fermentation, which is believed to have been invented some 400 years ago in North China. Primitive stills are used in batch distillation, and aging is done in earthen jars.

4.4.1 Ingredients

Many of the best-known white liquors use sorghum as the sole ingredient. Other ingredients include corn, rice, barley, sweet potatoes, and manioc. All starch-containing or sugar-containing substances could be used, and the list of "substitute ingredients" include rice chaff, sugarcane, beets, seeds of loquat, and more than 100 kinds of plant products, including acorns, mulberries, black dates, and many little known in the West. The famed five-grain nectar uses sorghum, rice, glutinous rice, maize, and wheat.

The chaff of sorghum has some tannin, which imparts a characteristic flavor to the final liquor. Rice imparts a characteristic mild flavor. The germ portion of corn contains excess fat and should be precluded. Barley has too much protein and cellulose, generating higher alcohols that give a pungent tartness to the liquor. Sweet potatoes tend to produce harmful methyl alcohol, as does manioc, which, in addition, has cyanogen, which must be removed in pretreatment. Since the process involves distillation, the requirement on water quality is not as stringent as in yellow wine but should be more or less neutral, with neither pollutants nor unwanted microorganisms.

4.4.2 Saccharification and Fermentation

Quality white liquor calls for the use of the slow-acting large *qu,* and the preferred method of saccharification and fermentation is by slow-acting solid-phase fermentation, using very little added water.

The ingredients are boiled into a paste and then cooled. They are mixed with powdered *qu,* yeasts, and a small amount of water and then sealed in an airtight vault for months. Within the vault, solid, liquid, and gaseous phases coexist, and the chemical actions are exceedingly complex. In general, the longer the fermentation period, the more fragrant substances are produced.

4.4.3 Distillation

While major distillers in the United States and elsewhere are using sophisticated, efficient multicolumn stills for continuous production, minimum supervision, and ease of component separation, the tendency in China is to stick to the tried-and-true, labor-intensive methods to guarantee product quality (Peng et al., 1995; Chang, 1995). In this way, they are no different from the small Highland Scotch malt whiskey manufacturers.

Most white liquors are made a batch at a time, using traditional simple stills called "sky pots" *(tianguo),* with a pot of cooling water above and heated water below; the mash from the fermentation vault plus the used mash from a previous pass (if any) are placed in the middle part, separated from the heated water below by a bamboo sieve. The upper pot comes in two styles. It may be concave upward, condensing the distillate at the bottom center, to be collected by a funnel, or it may be concave downward, allowing the condensed distillate to drip through a side tube leading to the collector outside. During distillation, the saccharification and fermentation processes continue while low-molecular-weight products evaporate and become the distillate.

As the saying goes, "Slow steaming for the liquor, and high steaming to chase the end." This means that the heat should be applied slowly initially and then built up steadily, without sudden fluctuations, but vigorous heat should be used near the end. Supervision by skilled operators is critical. The first part of the distillate, called "the liquor head" *(jiu tou),* is fragrant and tart to the palate, containing low-molecular-weight aldehydes, ketones, and esters of formic and acetic acids. It can be retained for later blending. The last part, called "the liquor tail" *(jiu wei),* contains organic acids, ethyl lactate, high-molecular-weight aldehydes, esters of fatty acids, and fusel oil, but much less alcohol; it is often oily and cloudy in appearance, fishy in smell, and slightly bitter in taste. It can be returned to the still to mix with the next batch or could be mixed with a new batch in the fermentation vault.

The midportion(s), called "the liquor body" *(jiu sheng),* is the main product. It contains the most ethyl alcohol and has esters of acids with 2–6 carbon atoms. It is fragrant, with a pleasant aftertaste.

Because both the solid-phase fermentation process and the distillation are relatively inefficient, the lees after a distillation run still contain incompletely fermented ingredients and valuable trapped substances. It can be added back to the fermentation vault or participate in the next batch of distillation until little extractable alcohol remains, when it is finally discarded as animal feed. The water remaining in the still bottom contains high-molecular-weight substances that could also be used in blending.

4.4.4 The Making of Moutai Liquor

Among the most elaborate distillation technique is that used in making Moutai liquor. The (initial) saccharification/fermentation takes a full month, before the mash is distilled. The first pass distillate is returned to the fermentation vault *in toto* "to nourish the vault." A month later, the looping passes of the distillation-revaulting cycle begin. In each pass a midportion distillate is accepted, but the liquor head and liquor tail are returned to the vault together with added *qu.* The production cycle of the raw liquor is complete only after looping six times, for a total of seven distillation passes over a span of 10 months. The last pass distillate also returns to the vault to join in the next cycle of manufacture. The second through the sixth midportion distillates are aged separately before blending.

4.4.5 Aging and Blending

The raw distillate is commonly kept in earthen jars for months to years, unlike the use of charred wood barrels in the West, with inherently more breathing yet much evaporation loss. During aging, the effervescent aldehydes disappear, and the white liquor, losing its tartness, turns mellow, with an enhanced bouquet. Then, highly skilled blending is done to ensure uniform quality and to bring out the personality of the brand. Another popular saying in the industry is, "Bouquet generation calls for fermentation, bouquet extraction takes distillation, and shaping requires blending." Moutai liquor takes 3 years before blending and then at least another year before release to consumers.

5. ALCOHOLIC BEVERAGES AS A CHINESE CULTURAL TRADITION

5.1 ALCOHOL BEVERAGES IN MEDICINE

In Chinese medicine, everything to be taken by the mouth has an intrinsic caloric value or humor (Anderson, 1988). A primary aim of Chinese medicine is to reach caloric equilibrium in the human body; any excess of a humor is redressed by administering a compensating humor through the mouth. In this regard wine lees, *qu,* and wine of low alcohol content are considered sweet, peppery, and warm; red *qu* is milder in taste and is considered sweet, warm, and nonpoisonous. Whiskeys are judged to be "very hot," undoubtedly because they improve circulation, warming the drinker all over. Chinese doctors often advise patients to take a moderate quantity of yellow wine or white liquor every day, as medicine or tonic. After childbirth, women are given chicken boiled in yellow wine and ginger to improve circulation and to replenish loss of blood. White liquor, often mixed with herbs, is also used externally as a body rub for stiff joints and rheumatism.

Alcohol is an excellent solvent and reaches the blood vessels quickly. It is therefore used to extract the essential ingredients from raw medicine, also a good agent [called "medicine inducer" *(yaoyin)*] to coax the body into receiving medicine suspended and/or dissolved therein.

The most common wine-medicine combinations are the medicinal wines, using yellow wine or white liquor as a solvent to dissolve the essences of herbs and animal products as if they were flavoring agents.

In Chinese medicine, many plants and animal parts are valued for their ability to reinforce *(bu)* the body, particularly for the aged, frail, or infirm, and up to a dozen such ingredients may be infused in white liquor or yellow wine to become tonic wine *(bu jiu)*. To prove authenticity, the major ingredients may be left inside the transparent bottle for buyer scrutiny. Some tonic wines are reserved for male drinkers only, many of which are said to strengthen the kidney (i.e., increase sexual potency). A few contain *dongquai (Angelica sinensis),* known in Chinese medicine for its ability to cure or relieve female disorders. Herbs are even basic ingredients of *qu,* participating in the alcoholic beverage-making process itself.

Many herbs used in medicinal wines are known in the West only through their scientific names. Notable ones are *ginseng (Panax ginseng), dangquai, lingzhi (Gandoderma lucidum) heshouwu (Polygonum multiflorum),* and *wujiapi (Acanthopanax gracilistylus).* The animal products used include mutton, chicken, black-boned chicken, *lurong* (the fluffy horns of young deer), tails of doe, turtle shells, snakes, geckos, sea horses, and dragon fleas (the beetle *Cybister japonicus*).

5.2 ALCOHOLIC BEVERAGES AND FOOD PREPARATION

Alcoholic beverages are usually consumed with food in the Orient and are an important ingredient of Oriental cuisine, imparting aroma while adding taste to food. Moreover, they dispel the unpleasant odors of fish and raw food.

Very broadly speaking, lees and low-proof wines easily affect soft and blander foods, but it takes liquor to assert its own presence in the company of strong-tasting foods to penetrate foods firm in texture or to neutralize the effect of excessive grease. Thus, in Chinese cuisine, wine lees and yellow wine are commonly cooked with fish, fowl, and pork, while white liquor is often used for beef, fatty foods, and the firm Chinese broccoli.

5.2.1 Wine Lees and Red *Qu*

Lees *(zao)* are extremely important in East China cuisine. They come in three kinds: sweet lees, white lees, and red lees. Sweet lees *(tian zao),* or "wine ferment" *(jiuniang),* are essentially glutinous rice fermented by yeast for 24 h. Both the fluffy lees and the liquid (which is sweet glutinous rice wine) are used together in cooking. Sweet lees are served with dumplings as a snack in Shanghai and are used extensively to marinate food, partly to prolong storage, but more importantly to create new flavors, as in lees-marinated duck's eggs, which keeps for months. In the Jiangsi Province, it replaces yeast as a levener to make wine-flavored wine lees buns. Aside from sweet lees, fragrant lees *(xiang zhao)* are the by-products of yellow wine making and are further divided into white lees (from the yellow *Shao* wine) and red lees (from yellow wine fermented using the red *qu*). Sweet lees are often substituted for white lees in fish and fowl dishes, but experts could tell the difference. Red lees have a telltale red color and cannot easily be mistaken. All three lees impart a pleasant aroma and a refreshing taste to food. The sweet lees are sweeter, and the red lees are milder and more colorful. Several special cooking techniques employ the lees: lees slipping, lees pan-frying, and lees blasting. The most common ingredients are fish, shrimp, chicken, duck, pork, and lamb and their internal organs.

Red *qu* can be ground into a powder (called red *qu* powder), and the powder can be suspended in water (called red *qu* water); both are used in cooking. Red *qu* and red lees dishes are a halllmark of the Fujian Province and Taiwan in the Southeast. But duck in red *qu* sauce, with the duck braised in red *qu* water in an earthen pot, is a dish from Yangzhou, just north of theYangtze River, 150 miles from Shanghai.

5.2.2 Yellow Wine and Beer

The most common wine used in cooking is *Shaoxing jiu,* a yellow wine from Shaoxing, the renowned capital of yellow wine. It has roughly the same alcoholic content as sherry, which is often used as a substitute outside China. It is used extensively in food processing; for example, it is added in the last stages of the manufacturing of bean cheese *(furu),* which is fermented from beancurd. A lower grade yellow wine especially for cooking is called *liaojiu* ("ingredient wine").

In a Chinese stir-fry, often yellow wine is whirled onto the heated wok containing the ingredients, producing a penetrating aroma and characteristic taste. Yellow wine is also added to fish soups to neutralize the fishy smell.

Yellow-wine shrimplets is a famed stir-fry dish from Hangzhou, about 25 miles from Shaoxing. Wine-cooked pork from Shanghai is slow-cooked in a sealed pot using yellow wine instead of water. A Cantonese specialty, *huadiao* chicken, is a whole chicken braised with yellow wine in an earthen pot. Three-cup chicken, a Taiwan invention, is a whole chicken braised in a cup of yellow wine, a cup of soy sauce, and a cup of water; a derivative is beer-braised duck, using beer instead of yellow wine and duck instead of chicken.

5.2.3 White Liquor and Flavored Wines

High-proof white liquor (notably *Fen* liquor) is a Cantonese favorite in stir-frying the firm-textured Chinese broccoli. *Fen* liquor also goes well with beef. In the United States, Chinese chefs often use Scotch whiskey when Chinese white liquor is not available.

To fight the winter chill, many homes serve air-dried meat under the general name *la.* Essentially, these are meat or fowl cured in salt, saltpeter, and *fen* liquor and air-dried in the winter sun. *La* sausage is known in the West as Chinese sausage; there are also *la* pork, *la* duck, and *la* chicken. The Chinese counterpart of beef jerky also may use white liquor. In Cantonese cuisine, rose dew liqueur is used to flavor soy sauce chicken.

5.2.4 "Drunken" Foods.

Many animal-based foods are marinated in yellow wine cooked and served cold, treasured for its wine flavor and crisp texture. Examples are drunken chicken and drunken squab. Marinated crustaceans may even be served live, as in drunken shrimp and drunken crab; this practice is dangerous, particularly for freshwater shrimps and crabs, because their parasites are known to cause a delayed-action attack on the human nervous system. Another exotic dish is flaming shrimp, with live drunken shrimp set aflame using high-proof white liquor, just before serving.

5.3 DRINKING

5.3.1 The Culture of Imbibing

The Chinese normally do not drink alcoholic beverages alone or without the accompaniment of food. The solitary drinker who drinks himself into oblivion without any solid nourishment is rarely found. Li Po (701–762) wrote,

Among the flowers, lo! a jug of wine,
I longed to drink but had no company.
I raised the cup to hail the shiny moon,
And then my shadow too; now we were three.

He drank alone that time, though clearly not by choice.

As late as the 1950s, wine shops displayed earthen jars of wine, to be ladled to order by customers who brought their own bottles for refilling. Most jars looked alike, and after a jar had been opened, there was a definite chance of dilution by unscrupulous merchants. To buy good wine, one must find a knowledgeable and honest shopkeeper, and to buy very good wine at any price, the customer has to be known to the merchant as a connoisseur deserving of the nectar being purchased (Cheng, 1954, pp. 117–129). With modern bottling practices, the retailing by ladling from poorly labeled jars is no longer common. But from time to time, one still hears of people poisoned by methyl alcohol in counterfeit white liquors bought under suspicious circumstances.

Yellow wines are usually served warm, after the container has been given a hot-water bath. White liquors are served at room temperature but may also be warmed on a cold day. Porcelain cups are used in drinking; the white liquor cups are small, holding about 20 ml. Yellow wine cups are much larger, and in taverns, they are actually shallow bowls.

Despite the increased prosperity in recent years, the per capita consumption of alcohol is very low by Western standards. Most Chinese drink tea rather than alcoholic beverages, except on special occasions, and most Chinese women are teetotalers. Religious rituals often call for alcoholic beverages, but the amount required is small.

5.3.2 Affinities with Food

Drinkers in the West tend to follow rules on pairing wine with food, such as "red wine goes with red meat, white goes with fish," although experts may stress that the real choice rests with the diner. The Chinese are not punctilious on this score and may use the same beverage, regardless of the food being served (Chang, 1977, pp. 308–309). The expertise in matching is more evident in the kitchen than at the dinner table. We can think of several reasons for this apparent lack of dining sophistication. Grain-based wines and liquors appear to blend well with a wider diversity of foods than grape-based wines and spirits. The improvished could afford only one kind of beverage; the well-to-do are presented with several food courses often on the table at the same time, already planned for internal harmony. Or the diner could stick to the food matching his (her) wine. Notwithstanding, the simple fact is that most Chinese have not been taught the idea of pairing. On the other side of the coin, there is the old wives' taboo against drinking *wujiapi* wine with crab, but scientific tests have proved it groundless.

The Western habit of having champagne, a cocktail, or an aperitif before a meal, wine during a meal, and then brandy, port, or liqueur after a meal is again not a Chinese tradition. The drinker may move from yellow wine to white liquor and vice versa but seldom takes three or more kinds of alcoholic drinks in one sitting.

5.3.3 Drunkenness and Capacity to Drink

The word *zui* is often translated as "drunk," but there is a major difference. F. T. Cheng wrote that it should really be translated as "up to, or exceeding, one's capacity (to drink)" (Cheng, 1954, pp. 117–129), and T.C. Lai employs the single word *tipsy* (Anderson, 1988. p.121). Public drunkenness as found in the West is not unknown but is quite rare; most people get roaring drink only once in their lives, namely on their wedding day, after the groom gamely submits to the repeated urging of "one more cup" by old friends. While people promise each other to "empty the cup" *(ganbei),* they usually do not. One often begs to be excused from further drinking by claiming, "I am already up to my capacity," or "My face is red already." It has been pointed out (Wolff, 1972; Chang, 1977, p. 343) that the faces of East Asians tend to turn red after drinking a small amount of alcohol. Actually, this characteristic is not universal. The Chinese, however, discern two types of drinkers among themselves. The faces of the majority turn red, yet about 20% turn a little bluish-green instead. The latter type seem to have a much higher capacity for alcohol, fortunately so, for they could not use a red face as an excuse to decline another drink.

5.3.4 The Milieu

In most Chinese households, wine and white liquor are used only for cooking, and the alcoholic beverage taken directly, if any, is beer. Nevertheless, in many homes, the elders of a family may drink yellow wine or white liquor, usually with meals and often in the company of nondrinkers, unless there are visiting friends and relatives to share with. As an alternative, drinkers in Shaoxing, the capital of yellow wine, may go to a tavern (called "hot wine shop") for warmed yellow wine, to be consumed with snacks such as stir-fried peanuts, cooked fava beans, cooked trapa nut *(lingjiao; Trapa natans* L.), dried spiced beancurds, and sea snails.

The Chinese hold banquets for business or political reasons, for celebration, or even for mourning, offering a choice of beverages, a full array of food, and company. Yet there are far too many rituals in formal banquets to make the imbiber really comfortable. Often, the main courses are presented one at a time, and the appearance of each is an occasion for a round of drinking. Often, one toasts the host and hostess and everyone at the table in turn. Both the initiator and the recipient of the toast are to drink for appearance sake, though teetotalers may use tea instead. In the People's Republic of China, yellow wine, white liquor, and tea are served at the same time. In Hong Kong these days, the typical alcoholic beverage served is cognac, and banqueting is such an important aspect of business life there that people have to learn to drink it in tumblers to get ahead.

Informal banquets such as wedding feasts and birthday parties are far more congenial, and participants may play a noisy, old, two-player guessing game called *huachuan* (gesturing with the fist). At an agreed-upon instant, player A may stretch N(A) fingers in one hand and simultaneously utter a guessed sum S(A); likewise, person B stretches N(B) fingers and utters S(B). The winner is the player who guessed the sum (N(A) + N(B)) correctly; if both win or both lose, the round ends in a draw, and the guess may start anew. What makes this game fascinating is the total absence of material reward for the winner and the nature of punishment: the loser is punished by taking a swallow. This clearly implies that reaching one's drinking capacity in public is something to be ashamed of. Also, the skillful player who is dying of thirst may have to find a way to lose, illustrating the Chinese strategy: "Advancing through retreats."

5.4 DRINKING AND ARTISTIC CREATION

Under the influence of alcohol, emotions get a free rein, senses become heightened, and the artist within the imbiber comes out, unfettered by conventions, as is true everywhere in the world. East Asian culture is unique in the modes of impromptu artistic expression available for the *literati* of the old school, in addition to good conversation and an occasional musical performance. After a few cups, the imbiber may write poetry directly on the tavern wall in flowing calligraphy, sometimes as payment for the wine consumed. Incidentally, much of Chinese poetry refers to wine and drinking; by contrast, precious little was written about food, wine's faithful companion.

Poetry clubs may meet regularly in restaurants for members to enjoy each other's company, to play word games requiring composing and/or reciting poems, and for "on the table swinging of the brush," upholding the tradition of broad-stroke painting, brush calligraphy, and poetry. Each, indeed all three, can be created at the spur of the moment, while the inspiration (and the effects of the wine) lasts. It is not unusual to see a painting (in Chinese ink or in color) with a poem as commentary beautifully written on brush, all on the same piece of paper, within a half hour, with several drinking companions pitching in or by a single erudite artist. Such impromptu art is still being created these days, but practitioners are few in number and usually advanced in years. The next generation simply has too many other things to do.

5.5 CHANGING TRENDS

The Chinese imbibing culture is in transition. Old traditions are constantly under challenge because historical social upheavels, improving economic conditions, mobility of the people, cultural globalization, and advances in science and technology.

The fastest growing segment of the market probably belongs to imported wine types. Beer is perhaps the most popular drink in China today, and there are now quite a few good brands. Grape wines are produced in quantity; some represent joint ventures with Western wine establishments, with clearly labeled varietal names and vintage. Vermouth, gin, rum, and vodka are produced, some (e.g., the vermouths from Qingtao and Yantai, both from Shandong Province) with a distinct Chinese herbal twist.

The new moneyed class adopts imported beverages with gusto, not always just because they are expensive. An unusual practice, from the Western standpoint, is the dropping of a salted plum into a glass of, say, Scotch or vintage Chateau Lafite-Rothschild. Hong Kong, now officially an autonomous administrative region of China, continues the long tradition of serving cognac in tumblers but is paying increasing attention to good grape wines. Western wine and spirit books are read, and there are some good books on wine selection and matching in Chinese. Cognac and imported wines are also popular in Taiwan. Under the Taiwan Tobacco and Wine Monopoly Bureau high-quality, grain-based wines and white liquors are being made for an appreciative public. The Taiwanese are also fond of their own excellent beer.

Traditional Chinese alcoholic beverages are more than holding their own in the market. It is true that, with advances in packaging, wine retailing by ladling has completely disappeared, and oldtimers may still lament that the best Moutai liquor has disappeared from the market. On the other hand, well-known yellow wines and white liquors are being distributed worldwide, and advances in fermentation research and engineering have guaranteed uniform quality, reduced wastage, lowered production costs, and allowed the enforcement of strict laws against excessive impurity. Furthermore, increased sharing of commercial secrets and better understanding of the physical, chemical, physiological, and psychological bases of the enjoyment of imbibing are paving the way for the next advance in alcoholic beverage technology. Some of the best white liquors were made after the 1940s. For example, five-grain nectar *(wuliangye)* began production only in 1951.

6. ACKNOWLEDGMENT

The writers acknowledge the contributions of Mr. Jose Puebla, Technical Center, Joseph E. Seagrams and Sons, Inc., and Pearl K. Chen, writer of Chinese cuisine.

7. REFERENCES

Anderson, E. N. 1988. *The Food of China.* Yale University Press, New Haven, CT.

Chang, K. C., Ed. 1977. *Food in Chinese Culture.* Yale University Press, New Haven, CT.

Chang G. Q. 1995. How to control distillations in white liquor making. *Brewing Technology* (6):21–22 (in Chinese).

Cheng, T. S. 1954. *Musings of a Chinese Gourmet.* Hutchinson, London, England.

Heilongjiang Business Institute and Beijing Municipal Sugar, Tobacco and Wine Company, 1980. *Zhongguo Jiu* (Chinese Alcoholic Beverages), The Chinese Finance and Economics Publishing Company, Beijing, China (in Chinese).

Hsu, K. C. 1975. *The Food Industry,* 2nd ed. The Hsu Foundation, Taipei, Taiwan (in Chinese).

Packowski, G. W. 1978. Beverage spirits, distilled. In: *Kirk-Othmer Encyclopedia of Chemical Technology,* Vol. 3, 3rd ed., John Wiley and Sons, New York, pp. 830–863.

Peng M. Q., Lu B., Lai, D. F., Hu S., Pan, Q. R., and Ding Z. X., 1995. The investigation of the mode of cooling in the unified distillation using ancient sky pots. *Brewing Technology* (6):26–29 (in Chinese).

Simoons, F. J. 1991. *Food in China: A Cultural and Historical Inquiry.* CPC Press, Boston, MA, pp. 448–454.

Temple, R., 1986. Brandy and whisky—seventh century A.D. In: *The Genius of China—3000 Years of Science, Discovery and Invention,* Simon and Shuster, Upper Saddle River, NJ, pp. 101–103.

Wang R. X. 1989. *Min Yi Shi Wei Tian* (People Treasure Food Above All Else), *Vol. 1.* Zhonghua Book Co., Hong Kong (in Chinese).

Wolff, P. H. 1972. Ethnic differences in alcohol sensitivity. *Science* 175(4020):449–450.

Yao, W. J., 1989. *Zhongguo Yinshi Wenhua Tanyuan* (Exploring the Origins of Chinese Food and Drink Culture). Guangxi People's Publishing Company, Nanning, Guangxi, China (in Chinese).

Zeng, Z. Y. 1980. *Zhongguo Mingjiu Zhi* (Records of the Famed Alcoholic Beverages in China). The Chinese Tourism Publication Company, Beijing, China (in Chinese).

Zhu, Q. P. 1991. The technological characteristics of famous chinese liquor brewing and its bouquet formation. In: *The First Int'l. Symposium on Chinese Dietetic Culture,* Beijing, China, pp. 339–345 (in Chinese).

CHAPTER 14

Traditional Chinese Functional Foods

YAO-WEN HUANG
CHUNG-YI HUANG

1. INTRODUCTION

A functional food is any food or part of a food that has a medicinal effect and provides health benefits, including the prevention and treatment of a disease. A functional food is also often called a nutraceutical, designer's food, or pharmafood. Such products may include original food products, isolated components, genetically engineered foods, herbal products, and processed foods, including beverages.

Since ancient times, the Chinese have believed that foodstuff and drugs come from the same source (Li, 1578). For the Chinese, food is not only to be enjoyed for the taste, but also to be appreciated for its medicinal values. Many Chinese cuisines were prepared by using traditional functional foods for preventing or healing various diseases.

Traditional Chinese medicine is emphasized for the prevention of diseases and promotion of health. In Chinese medicine, foodstuffs having actively healing effects are categorized as drugs, and those having mild effects are categorized as foods. Chinese herbalists use an appropriate combination of food ingredients and proper cooking methods to enhance the medicinal effect. Using food for health purposes, it is crucially important to provide, not only medicinal value, but also a pleasant palate.

2. HISTORY OF CHINESE FOOD USED AS MEDICINES

Throughout history, the Chinese have developed knowledge of medicine for 4000 years. In the Chou Dynasty, the "Doctor of Food" was established in the palace along with other doctors such as internal medicine, surgeon, and veterinarian. The Doctor of Food played an important role as a dietitian for the royal family. In the West-Han Dynasty (206–8 B.C.), the first Chinese medicine book *Huang-Ti-Nei-Chin* (Yellow Emperor's Classic of Internal Medicine) was published. This book includes a chapter "Su-Wen-Pein" that discusses the importance of balanced foods. In the East-Han Dynasty (25–220 A.D.), the earliest book discussing drugs was called *Seng-Non-Ben-Tseu-Chin* (Seng Nong Herbal Essentials). In the Tang Dynasty (618–907 A.D.), an important book entitled *Cheng-Chin-Yau-Fong* (Effective Emergency Treatments) discusses the uses of foods instead of drugs to treat diseases. Currently, people are familiar with the uses of fats, proteins, vitamins, and minerals as sources of nutrients but tend to forget the classification of Chinese medicine.

3. THE BASIC PRINCIPLES BEHIND CHINESE MEDICINES

In the past, the Chinese used the *yin* and *yang* as two forces to explain how the universe worked (Williams, 1995). *Yin* represents a negative force, coolness, and the feminine side of nature, while *yang* represents a positive force, heat, and the masculine principle. The forces are complementary, and neither can exist without the other.

Later, the "Five Elements" were derived from the *yin-yang* concept. Chou Tun-I, the great neo-Confucian philosopher of the Sung Dynasty (964–1279 A.D.) stated that, by the transformations of the *yang* and the union therewith of the *yin,* water, fire, wood, metal, and earth are produced. These five elements become diffused in harmonious order, and the four seasons proceed in their course. The use of food as medicine is based on this five elements concept. Traditionally, the Chinese categorize foods into five tastes: sourness, sweetness, bitterness, acridness, and saltiness. The foods used in Chinese medicines are based on these five tastes.

4. GENERAL PROPERTIES OF CHINESE FOODS AND DRUGS

The Chinese believe that foods and drugs with the same taste should have the same medicinal effect. From the reaction of foods on the human body, foods can be classified into five categories: hot, warm, neutral, cool, and cold. This is called "Five Properties." This concept is the same for Chinese drugs. In Chinese medicine theory, the nature (energy, *qi* or *chi*) and its taste *(wei)* of food or drug are interrelated (Anon., 1977). The "four energies" are varied in degree of energy, namely cold *(han),* cool *(liang),* warm *(wen),* and hot or fever *(jeh).* The balanced one that does not lean toward warmth or coolness is called neutral *(ping).*

The five tastes include the sour, sweet (pleasant), bitter, acrid (peppery), and salty flavors. Some bland and tasteless ones are also included in the five tastes.

Generally, these tastes relate to the medicinal effect. Sour-tasting items exert an astringent effect; sweet-tasting ones, a soothing and tonic effect; bitter-tasting ones, an antipyretic and moisture-drying effect; acrid-tasting ones, a dispersing and stimulating effect; salty ones, a purgative effect; and bland-tasting ones, a diuretic effect. Based on the principles posed by the four energies and five tastes, one can easily understand the medicinal action of Chinese functional foods and herbs.

In general, foods are traditionally categorized as follows: (1) grain—wheat flour and sweet rice are warm in nature, while barley and Chinese pearl barley are cool in nature; (2) fruits—date, walnut, almond, peach, cherry, and plum are warm in nature, while watermelon, banana, grapefruit, citrus, and pear are cool in nature; (3) vegetables—pumpkin, celery, Chinese onion, and carrot are warm in nature, while turnip, cucumber, bamboo shoot, bitter melon, bean sprout, tomato, winter melon, spinach, and water lily root are cool in nature; (4) meats—lamb, chicken, deer, and ham are warm in nature, while rabbit and horse are cool in nature; (5) seafood—sea cucumber and abalone are warm in nature, while clam, crab, saltwater eel, mussel, and snail are cool in nature; and (6) miscellaneous—chicken egg and spices (ginger, Szechuan pepper, chili pepper, mustard, mint, and anise) are warm in nature, while honey, duck egg, kelp, seaweed, tea, salt, and soy sauce are cool in nature. Using the right food property to promote human health has been applied in both East and West. Much evidence supports this theory (Carper, 1989; Polunin and Robbins, 1992; Lu, 1994). Chinese believe the best preventive medicine is the use of proper food in diet. The second way is by taking herb remedies. The last resort then is chemical drugs.

5. NEW CATEGORY OF THE CHINESE FUNCTIONAL FOODS

Many Chinese food items contain medicinal effects, and the Chinese health authority has listed those foods in a special category called "Item with both food and drug properties." It can also be called functional foods. These food items have been consumed for thousands of years and at the same time have been recorded in books as drugs (Lin, 1994). Because of the confusion of food or drug properties, many food and beverage manufacturers have recently been producing products using ingredients with both drug and food properties. Health claims have also been stated on the labels.

In order to regulate the uses of drugs in food formulas, the Chinese health authority, the National Chinese Medicine Administration, and the Ministry of Health issued the regulations for preventing the addition of drugs into food products on October 22, 1987. The first 33 traditional items, listed in both the *The People's Republic of China Pharmacopoeia* (the 1985 version) and the *Food Composition Tables* (the 3rd edition published in 1981 by the Institute of Hygiene Research of the China Medical Science Academy), were included in the regulations appendix. These items are Flos aurantii *(Dai dai hua),* Flos carthami *(Hong hua),* Flos caryophyli *(Ding xian),* Flos chrysanthemi *(Jiu hua),* Fructus amomi *(Sha ren),* Fructus cannabis *(Huo ma),* Fructus citri *(Xiang yuan),* Fructus gardeniae *(Zhi zi),* Fructus hippophae *(Sha ji),* Fructus momordicae *(Lou hang guo),* Fructus mume *(Wu mei),* Exocarpium citri rubrum *(Ju hong),* Pericarpium citri reticulatae *(Chen pi),* Semen cassiae *(Jue ming zi),* Semen ginkgo *(Bai guo),* Semen myristicae *(Rou dou kou),* Semen pruni *(Yu li ren),* Semen raphani *(Lai fu zi),* Semen ziziphi spinosae *(Suan zao ren),* Folium perillae *(Zi su ye),* Herba agastachis *(Huo xiang),* Herba elsholtziae *(Xiang ru),* Herba menthae *(Bo he),* Radix angelica dahuricae *(Bai zhi),* Radix glycyrrhizae *(Gan cao),* Rhizoma galangae *(Gao liang jiang),* Bulbus allii macrostemi *(Xie bai),* Cortex cinnamomi *(Rou gui),* Tuckahoe *(Fu ling),* Concha ostreae *(Mu li),* and Zaocys *(Wu shao she).*

The Ministry of Health added an additional 29 items to the list on April 16, 1988. The addition includes Fructus anisi stellati *(Ba jiao hui xiang),* Fructus chaeomelis *(Mu gua),* Fructus citri sarcodactyli *(Fo shou),* Fructus crataegi *(Shan zha),* Fructus canarii *(Qing guo),* Fructus euphori *(Long yuan),* Fructus foeniculi *(Xiao hui xiang),* Fructus jujubae *(Zao),* Fructus lycii *(Gou qi zi),* Fructus mori *(Sang shen),* Fructus piperis *(Hu jiao),* Pericarpium zanthoxyli *(Hua jiao),* Semen armeniaceae amarum *(Ku xing ren), Semen canavaliae (Dao dou),* Semen coicis *(Yi yi ren),* Semen euryales *(Qian shi),* Semen lablab album *(Bai bian dou),* Semen loti *(Lian zi),* Semen persicae *(Tao ren),* Semen phaseoli *(Chi xiao dou),* Semen sesami nigrum *(Hei zhi ma),* Semen sojae preparatum *(Dan dou chi),* Semen torreyae *(Fei zi),* Herba cichorii *(Ju ju),* Rhizoma diosacoreae *(Shan yao),* Rhizoma zingiberis *(Jiang),* Bulbus lilii *(Bai he),* Thallus laminariae (eckloniae) *(Hai dai),* and Mel *(Feng mi).*

In 1991, the Ministry issued a second list with eight items: Fructus hordei germinatus *(Mai ya),* Semem sinapis *(Huang jie zi),* Folium loti *(He ye),* Folium mori *(Sang ye),* Herba portulaceae *(Ma chi xian),* Rhizoma imperate *(Mao gen),* Rhizoma phragmitis *(Lu gen),* and Corium stomachichum galli *(Ji nei jim).* In March 1998, another eight items were added to the category of items with both food and drug properties. They are Flos lonicerae *(Jin yin hua),* Fructus alpiniae oxyphyllae *(Yi zhi),* Fructus phyllanthi *(Yu gan zi),* Semen sterculiae lychnophorae *(Pang da hai),* Herba houttuyniae *(Yu xing cao),* Herba lophatheri *(Dan zhu ye),* Herba taraxaci *(Pu gong ying),* and Radix puerariae *(Ge gen).* Currently, a total of 78 traditional Chinese foods with medicinal effects will be treated as food and not regulated by drug standards. Selected functional foods with medicinal effects are listed in Table 14.1.

6. MANNER OF USE FOR FUNCTIONAL FOODS

Chinese functional foods can be prepared as herbal diets using steaming, stewing, simmering, boiling, stir-frying, and tea-making techniques. Two

TABLE 14.1. Selected Chinese Functional Foods with Medicinal Effects.

A. Antianginal Effect	
1. *Ge gen*	*Pueraria lobata*
2. *Ju hua*	*Chrysanthemum morifolium*
B. Antihypercholesterolsterolemic Effect	
1. *Shan zha*	*Crataegus pinnatifida*
2. *Jue ming zi*	*Cassia obtusifolia*
C. Antishock Effect	
1. *Zhi shi*	*Citrus aurantium*
D. Sedative and Hypnotic Effect	
1. *Suan zao ren*	*Ziziphus spinosa*
E. Stomachic and "Wind"—Dispelling Effect	
1. *Ba jiao hui iang*	*Illicium verum*
2. *Xiao hui xiang*	*Foeniculum vulgare*
3. *Chen pi*	*Citrus reticulata* Blanco
4. *Sheng jiang*	Zingiber officinale Rose
5. *Rou dou kou*	*Alpinia katsumadii* Hayata
6. *Huo xiang*	*Pogostemon cablin* Benth.
7. *Cao guo*	*Amonum tsaoko*
8. *Sha ren*	*Amomum villosum* or *A. longiligulare*
9. *Gao liang jiang*	*Alpinia officinarum*
F. Promoting Digestion	
1. *Mai ya*	*Hordeum vulgare*
2. *Ding xiang*	*Eugenia caryophyllata*
3. *Shan zha*	*Crataegus pinnatifida Major*
4. *Ji nei jin*	*Gallus gallus domesticus* Brisson
G. Antiacid and Antiulcer Effect	
1. *Mu li*	*Ostrea gigas*
H. Laxative Effect	
1. *Huo ma ren*	*Cannabia sativa*
2. *Yu li ren*	*Prunus humilis* Beg.
I. Antidiarrheal Effect	
1. *Rou dou kou*	*Myristica fragrans* Houtt.
2. *Qian shi*	*Euryala ferox*
3. *Wu mei*	*Prunus mume* Sieb. et Zucc.
4. *Bai guo*	Ginkgo biloba L.
J. Choleretic and Antihepatitis Effect	
1. *Zhi zi*	*Gardenia jasminoides*
2. *Yu jin*	*Curcuma aromatica*
K. Tonics and Supporting Effect	
1. *Gou qi zi*	*Lycium barbarum*
2. *Bai he*	*Lilium lancifolium*
L. Antitussive Effect	
1. *Xiang ren*	*Prunus armeniaca*
2. *Yu gan zi*	*Phyllanthus emblica*
M. Expectorants	
1. *Ju hong*	*Citrus reticulata*
2. *Luo han guo*	*Momordica grosvenori*

(continued)

TABLE 14.1. (continued)

N. Antiasthmatic Effect	
1. *Hai dai*	*Laminaria saccharrina* (L) Lamour
O. Diuretic Effect	
1. *Bai mao gen*	*Imperata cylindrica*
P. Foods Affecting the Uterus	
1. *Hong hua*	*Carthamus tinctorius*
Q. Foods Affecting the Thyroid Glands	
1. *Hai dai*	*Laminaria saccharrina* (L) Lamour
R. Foods Affecting the Adrenal Cortex	
1. *Gan cao*	*Glycyrrhiza uralensis*
S. Antbacterial, Antiviral, and Antifungus Effect	
1. *Jin yin hua*	*Lonicera japonica*
2. *Pu gong ying*	*Taraxacum mongolicum*
3. *Yu xing cao*	*Houttuynia cordata*
4. *Ma chi xian*	*Portulaca oleracea*
5. *Zhi zi*	*Gardenia jasminoides*
T. Anthelmintics	
1. *Wu mei*	*Prunus mume*
2. *Hua jiao*	*Zantoxylum schinifolium*
U. Anticancer Effect	
1. *Yi yi ren*	*Coix lachryma*

popular methods for preparation are making gruel (also called congee or porridge) and tea (Figure 14.1). When making gruel, there are two ways of preparation: (1) direct cooking and (2) decoction and cooking.

(1) Direct cooking: A sufficient amount of water is added to the rice and cooked with the item directly. Sometimes, two or more other items may be combined (Figure 14.2).
(2) Decoction and cooking: Decoct the item(s) in water and sift the liquid from the dregs; then add rice and water to cook.

For tea making, there are three methods of preparation: (1) decoction, (2) infusion, and (3) maceration. Of these, infusion is the most often used. On many occasions, a combination formula of several items is used for medicinal purposes (Figure 14.3):

(1) Decoction: The required amount of the item is placed in cold water and boiled for 5 min.
(2) Infusion: Boiling water is poured over the prescribed amount of the item in a container that is then covered for 5 min. This method is commonly used for flowers, leaves, and whole leafy plants. It is also used for comminuted roots and barks.

Figure 14.1 Tea is the most common method for using traditional Chinese functional foods. From left to right: chrysanthemum tea, tea of chrysanthemum and Frutus Momordicae, and *Gang he* tea made of more than 10 functional foods and Chinese herbs.

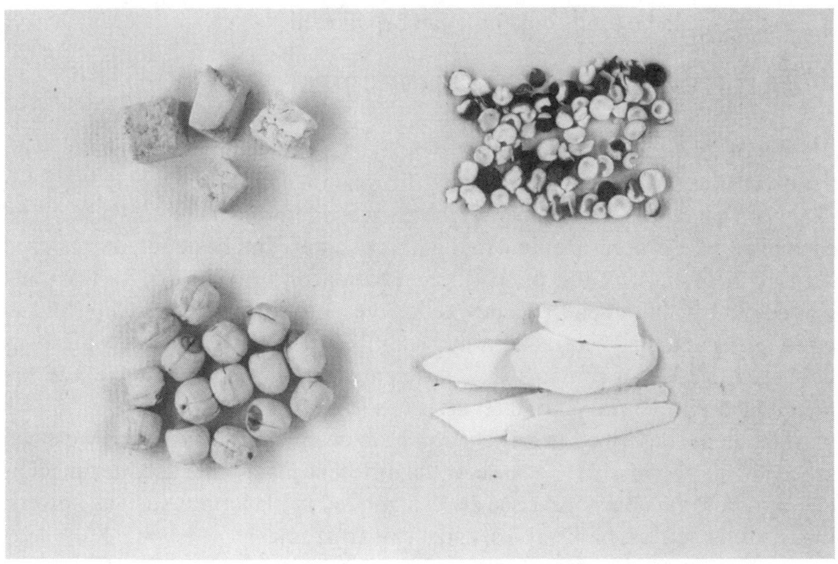

Figure 14.2 Combination of more than two items is a common practice for usage.

Figure 14.3 Example of a drink made of five functional foods—*Rhizoma diosacoreae, Semen coicis, Bulbus lilii, Semen euryales* (center), *Semen loti*—and one herb (polygonatum).

(3) Maceration: The required amount of cold water is poured over the prescribed amount of the item and allowed to stand at room temperature for 3–5 h. It can be drunk cold or it can be warmed.

7. MAJOR SOURCES OF BASIC INFORMATION

Most of the listed food items with a medicinal value come from plants, with the exception of five items originating from animals. Those originating from plants are categorized by the part of the plant used for the medicinal effect, and each plant part is described in detail in this chapter. The basic references used here are the *Ben Cao Gong Mu* (Chinese Pharmacopoeia) (Li, 1578). The book was edited by Shih-chen Li. Li devoted 27 years to study the *Ben Cao* (Material Medica) and made corrections and additions. The book of 25 volumes was published in 1578. It contains 1892 different kinds of medicinal materials that are divided into 16 classes and 60 divisions (Figure 14.3).

Many items described in this chapter have been naturalized in the West and are available worldwide. Because of the different plant parts used for medicinal purposes, the items discussed are categorized in plant parts such as flowers, fruits, seeds, leaves, barks, roots, and stems (rhizomes).

The information on each functional food item is based on *Ben Cao Gong Mu* (Li, 1578). However, two books, *A Barefoot Doctor's Manual* (Anon.,

1977) and *The People's Republic of China Drug Book* (Anon., 1991), were used for confirmation. Citations not specifically mentioned are from the *Ben Cao Gong Mu*. However, other books were also used as supplements. Citation are given when the information was different from the previous books. The chemical components of each functional food are taken from *The Pharmacology of Chinese Herbs* (Huang, 1993a) and *Herbal Drugs and Phytopharmaceuticals* (Bisset, 1994).

For safety classification, the *Botanical Safety Handbook* (McGuffin et al., 1997) was used, although the book did not include all the items. The classification of the safety of each item is based on data that are associated with the use of a specific item in isolation and in quantities generally consumed for a therapeutical effect. Four classes are defined as follows:

(1) Class 1: items that can be safely consumed when used appropriately
(2) Class 2: items for which the following use restrictions apply, unless otherwise directed by an expert qualified in the use of the described substance:
  - 2a: for external use only
  - 2b: not to be used during pregnancy
  - 2c: not to be used while nursing
  - 2d: other specific use restrictions as noted
(3) Class 3: items for which significant data exist to recommend the following labeling: "To be used only under the supervision of an expert qualified in the appropriate use of this substance." Labeling must include proper use information: dosage, contraindications, potential adverse effects and drug interactions, and any other relevant information related to the use of this substance.
(4) Class 4: items for which insufficient data is available for classification

8. PLANT ORIGIN

The flower, fruit, peel, seed, leaf, whole plant, and root are discussed here. Each item is listed in alphabetical order.

8.1 FLOWERS

8.1.1 Flos Aurantii *(Dai Dai Hua)*

Family: Rutaceae.

Scientific Name: *Citrus aurantium* L. var. *amara* Engler. (*C. daidai* Sieb.).

Related Species: *Citrus aurantium* 'Huangpi,' *C. aurantium* 'Chuluan,' and *C. aurantium* 'Tangcheng.'

Other Common Names: bitter orange, Seville orange, or Bigarade or Neroli flowers.

Properties: cold in nature; sour and bitter, yet acrid, to taste. Safety class: 1 (listed for the peel).

Chemical Components: 0.2–0.5% essential oil containing monoterpenes (linalyl acetate, α-pinene, limonene, linalool, nerol, geraniol); methyl anthranilate is a characteristic constituent. Bitter substance and flavonoids are also present (Bisset, 1994). Two flavones are tangeratin and nobiletin (Huang, 1993a).

Medicinal Action: improves coronary circulation and cerebral blood flow and inhibits intestinal smooth muscle contraction (Huang, 1993a).

Therapeutic Uses: for the treatment of nervousness and disturbed sleep as a sedative agent (Bisset, 1994); also used for fullness of lungs, poor appetite, vomiting, soothing the stomach.

Manner of Use: commonly used as a tea by infusion or decoction.

8.1.2 Flos Carthami *(Hong Hua)*

Family: Compositae.

Scientific Name: *Carthamus tinctorius* L.

Other Common Names: *hong lan hua* (red orchid), *huang lan* (yellow orchid); safflower.

Properties: a dried flower that is warm in nature; acrid to taste. Safety class: 2b.

Chemical Components: 0.3–0.6% glycoside carthamin (a yellow substance), which converts into carthamone and carthemidin by enzymes. Neocarthanmin and saflor yellow (20–30%) have also been isolated from the flower (Huang, 1993a).

Medicinal Action: causes a rhythmic contraction of the uterus; stimulates the heart; lowers blood pressure; prolongs the blood coagulation time; lowers plasma cholesterol and triglyceride levels (Huang, 1993a); resolves bruises and stimulates tissue regeneration; activates and clears meridian channels.

Therapeutic Uses: for preventing amenorrhea, unexpelled dead fetus, prolonging postpartum discharge; dysmenorrhea, menopause, and angina pectotis; also used externally in treating traumatic injuries and painful bruises (Huang, 1993a).

Manner of Use: rehydrated first and then used for the preparation of a Chinese dish "crabs stir-fried with carthamus." The dish functions by cleaning the blood and cures women's diseases such as menstrual aberration and cold hands and feet (Su, 1993). It is also used commonly to enliven the hues (red color) of cream soups, marinades, pale sauces, salad dressings, basting liquids, flavored vinegars, pasta salads, and curries (Kowalchik and Hylton, 1998). It can be substituted for saffron *(Crocus sativus)* for coloring food. Egypt also has a long history to use safflower oil in cooking (Kowalchik and Hylton, 1998).

8.1.3 Flos Caryophylli *(Ding Xian)*

Family: Myrtaceae.

Scientific Name: *Eugenia caryophyllata* Thunberg (*Caryophyllus aromaticus* L.).

Other Common Names: caryophyllum, clove.

Properties: dried flower bud that is warm in nature; acrid to taste. It warms the body center and alleviates pain and curtails any bad effects. Safety class: n/a.

Chemical Components: 15–20% essential oils (eugenol, acetyl-eugenol, α- and β-caryophylline, ylangene, chavicol, and humulene). Also contains flavonoids (rhamnetin, kaempferol, oleanolic acid, eugenitin, and isoeugenitin) (Huang, 1993a); tannins and phenolic carboxylic acids (gallic acid, protocatechuic acid, etc.) (Bisset, 1994).

Medicinal Action: stimulates gastric secretions to increase digestion and dispels the gases; it is also used as a potent ascaricide (Huang, 1993a).

Therapeutic Uses: for treatment of han (cold-type) stomach vomiting, hiccups, pains in the heart and abdomen, hernia (Anon., 1977); also used for hiccups as an antiemetic agent, for externally treating fungal infections and skin infections (Huang, 1993a).

Manner of Use: mixed with ginger juice and milk for drinking (Zhang et al., 1988). The Chinese dish "clove duck" is an example of an herbal diet dish in which alpiniae and cinnamon are also used to simmer the duck. The dish is good for the stomach and kidneys. It effectively alleviates coughing and vomiting (Su, 1993).

8.1.4 Flos Chrysanthemi *(Jiu Hua)*

Family: Compositae.

Scientific Name: *Chrysanthemum morifolium* Ramat.

Related Species: *Chrysanthemum indicum* L. *(yie jiu hua), C. boreale* Mak., and *C. lavandulaefolium* (Fisch.) Mak.

Other Common Names: *jiu hua, hang jiu* (Hang-chow chrysanthemum), *huang jiu hua* (yellow chrysanthemums), *cha jiu* (tea chrysanthemum); chrysanthemum (Figure 14.4).

Properties: dried flower that is slightly han (cold) in nature; bitter, yet pleasant in taste. Safety class: n/a.

Chemical Components: chrysanthenone, camphor, the alkaloid stachydrine, glucosides (acacetin-7-rhamnoglucoside, cosmosiin and acacetin-7-glucoside, diosmetin-7-glucoside), adenine, choline, vitamin B- and A-like substances, and bornol (an essential oil) (Huang, 1993a).

Medicinal Action: increases coronary vasodilatation and coronary blood flow (Huang, 1993a); acts as a carminative, antipyretic, antibacterial, or detoxifying agent (Huang, 1993a).

Figure 14.4 *The Ben Cao Gang Mu,* or *The Chinese Pharmacopeia,* by Shih-chen Li, lists 1892 different kinds of medicinal materials.

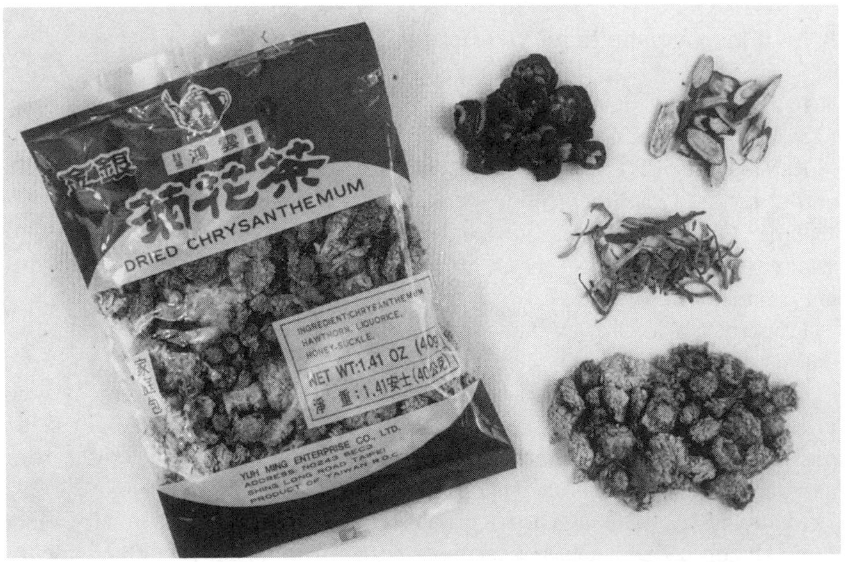

Figure 14.5 Example of tea made of chrysanthemum.

Therapeutic Uses: for treatment of the common cold, headache, hypertension, dizziness associated with wind-caused fever, tinnitus, conjunctivitis, boils, and abscesses.

Manner of Use: commonly taken as a tea (decoction or infusion). It is also used as a composite drink by mixing with Hawthorn fruit or with Sophora flower and green tea. The fresh flower is also used to cook with fish, especially perch, as a traditional herbal diet (Su, 1993).

8.1.5 Flos Lonicerae *(Jin Yin Hua)*

Family: Caprifoliaceae.

Scientific Name: *Lonicera japonica* Thunb., *L. hypoglauca* Miq., *L. confusa* DC., *L. dasystyla* Rehd., *L. brachypoda, L. chinensis, L. flexuosa.*

Related Species: *L. periclymenum* (Potterton, 1983).

Other Common Names: *yen hua teng* (silver-flower vine), *yen yang hua* (Mandarin duck flower), *ren dong hua,* Chinese honeysuckle, Japanese honeysuckle, lonicera, woodbine.

Properties: dried flower bud cool in nature; bitter, yet pleasant in taste. Safety class: 1.

Chemical Components: luteolin, inositol, saponins, and chlorogenic acid (an antibacterial agent).

Medicinal Action: clears fevers and detoxifies. It has antibacterial effects against *Staphylococcus aureus,* streptococci, pneumococci, *Bacillus dysenterii, B. typhoid,* and paratyphoid (Huang, 1993a). It is an antiviral agent. It has antilipinemic actions, interfering with lipid absorption from the gut (Huang, 1993a).

Therapeutic Uses: for treatment of colds, laryngitis, bacterial dysentery, enteritis, infected boils, skin sores, lymphadenitis, or rheumatism.

Manner of Use: prepared as a tea (decoction). Drink of *jin yin hua* and *xing ren* (apricot kernel) includes mulberry leaf, platycodon root, and licorice.

8.2 FRUITS

8.2.1 Fructus Alpiniae Oxyphyllae *(Yi Zhi)*

Family: Zingiberaceae.

Scientific Name: *Alpinia oxyphylla* Miq. (Anon., 1991).

Properties: the dried fruit of ginger plant with a warm and acrid taste. The safety classification is not available, but other species *(A. galanga, A. officinarum)* are listed as Class 1.

Chemical Components: essential oils.

Medicinal Action: nourishes the spleen and kidneys; stops diarrhea; reduces urine.

Therapeutic Uses: for treatment of a cold spleen caused by diarrhea, cold-caused abdominal pain, polyuria, enuresis, or spermatorrhea.

Manner of Use: parched with vinegar and ground into fine powder and then used as tea by infusion (Zhang et al., 1988).

8.2.2 Fructus Amomi *(Sha Ren)*

Family: Zingiberaceae

Scientific Name: *Amomum xanthioides* Wallich. *A. villosum* Loureiro; *A. longiligulare* T.L. Wu.

Other Common Names: bastard "cardamom."

Properties: warm in nature; acrid to taste. The safety class is not available, but other related species [*A. tsao-ko* crev. et Lem. *(cao guo)* and *A. melegueta* Roscoe (grains of paradise)] are all listed as Class 1.

Chemical Components: n/a.

Medicinal Action: stimulates energy circulation; regulates the center of the body; alleviates pain; arrests diarrhea; prevents miscarriage.

Therapeutic Uses: for indigestion and gas collection in stomach-spleen, cramps, abdominal distension, hiccups, vomiting, han (cold) dominant diarrhea and dysentery.

Manner of Use: commonly decocted and used as a drink.

8.2.3 Fructus Anisi Stellati *(Ba Jiao Hui Xiang)*

Family: Illiciaceae (McGuffin et al., 1987)

Scientific Name: *Illicum verum* J.D. Hooker

Other Common Names: star anise.

Properties: warm in nature; peppery to taste. Safety class: 1. However, *I. verum* should not be confused with the toxic Japanese or bastard star anise, *I. lanceolatum,* or *I. anisatum* (syn. *I. religiosum*) (McGuffin et al., 1987).

Chemical Components: essential oils, including anethol, anisaldehyde, safrole, and anisic ketone.

Medicinal Action: warms the central organs; dispels han (cold); corrects energy; alleviates pain.

Therapeutic Uses: for treatment of hernia, gaseous belching, han (cold) stomach and vomiting, and han (cold) abdominal pain.

Manner of Use: commonly used as a spice in cooking.

8.2.4 Fructus Canarii *(Qing Guo)*

Family: Burseraceae.

Scientific Name: *Canarium album* Raeusch; *C. sinense; Pimela alba.*

Other Common Names: white Chinese olive.

Properties: neutral in nature; pleasant, yet acrid to taste. Safety class: 1.

Chemical Components: n/a.

Medicinal Action: removes the fever; purifies the lungs; eliminates apprehension; stimulates appetite; promotes salivation; detoxifies the body.

Therapeutic Uses: for treatment of sore throat, thirst, restlessness, globefish poisoning, and alcohol intoxication.

Manner of Use: commonly made into preserved fruit products.

8.2.5 Fructus Cannabis *(Huo Ma)*

Family: Cannabinaceae.

Scientific Name: *Cannabis sativa* L.

Other Common Names: *hou ma ren* (fiery hemp seed); hemp; marijuana.

Properties: dried fruit and seed that is neutral in nature; pleasant to taste. Safety class: n/a. However, a large dose can cause cholinergic intoxication, manifested as nausea, vomiting, diarrhea, convulsions, and coma (Huang, 1993a).

Chemical Components: trigonelline, *l(d)*-isoleucine betaine, cannabinol, tetrahydro-cannobinol, and cannabidiol; muscarine, choline, vitamins B_1 and B_2 (Huang, 1993a).

Medicinal Action: moistens "fire"; stimulates intestinal mucosa (laxative).

Therapeutic Uses: for treatment of overly "hot" intestines, constipation, and coughing.

Manner of Use: commonly decocted.

8.2.6 Fructus Chaeomelis *(Mu Gua)*

Family: Rosaceae.

Scientific Name: *Chaenomeles speciosa* (Sweet) Nakai. (Anon., 1991; McGuffin et al., 1997); C. *lagenaria* Koidz, C. *sinensis* (Thouin) Koehue.

Other Common Names: *du mu gua, hsuan mu gua,* flowering quince.

Properties: warm in nature; sour and biting in taste. Safety class: 1. However, excessive use can damage teeth and bones (McGuffin et al., 1997).

Chemical Components: not available.

Medicinal Action: fortifies the spleen; resolves moisture; loosens up the sinews; activates the muscles.

Therapeutic Uses: for treatment of vomiting and diarrhea, cholera-associated cramps, joint pains in the back and knee, and numbness in cases of beriberi.

Manner of Use: served as a drink.

8.2.7 Fructus Citri *(Xiang Yuan)*

Family: Rutaceae.

Scientific Name: *Citrus medica* L.; *C. wilsonii* Tanaka; *C. grandis* Oseck var. *shangyuan* Hu.

Other Common Names: citron.

Properties: warm in nature; bitter, sour, and acrid to taste. Safety class: n/a. However, many other related species of citrus are listed as Class 1.

Chemical Components: n/a.

Medicinal Action: corrects energy circulation; strengthens the liver; counteracts moisture (excessive in the body); and resolves phlegm.

Therapeutic Uses: for treatment of fullness in chest and abdomen, regurgitation and vomiting, chest and abdominal pains, poor appetite, productive coughing, indigestion, or diarrhea.

Manner of Use: commonly made into preserved fruit products.

8.2.8 Fructus Citri Sarcodactyli *(Fo Shou)*

Family: Rutaceae.

Scientific Name: *Citrus medica* L. var. *sarcodactylus* Swingle.

Other Common Names: Bergamot.

Properties: warm in nature; bitter, sour, and acrid to taste. Safety class: n/a. However, many related species are classified as Class 1.

Chemical Components: n/a.

Medicinal Action: corrects energy circulation; strengthens the liver; soothes the stomach, relieves the pain.

Therapeutic Uses: for treatment of fullness in chest and abdomen, regurgitation and vomiting, chest and abdominal pains, poor appetite, productive coughing, indigestion, and diarrhea.

Manner of Use: slices added to white wine and stored for 1 month to make Bergmot wine (Su, 1993). It stimulates the function of the liver, stomach, and spleen.

8.2.9 Fructus Crataegi *(Shan Zha)*

Family: Rosaceae.

Scientific Name: *Crataegus cuneata* Sieb. et Zucc., *C. pinnatifida* Bunge var. *major, C. pinnatifida.*

Related Species: *C. monogyna* (common names are hawthorn, may blossom, and quickthorn) (Potterton, 1983).

Other Common Names: crab apple, hawthorn seed.

Properties: slightly warm in nature; sour, yet pleasant to taste. Safety class: 1. However, according to Huang (1993a), nausea and vomiting may be observed if large quantities are consumed.

Chemical Components: chlorogenic acid, caffeic acid, citric acid, crataegolic acid, maslinic acid, ursolic acid, and saponins.

Medicinal Action: helps digestion; stimulates blood circulation; stops diarrhea; lowers blood cholesterol; smoothes the surface of the atherosclerotic area; increases blood flow in heart; increases the myocardial contractibility; lowers blood pressure (Huang, 1993a).

Therapeutic Uses: for treatment of indigestion, infantile marasmus; menstrual cramps, diarrhea, and dysentery; and hernia. Hyperchole sterolemia, agina pectoris, and hypertension (Huang, 1993a). Today, it is often used as a cardiac tonic, and the blossoms are also effective (Potterton, 1983).

Manner of Use: The "sweet and sour pork with hawthorn" is a dish in which hawthorn and licorice are first cooked. It is used as a sauce for deep-fried pork (Lee, 1990). It can be consumed as a snack food such as hawthorn cookies and hawthorn cake.

8.2.10 Fructus Euphori *(Long Yuan)*

Family: n/a

Scientific Name: *Euphoria longan* (Lour.) Steud.

Other Common Names: longan.

Properties: neutral in nature; pleasant to taste. Safety class: n/a.

Chemical Components: vitamin B, glucose, sucrose, and tartaric acid.

Medicinal Action: nourishes the spleen; cultivates the heart; supplements the intellect.

Therapeutic Uses: for anemia, hyperactive mental activity, and forgetfulness.

Manner of Use: simmered with meat and chicken.

8.2.11 Fructus Foeniculi *(Xiao Hui Xiang)*

Family: Umbelliferae; but classified as Apiaceae by McGuffin et al. (1997)

Scientific Name: *Foeniculum vulgare* Miller.; *Anethum foeniculum.*

Other Common Names: *xia hui* (ragrant fennel); sweet fennel, cumin.

Properties: warm in nature; sharp and sweet tasting. Safety class: 1.

Chemical Components: essential oils, including anethol, *d*-fenchone, methylchavicol, and anisaldehye.

Medicinal Action: stimulates energy; promotes digestion; resolves phlegm; stimulates milk production. It also restores normal functioning of the stomach, warms taste sensations, dispels the colds, normalizes the flow of *chi* (or *qi*), and relieves pain (Huang, 1993a).

Therapeutic Uses: for treatment of gastroenteritis, hernia, indigestion, and abdominal pain.

Manner of Use: commonly decocted and used as a drink.

8.2.12 Fructus Gardeniae *(Zhi Zi)*

Family: Rubiaceae.

Scientific Name: *Gardenia jasminoides* Ellis.

Other Common Names: garden cardenia; cape jasmine; gardenia fruits.

Properties: cool in nature; bitter to taste. Safety class: 1.

Chemical Components: gardenin, gardenoside, shanzhiside, usolic acid, crocin, and crocetin.

Medicinal Action: clears fevers and purges fire; cools the blood; has antipyretic and detoxification functions, stimulating biliary secretion and reducing plasma bilirubin levels; displays sedative, hypnotic, anticonvulsant, antibacterial, and anthelmintic properties; also used as an effective choleretic (Huang, 1993a).

Therapeutic Uses: for treatment of icteric hepatitis with jaundice, high fevers associated with influenza; styes, canker sores, toothache, mastitis; epistaxis, hematemesis, hematuria; bacterial dysentery; snakebites.

Manner of Use: normally decocted.

8.2.13 Fructus Hippophae *(Sha Ji)*

Family: n/a

Scientific Name: *Hippophae rhamnoides* L.

Other Common Names: n/a

Properties: warm in nature; sour and astringent to taste. This item is commonly used by Mongolians and Tibetans in North and Southwest China, respectively. Safety class: n/a.

Chemical Components: n/a.

Medicinal Action: has antitussive effect and used as an expectorant.

Therapeutic Uses: for treatment of coughing, indigestion, and abdominal pain.

Manner of Use: normally decocted.

8.2.14 Fructus Hordei Germinatus *(Mai Ya)*

Family: Gramineae (Poaceae: McGuffin et al., 1997)

Scientific Name: *Hordeum vulgare* L.

Other Common Names: germinated barley.

Properties: neutral in nature; pleasant to taste. Safety class: 2b.

Chemical Components: enzymes (invertase, amylase, and proteinase), lipid, vitamins B and C, maltose, and dextrose (Huang, 1993a).

Medicinal Action: helps digest carbohydrates and protein; balances the central organs; controls lactation.

Therapeutic Uses: for treatment of indigestion, fullness in chest and abdomen, loss of appetite, and fullness of lactating breasts.

Manner of Use: boiled with duck gizzards and preserved dates. Then switched to low heat and parboiled for 2 more hours (Lee, 1990). The soup can help digestion and increase appetite.

8.2.15 Fructus Jujubae *(Zao)*

Family: Rhamnaceae.

Scientific Name: *Zizyphus vulgaris* var. *spinosa* (*Ziziphus jujuba* Mill.: McGuffin et al., 1997)

Other Common Names: *da zao, jujube,* Chinese *jujube.*

Properties: warm in nature; pleasant to taste. Safety class: 1.

Medicinal Action: strengthens the spleen and stomach; moisturizes the heart and lungs; regulates various medications.

Therapeutic Uses: for treatment of weak stomach and spleen, anemia, inadequate energy (fatigue), and salivation.

Manner of Use: cooked with sufficient water and made into gruel.

8.2.16 Fructus Lycii *(Gou Qi Zi)*

Family: Solanaceae.

Scientific Name: *Lycium chinense* Miller; *L. barbarum* L.

Other Common Names: lycium, medlar, wolfberry, matrimony vine.

Properties: neutral in nature; pleasant to taste. Safety class: 2b.

Chemical Component: betaine, zeaxanthin, physalein, and vitamins (carotine, nicotinic acid, and vitamin C).

Medicinal Action: strengthens the kidneys; restores semen; nourishes the liver; clears vision.

Therapeutic Uses: for treatment of nutritional deficiency eye diseases, diabetes, inadequate liver and kidney function, and seminal emission.

Manner of Use: A dish called "pork kidney with lycium fruit" (other ingredients include squid and lycium bark) can energize the body and supplement the blood. It can be a mild treatment for diabetes and vision defects (Su, 1993). Lycium can be cooked with chicken or rice (Chen, 1997). It can be decocted as a tea for drinking.

8.2.17 Fructus Momordicae *(Lou Han Guo)*

Family: Cucrurbitaceae.

Scientific Name: *Momordica grosvenori* Swingle.

Properties: cold in nature; pleasant to taste. Safety class: n/a. This item is limited to regional use in southern China and never served as a traditional fruit (Huang, 1993a).

Chemical Components: glucoside esgoside (0.1% of total content) whose hydrolysis product is fructose.

Medicinal Action: clears heat and dryness in mouth; soothes the lungs.

Therapeutic Uses: for treatment of whooping cough, sore throat, voice loss, and constipation.

Manner of Use: commonly made into a tea.

8.2.18 Fructus Mori *(Sang Shen)*

Family: Moraceae

Scientific Name: *Morus alba* L.

Other Common Names: white mulberry, mulberry achenes.

Properties: cold in nature; sour and tart, yet pleasant to taste. Safety class: 1.

Chemical Components: n/a. However, morin, dihydromorin, dihydrokaempterol, 2,4,4',6-tetrahydroxybenzophenone, maclurin, mulberrin, mulberrochromene, and cyclomulberrochromene are isolated from mulberry twigs (Huang, 1993a).

Medicinal Action: strengthens kidneys; aids vision; nourishes blood and *yin* element.

Therapeutic Uses: for treatment of agitation and insomnia, deafness and blurred vision, white patches in hair and beard, "hot" intestines and constipation, pain in back and knees, and stiffness of muscles and joints. (In addition to fruits, the twigs, leaves, and root bark also have medicinal effects.)

Manner of Use: A famous mulberry gruel is made with mulberry fruits, rice, chicken, and other ingredients, including red *jujubes,* lotus seeds, and pine seeds. The congee is very effective for bronchitis, sinusitis, and asthma; it strengthens the lungs and is used for antitussiveness (Chen, 1997). It is also consumed as a fruit beverage.

8.2.19 Fructus Mume *(Wu Mei)*

Family: Rosaceae.

Scientific Name: *Prunus mume* Sieb. et Zucc.; *P. armeniaca* (Thunb).

Other Common Names: black "prune," Japanese apricot.

Properties: The unripe fruit is neutral in nature; sour and tart to taste. Safety class: 1.

Chemical Components: glucoside prudomenin, malic acid, and succinic acid.

Medicinal Action: acts as an astringent, antipyretic, and vermicidal; stimulates contraction of the muscles of intestinal parasites and also gallbladder; causes relaxation of the bile duct (Huang, 1993a). It is also an antimicrobial agent.

Therapeutic Uses: for treatment of chronic diarrhea and dysentery, feverish thirst, achlorhydria, no appetite, residue coughing, chronic malaria, bliary ascariasis, hookworms, abdominal pain, cholecystitis, and gallstones.

Manner of Use: commonly made into a beverage or wine.

8.2.20 Fructus Phyllanthi *(Yu Gan Zi)*

Family: Euphorbiaceae.

Scientific Name: *Phyllanthus emblica* L.

Other Common Names: emblic, ambal.

Properties: cool in nature; sour and astringent, yet pleasant to taste. Safety class: n/a. The item is used by Tibetan people as a medicine.

Chemical Components: n/a.

Medicinal Action: clears fevers; cools blood; reinforces the stomach; stops cough.

Therapeutic Uses: for treatment of liver and bile diseases, indigestion, abdominal pain, coughing, sore throat, and dry mouth.

Manner of Use: decocted and ingested as a drink.

8.2.21 Fructus Piperis *(Hu Jiao)*

Family: Piperaceae

Scientific Name: *Piper nigrum* L.

Other Common Names: black pepper (with cortex), white pepper (without cortex), pepper, black peppercorns.

Properties: hot in nature; acrid to taste. Safety class: 1.

Chemical Components: alkaloids, including piperine, chavicine, and piperamine; 0.8% essential oils, including piperonal, dihydrocarverol, caryophyllene, and cryptone.

Medicinal Action: warms the central organs and dispels cold; eliminates abdominal distension and alleviates pain; used as an anticonvulsive and a sedative (Huang, 1993a).

Therapeutic Uses: for treatment of vomiting and diarrhea caused by cold stomach, "cold" putum and abdominal fullness, "cold" energy dominance, "cold" diarrhea and dysentery, "*yin*-cold" abdominal pain, and food (fish, meat, crab, mushroom) poisoning.

Manner of Use: used as a spice in cooking or used as drink (decoction).

8.3 PEEL

8.3.1 Exocarpium Citri Rubrum *(Ju Hong)*

Family: Rutaceae.

Scientific Name: *Citrus reticulata Blanco; C. grandis* (L.) Osbeck (Huang, 1993b), *C. grandis* Osbeck var. *tomentosa* Hortorum.

Other Common Names: red tangerine peel.

Properties: red-colored external layer of the pericarp warm in nature; bitter and acrid to taste. Safety class: 1.

Chemical Components: citral, geraniol, linalool, methylanthranilate, stachydrine, putrescine, and apyrocatechol. Also, contains glucosides (naringin, poncirin, hesperidin, neohespiridin, and nobiletin).

Medicinal Action: corrects energy circulation; strengthens the lungs; and resolves phlegm.

Therapeutic Uses: for treatment of fullness in chest and indigestion, eliminating sputum and coughing.

Manner of Use: commonly used as a tea; a red tangerine peel gruel is made by decocting and cooking. The ingredients include red tangerine peel and bitter apricot kernel. The peel can be steamed with chicken and wine (Huang, 1993b).

8.3.2 Pericarpium Citri Reticulatae *(Chen Pi)*

Family: Rutaceae.

Scientific Name: *Citrus reticulata* Blanco and *C. sinensis* Osbeck; *C. aurantium* L. Subsp. *aurantium.*

Related Species: *C. reticulata* 'Chachi,' *C. reticulata* 'Dahongpao,' *C. reticulata* 'Unshiu,' *C. reticulata* 'Tangerina;' *C. tangeriana* Hort et Tanaka; *C. erythrosa* Tanaka.

Other Common Names: tangerine peel, mandarian orange peel.

Properties: dried peel warm in nature; bitter and acrid to taste. Safety class: 1.

Chemical Components: bitter-tasting flavone glycosides (neohesperidin and naringin, neohesperidose); nonbitter flavonoids (hesperidin, rutoside, sinensetin, nobiletin, tangeratin); 1–25 essential oils (limonene); pectin (Huang, 1993a).

Medicinal Action: corrects energy circulation; strengthens the spleen; counteracts excessive moisture in the body; resolves phlegm.

Therapeutic Uses: for treatment of fullness in chest and abdomen, regurgitation and vomiting, chest and abdominal pains, poor appetite, productive coughing, indigestion, and diarrhea.

Manner of Use: served as a famous dish called "stir-fried chen pi beef" and as preserved fruit products to enhance the digestive system and blood circulation.

8.3.3 Pericarpium Zanthoxyli *(Hua Jiao)*

Family: Rutaceae.

Scientific Name: *Zanthoxylum bungeanum* Max. *Z. schinifolium* Sieb. et Zucc.

Other Common Names: *qin jiao, chuan jiao,* Szechuan pepper.

Properties: fruit rind warm in nature; acrid tasting and toxic. Safety class: 2b.

Chemical Components: estragol, citronellol, phellandrene, and essential oil zanthoxylene.

Medicinal Action: warms the central organs; dispels chill cold; counteracts moisture; acts as a vermicide against parasites.

Therapeutic Uses: for treatment of chills and pains in the abdomen, vomiting, cold-damp diarrhea and dysentery, ascariasis-caused abdominal pain, and moist sores on the skin.

Manner of Use: used as a spice in cooking or for drinking by being pulverized and mixed with water.

8.4 SEEDS (Figure 14.6)

8.4.1 Semen Armeniaceae Amarum *(Ku Xing Ren)*

Family: Rosaceae.

Scientific Name: *Prunus armeniaca* L. var. *ansu.* Maxim., *P. sibirica* L., *P. manshurica* (Maxim.) Koehne, *P. armeniaca* L. (Anon., 1991; Huang, 1993a).

Related Species: *Amygdalus communis, Prunus amygdalus, P. communis* (Anon., 1977).

Other Common Names: "bitter almond" used for *P. dulcis* (Mill.) D.A. Webb var. *Amara* (DC.) H.E. Moore, while apricot, or *xingren,* is used for *P. armeniaca* L.

Properties: warm in nature; bitter to taste. Safety class: 3 (Chang and But, 1986).

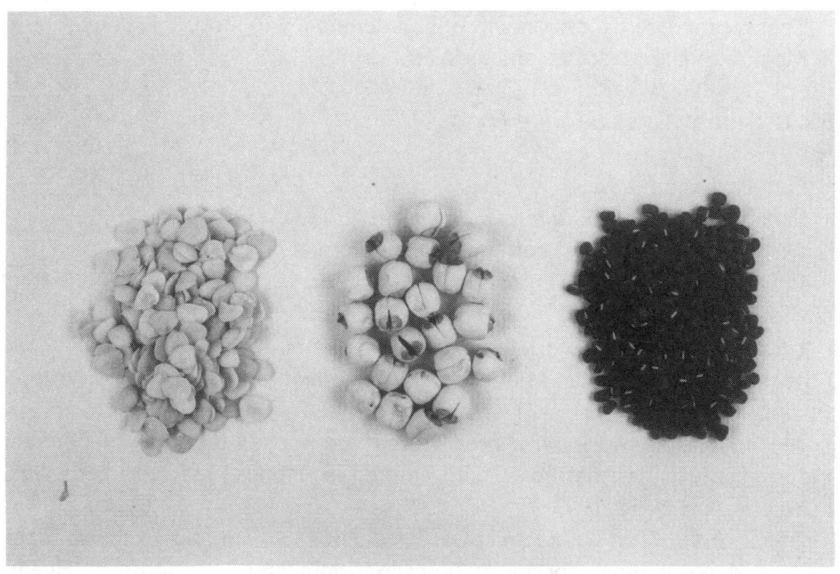

Figure 14.6 Example of seeds: *Semen Armeniaceae amarum, Semen loti,* and *Semen Semen phaseoli* (left to right).

Chemical Components: amygdalase and amygdalin (cyanogenic glucoside, up to 8.0%) (Bensky and Gamble, 1986; Huang, 1993a; Yeung, 1985).

Medicinal Action: resolves phlegm; quiets cough; lowers excessive energy; the fat from almonds can lubricate the intestines. A small quantity of cyanic acid (HCN), produced from amygdalin by amygdalase and pepsin of gastric juice, can stimulate the respiratory center and produce a tranquilizing effect (i.e., antitussive and antiasthmatic effects) (Huang, 1993a). Overdosage can cause cyanide intoxication, especially in children.

Therapeutic Uses: for treatment of cold and coughing, unproductive coughing, constipation, dyspnea, and asthma.

Manner of Use: prepared in decoction and used as a drink.

8.4.2 Semen Canavaliae *(Dao Dou)*

Family: Leguminosae.

Scientific Name: *Canavalia gladiata* (Jacquin) De Candolle., *C. gladiata, Dolichos ensiformis.*

Other Common Names: droad bean; jack bean.

Properties: warm in nature; pleasant to taste. Safety class: n/a.

Chemical Components: canavaline, canavanine, and urease (20% in total); gibberelin A_{21} and A_{22}, canavalia gibberellin I, and canavalia gibberellin II.

Medicinal Action: warms the central region of the body; suppresses evil *qi;* corrects circulation.

Therapeutic Uses: for treatment of han deficiency to tonify the body.

Manner of Use: decocted and used as a drink.

8.4.3 Semen Cassiae *(Jue Ming Zi)*

Family: Leguminosae.

Scientific Name: *Cassia tora* L., *C. obtusifolia* L.

Other Common Names: coffee weed, sickle-pod.

Properties: slightly han in nature; salty; bitter, yet pleasant to taste. Safety class: n/a.

Chemical Components: chrysophenol, emodin, aloe-emodin, rhein, physcion, obtusin, aurantio-obtusin, chrysobtusin, rubrofusarin, norrubrofusarin, and toralactone.

Medicinal Action: removes "heat" from liver; purifies the liver; supports the kidneys; expells gas; clarifies vision. Emodin is responsible for the herb's laxative effects (Huang, 1993a).

Therapeutic Uses: for treatment of headache, swollen and red eyes, dizziness, and pterygiums; for hypercholesterolemia and hypertension (Huang, 1993a).

Manner of Use: decocted and used as a drink.

8.4.4 Semen Coicis *(Yi Yi Ren)*

Family: Gramineae.

Scientific Name: *Coix lachryma-jobi* L., *C. lachryma-jobi* L. var. Ma-yuen (roman.) Stapf, *C. agrestis, C. chinensis, C. lacryma.*

Other Common Names: Chinese pearl barley, adlay, Job's tears.

Properties: slightly cold in nature; pleasant and bland to taste. Safety class: 2b (Bensky and Gamble, 1986; Yeung, 1985).

Chemical Components: α,β-sitosterol, amino acids, and lipids (Huang, 1993a).

Medicinal Action: strengthens the spleen; converts moisture; promotes diuresis.

Therapeutic Uses: for treatment of lung abscess, lobar pneumonia, appendicitis, rheumatoid arthritis, beri-beri, diarrhea, edema and difficult urination; also for lung and cervix cancers and chorionic epithelioma (Huang, 1993a).

Manner of Use: made into gruel.

8.4.5 Semen Euryales *(Qian Shi)*

Family: Nymphaeaceae.

Scientific Name: *Euryale ferox* Salisb.

Other Common Names: *chi-toulien* (chicken-head water lily), euryale seed, fox nut.

Properties: neutral in nature; pleasant, yet acrid to taste. Safety class: 1.

Chemical Components: protein, lipid, carbohydrate (starch), and vitamins.

Medicinal Action: fortifies the spleen; strengthens the kidneys; stops diarrhea and seminal emission.

Therapeutic Uses: for treatment of diarrhea, incontinence of urine, seminal emission, leukorrhea, joint pains in lower extremities, and backache.

Manner of Use: used as an herbal diet called "honey elixir" in which euryale seeds are simmered with lotus seeds, ginkgo, red *jujubes,* and longan flesh; then stir in honey and serve (Chen, 1997). It can be cooked with duck gizzard to make a soup (other ingredients include lotus seed, dried lily bulb, dates, and dried tangerine peel) (Lee, 1990).

8.4.6 Semen Ginkgo *(Bai Guo)*

Family: Ginkogaceae.

Scientific Name: *Gingko biloba* L.

Other Common Names: ginkgo nut, maidenhair tree.

Properties: The seed has neutral properties; pleasant, yet bitter and biting to taste. Safety class: 2d. Do not exceed recommended dose (no more than 10 seeds a day for the boiled or roasted seeds; the standard dose for the prepared ginkgo seed is 4.5–15.0 g/day); not for long-term use (Bensky and Gamble, 1986; Chang and But, 1986; Hsu, 1986; Leung and Foster, 1996). Canadian regulations do not allow ginkgo seeds in foods (Welsh, 1995).

Chemical Components: ginkgolic acid and bilobol (Huang, 1993a).

Medicinal Action: exerts an astringent effect on pulmonary energy; stops coughing and asthma; stabilizes spermatogenesis, stops leukorrhea. It has antitussive, antituberculous, and antibacterial functions (Huang, 1993a).

Therapeutic Uses: for treatment of pulmonary tuberculosis, seminal emissions, leukorrhea, and frequent micturition.

Manner of Use: A dish "ginkgo and sweet rice ball soup" is drunk for nourishing the lungs and curing severe coughing (other ingredients, including dates and assorted fruits, are also added).

Note: Ginkgo leaf extracts (concentrated to 24% flavone glycoside and 6% terpenes) are among the most actively studied of the modern phytopharmaceutical substances (McGuffin et al., 1997). The leaf contains kaempterol-3-rhamnoglucoside, ginkgetin, isoginketine, and bilobetin. Other substances include isorhamnetin, shikimic acid, D-glucaric acid, and anacardic acid (Huang, 1993a).

8.4.7 Semen Lablab Album *(Bai Bian Dou)*

Family: Leguminosae

Scientific Name: *Dolichos lablab* L., *D. albus, D. perinans, D. purpureus, Lablab nankinensis, L. niger, L. vulgaris.*

Other Common Names: *bian dou,* bonavista bean, lablab.

Properties: slightly warm in nature; pleasant to taste. Safety class: n/a.

Chemical Components: n/a.

Medicinal Action: moderates the central organs; dispels moisture; cools and clears summer heat; and detoxifies.

Therapeutic Uses: for treatment of summer moisture and cholera, vomiting, diarrhea and thirst, leukorrhea, gonorrhea, alcoholic intoxication, and globefish poisoning.

Manner of Use: made into gruel.

8.4.8 Semen Loti *(Lian Zi)*

Family: Nymphaeaceae.

Scientific Name: *Nelumbo nucifera* Gaertner, *Nelumbium nelumbo, Nelumbium speciosum, Nymphaea nelumbo.*

Other Common Names: lotus seed.

Properties: The seed (including plumule and radicle may taste bitter and cold in nature) is neutral in nature; pleasant, yet astringent to taste. Safety class: 2d (but the plumule is classified as Class 1).

Chemical Components: alkaloids, including liensinine, isoliensinine, neferine, lotusine, methylcorypalline, and demethyl-coclaurine.

Medicinal Action: strengthens spleen; cultivates the heart; controls peristalsis; stabilizes sperm (seminal control); removes "heat"; has antihypertensive and tranquilizing effects (Huang, 1993a).

Therapeutic Uses: for treatment of spleen-deficient diarrhea, excess dreaming and seminal emissions, metrorrhagia and leukorrhea.

Manner of Use: used for cooking or made into gruel.

8.4.9 Semen Myristicae *(Rou Dou Kou)*

Family: Myristicaceae.

Scientific Name: *Myristica fragrans* Houtt.

Other Common Names: nutmeg.

Properties: warm in nature; acrid to taste. Safety class: 2b. More than 5.0 g of powder nutmeg affect the central nervous system, producing hallucinations, headache, dizziness, and nausea (Leung and Foster, 1996).

Chemical Components: α-pinene, *d*-camphene, and myristicin

Medicinal Action: warms the central organs; moistens intestines; expels gas; is central depressant (Huang, 1993a).

Therapeutic Uses: for treatment of diarrhea, vomiting, and abdominal ache.

Manner of Use: used as a spice in cooking or decocted and used as a drink.

8.4.10 Semen Persicae *(Tao Ren)*

Family: Rosaceae.

Scientific Name: *Prunus persica* (L.) Batsch, *P. davidiana* (Carr.) Franch.

Other Common Names: peach kernel.

Properties: neutral in nature; bitter, yet pleasant to taste. Safety class: 2b or 3.

Chemical Components: cyanogenic glycoside (amygdalin, 2.0–6.0%) (McGuffin et al., 1997).

Medicinal Action: reduces phlegm; resolves clots; moistens and lubricates the intestines.

Therapeutic Uses: for treatment of constipation, painful urination, and hematemesis.

Manner of Use: used as an ingredient in cake.

8.4.11 Semen Phaseoli *(Chi Xiao Dou)*

Family: Leguminosae.

Scientific Name: *Phaseolus angularis* Wight., *P. calcaratus* Roxb.

Other Common Names: *chuan hong dou* (scarlet bean), *you lu dou* (wild mung bean), *hsueh dou* (blood bean); *adzuki bean.*

Properties: neutral in nature; pleasant tasting. Safety class: n/a.

Chemical Components: n/a.

Medicinal Action: promotes diuresis; stimulates blood circulation; reduces swelling; drains pus.

Therapeutic Uses: for treatment of edematous beri-beri, dysentery, sores, and abscesses.

Manner of Use: cooked into gruel or soup.

8.4.12 Semen Pruni *(Yu Li Ren)*

Family: Rosaceae.

Scientific Name: *Prunus humilis* Bge., *P. japonica* Thunberg, *P. pedunculata* Maxim.

Other Common Names: Japanese plum; dwarf flowering cherry.

Properties: neutral in nature; acrid and bitter, yet pleasant to taste. Safety class: n/a.

Chemical Components: n/a.

Medicinal Action: expels flatus; supports energy.

Therapeutic Uses: for treatment of sluggish colon, constipation, edema, and inadequate elimination of urine.

Manner of Use: decocted and used as drink.

8.4.13 Semen Raphani *(Lai Fu Zi)*

Family: Cruciferae

Scientific Name: *Raphanus sativus* L. var. *macropodus* Makino.

Other Common Names: turnip seed.

Properties: neutral in nature; acrid, yet pleasant to taste. Safety class: n/a.

Chemical Components: n/a.

Medicinal Action: resolves phlegm; cures intestinal parasites.

Therapeutic Uses: for treatment of indigestion, abdominal distension, hiccups, excessive sputum, asthma, abdominal pain, and diarrhea.

Manner of Use: concocted brew.

8.4.14 Semen Sesami Nigrum *(Hei Zhi Ma)*

Family: Pedaliaceae.

Scientific Name: *Sesamum indicum; S. orientale.*

Other Common Names: black sesame seeds.

Properties: neutral in nature; pleasant to taste. Safety class: 1.

Chemical Components: n/a.

Medicinal Actions: strengthens the liver and kidneys; moistens the five viscera (organs).

Therapeutic Uses: for inadequate liver and kidney function, head-cold dizziness, numbness and paralysis, and constipation.

Manner of Use: used as an ingredient in the preparation of candies and cakes.

8.4.15 Semen Sinapis *(Huang Jie Zi)*

Family: Brassicaceae.

Scientific Name: *Brassica juncea* (L.) Czern. et Coss.

Related Species: *Sinapis alba* L. *(Bai Jie Zi).*

Other Common Names: *jie.*

Properties: warm in nature; acrid to taste. Safety class: 1.

Chemical Compoents: sinalbin.

Medicinal Action: warms the lungs; stops pain; has an antitussive and expectorant effect.

Therapeutic Uses: for treatment of asthma and coughing.

Manner of Use: decocted and used as drink.

8.4.16 Semen Sojae Preparatum *(Dan Dou Chi)*

Family: Leguminosae.

Scientific Name: *Glycine* max (L.) Merr.

Other Common Names: fermented black beans.

Properties: cold in nature; bitter and acrid to taste. Safety class: n/a.

Chemical Components: n/a.

Medicinal Action: relaxes muscles; promotes perspiration; clears fever; eliminates apprehension.

Therapeutic Uses: for treatment of fever, headache, and restlessness connected with colds, chest discomfort, macula, and measles.

Manner of Use: used as an ingredient for cooking.

8.4.17 Semen Sterculiae Lychnophorae *(Pang Da Hai)*

Family: n/a.

Scientific Name: *Sterculia lychnophora* Hance.

Properties: cold in nature; pleasant and bland to taste. Safety class: n/a.

Chemical Components: n/a.

Medical Action: clears fevers; detoxifies; keeps bowels open.

Therapeutic Uses: for treatment of dry coughs, hoarseness, aches, pains in bones, nosebleeds, pink (red) ear, wind-fire dominant toothache, and hemorrhoids.

Manner of Use: decocted and used as a beverage.

8.4.18 Semen Torreyae *(Fei Zi)*

Family: Taxaceae.

Scientific Name: *Torreya grandis* Fort., *T. nuifera* Sieb. et Zucc.

Other Common Name: Japanese torreya.

Properties: Neutral in nature; pleasant to taste. Safety class: n/a.

Chemical Components: essential oils, oxalic acid, glucose, tannin, glyceride of linoleic acid, and lipids, including palmitic acid, stearic acid, and oleic acid (Zhang et al., 1988).

Medicinal Action: kills worms; eliminates stoppages (marasmus caused); exerts laxative action.

Therapeutic Uses: for treatment of abdominal pain caused by intestinal worms and constipation associated with hemorrhoids.

Manner of Use: to be eaten parched (Zhang et al., 1988).

8.4.19 Semen Ziziphi Spinosae *(Suan Zao Ren)*

Family: Rhamnaceae.

Scientific Name: *Ziziphus jujuba* Mill. var. *spinosa* (Bunge) Hu ex H.F. Chou, *Z. spinosa* (Huang, 1993a).

Other Common Names: spiny *jujuba.*

Properties: neutral in nature; sour, yet pleasant to taste. Safety class: 2b (it is listed as emmenagogues and uterine stimulants).

Chemical Components: betulin, betulic acid, and the glycosides jujuboside A and B, which on hydrolysis produce jujubogenin. Sanjoinines, the peptide alkaloids (14 in all) have also been isolated (Huang, 1993a).

Medicinal Action: nourishes the liver; calms the mind; preserves *qi,* nourishes the muscle; enriches the bone marrow; has tranquilizing and hypnotic effects.

Therapeutic Uses: for treatment of nervous exhaustion, insomnia, neurasthenia, and irritation.

Manner of Use: decocted and used as a drink.

8.5 LEAVES

8.5.1 Folium Loti *(He Ye)*

Family: Nymphaeaceae.

Scientific Name: *Nelumbo nucifera* Gaertner.

Other Common Names: waterlily leaves.

Properties: neutral in nature; bitter to taste. Safety class: 1.

Chemical Components: alkaloids, including nuciferine, roemerine, *o*-nornuciferine, anonaine, lirodenine, dihydro-nuciferine, pronuciferine, anneparine, *N*-methylcoclaurine, and *N*-methylisococlaurine.

Medicinal Action: raises stomach energy; loosens up clots; and stops bleeding; disperses body heat during summer; increases body energies (Huang, 1993a).

Therapeutic Uses: for treatment of summer moisture and diarrhea, hematemesis, epistaxis, metrorrhagia, bloody stools, and bloody vaginal discharge.

Manner of Use: a famous southern Chinese dish "rice wrapped in leaf" served in summertime.

8.5.2 Folium Mori *(Sang Ye)*

Family: Moraceae.

Scientific Name: *Morus alba* L., *M. constantinopalitana, M. indica.*

Other Common Names: Chinese white mulberry, white mulberry, mulberry leaves.

Properties: cool in nature; bitter yet pleasant to taste. Safety class: 1.

Chemical Components: n/a.

Medicinal Action: acting as a carminative and antipyretic.

Therapeutic Uses: for treatment of fever from colds, headache, bloodshot eyes, "hot" pulmonary coughing, sore throat, and toothache.

Manner of Use: concocted brew.

8.5.3 Folium Perillae *(Zi Su Ye)*

Family: Labiatae.

Scientific Name: *Perilla frutescens* (L.) Britton var. *crispa* Decaisne.

Other Common Names: *yeh su, zi su,* bushy perilla.

Properties: warm in nature; acrid to taste. Safety class: n/a.

Chemical Components: *l*-perilla aldehyde and alcohol.

Medicinal Action: dispels cold; corrects energy balance; relieves asthma; quiets restless fetus; and detoxifies.

Therapeutic Uses: for treatment of vomiting, abdominal distension and flatulence, restless fetus, and seafood (fish and crab) poisoning and can be used as an antitussive agent for epidemic influenza, colds, and malaria.

Manner of Use: concocted brew.

8.6 WHOLE LEAFY PLANTS

8.6.1 Herba Agastachis *(Huo Xiang)*

Family: Labiatae (Anon., 1977), Laminaceae (McGuffin et al., 1997).

Scientific Name: *Agastache rugosa* (Fisch. et Mey.) O. Kuntze.

Other Common Names: *hsing jen hua* (almond blossoms), bishop-wort, agastache.

Properties: warm in nature; pleasant, yet acrid to taste. Safety class: 1.

Medicinal Action: clears fever; resolves moisture; strengthens the stomach; stops vomiting.

Therapeutic Uses: for treatment of wound injuries, summer moisture, vomiting, diarrhea, and angina pains.

Manner of Use: decocted and used as a drink.

8.6.2 Herba Cichorii *(Ju Ju)*

Family: Asteraceae.

Scientific Name: *Cichorium glandulosum* Boiss. et Hout, *C. intybus* L. (Anon., 1991; Potterton, 1983)

Other Common Names: succory, chicoy

Properties: cool in nature; salty and slightly bitter to taste. Safety class: 1 (root is also commonly used). It is used by the Chinese Muslims in Northwest China as a popular herb drug).

Chemical Components: n/a.

Medicinal Action: clears liver; reinforces gallbladder; stimulates urine flow; removes edema fluid.

Therapeutic Uses: for moist heat and jaundice, abdominal pain, edema, and oliguria. It is also effectual for sore eyes that are inflamed or for nursing mothers whose breasts hurt because of the abundance of milk (Potterton, 1983).

Manner of Use: boiled in wine or water and drunk; used most as a salad vegetable: "garden endive" (Potterton, 1983).

8.6.3 Herba Elsholtziae *(Xiang Ru)*

Family: Labiatae.

Scientific Name: *Elsholtzia splendens* Nakai ex F. maekawa.

Related Species: *E. patrini* Garcke.

Other Common Names: *xiang ru.*

Properties: slightly warm in nature; acrid to taste; fragrant (Anon., 1991). Safety class: n/a.

Medicinal Action: breaks up clots; alleviates pain; resolves gonorrheal discharge; promotes diuresis; neutralizes heat (Anon., 1991).

Therapeutic Uses: for treatment of traumatic bruises and pain, poisonous snakebites, heatstroke fever, edema, and itching from oozing rash.

Manner of Use: used as a spice in cooking.

8.6.4 Herba Houttuyniae *(Yu Xing Cao)*

Family: Saururaceae.

Scientific Name: *Houttuynia cordata* Thunb.

Other Common Names: *chou mu tan* (stinky peony), *la-tzu cao* (pepper grass), *nai-tou cao* (nipple grass).

Properties: cool in nature; acrid to taste. Safety class: n/a.

Chemical Components: n/a.

Medicinal Action: clears fevers; detoxifies; promotes diuresis; reduces swelling.

Therapeutic Uses: for treatment of upper respiratory tract infections, lung abscesses, inflammatory conditions and other infections of the urinary tract, and poisonous snakebites.

Manner of Use: cooked as gruel.

8.6.5 Herba Lophatheri *(Dan Zhu Ye)*

Family: Gramineae.

Scientific Name: *Lophatherum gracile* Brongn., *L. elatum, L. japonicum, L. lehmannii, Centrotheca affine.*

Other Common Names: *zhu ye mai dong* (bamboo leafed lily turf), *zhu ye cao* (bamboo grass), *shu zhu ye* (aquatic bamboo leaf).

Properties: dried aerial part of the plant cold in nature; bitter to taste. Safety class: n/a.

Chemical Components: arundoin, cylindrin, and friedelin.

Medicinal Action: clears fever; dispels feelings of agitation and apprehension; promotes diuresis.

Therapeutic Uses: for treatment of measles, influenza, heatstroke, restlessness and insomnia, thirst, sore throat, and painful and difficult urination.

Manner of Use: boiled with water and used as a beverage.

8.6.6 Herba Menthae *(Bo He)*

Family: Ladiatae.

Scientific Name: *Mentha haplocalyx* (Huang, 1993a); *M. arvensis* L.

Common Names: peppermint.

Properties: cool in nature; acrid to taste; aromatic. Safety class: 1.

Chemical Components: essential oils, methol (70–90%), menthone (10–20%), and menthyl acetate (Huang, 1993a).

Medicinal Action: expels flatulence; clears fevers, promotes perspiration and measles rash appearance; reduces swelling; relieves itching; stimulates gastrointestinal tract motility; promotes the forward movement of luminal contents (Huang, 1993a).

Therapeutic Uses: for treatment of influenza (to dispel "wind" and "heat"), headache, cough, incomplete measles eruption, pharyngitis, conjunctivitis, beri-beri edema, and pruritus.

Manner of Use: used in making sauces, jellies, tea, salad, and meat products.

8.6.7 Herba Portulaceae *(Ma Chi Xian)*

Family: Portulacaceae.

Scientific Name: *Portulaca oleracea* L.

Other Common Names: *ma chi cao, ma xian, lao-shu erh* (mouse's ear), purslane, pigweed.

Properties: dried aerial parts cold in nature; sour and bitter to taste. Safety class: 2b or 2d. Individuals with a history of kidney stones shoud use it cautiously.

Chemical Components: a great deal of noradrenaline, sylvite, dihydroxyphenyl ethylamine, dihydroxy-phenylpropylamine, malic acid, citric acid, glutamic acid, aspartic acid, alanine, sucrose, glucose, fructose (Zhang et al., 1988), and oxalates (up to 1.7% oxalic acid) (Huang, 1993a).

Medicinal Action: clears away heat and toxic materials; promotes blood circulation; relieves swelling.

Therapeutic Uses: for treatment of dysentery, enteritis, urinary tract infections, leukorrhea, hemorrhoids, erysipelas, boils, and ulcers.

Manner of Use: cooked with rice and a sufficient amount of water to make gruel and served on an empty stomach (Huang, 1993b). It can also be eaten as a stir-fried dish.

8.6.8 Herba Taraxaci *(Pu Gong Ying)*

Family: Compositae; but classified as Asteraceae by McGuffin et al. (1997)

Scientific Name: *Taraxacum mongolicum* Hand.-Mazz., *T. sinicum* Kitag., *T. ceratophyllum, T. corniculatum, T. dens-lionis, T. officinale, T. sinense.*

Other Common Names: huang hua ti ting (yellow-flowered violets), *ju chi cao* (milky grass), Asian dandelion, common dandelion, Mongolian dandelion.

Properties: dried aerial part of the plant cold in nature; bitter, yet pleasant to taste. Safety class: 1. However, the root has also been used and belongs to Class 2b (McGuffin et al., 1997).

Chemical Components: taraxasterol, taraxerol, taraxacerin, taraxacin, and vitamins A, B, and D (Huang, 1993a).

Medicinal Action: clears fevers; detoxifies, breaks up congestion; strengthens the stomach; stimulates milk flow. It also has antibacterial action against *Staphylococcus aureus, Streptococcus hemolyticus,* typhoid bacilli, dysentery bacilli, tuberculosis, and most Gram(+) bacteria (Huang, 1993a). It has antispirochetic and antiviral effects and acts as a choleretic agent for the protection of liver functions.

Therapeutic Uses: for treatment of mastitis, boils and abscesses, stomachache, and inadequate milk supply. It is also used for hepatitis and upper respiratory infections, such as tonsillitis and laryngitis (Huang, 1993a).

Manner of Use: boiled in water and used as a drink.

8.7 ROOTS

8.7.1 Radix Angelica Dahuricae *(Bai Zhi)*

Family: Umbelliferae (Anon., 1977); Apiaceae (McGuffin et al., 1997)

Scientific Names: *Angelica anomala* Lallement (Anon., 1977), *A. dahurica* (Fischer ex Hoffm.) (McGuffin et al., 1987), or *A. taiwaniana* Boiss. (Huang, 1993a).

Other Common Names: Chinese Angelica, fragrant Angelica.

Properties: warm in nature; acrid to taste; slight bitterness. Safety class: 1. However, *A. arhangelica* L. and *A. atropurpurea* L. belong to Class 2b and 2d, respectively.

Chemical Components: The whole plant contains essential oils, and the root contains various furocoumarin derivatives: 0.2% Byak-angelicin, 0.2% Byak-angelicol and oxypeucedanine, imperatorin, isoimperstorin, phellopterin, xanthotoxine, marmesin, scopoletin, neobyakangelicol, etc. (Huang, 1993a).

Medicinal Action: dialates the cornary vessels; increases coronary circulation (Huang, 1993a); clears the inner organs and dispels gas; warms (excessive) hydration; promotes pus drainage (Anon., 1991). It possesses antipyretic, analgesic action and an antibacterial effect.

Therapeutic Uses: for treatment of colds, toothache, headache, head "stuffiness" and coryza, "spotting," leukorrhea, and boils and abscesses.

Manner of Use: combined with cuidium *(chu-kung)* to stew pig brain. The dish is good for both men and women, but especially for expelling woman's head "wind" (Lee, 1990).

8.7.2 Radix Glycyrrhizae *(Gan Cao)*

Family: Leguminosae (Anon., 1977); Fabaceae (McGuffin et al., 1997).

Scientific Names: *Glycyrrhiza uralensis* Fischer; *G. inflata,* or *G. glabra; G. echinata* L. (McGuffin et al., 1997)

Other Common Names: licorice.

Properties: neutral in nature; pleasant to taste. Safety class: 2b, 2c, or 2d.

Chemical Components: glycyrrhizin (Ca and K salt of glycyrrhinic acid), three glycosides (liquiritin, isoliquiritin, and neoliquiritin) (Huang, 1993a).

Medicinal Action: revitalizes the center; supplements energy; detoxifies, loosens phlegm (Anon., 1977); stimulates the adrenal cortex; increases mineralocorticoid secretion (Huang, 1993a).

Therapeutic Uses: for splenic and gastric imbalance, abdominal pain, vomiting and diarrhea, productive cough, parched and sore throat, and swollen abscesses (Anon., 1977). Addison's disease (hypocorticosteroidism), bronchitis, tuberculosis, and peptic ulcers (Huang, 1993a). It is commonly used as an adjuvant to other herbs to smooth their taste and reduce their side effects.

Manner of Use: used as a sweetener for preserved fruits and candies. A dish, "chicken wings with quail eggs and soybean," has licorice for improving taste. The licorice in the dish will help treat pain, cough, stomachache, and sore throat (Lee, 1990).

8.7.3 Radix Puerariae *(Ge Gen)*

Family: Leguminosae; also classfied as Fabaceae by McGuffin et al. (1997).

Scientific Name: *Pueraria lobata* (Willd.) Ohwi, *P. thomsonii* Benth., *P. pseudohirsuta* Tang et Wang.

Other Common Names: *ge, fen ge* (starchy arrowroot), *ge teng* (arrowroot vine); mealy kudzu *(P. thomsonii).*

Properties: cool in nature; acrid, yet pleasant to taste. Safety class: 1.

Chemical Components: alkaloid (kassein), glucosides including daidzin, daidzin-4,7-diglucoside, puerarin, and xylopurarin.

Medicinal Action: relieves hunger; lowers fever; stops diarrhea. Also improves coronary circulation; lowers myocardial oxygen consumption; improves cerebral ciculation (Huang, 1993a).

Therapeutic Uses: for exposure-caused colds and fever, incomplete measles rash breakout, diarrhea, dysentery, enteritis, angina pectoris, hypertension, deafness, and optic nerve atrophy or retinitis.

Manner of Use: used as tea by decoction.

8.8 RIZOME/STEM TUBERS (Figure 14.7)

8.8.1 Rhizoma Diosacoreae *(Shan Yao)*

Family: Dioscoreaceae.

Scientific Name: *Dioscorea opposita* Thunb. (Anon., 1991; Huang, 1993b), *D. batatas* Decaisne (Anon., 1977).

Other Common Names: *huai shan, huai shan yao, shan shu* (mountain tuber); wild yam, Chinese yam.

Properties: neutral in nature; pleasant to taste. Safety class: 1.

Chemical Components: saponin, phlegm (mannan and phytic acid), starch (16%), glucoprotein, amino acids, *d*-abscisin II, polyphenoloxidase, vitamin C, and 3,4-dihydroxyphenylethylamine (Zhang et al., 1988).

Medicinal Action: strengthens the spleen; tonifies the lungs; reinforces the kidney.

Therapeutic Uses: for treatment of chronic enteritis, dysentery, poor indigestion, asthma, "wet dreams," excessive perspiration, leukorrhea, and neurasthenia.

Manner of Use: stir-fried after being peeled. It can also be decocted as tea or made into gruel by adding red *jujubes.*

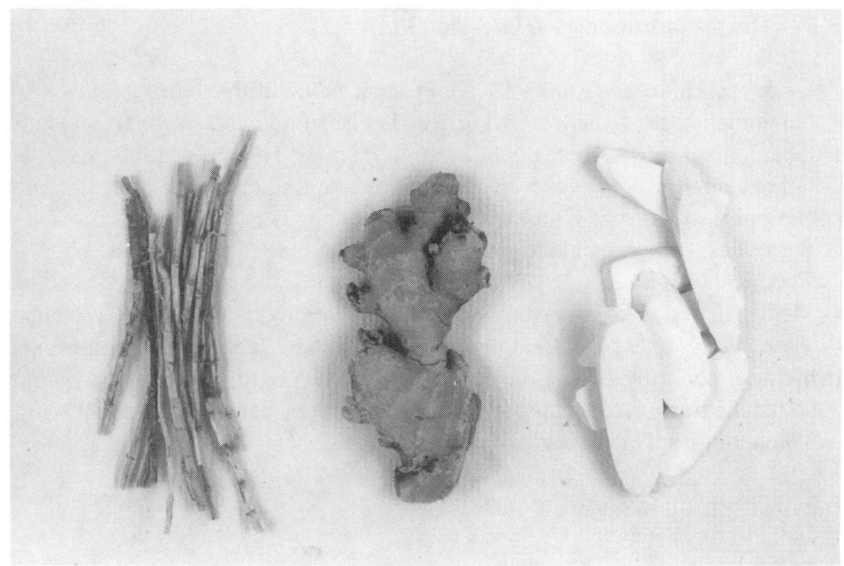

Figure 14.7 Example of rhizoma: *Rhizoma imperate, Rhizoma zingiberis,* and *Rhizoma diosacoreae* (left to right).

8.8.2 Rhizoma Galangae *(Gao Liang Jiang)*

Family: Zingiberaceae (Bisset, 1994).

Scientific Name: *Alpinia officinarum* Hance (Bisset, 1994).

Related Species: *A. galanga* (L.) Willd. (McGuffin et al., 1997) (interchangeable with *A. officinarum*).

Other Common Names: galanga, galangal, colic, East Indian root, Chinese ginger. *A. galanga* is called "greater galangal," while *A. officinarum* is called "lesser galangal" (McGuffin et al., 1997).

Properties: hot in nature; acrid to taste. Safety class: 1.

Chemical Components: 0.5–1.0% essential oil (sesquiterpene hydrocarbons and alcohols), eugenol, pungent substances (diarylheptanoids and gingerols), flavoinds (quercetin and kaempferol), sterols and sterol glycosides (Bisset, 1994).

Medicinal Action: warms the central region; dispels han (cold); digests foods; relieves the pain.

Therapeutic Uses: for treatment of stomach pain and "cold," vomiting due to stomach han, and heartburn.

Manner of Use: made into a tea by simmering three or four slices of the steam tuber in a pan of water for 15–20 min. In the United States, *A. galanga* is allowed to be used only as a flavoring agent in alcoholic beverages (McGuffin et al., 1997).

8.8.3 Rhizoma Imperate *(Bai Mao Gen)*

Family: Gramineae (Anon., 1977), Poaceae (McGuffin et al., 1997).

Scientific Name: *Imperata cylindrica* (L.) Beauvois var. *major* (Nees) C.E. Hubbard, *I. cylindrica* Beauv. var. *koenigii* Durand (Huang, 1993a).

Other Common Names: *bai mao gen* (white cogongrass), *ssu maogen* (silky cogongrass), *yeh lu hua* (wild reed-flower), couch grass root.

Properties: cold in nature; pleasant and bland to taste. Safety class: 1.

Chemical Components: cylindrin and simiarenol.

Medicinal Action: clears fevers; stops bleeding; promotes diuresis; allays thirst.

Therapeutic Uses: for treatment of hemoptysis, hematemesis, epistaxis, hematuria, nephritic edema, urinary tract infections, high fevers, and thirst. Also for the treatment of acute toxic hepatitis and measles (Huang, 1993a).

Manner of Use: made into a tea by decoction.

8.8.4 Rhizoma Phragmitis *(Lu Gen)*

Family: Gramineae.

Scientific Name: *Phragmites communis* (L.) Trinius.

Other Common Names: *wei gen* (roots of bullrushes), *lu chu ken, lu mao ken* (hairy roots of reed), *lu wei, carrizo,* common reed, reed grass, water reed.

Properties: cold-natured, but pleasant to taste. Safety class: n/a.

Chemical Components: n/a.

Medicinal Action: clears fevers; detoxifies; promotes salivation to stop thirst; promotes diuresis.

Therapeutical Uses: for treatment of thirst in fevers, nausea, vomiting, hematuria, and burning urination.

Manner of Use: made into a tea (decoction).

8.8.5 Rhizoma Zingiberis *(Jiang)*

Family: Zingiberaceae.

Scientific Name: *Zingiber officinale* Roscoe.

Other Common Names: *gan jiang* (dried ginger), *sheng jig* (fresh ginger).

Properties: slightly warm in nature; acrid to taste. Safety class: 2b and 2d. It should be avoided by those who manifest the symptoms of interior heat caused by *yin* deficiency or bleeding caused by blood-heat (Zhang et al., 1988).

Chemical Components: 2.5–3% essential oil, sesquiterpenes [zingiberene, curcumene, β-bisabolone, and (ϵ)-a-farnesene] predominate.

Medicinal Action: warms the central region; dispels han (cold); helps the circulation in the body.

Therapeutic Uses: for treatment of *yang* deficiencies, slow pulse, cold extremities, deficient-cold stomach-spleen, diarrhea, moderate cold-caused abdominal pain, and moist coughs.

Preparation: used as a seasoning in cooking; also to be made into a sweet ginger soup used as a stomach tonic and used as a digestant in sub-acid gastritis and for lack of appetite.

8.9 BULB

8.9.1 Bulbus Allii Macrostemi *(Xie Bai)*

Family: Liliaceae.

Scientific Name: *Allium macrostemon* Bge., *A. fistulosum.*

Other Common Names: *cong jing bai, cong bai tou.*

Properties: warm in nature; bitter, yet acrid to taste. Safety class: 1.

Chemical Components: volatile oils (allicin, dially sulfide), nicotinic acid, lipid, mucus, and vitamins C, B_1, B_2, and A.

Medicinal Action: removes superficies syndrome; activates *yang;* clears away toxic materials.

Therapeutic Uses: for treatment of chills, fever, and headache casued by exogenous febrile diseases, abdominal pain caused by *yin*-cold, ascaris intestinal obstruction, constipation and uroschesis, dysentery, carbuncle, and swelling.

Manner of Use: decocted or boiled in aloholic drinks.

8.9.2 Bulbus Lilii *(Bai He)*

Family: Liliaceae.

Scientific Name: *Lilium brownii* F.E. Brown var. *viridulum* Baker, *L. lancifolium* Thunb., *L. lumilum, L. pumilum* D.C., *L. candidum, L. odorum.*

Other Common Names: lily root (Figure 14.8).

Properties: slightly cold in nature; bitter to taste. Safety class: 1.

Chemical Components: n/a.

Medicinal Action: moistens the lungs to arrest coughing; clears fevers; calms nerves; promotes diuresis; increases the leukocyte count.

Therapeutic Uses: for treatment of coughing, hematemesis nerves, anxiety, and indigestion. It also promotes uiuresis.

Manner of Use: used in cooking or is cooked with water for drink.

8.10 BARK

8.10.1 Cortex Cinnamomi *(Rou Gui)*

Family: Lauraceae.

Scientific Name: *Cinnamomum cassia* J. Presl., *C. aromaticum.*

Other Common Names: cassia, cassia cinnamon, cassia lignea, cinnamon bark.

Figure 14.8 Dried *Bulbus lilii.*

Properties: very "hot" in nature; pleasant, yet acrid to taste. Safety class: 2b. (Twig part is also used and categorized as Class 1.)

Chemical Components: 1–2% cinnamic oil (75–90% cinnamic aldehyde, a little cinnamyl acetate, and phenyl propyl acetate) (Zhang et al., 1988).

Medicinal Action: warms the kidneys; supplements the body fire; dispels cold; alleviates pain.

Therapeutic Uses: for treatment of cold visceral organs, chronic diarrhea, cold and pain in heart and abdomen, inadequate yang element in kidneys, chilled lungs, coughing and wheezing, and lumbago; also used for vasodilatation.

Manner of Use: cooked together with vegetables or used as an ingredient in a recipe.

8.11 FUNGUS

8.11.1 Tuckahoe *(Fu Ling)*

Family: Polyporaceae.

Scientific Name: *Poria cocos* Wolf.

Other Common Names: China root.

Properties: the dried sclerotium of the fungus is neutral in nature; pleasant and light to taste. Safety class: n/a.

Chemical Components: pachymarose (a polysaccharide) and organic acids, including pachymic acid, tumulosic acid, eburicoic acid, and pinicolic acid.

Medicinal Action: breaks down moisture; promotes diuresis; benefits the stomach and spleen; settles nerves (tranquilizing effects); lowers the blood sugar level; is used as a cardiotonic; increases the immune response of the body to cancer cells (Huang, 1993a).

Therapeutic Uses: for treatment of moisture dominance in kidney deficiency, edema, pulmonary congestion, vomiting and diarrhea, difficult urination, apprehension, and insomnia. Outer covering tends to promote diuresis and reduce edema; scarlet *fu ling* tends to circulate moisture and to reduce moisture-based heat.

Manner of Use: used in cooking.

8.12 MARINE PLANTS

8.12.1 Thallus Laminariae (Ecklonae) *(Hai Dai* or *Kun Bu)*

Family: Laminariaceae.

Scientific Name: *Laminaria saccharina* (L.) Lamour.

Other Common Names: *hai ma lin, hai cao* (sea grass), eel-grass, kelp.

Properties: cold in nature; salty to taste. Safety class: n/a. May have headache and heart palpitation as adverse effects. Because of a high iodine content, it should be used cautiously in cases of acute tuberculosis and chronic bronchitis (Huang, 1993a).

Chemical Components: iodine (0.34% of total weight), K, Ca, laminarin, laminine, algin, and amino acids (Huang, 1993a). It also contains 8.82% of pentosan and zosterin, galacturonic acid, galactose, arbinose, xylose, α-methyl-xylose, tannin, and vitamin B_2 (Zhang et al., 1988).

Medicinal Action: softens the "hard," resolving phlegm; induces diuresis; expels pathogenic heat; lowers blood pressure; is hemostatic; has a hypolipemic effect (due to laminarin) (Huang, 1993a).

Therapeutic Uses: for treatment of enlarged lymph nodes, tumors, edema, congestion, painful testicles, goiter, tuberculosis, and hypertension.

Manner of Use: used as a food or snack.

9. ANIMAL ORIGIN

9.1 CONCHA OSTREAE *(MU LI)*

Family: Ostreae.

Scientific Name: *Ostrea gigas* Thunberg., *O. talcenuhanesis* Crosse, *O. rivularis* Gould.

Other Common Names: Oyster shell.

Properties: shell cool in nature; salty to taste. Safety class: n/a.

Chemical Components: 80–95% calcium carbonate and some calcium sulfite and aluminum (Huang, 1993a).

Medicinal Action: moderates the *yang;* stabilizes body chemistry; clears fever; breaks up congestion.

Therapeutic Uses: for treatment of hypertension, dizziness, seminal emission, tuberculous cervical nodes, hidrosis. Used as an antacid and astringent.

Manner of Use: used as calcium supplements.

9.2 CORIUM STOMACHICHUM GALLI *(JI NEI JIM)*

Family: Phasianidae (Anon., 1977).

Scientific Name: *Gallus gallus domesticus* Brisson.

Other Common Names: chicken giblets.

Properties: neutral in nature; pleasant to taste. Safety class: n/a.

Chemical Components: ventriculin and vitamins.

Medicinal Actions: increases gastric secretion; tonifies stomach; helps digestion.

Therapeutic uses: for treatment of overeating, vomiting, diarrhea, and marasmus in children.

Manner of Use: mixed with vinegar and cooked.

9.3 MEL *(FENG MI)*

Family: Apidae.

Scientific Name: *Apis cerana* Fabricius; *A. mellifera* Linnaeus.

Common Name: honey.

Properties: neutral in nature; sweet and pleasant to taste. It should be avoided by those who have symptoms of accumulation of phlegm-dampness in the body, abdominal distension, and lingering diarrhea. Safety class: n/a.

Chemical Components: fructose, glucose (70%), a little sucrose, maltbiose, dextrin, gum, nitrogen-containing compounds, organic acid, volatile oil, pigment, wax, plant bits, and yeast (Zhang et al., 1988).

Medicinal Action: moistens dryness; relieves spasm; clears away toxic materials.

Therapeutic Uses: for treatment of constipation, cough, epigastralgia, rhinorrhea with turbid discharge, aphthae, a scald, or burn.

Manner of Use: drink soon after being infused.

9.4 ZAOCYS *(WU SHAO SHE)*

Family: Colubridae.

Scientific Name: *Zaocys dhumnades* Oshimai (Stejneger).

Other Common Names: green snake.
Properties: neutral in nature; pleasant to taste. Safety class: n/a.
Therapeutic Uses: for dispelling moisture.
Manner of Use: cooked in soup or stored in wine.

10. CONCLUSION

Although there are many other Chinese foods having medicinal effects, such as wood ear, mushrooms, garlic, soybeans, etc., the items discussed here are limited to the official list issued by the Chinese government. In Taiwan, the Bureau of Food Hygiene and the Department of Health recognized that the traditional Chinese herbs listed in the *Ben Cao Gong Mu* have been traditionally used as foods and can still be treated as foods (Chen, 1998). In Japan, functional foods have been regulated by the govermental health authority. A special category for foods with a medicinal effect is listed as Foods for Specified Health Use (FOSHU) in the "Nutrition Improvement Act" (Anon., 1997). In the United States, the Dietary Supplement Health and Education Act (DSHEA) was passed by Congress in 1994 (Anon., 1994). The Act expanded the traditional dietary supplements as essential nutrients to substances like ginseng, garlic fish oils, and so forth. As a result of the Act, dietary ingredients used in dietary supplements are no longer subject to the premarket safety evaluations required of other new food ingredients or new uses of old food ingredients (Anon., 1994a). The new law will help to regulate the proper use of functional foods.

11. REFERENCES

Anon. 1977. *A Barefoot Doctor's Manual: A Guide to Traditional Chinese and Modern Medicine.* Prepared by The Revolutionary Health Committee of Hunan Province. Madrona Publishers, Seattle, WA.

Anon. 1991. *People's Republic of China Drug Book.* Department of Health, People's Hygiene Publisher, Beijing, China (in Chinese).

Anon. 1994. The Dietary Supplements Health and Education Act of 1994 (DSHEA). Food and Drug Administration, U.S. Department of Human Resource and Service, Washington, D.C.

Anon. 1997. *A Guideline Handbook on Foods for Specified Health Use (FOSHU).* Japan Health Food and Nutrition Food Association, Ministry of Health and Welfare. Tokyo, Japan.

Bensky, D. and Gamble, A. 1986. *Chinese Herbal Medicine.* Eastland Press, Inc., Seattle, WA.

Bisset, N. G., ed. 1994. *Herbal Drug and Phytopharmaceuticals* (Max Wichtl, editor for German edition). Medpharm Scientific Publishers, Stuttgart, Germany and CRC Press, Boca Raton, FL.

Carper, J. 1989. *The Food Pharmacy.* Bantam Books, New York.

Chang, H. M. and But, P. P. H. 1986. *Pharmacology and Applications of Chinese Materia Medica.* World Scientific, Philadelphia, PA.

Chen, S. K. 1998. Personnel communication. Bureau of Food Hygiene, Department of Health, Republic of China Executive Yuan, Taipei, Taiwan.

Chen, W. C. 1997. *Chinese Herb Cooking for Health.* Chin-Chin Publishing Co., Ltd., Taipei, Taiwan.

Hsu, H. 1986. *Oriental Material Medica: A Concise Guide.* Oriental Healing Arts, Long Beach, CA.

Huang, K. C. 1993a. *The Pharmacology of Chinese Herbs.* CRC Press, Inc., Boca Raton, FL.

Huang, S. C., ed. 1993b. *Shiyong Zhongcaoyao Caise Tupu* (Chinese Medicinal Herbal Color Book). Gaunsi Science and Technology Publisher, Naning, Gaundong (in Chinese).

Kowalchik, C. and Hylton, W. H., eds. 1998. Safflower. In: *Rodale's Illustrated Encyclopedia of Herbs,* Rodale Press, Emmaus, PA. pp. 436–437.

Lee, Y. H. 1990. *Home-Cook Nutritional Recipes.* Haibin Publishing Co., Hong Kong.

Leung, A. Y. and Foster, S. 1996. *Encyclopedia of Common Natural Ingredients Used in Food, Drugs and Cosmetics,* 2nd ed. John Wiley & Sons, Inc., New York.

Li, S. C. 1578. *Ben Cao Gong Mu* (The Chinese Pharmacopoeia) Volumes 1 and 2 (new audited version published in 1991). People's Hygiene Publisher, Beijing, China (in Chinese).

Lin, R. I. 1994. Phytochemicals and antioxidants. In: *Functional Foods: Designer Foods, Pharmafoods, Nutraceuticals,* I. Goldberg, ed., Chapman & Hall, Inc., New York, pp. 453–467.

Lu, H. C. 1994. *Chinese Foods for Longevity.* Yuan-Liou Publishing Co., Ltd., Taipei, Taiwan (in Chinese).

McGuffin, M. Hobbs, C., Upton, R. and Goldberg, A., eds. 1997. *American Herbal Products Association Botanical Safety Handbook.* CRC Press, Boca Raton, FL.

Ortiz, E. L. 1992. *The Encyclopedia of Herbs, Spices, and Flavors.* Dorling Kindersley, Inc., New York.

Potterton, D., ed. 1983. *Culpeper's Color Herbal.* Sterling Publishing Co., Inc., New York.

Polunin, M. and Robbins, C. 1992. *The Natural Pharmacy.* Macmillan Publishing Co., New York.

Su, W. L. 1993. *Oriental Herbal Cook Book for Good Health (I).* Shun An Tong Corp., Flushing, NY.

Tang, W. and Eienbrand, G. 1992. *Chinese Drugs of Plant Origin.* Springer-Verlag, New York.

Wiss, R. F. 1988. *Herbal Medicine.* Beaconsfield Publishers Ltd., Beaconsfield, England.

Welsh, F. W. 1995. Letter to Canadian Health Food Association on letterhead of Health Protection Branch, Health Canada, dated February 6, 1995, and Enclosure.

Williams, T. 1995. *Chinese Medicine.* Element., Rockport, MA.

Yeung, H. 1985. *Handbook of Chinese Herbs and Formulas, Volume 1.* Institute of Chinese Medicine, Los Angeles, CA.

Zhang, W., Jia, W., Li, S., Zhang, J., Qu, Y., and Xu, X. 1988. *Chinese Medicated Diet.* Publishing House of Shanghai College of Traditional Chinese Medicine, Shanghai, China.

CHAPTER 15

Cultural Aspects of Asian Dietary Habits

JACQUELINE M. NEWMAN

1. INTRODUCTION

ASIA is one of the fastest growing continents in terms of both population and economy. In the age of globalization, as Asians are coming to countries in the West, more and more people from Western countries, including the United States, are traveling to Asia. Some go as tourists, others to participate in commercial ventures. A growing number of those going to Asia are the legions of businesspersons seeking out and making foreign investments, exploring and consummating business deals, and coming and going to enlarge and enhance the rapidly growing Asian export trade.

When people cross the oceans, no matter the direction, their visits can be enhanced if, before they go, they study the food habits to be encountered. Should their travel be for business, learning can maximize business potential because understanding food habits is particularly important in cultures where social and business arrangements often are initiated or finalized at the table; such is the case among most people in Asia. Therefore, for business and pleasure, it is important to understand Asian people's foodways* and to know that their food habits are characterized by unity of continent but diversity of culture. It is important to explore the foodways of their individual cultures, their home country beverage and food consumption, their social concerns, availability of resources, religious sanctions and taboos, and more.

*Foodways means "food choice—methods of eating, preparation, number of meals per day, time of eating and the size of portions eaten, are an integrated part of a coherent pattern in which each custom and practice has a part to play" (Fieldhouse, P., 1986. *Food and Nutrition: Customs and Culture.* Croom Helm, London.)

The purpose of this chapter is to initiate a general understanding of some Asian food behaviors and to provide data about beliefs and practices of those populations whose flavors are increasingly a part of the world's diet. Each chapter section investigates each of the larger Asian population groups living in or from the countries of, in an alphabetical order, Bangladesh, Cambodia, China, India, Indonesia, Japan, North and South Korea, Laos, Malaysia, Nepal, Pakistan, the Philippines, Thailand, and Vietnam. The people and the food behaviors will be discussed in terms of geography, language, population, and religious data under headings that speak for themselves, namely: "Food Selection," "Eating Behaviors," "Intracultural Differences," and "Feasts, Holidays, and Special Foods." Some smaller Asian populations are also offered. Differences in the length of each section are somewhat arbitrary but, for the most part, are based on the size of population and/or representative of the population in the Americas, and/or on the number of people that visit or do business with them.

This chapter is about more than one-quarter of the world's population. It is, therefore, only a small taste of the differences between and within Asian peoples. It cannot be, nor is it intended to be, all-inclusive. And if it is not what you know about one or more population groups, keep in mind that there is a myriad of differences because of availability, culture, and practices whether they are between or within regions, countries, and populations. Therefore, after reading this chapter, consult the References and Recommended Readings at the end, which is divided into two sections: "Food Behaviors, Customs, and Holidays" and "Recipe Resources." Several good additional readings are, in alphabetical order: Barer-Stein (1979); Bryant, Courtney, Markesbery, and DeWalt (1985); Dickenson (1993); Far Eastern (n.d.); Kittler and Sucher (1998); Tannahill (1973); Toussaint-Samat (1992); Trager (1995); and Zibert, Stevens, and Vermont (1995), to name but a few. They discuss various cultures or cultural aspects as can individual embassy and cultural attaches, other books and magazines, and the vertical files in libraries and more. All are means of appreciation and understanding. All can serve as a knowledge base and ways to remove misunderstandings.

2. FOODS OF BANGLADESH

Bangladesh (once known as East Pakistan) has foods and food behaviors similar to those in Pakistan and India (which follows in alphabetical order). The major identifying factor is the Muslim heritage of the people. Thus, the eating habits of most Bangladeshi people mirror those of the Muslims in India. The geography of the country is such that rice is harvested three times a year but consumed in such huge quantities that additional imports are needed to meet the demand. In Bangladesh, there is a variety of medicinal/food systems borrowed from Arabic practices, homeopathic systems, and regional and local practices and considerable adherence to *Ayurvedic* beliefs and food practices. Tea is the beverage of choice, and foods are similar to both India and Pakistan, as long as they adhere to the Muslim practices of *halal* meats, no stimulants, and no pork.

3. CAMBODIAN FOODS

The foods of Cambodia have strong Chinese influences, along with their own character. Rice is the main grain served at the main meal, and noodles are a breakfast staple. Unlike China, tea consumption is low because the mealtime beverage of choice is often warm or hot water, boiled or not. Water is the preferred between meal drink, too. Fish is the main protein food, with fresh varieties more popular than those of the sea. All meats are consumed, except mutton; poultry is not popular, and fish sauces and peanuts are used in most dishes. Black or white pepper is the most important condiment, with ginger next in importance. Most Cambodians prefer two meals a day, the first late in the morning, the second near dusk. At meals, knives never appear at table, soup is invariably served, and foods are eaten with forks and spoons; rarely are chopsticks used. Snacks are very popular before, after, and between these meals. The main holidays are the lunar New Year and a New Year celebrated in April. Barbecued or roasted meats, fried chicken, rice, many vegetables, and dessert are served and there may be a *baci* ceremony. *Baci* is an important, but informal, ceremony used to mark any life-cycle, business, or personal occasions. It has elements of Buddhism and spirit worship, eggs have symbolic meaning, prayers and a benediction are intoned, the table centerpiece is to look tree-like, and strings wrapped around wrists are used to express specific good wishes; they should not be cut for at least 3 days.

4. CHINESE FOODS

4.1 FOOD SELECTION

The Chinese cuisine we know evolved during the Ming (1368–1644) and Qing (1644–1911) Dynasties. It has many Manchu influences, along with lesser borrowing from the Japanese during their sojourns in Ming and later times. Wen (1974) and others have indicated that this cuisine and food in general is one of the people's greatest pleasures. Perhaps this is based upon an old Chinese proverb that says: Food for the people equals heaven (Sun, 1987), or perhaps it is because the Chinese respect food, talk about it frequently, and share it often. Chinese foods may seem alike to those unfamiliar with them; however, there are major and minor differences, depending upon region of ancestry, county and country of origin, intermarriage with those from other regions within China, residency or marriage outside of it, and the other land(s) that they or their ancestors visited or lived in.

The most important Chinese food habit and heritage difference is *fan-tsai* where *fan* is the grain and *tsai* the vegetables, meats, and other accompaniments of a meal (Chang, 1977). This main difference is a north/south dichotomy with respect to the primary *fan,* or grain, used. It is important because Chinese people get from 60% to 80% of their calories from grain (Anderson, 1988). The larger figure is common among people in China, the lesser amounts more common among Chinese who live in Asian or other countries. The variation among the grains selected is that for those whose heritage is south of the Yangtze River, the major grain is rice. Those living north of this river use more wheat, millet, sorghum *(kaoling),* and other lesser well known grains; they eat very little rice (Anderson, 1988; Chang, 1977).

There are other universalities in the cuisine. These include, but are not limited to, parents who tell their children to leave the table 70% full and providing respect for personal dietary control. Another is a concern that there be a plethora of textures, temperatures, and tastes at all meals, assuring that people serve a variety of dishes to season their *fan.* Yet another commonality is the belief that food is medicine. This is especially true among Cantonese but not limited to those who classify some foods as *pu,* or remedial, and others as *hsieh,* or damaging. Food as medicine is also seen in complementary forces that affect life, including the duality of *yin* and *yang.* Foods and diseases are labeled with one of these forces that translates thus to the table: People use a *yin* food to treat a *yang* condition or a *yang* food when the illness is considered *yin* (Koo, 1982).

The diet of most Chinese includes the *fan* foods mentioned above and also corn, oats, potatoes, yams, and other roots and tubers. The main meat is pork, and the word *meat* and the word *pork* are one and the same in many parts of China. In addition to pork, there is considerable use of lamb by Muslim and other minority populations who, for religious reasons, do not eat pork and by others living in northern areas. Very little beef is available because of minimal grazing land. The Chinese eat all other land and sea creatures and all parts of them. Vegetable use is extensive, with the more common ones being soybeans and other leguminous plants, *bok choy* and all members of the cabbage or Brassica family, eggplant, members of the Allium family (such as garlic and onions), other green vegetables, and herbs and grasses. Fruits and nuts are eaten frequently, and persimmons, pomegranates, peaches, pears, apples, grapes, Chinese dates called *jujubes,* chestnuts, walnuts, and peanuts are popular (Simoons, 1991). Consumption and variety of fruits and vegetables depend upon geographic location and season because in China there is limited transportation of common foods.

4.2 EATING BEHAVIORS

Most Chinese eat three meals a day, with lots of rice, noodles, other grain dishes, and/or steamed bread, along with several stir-fried, steamed, red-simmered, or deep-fried dishes. The number of dishes relates to the number of diners and dollars. Usually three or more dishes are served, but there can be as many as a dozen. At meals, everyone gets their own large bowl of *fan* and chooses *tsai* from the accompanying serving dishes. Ingredients are cut before cooking, dishes are served together, and foods are eaten with chopsticks. No knives come to the table because they are considered weapons. Soup or a soup-like dish such as *congee* is the beverage of choice, eaten with a ceramic or plastic soup spoon or sipped from the bowl. Soup is most often eaten during and at the end of the meal. Desserts are not common at main meals, but pastry-type items are available at *dim sum* meals (see below), and they are popular for special occasions and banquets. Tea is not common at meals, except among the Cantonese, and it is always served to visitors as soon as they enter a home. Sweets and nuts might accompany this tea as might small dishes such as dumplings. Alcohol is rarely served with ordinary meals and almost always is served at banquets or at special meals such as those celebrating life-cycle events. Alcoholic beverages are never consumed by Buddhists, Muslims, and some minority populations.

4.3 INTRACULTURAL DIFFERENCES

In the south of China, they grow and consume large quantities of rice and other foods thanks to two or three growing seasons. Further north there is only one, and in some northern areas they have a minimal ability to produce any food. Dairy products are almost nonexistent in the South. There is some use of milk, yoghurt, and other milk products in the North, primarily in Tibet and Mongolia, where the main milk is yak milk. Some yoghurt is available in and around Beijing.

There is much conversation about and cookbooks discuss different numbers of culinary regions. Some say there are the four regions, named for the points of a compass: South around Guangdong (Canton), East in and near Shanghai, West around Sichuan and Hunan, and North in Beijing and environs. Still others talk about four but refer to them as North or Peking cuisine; East, including the entire lower Yangtze River valley; Central, meaning Sichuan, Yunnan, and Hunan; and Southeast, around Guizhou and Fujian and Taiwan. Still others go so far as to say there are 20 or more regional cuisines. To simplify at least in terms of mealtime, people of southern heritage with roots south of the Yangtze River eat a breakfast and a lunch that are similar. Those north of this river have lunches and dinners more similar and a breakfast that is different.

Southern morning or early afternoon meals usually include *dim sum.* This translates as a "dot the heart" meal that is often eaten in a tea house or restaurant. There it is served from early morning until sometime past midday. Foods are eaten outside, if one can, because they include labor-intensive items such as dumplings and other small dough-wrapped items, long-cooked foods, and little foods served steamed, baked, fried, and roasted. Literally hundreds of choices are available to "dot the heart" and begin the day. *Congee,* a rice gruel or porridge served then or at home is Cantonese and is known as *juk* in Mandarin. This rice porridge is eaten with pickled vegetables, with tiny bits of meat or legumes, or even with the previous day's leftovers in or served with it. Southerners drink tea at *dim sum* meals, choosing among green, oolong, or red varieties. This refers to the color of the tea infusion; the color of the leaf is called green, oolong, or black. The Cantonese also eat all manner of animals, some only in colder seasons. They have a love of exotic fruits and foods, enjoyed centuries of trade that brought them all manner of delicacies, have an obsession about peak freshness and flavor, and eat lots of fish and sea creatures from their long coastline. For them, soups and seafood are omnipresent, and tea is consumed frequently. Popular dishes are many and often include black beans, light soy sauce, oyster sauce, shrimp paste, Chinese sausage, and roast pork. All are made by what some might say is undercooking the freshest of ingredients.

While Southerners and many Easterners eat similarly, people from Shanghai and environs prefer their foods red-cooked, often indulge in freshwater eels, and eat all their foods a bit sweeter than their southern, western, or northern neighbors. Those from western areas such as Sichuan and Hunan like their main dishes more piquant than do Chinese from other regions, and they use a lot of fagara, a dried Sichuan peppercorn flower. The western region has little access to internal waterways or the coastline, so they eat little fish. Instead, they enjoy a plethora of preserved vegetables, make a great ham, double cook many foods, and use hot peppers as well as fagara.

For breakfast, Northerners prefer warm soymilk and fried wheat-dough crullers. Sometimes they make or buy steamed bread and have it with one or more stir-fried, steamed, or red-simmered dishes. They like boiled peanuts, a few pickled vegetables, and some sugar with their crullers in the morning, and they dip them in sugar and warm milk before biting off a piece. At other meals, people of northern heritage like to drink the water their noodles have been cooked in—plain or flavored. In addition to pork as the main meat, there is considerable use of lamb in the North. Since Muslims never eat pork, lamb is always their main meat. Dairy products in Tibet, Mongolia, and Manchuria are made from yak milk, and Tibetans like milk with their tea, along with salt, sugar, or both. Vegetarians, be they from north or south, get their protein from a sophisticated cuisine based upon considerable use of gluten and *tofu,* and they make it to taste like and imitate animal food.

4.4 FEASTS, HOLIDAYS, AND SPECIAL FOODS

Most Chinese observe the major holidays of the lunar year. Of these, New Year *(Yin Li)* is the one observed with the greatest enthusiasm (Newman et al., 1988; Latsch, 1988). It starts just before the first day of the first moon, late at night on the last day of the old year. Before that, homage was paid to *Tsao Chun,* the kitchen god, before he leaves to report about behaviors of family members. New Year involves 3–5 days of celebration and festivities, uniting families, settling debts, patching up quarrels, and preparing a large variety of dishes. Favorite gifts are oranges, tangerines, or other citrus fruits that wish abundant happiness, peace (apples), long life (peaches), fertility (pomegranates), ingots of gold (olives), or that there be no separation among friends (pears). People eat many, many New Year dumplings *(jiao tze),* traditionally round, small, and stuffed full with sesame paste, red bean paste, and preserved cassia blossoms as symbols of good fortune; they might also be filled with pork or vegetables or both. Spring rolls, looking like rolled gold and glutinous rice cakes *(nien gao),* sounding like "going high," are popular New Year good luck symbols. The New Year officially ends 15 days later at the Lantern Festival, a time to help wandering souls go to judgment and reincarnation.

Other important holidays include the Clear Brightness Festival, known to some as the Remembrance of the Ancestors and to others as the Feast of the Dead *(Qing Ming).* This holiday starts 2 weeks after the vernal equinox (very early in April) when it is considered obligatory to visit ancestral spirits and sweep their tombs. The entire family goes together and brings food to the ancestors, often staying and eating it and more thereafter. Dragon Boat Festival *(Duan Wu),* on the fifth day of the fifth lunar month, has multiple meanings, the most recent of which is to appease the soul of an ancient minister with triangular, leaf-wrapped rice cakes dropped into the sea or river. After that rite, dragon boat races, family reunions, and feasting can begin. The Moon Festival, sometimes called The Moon's Birthday, is on the fifteenth day of the eighth lunar month. It is at harvest time when families get together and give thanks for the past year's crops; they also wish for a great and abundant coming year. Moon cakes *(yue bing),* heavy with fruits and nuts, identified on top with a specially designed stamp, are purchased, presented, and enjoyed by all on this holiday (Stepanchuk and Wong, 1991).

5. INDIAN FOODS

5.1 FOOD SELECTION

From a culinary perspective, the cuisines of this subcontinent are as varied as the regions, ethnicities, languages, and religions; the latter are mostly Hindu and Sunni Muslims, Christians, Buddhist, Jain, Sikh, and Parsi (Chakravarti,

1974). Most people use large quantities and varieties of cereals and legumes; fairly large amounts of dairy products, though not necessarily as beverages; and smaller amounts of vegetables and fruits that often come to the table as chutneys, pickles, or condiments. However, about 30% of the population is vegetarian, with assorted sets of eating taboos such as those who only consume all kinds of fruits, vegetables, legumes, and nuts; others who eat these and dairy products but avoid eggs; and still others who do not eat eggs, garlic, and onions but do consume dairy products (Chakravarty, 1972). Meat, poultry, and fish are used in lesser amounts, and some vegetarians of the Hindu faith eat fish because they consider them fruits of the sea. Sweets are commonplace, and everyone eats dals and lentils of all colors: chickpeas, black-eyed beans, red kidney beans, split peas, and other pulses not available outside of India. Coriander, cumin, curry leaves, fenugreek, asafetida (a plant resin related to fennel and called *heeng*), mango powder, mustard seeds of different colors, pistachio nuts, poppy and sesame seeds, saffron, tamarind, and a piquant spice mix called *garam masala* are the more popular flavoring ingredients. These are used along with a plethora of oils, including, but not limited to, coconut, peanut, sesame, and mustard oil (Achaya, 1994).

5.2 EATING BEHAVIORS

Most Indian people eat two meals a day, times varying from region to region. In the morning they might have *uppama,* a popular breakfast porridge made from semolina, onions, ginger, chilies, and black mustard seeds, or they might eat long-cooked rice similarly seasoned and have tea or coffee laced with sugar and boiled milk. Fruits and pickles can be served then and are always available at the evening meal, along with one or more curries, lots of grain foods, legumes, breads, a very sweet milk-based dessert such as *gajjur ks halwa* (carrot halva), *jallebi* (batter-dipped fried sweets in syrup, *kheer* (rice pudding), *barfi* (sweet toffee), *ras malai* (cheeseballs in milk), and/or a buttermilk beverage. *Dosa* (a pancake used as a wrapper) and *idli* (a steamed rice cake) are rarely served alone or plain; instead, they are filled with or eaten with a bit of spicy vegetable, a mite of lentil curry, or some chutney. Other popular bread foods are *wellappam, idyappam, paratha,* and *payasam*—rice-based batter bread with coconut, little baskets of rice noodles, griddled wheat bread, and a slow-cooked cardamon-flavored, milk-based pudding, respectively (Achaya, 1994). Their foods are served on a platter called a *thali* or on large leaves and put on the floor or low tables. They are eaten, after thorough washing of hands, faces, and legs, with the first three fingers of the right hand tearing off pieces of bread or making clumps of rice or other grains to get their curries or whatever else they are eating to their mouths from their own *thali* or the one *thali* served for everyone.

As indicated, vegetarian practices vary and are adhered to in differing amounts. Males and females of the Hindu faith are vegetarians as are most Buddhists and Jains, while more male Brahmins and Vaisays eat meat. Though Muslims don't eat beef, they have no such sentiment about buffaloes, goats, and sheep. Also, most Brahmins and Jains do not eat onions, garlic, coffee, or tea because these are considered sexually stimulating foods. Other than these generalities, illness, pregnancy, and lactation are times when strict adherence to these and other ancient medicinal beliefs, including hot and cold food differentiations, impact food consumption (Storer, 1977). Many people are unaware or even don't know these ancient health considerations, yet they practice them by habit, cultural association, or heritage. One such is *ayurveda,* a Sanskrit 4000-year-old Vedic conceptual system of nourishment and daily life. It is popular with more than 300,000 medical practitioners teaching and practicing this natural healing system as they determine body constitutions *(vata, pitta,* or *kapha).* They do this to assure that foods used are complementary and soothing to body and mind. A *dosha,* or constitution, is present in every individual, as are seven types of body tissues and many *srotas,* or body channels. In addition, there are five elements, three food types *(sattva, rajas,* and *tamas),* and various integrating concepts. *Ayurvedics* focus on their food when eating, do not eat in a hurry, and do not eat fried foods. They only eat on an empty stomach, rotate their diets, meet polarity considerations, and use positive accumulated energy. In addition, they serve food considering the six tastes of sweet, sour, salty, pungent, bitter, and astringent (Morningstar and Desai, 1990).

5.3 INTRACULTURAL DIFFERENCES

From a culinary perspective, India has been divided into wheat, rice, and rice-and-wheat consuming areas; it has also been divided by province, religion, and so forth. In the North with a cuisine influenced by the Moguls, pork is rarely consumed, and foods are richer, creamier, and heavier than in the South. There, wheat is the primary grain for breads such as *chapatti* baked in a tandoori oven and *paratha* and *naan* made on flat griddles. These breads and other grain foods are the staples of the northern diet. Main dishes accompany them, if affordable, made with chicken, vegetables, other meats, or combinations. Rich yoghurt-based sauces are used with or in them from the most elaborate to the simplest, and they can be used on salads, which are called *chaats.* The main cooking fat is a clarified butter called *ghee.* Not all Pujabi and Kashmiri peoples and Hindus and Muslims avoid eating meat. Lamb is popular for Muslims who do not eat pork and those of the Hindu faith who do not consume beef. Tandoori cooking is popular, and breads are made in a tandoor oven; the meats are roasted there after application of a paste made with turmeric and chili powders, garum masala, and salt mixed into milk, cream, or yoghurt. Also common and adored is *dal* made with black beans and *kulfi* made with nuts; *kulfi* is an Indian-style ice cream.

In the South, rice is the staple grain. The main course dishes are more highly spiced but lighter than those in the North. In this region, breads made from rice or potato flours such as *idli* and *dosa* are main grain foods; people like the latter with lentils and chutney. Popular foods include *lamb biriyani* made with rice and *baghare baigan,* a spiced eggplant. *Raita,* a spiced yoghurt and *mirchi ka salan,* a very hot dish of chili curry, are popular as are *sambars* and the lentils and vegetables in them. All of these are made dozens, if not hundreds, of different ways. Many Indians in the South are vegetarians with a highly developed meatless-curry-culinary repertoire. Their main cooking fats are coconut or other vegetable oils, and milk is an important ingredient consumed as yoghurt, in *paneers* (cheeses), and in sweet desserts.

North and south, curried dishes are made with simple or complex blends of spices mixed with the meats, poultry, fish, and/or vegetables, along with onions, garlic, ginger, tomatoes, okra, coconut, nuts, and/or yoghurt. Indian cooks carefully grind selected spices daily. They use a lot of cumin, coriander, turmeric, chilies, ginger, cinnamon, various mustard seeds, cardamon, fenugreek, or laos (a dried ground version of galangal), along with asafetida, mango powder, tamarind, and other ingredients. With their breads and pulses, they eat a variety of gourds, greens, roots, and other vegetables, as well as meats, fish and prawns, eggs, and chicken, and they consume wheat, maize, barley, millet, *ragi,* and *bajra* for porridge and more. Leguminous pulse and seed crop use varies by region, season, and heritage and can include Bengal gram, black gram, cow gram, green gram, lentil, moth beans, peas, and red grams. These are the principal protein for most people, vegetarian or not.

5.4 FEASTS, HOLIDAYS, AND SPECIAL FOODS

Indian festivals can be religious and/or follow the agricultural or Western calendars or mythological tales. *Divali,* the Hindu Festival of Lights, honors the religious New Year and is one of the most important of holidays occurring in late fall when it welcomes *Lakshmi,* the goddess and giver of wealth. Other New Year festivals are regional, such as *Godi Padwa, Nav Warih,* or *Varudu Pinappu,* in Maharashtia, Kashmir, or Tamil Nadu, respectively. The festival *Dashera* is in the fall and celebrates a Maharajah's birthday, remembering the warring past. *Teej* is in late summer and is a woman's festival dedicated to the goddess *Parvati. Raakhi* is in August when women with brothers seek their titular, if not real, protection. Other holidays commemorate birthdays of gods, saints, or honored persons; are regional celebrations or fairs; or are other types of festivals, some of which include tests of faith such as abstinence, long walks made barefoot, and so on. The Muslim festival that breaks *Ramadan,* though religious in nature, is an evening food feast on the last of a 30-day, sunup to sunset fasting period. Other religious observances can include temporary abstinence from food or avoidance of meats. On holidays and every day, observant Muslims eat only *halal* or Islamic-approved ritually slaughtered meats and abstain from alcohol, coffee, tea, or other stimulants.

6. INDONESIAN FOODS

The dietary of the people of this equatorial island archipelago of Indonesia, whose national motto: *Bhinneka tunggal ika* means "unity in diversity," is just that, a very diverse mix of Malay, Java, Sumatra, the Arab world, and others, a plethora of religious influences (mainly Muslim), and many local traditions. Rice is the main carbohydrate, made in dozens of ways and consumed two or three times daily. *Nasi goreng* (fried rice) is the best known of the rice dishes outside of the archipelago, and though noodles and noodle dishes are consumed, they are less well known. Saltwater fish and freshwater varieties raised in ponds are the main animal protein. They are eaten frequently, both fresh and dried, the anchovy-like *ikan teri* being the most popular. Vegetables are used at most meals but are not a major portion of the diet. Tubers such as sweet potatoes are loved and white potatoes, manioc, tapioca, and other tubers less so. Cucumbers, pumpkins, eggplant, corn, coconut, and tomatoes are enjoyed, as are local leafy vegetables such as *salam, jeruk purut,* and *sereh.* These islands grow large amounts of popular spices, including nutmeg, mace, cloves, chilies, coriander, cumin, and turmeric. Traders from the Middle East, India, and other countries sought out these islands to buy the spices; they also brought their own indigenous spices, many of which became popular, flavoring Indonesian foods with items such as lemongrass, galangal, peppercorns of all colors and intensities, garlic, onions, limes, peanuts and candlenuts, shrimp paste, tamarind, and soy sauce. In spite of, or a few people say because of, the intensity of flavor mixtures, foods are preferred sweet; even the local soy sauce, *kecap manus,* is fermented sweetened.

Fruits, including soursop, avocado, banana, and orange, are found in beverages more frequently than they are consumed raw; they are loved cut up, consumed with a spoon, and mixed with colorful pieces of gelatin or noodles made of bean flour, dried red beans, kernels of sweet corn, flavored syrups, sweetened condensed milk, and shaved ice. Dairy foods are not popular, cheese is generally not known, and butter is used rarely. Vegetable oils, coconut milk, and cream are used in their place. Coffee and tea, when consumed, are served hot and sweetened with either sugar or condensed milk. Fruit and alcoholic beverages such as beer, hard liquor, and less potent local brews are available made from palm, sago, rice, and coconut; however, they are not popular because the largest religious group, Muslims, are forbidden their use (Peterson and Peterson, 1997). A typical Indonesian meal is based on rice cooked alone or prepared with a *sambal,* which is a hot chili condiment; it is served with *krupuk* (cracker-like wafers made of rice, other grains, or vegetables) and fried shallots or onions. Along with these come main dishes, many small portions of each. Main dishes are made with fish, meat, poultry, fruits, and/or vegetables. Of course, not everyone eats rice, though most people do; others use cassava, corn, sago, or any combination of these. The main meal of the day is consumed in the evening, breakfast can be early, and multiple snacks are eaten between breakfast and dinner, making snacking popular at home and from street vendors whose noodle dishes are sought after as soups or without broth but with fried toppings and many seasonings. Peanuts, fritters, pancakes, and puddings are popular, too.

In Indonesia, spices are preferred rubbed or ground before use; cleavers are used to cut up meats, vegetables, and fruits; and a wok is used to cook them all. If not a wok, a *belanga* is used; it is round-bottomed, earthenware, and deeper than a wok. Many foods are served wrapped and eaten with fingers. Banana and other leaves and an assortment of rolled dough are popular packaging for raw, steamed, and grilled foods. Main dishes are preferred cooked in coconut milk, sweetened, and given texture with ground candlenuts, peanuts, tapioca, fried shallots, and the like. Forks and spoons are the utensils of choice; second are the use of fingers. They are the means to eat one or more *sambal* varieties at meals enhanced with chayote, mango, shallots, or jicama, and raw or cooked vegetables called *lotek* are popular in a peanut sauce. Fruits are also served, as is *rujak,* a popular fruit salad sauce made with jicama, papaya, mango, tart apple, peaches, or whatever is in season. Also commonplace at meals are *gado gado* (a mixed salad), stuffed pancakes called *martabak,* curries as soups or stews, rice dishes, marinated bite-sized meat cubes threaded on bamboo skewers and barbecued called *satays, acars* (sour pickle mixtures), sweet chilled beverages, and steamed or fried fruits. These foods are also served at *rissjtafels* when dozens of these small dishes are served with rice. *Rissjtafel* is known as a rice table. Main dishes at it and all meals must have a variety of flavor, texture, color, and cooking method and an assortment of interesting balances and startling contrasts.

Almost all Indonesians celebrate ritual events with a communal feast called *selamatan.* One such is when vegetables are carried in the form of a sacred mountain during the Yogyakarta Palace Festival. This festival and others have elements of ancient local Buddhist and Hindu celebrations. At the meal portion of these celebratory gatherings, people enjoy mounds of yellow rice and a dozen or more dishes to accompany it. Many of these accompanying foods are fried or boiled; others are served cold and raw, as are salads. Bean sprouts, *sambal,* and *krupuk* grace tables, along with many small main dish food items. For most celebrations, palm wine called *tuak* is popular; it and other alcoholic and nonalcoholic beverages are served before and during the meal. Muslims, the largest religious group on these islands, celebrate *Lebaran* at the end of the fasting month of *Ramadan.* The evening meal that breaks their fast is much like other *selamatans,* except that they do not consume alcoholic beverages. Incidentally, there is a difference between a *selamatan* and a *rijsttafel;* the former is the precursor of the latter and always has both prayer and feast. Today, *selamatans* are held in honor of any religious festival, ritual, life-cycle event, rice harvest, house raising, or even a community event. Indonesians enjoy many dishes at all of them and celebrate in temples, streets, homes, small shops, and large restaurants. All of the foods already mentioned are used for these celebrations and at main meals every day. There is one difference, however. For holidays, yellow and not plain rice is served. It is made yellow with cumin and turmeric. Sometimes the rice, yellow or not, is also sweetened with coconut and condensed milks and served as a sweet cooling dessert after the meal.

7. JAPANESE FOODS

7.1 FOOD SELECTION

The Japanese dietary includes Shinto inspiration, beauty, and simplicity (Seligman, 1994) that came of age in the Kamakura Period (circa 1185–1333), matured during the Edo Period (1603–1867), and had external influences from Chinese neighbors during the Nara Period (710–794). This dietary is the most spice-free dietary of any in Asia. Peppercorns are used today along with a few other and newer seasonings such as *togarashi,* a spiced powder with red pepper flakes. When foods are flavored, it is with *kinome* leaves, *sansho* berries of the prickly ash, *wasabi* (a green horseradish paste), *dashi* (an all-purpose stock), *konbu* (made of kelp), and *bonito* in the form of flakes of this tuna. Japanese people prize individual ingredients, and when preparing foods, they prefer their composition presented on the plate and not in the pot. In general, their meals are light, use limited amounts of fats and oils, and include little starch, except for rice, which is eaten daily and preferred plain. There is some use of *udon* and *soba,* both noodles but from different flours. *Udan* is made from wheat and *soba* from buckwheat; both are popular in soups. A tuber called *konyyakku* is eaten raw, boiled, or made into a flour for a gelatinous noodle eaten plain or in stir-fried dishes. It is rarely used in soups.

Japanese soy sauce, called *shoyu,* is lighter, sweeter, and less salty than most Chinese varieties and is used sparingly in soups and simply prepared main dishes. *Miso,* a fermented soy paste is also used in soups and stocks and is often preferred mixed with *dashi. Tofu* (soybean curd) is simmered, steamed, grilled, and even fried, and when cut into small pieces, it is put in soups. Soups are frequently eaten before raw fish, be it *sushi* or *sashimi. Sushi* is fish served over *wasabi,* which is a horseradish paste served on vinegar-flavored rice. Fish can also be rolled in thin sheets of *nori* (purple or black seaweed) and served alongside a dollop of *wasabi* paste and some *beni shoga* (vinegared red ginger). *Sashimi* is pieces of raw fish served with both of the above. Kelp and *wakame* are other popular seaweeds used to make *sushi,* and they are also used in other ways (Japan Travel Bureau, 1991). Other seasonings include black and white sesame seeds *(goma)* and *gomasio* and a mixture of black seeds and salt used as a common table condiment. Sesame oil is popular for flavoring but not for cooking. *Mirin,* a sweetened rice wine, is used in marinades, dips, soups, and main dishes. It and pickled or marinated vegetables are ubiquitous and served at most meals.

The Japanese are known for *kobe* and *matsuzaka* beef from cows that are fed a lot of beer and massaged to evenly distribute accumulated fat. Oily flesh fish are prized whether they are tuna, mackerel, blue fish, or salmon, as are squid, octopus, eel, and shrimp. Blowfish is a delicacy even though its liver, if consumed, is poisonous and can be fatal. A large variety of vegetables are eaten, such as broccoli, cauliflower, carrots, burdock, scallions, onions, tomatoes, eggplant, spinach, sweet potatoes, squash, green peppers, mushrooms, and the tubers mentioned above. Fruits, popular for snacks or desserts, rarely appear in main dishes. Rather, oranges and other citrus items, apples, pears, and melons are popular plain as desserts. *An-pan* (creamed buns), *senbei* (sweet rice cookies), and *kanten* made into sweet gelatin are other frequently consumed desserts. Popular dishes include *sukiyaki* (meats and vegetables braised or boiled in liquid), *tempura* (vegetables and shrimp or other foods of the sea battered and deep-fried), *teriyaki* (glazed and broiled meats or fish), *umi boshi* (brined sour plums), and a bubbling fondue-type main course soup-meal called *shabu-shabu.*

7.2 EATING BEHAVIORS

Classical five-item meals are preferred in the evening, and breakfasts and lunches are becoming westernized. At night, five dishes are served, each prepared one of five different ways (raw, simmered, steamed, barbecued, and fried) or a food or foods of five different colors (red, green, yellow, black, and white). These colors and cookery preparations are used in fish, vegetables (pickled, fresh, or both), other pickled foods, rice, soup, and desserts. Salads, when part of a main meal, are served before and/or after the main course. All foods are consumed with chopsticks called *hashi,* which are usually made of wood; they are shorter than those of the Chinese and are usually decorated beautifully (Shimoda, 1992). Japanese breakfast can be food left over from the night before or plain rice or porridge served with one or more pickled items. Lunches include *menrui,* or noodles, in a meat-, fish-, or vegetable-flavored soup, and more pickled small dishes. Formal meals and those for guests have *sake,* a rice wine served in small handle-less cups and consumed throughout the meal. Lunch and dinner end with green tea, thin or thick *(ususicha* or *koicha),* served in a cup without a handle, which is about twice the size of those used for *sake.* Foods are almost always served on lacquered trays, and soups in lacquered bowls. The dishes used are made of a variety of ceramic materials, each selected for beauty and the way it harmonizes with the foods to be served on it. Together, food and plate are meant to be a feast for the eye.

7.3 INTRACULTURAL DIFFERENCES

There are a few regional differences. The main consideration is differences in noodle consumption. *Udan* is favored in the South, while *soba* is preferred in the North. Religious considerations are about meat and alcohol use; the latter is not allowed in the Buddhist persuasion. Other differences are minor, such as the shape of rice cakes or their color.

7.4 FEASTS, HOLIDAYS, AND SPECIAL FOODS

The Japanese New Year, called *Oshogatsu,* begins January first; it is celebrated for a week. People exchange New Year greetings and drink cups of *fuka-cha,* a green or seaweed tea of good fortune. It has a pickled plum in it to protect against illness. They also drink *toso,* a special spiced rice wine, followed by a soup called *zoni* that is made of a bean mash with chicken, greens, and toasted pieces of *mochi,* the New Year's rice cake (Japan Travel Bureau, 1989). These rather chewy cakes are used to begin the year on the "right foot," and the vegetables used differ in some regions. In Tokyo, *mochi* are always rectangular in shape; in Osaka they are round. Nowadays, *mochi* are served year-round; traditionally, they were reserved for this one holiday. The New Year ends with a rice porridge seasoned with seven herbal grasses; the ones used vary by region. Other holidays include *Hinamatsuri* (Girl's Day) when a special sweet and mild *sake* is drunk and girls play hostess to boys who come to admire them and their dolls. Boys have a festival day, too, and some call it Children's Day when traditionally, bright carp-shaped banners are mounted on poles outside boys' homes; inside they display their or their family's figures and implements of war. Other holidays include *O-Higan,* celebrating the vernal equinox; *Hanamatsuri* to honor Buddha's birthday; and *Obon,* which is a memorial holiday to honor ancestors. Special cakes made from rice or adzuki beans and various noodle dishes are popular for different holidays in different regions. The tastes and shapes are regional, as are the colors; even some holidays are regional, such as *Namahage,* celebrated in the city of Akita as a mixture of harvest, ancient scaring rites, house visits, rice cakes, and *sake.*

8. KOREAN FOODS

8.1 FOOD SELECTION

Korean food habits have similarities with their Chinese neighbors, coupled with considerable Japanese influence, indigenous traditions (Nelson, 1993; Korea National Tourism Corporation, 1994a), and a preoccupation with food and health. *Kimchi* is their national dish, and it is consumed at every meal. It is exceptionally piquant and made of one or more different pickled or fermented vegetables that can be complemented with fresh and dried salted fish, meat, a combination thereof, and other seasonings; *kimchi* made with Chinese cabbage *(paechu)* is the most popular (Korea National Tourism Corporation, 1994b). Two or three kinds of *kimchi* are eaten at a typical meal, along with soup, lots of rice, and other dishes. The main meat is beef, a favorite particularly when marinated with soy sauce, garlic, ginger, and sesame oil. It is enjoyed grilled over charcoal, a Mongolian influence called *pulgogi,* and is eaten plain or in soups and stews. It is even loved fried; it is popular to fry an American canned, chopped meat product called Spam® (Lewis, 1994). Most meats and other foods are preferred flavored with *denjang,* a soybean paste that has been fermented with *koji* as its microbial starter. Dried anchovies and other dried and preserved foods are an important part of the cuisine, as are soy or fish sauces; *mukhuli,* which is rice wine; and *gochu jang,* a very hot fermented chili paste (Millon and Millon, 1991).

Meals include rice served alone or mixed with other grains. Koreans are particular about their rice; they prefer it wet and sticky; and they take pains to obtain new rice, which is preferred for its fresher taste and whiter color. Some millet is used plain or in place of rice, and wheat and barley are also used but mostly for noodles, though barley is also used as a tea-like beverage. Sweet potatoes and their vines are used in soups made in *tukbaeges* or round clay pots. Stir-fried dishes are made in woks called *sots.* Vegetables of many kinds, lightly seasoned and served raw or steamed, including cabbages, radishes, mustard greens, garland chrysanthemum, spinach, and watercress, are used pickled and stir-fried and referred to as *namul.* Vegetables, *tobu* (tofu), many known and lesser known greens, fish and other seafood of every variety, and beef are parts of meals but also used as side dishes referred to as *panchan.* Higher socioeconomic classes eat many different *panchan;* poor people need to be content with just *kimchi* and rice and maybe a soup called *guk.* Milk and milk products are not part of the traditional diet; neither are prepared desserts, though the latter are always served to guests when they come to dinner and served on festive occasions. For those in the countryside, breakfast is the main meal of the day; for those in the cities, it is the evening meal. In cities, breakfasts and lunches are quite similar; they have soups served with noodles and several side dishes. Dinners include rice, soup, several *kimchi,* and *panchan* selections. In summertime, evening meals almost always end with fresh fruit, the native pear quite popular, along with apples, persimmons, and other fruits. Sweets are not popular at meal time but are loved as snacks, as are dried rice cookies, dried fruits, seeds, and nuts (Millon and Millon, 1991).

In wooded and mountainous regions, use of wild and dried vegetables and roots such as *chui* (a mountain-type watercress), *kosari* (fern stems and shoots), *todok* (a hairy root), and *turup* (tender shoots of an angelic bush) are popular *namul* foods. Even city dwellers manage to obtain and enjoy them.

Garlic, scallions, vinegar, ginger root, ginkgo nuts, dates, and ginseng are used for seasonings and for brewing tea. Noodles are loved and made from mung bean or yam or other flours. Red chili peppers are used everywhere, not just in *kimchi* and other preserved foods. Sesame, as seed or oil, and seaweed are used in or on many dishes. The Korean beverage of choice is *poricha,* a barley tea; the second favorite is a sweet punch made with persimmons and flavored with ginger and cinnamon. Milk and other dairy products are not popular, nor are they part of the traditional Korean diet. Egg use is limited, and foods cooked in oil are usually fried in sesame or bean oils, brown or red.

8.2 EATING BEHAVIORS

Breakfasts are big; they are particularly sumptuous in guest houses and the homes of the middle and upper classes. It can be as little as *kimchi,* with mounds of steamed white rice. For those who can afford, it will be many kinds of *kimchi,* soup, and one or more *panchan.* Lunch and dinners are larger meals, dinner being the main meal of the day. Street foods are very popular, the most common being *pajon,* a scallion pancake made with or without other vegetables and shrimp and served with a vinegar-like dipping sauce. Traditionally, and still preferred, are foods cooked over coals. More recently, electric rice cookers and tabletop burners have become popular tableside. *Bulgogi* and *kalbi* are meats of choice served grilled and on the bone. *Bulgogi* pans with ridges allow juices and fat to move away from meat. Fish stews are popular made spicy and with *tofu,* garlic, green onions, and chili paste as is *pibimpap,* a mixture of rice, vegetables, and chili sauce that is often served with egg. All dishes are well seasoned with pride, and at a dinner table there is even more pride if foods appear in five colors (black, white, red, green, and yellow) and a variety of textures such as those using *kosari* (bracken fern) in salads. Popular foods include this fern and *kim* (a seaweed sheet) brushed with sesame oil and then grilled, salted, and cut into rectangles. Also popular is Spam® served with *kimchi* as a sushi-type dish called *kimpap* or used in a stew of pollack roe, *tofu, kimchi,* green onions, and egg. Chopsticks are used to bring these and other foods to the mouth. A long-handled spoon is served with them to reach for foods from the serving plates and for soup at the end of the meal; both are made of metal. Though rice and soup bowls are used, Koreans do not hold bowls or dishes in their hands. They eat rice and soup with their spoons and side dishes with their chopsticks. When finished eating, they signal that by arranging their spoon and chopsticks as they were originally set (Chung, Haffner, and Kaplan, 1990).

8.3 Intracultural Differences

Pork is more popular in the North, as is millet made into *shikke* and fermented with fish and turnips. There are other regional differences, with the South known for spicier, saltier, and more piquant foods. Some *kimchi* flavors are also regional, as are many specific dishes. For example, *nakchikui,* or broiled octopus tentacle, seasoned and wrapped in straw is very popular in Chollado in the fertile Hinam Plain region. Chejudo Island is known for abalone porridge called *chonbokchuk* and dishes with *pyogo,* a local mushroom. The people of the Chungchong Province love pickled dishes such as a crab stew called *kkotkejang* (Millon and Millon, 1991). In addition, there are minor regional differences, particularly in folk remedies. One recent one is *chugyom,* known as bamboo salt (Korea National Tourism Corporation, 1993). Though the idea is not new, manufacture is of salt packed into pieces of bamboo and then baked at exceptionally high temperatures that liquify and reharden. It is used in ginger and fruit teas, even in *kimchi* and soy sauce and is a traditional medicinal cure. Other health-type foods are dog, brews of herbs, tiger bones, and so on (WuDunn, 1997a, 1997b).

8.4 FEASTS, HOLIDAYS, AND SPECIAL FOODS

Koreans love celebrations. New Year is in January, and they also celebrate the Lunar New Year, called Spring Festival, in late January or February. *Samil,* a folk festival, is in late winter, the Cherry Blossom Festival is in spring, Liberation Day in August, and Harvest Festival is in September. There are many more celebrations and festive life-cycle and other occasions that are times for *shinsol-lao.* When a brass pot of the same name with chimney for coals dish is loved and used to cook meats and vegetables in broth. Once only for royalty, it is now commonplace for birthdays and other life-cycle events and religious holidays. Sweets are popular for ceremonial purposes, too, as is a special buckwheat cake, baked rice, and molasses cakes. Strings of dates are special at weddings, and they and Korean ginseng, referred to as *insam,* make great gifts, as does Spam®. Rice and buckwheat cakes are used for birthdays, weddings, the New Year, and other important feasts. They are colored intensely; attractively made; and fried, steamed, or cooked in a saucepan and sweetened with sugar, honey, or molasses, pine nuts, and/or sesame seeds. These cakes are served with plum beverages, wine, or barley tea. Hard liquor is reserved for men and most often consumed at special inns or brewhouses where the alcoholic beverages are made from rice, rye, wheat, barley, plums, magnolia flowers, or sweet potatoes. *Insam* is popular at other times, is considered one of the world's best ginsengs, and is both herb and food since Koreans do not distinguish between them, and is thought to be good for the body. It is so popular that there are many restaurants that specialize in foods made with ginseng. Also popular and used as curatives are garlic and moxa, an herb known for cautery and other healing purposes.

9. LAOTIAN FOODS

The food habits of Laotians have roots in China and are similar in many ways to those of Cambodians. Their diet includes rice, and they prefer the sticky, short-grain variety and most often like it plain. They eat it at all meals and sometimes combine it with yams or black beans. Other grains consumed are sweet potatoes, corn, taro root, yams, and other starches. They also eat soybeans, green onions, eggs, fruit, and some fish, and they drink tea and coffee; the latter is black or sweetened with condensed milk. Pork and chicken are their main meats, often boiled in stews or soups. Popular seasonings include lemongrass and coriander; they are in almost every dish. Other seasonings are hot chilies, mint, fish and shrimp sauces, curries, and coconut milk. Many Laotian dishes require long cooking times, and they rely on fresh, rather than preserved or pickled, ingredients. Forks are eating utensils of choice, along with spoons—the former the pusher and the latter the carrier of food to the mouth. In the countryside, palm leaves can be utensils. Some people use chopsticks and the above utensils. The main holiday in Laos is the Lunar New Year, and for that and other holidays and life-cycle events, *lup* is served. It is meat cooked with chilies, onions, mint, onions, fish sauce, fried rice, and vegetables. Overall, people eat three meals a day and not always at regular hours because it is common to eat when hungry, and if eating a large meal, they'll drink tea or coffee, though more often the latter black or with sugar and cream.

10. MALAYSIAN FOODS

Malaysia, two islands with those who are descendants of Malays and local tribal groups such as the Bidayuh, Ibans, Dyaks, and other indigenous peoples, have a lush, alluvial plain on the west coasts with many orchards of durian, rambutan, mangosteen, carambola, *langsat, jambu,* and other fruits. Market gardens grow water spinach, bamboo shoots, sweet potatoes, eggplant, taro, cabbages, carrots, peppers, and broccoli, and the mountainous areas offer a plethora of wild foods. There are also tea plantations, dairy and poultry farms, ranches raising cows or pigs but never both, and many paddy lands for rice. The sea coast provides shrimp, fish, and many other sea creatures; in the marketplace, they are available fresh and dried year-round. Most people's staple food is rice, with noodles popular in the mornings and for lunch eaten with available vegetables and meats.

The food of the Malays are a blend of Malay, Chinese, Indian, Cambodian, Laotian, and Thai. Family meals generally consist of one meat or fish dish and several dishes made with vegetables, rice, and soup. They eat their foods with three fingers of their right hand and use a spoon for their soup. Before they do, it is popular to put a pinch of salt or a small piece of a date on the tongue. Shallots, *pandan* leaf from the screwpine known as *daun pandan,* tamarind called *asam,* sour carambola, limes, a wild ginger bud called *bunga kantan,* and hot chili peppers are common in dishes, as are coconut milk, lemongrass, kaffir lime, turmeric, pepper, cardamon, star anise, fenugreek, and a local rhizome *cekur.* Fish is popular, and dried anchovies, dried shrimp, and a shrimp paste called *belacan* are used in many dishes. Curried dishes are popular, too.

One unique feature of Malaysia is the Chinese who came to work in the tin mines. Many married Malay women, called Nonyas, and their families are referred to as Peranakan or Straits Chinese, and their cooking is called Nonya Cuisine. Straits Chinese live mostly in Penang, Malacca, and Singapore; practice Buddhism; and cook their Chinese food with a Malay influence. Their Nonya cuisine is eaten communally, all dishes are served at once, and each person at the table takes some of them to go with their grain, be it rice or noodles eaten from their own bowl. Their dishes are made with soy sauce, bean sprouts, sesame seeds, dried mushrooms, a lemon and a pineapple sauce, and/or bean curd, as well as curry leaves, coriander, cumin, worcestershire, and tomato sauces. They also use a very pungent black shrimp paste and black rice, lemongrass, and galangal known as *lengkaus.* Every person is served a rice bowl, a tiny dish for sauce, chopsticks, and a soup spoon, traditionally made of porcelain to match the other dishes. Foods go from serving dishes to rice bowl and then to the mouth. Soup is often consumed at the end of the meal, and tea is sometimes served during it. Only breakfast is different. Common, then, is a porridge called *congee* or noodles in soup, both eaten with bits of leftovers in to flavor them, or *dim sum,* which are small dumplings and other delicacies.

Indians in Malaysia are mostly Hindu and vegetarian; others are Muslim, Sikh, Bengali, Hindu, or Christian. There is a small Ceylonese population whose foods differ only slightly from local Indian cuisine. They and Malays generally eat their main dishes and their desserts cold. What they make for lunch is generally eaten again for dinner, with several different dishes prepared each day and placed on a table under nets for family and friends to help themselves. Among these dishes Malays would include a pudding, but for the Hindu people in Malaysia, desserts or puddings are not common at meal's end. Rice and seafoods are dietary staples for these populations, and there is an ample supply of fruits and vegetables. They also eat foods consumed by the Chinese but add *ghee* and mustard seed when they cook them and they eat various *dahls,* steamed breads *(idli),* pancakes *(dosai),* and *roti canai* and *murtabak,* both Indian breads. Actually, these last two breads are loved by all Malaysians. Breakfast is often *dosai* with a lentil curry or another bread called *chappati* served with a vegetarian curry. For lunch and dinner, the foods are on a *thali* or large brass platter. This tray is loaded with several curries and rice and perhaps a chutney or two for everyone to share. Indians, like the Malays, generally eat with the first three fingers of their right hand. For those without a *thali,* banana leaves were traditionally used; they still are at weddings and other festive occasions.

People who live in the mountainous regions eat many ferns before they open, wild boar, wild pig, deer, and other animal and vegetable foods in their areas. They also like smoked foods, dried fish, many herbs, and a local sour fruit called *asam gelugur.* This and other intracultural differences in Malaysia are mainly regional and/or religious. Islam prohibits practitioners from eating pork; regions where they are the majority often have very little of it. Other ethnic differences have been delineated above, and regarding holiday and feast days, for Malays and others who are Muslims, the Mohammed's birthday and *Hari Raya,* the holiday that marks the end of Ramadan's 30-day fast from sunrise to sunset, are the most important holidays. At the end of Ramadan, the feast or *kenduri* is celebrated with special cakes called *kuihsi* and other delicacies. For the Chinese whose New Year may overlap *Hari Raya,* they celebrate their important holiday with traditional cakes and elaborate meals. For most Indians, there are three important holidays: *Thaipusam,* celebrated in January when only vegetarian foods are consumed; *Ponggal* or the harvest festival when on the first of three days, newly harvested foods—mostly rice and vegetables—are consumed; and *Deepavali,* also known as the Festival of Lights, when elaborate vegetarian and non-vegetarian meals are made and trays of sweets sent to family and friends. Also for *Deepavali,* rice flour is used to draw decoratively on the entrances of houses to welcome Lord Krishna and the Goddess Lakshmi.

11. NEPALESE FOODS

The cuisine of this country has been influenced by the early development of the Kathmandu Valley peoples and influences from India, Tibet, and China and more recently even some from Europe. Much harsh topography in mountain regions resulted in a need for many fermented foods, including leafy greens *(gundruk)* and bamboo shoots *(tama).* Many vegetable and animal foods are sun-dried, including *maseura,* which are balls of lentils and other vegetables; *sankkhtra,* or dried citrus pickles; and *machha selo* and *sukuti,* or dried fish and dried meat. The Nepali diet is based upon large amounts of rice, corn, millet, wheat, lentils, and other legumes and more recently, potatoes flavored with chilies, garlic, ginger, fagara, cilantro, and scallions. To these are added leafy and other greens and fresh and fermented vegetables and chutneys. On special occasions such as the Festival of Tihar or the Festival of Dasain, lavish meat dishes and many sweets are served. During Janai Purnima, a soup called *quantee* is essential; it is made from sprouted soy, chick pea, black beans, and other leguminous seeds and cumin, lovage seeds, curry powder, garlic, and ginger. Nepali foods use a lot of asafetida (a dried gum resin also popular in India), black cumin seeds, cardamon, curry powder, fenugreek, and tamarind. Dumplings *(mo mo)* are loved and are often filled with turkey or other poultry, lamb, or pork.

12. FOODS OF PAKISTAN

The foods of the Republic of Pakistan share a common heritage with India with major culinary influences of Arabic, Iranian, and Turkish origins, as well as the main ethnic populations of Punjabi, Sindhi, Pathan, and Baluchi. There is considerable rice eaten, preferred *biryani* style, which is cooked in a meat sauce with cloves, cardamon, cinnamon, and yoghurt. They prefer wheat breads, including *roti,* a whole wheat flat bread; *shirmal,* which is a wheat bread cooked with milk and eggs; and *roghnii,* a sweet wheat bread used to eat the meat dishes with little pork consumed because most are of the Islamic faith. Mutton, goat, chicken, lamb, and beef are the main meats served as kebabs, curries, stews, and in stuffing. Meats and eggs are cooked in *ghee* and served with salads. A popular snack food is *pan,* a tobacco-type paste mixed with various spices, betel nut, and coconut put on a green leaf and then spread with a lime paste. It is chewed during the day and also consumed with evening tea when it is served along with *samosa, jalibi,* and *mathi.* Rarely is alcohol consumed, and tea is the favorite beverage at and between meals and later in the evenings. Festivals include *Ed al-Fitter,* the feast to celebrate the end of Ramadan; *Eid al-Azha,* commemorating Abraham's offer to sacrifice his son; *Shab-i-Barat;* and *Maulud n Nabi,* the day of Mohammed's birthday.

13. PHILIPPINE FOODS

The foods in this island archipelago are a mix of Malay and Mongoloid influences, along with those of European, Chinese, and American. Rice, the principal food grain, is eaten along with foods such as corn, coconut, sugar cane, banana, pineapple, papaya, mango and other tropical fruits, sweet potatoes, cassava, and some fish, poultry, and pork. Everyday foods include three meals prepared simply; the main dish can be stuffed, sauteed, broiled, baked or, more likely, boiled *adobo*-style with vinegar, soy sauce, and garlic, or *sinigang* with sour fruits such as guava and/or tamarind or vegetables. Meats and seafoods can also be marinated in soy and fish sauces and coconut milk. They might also be cooked *adobo*-style.

On special occasions, foods are given lavish attention. At all times, sweets are loved, some foods being made exceptionally sweet, even prepared with a burnt sugar sauce called *estofado.* A variety of desserts and snacks such as rice cakes, all called *kakanin,* are served; *bibingka,* a flat cake most often made with coconut milk, eggs, cheese, grated coconut, and lots of sugar, is the most popular. Breakfast is often the largest meal of the day; lunch and dinner are smaller because mid-morning and mid-afternoon snacks are popular. Condensed milk is added to coffee, sometimes even to tea, and it is consumed very sweet; this form of milk is also popular in desserts. At festivals, *tigulasug* is served usually made with beef cooked in liquid with roasted coconut and a mix of salt, garlic, and an herb called *pamapa.* Favorite desserts are *tiyateg* and *dolol.* There are many ancient medical beliefs, an example of which are the peppers or herbs hanging outside homes to ward off evil spirits believed to cause sickness and death (Claudio, 1994). In the Philippines, there are also many customs, one of which dictates that, when entertaining guests, more food than needed should be prepared; politeness dictates leaveing something on the plate at the end of the meal.

14. THAI FOODS

14.1 FOOD SELECTION

The cooking in Thailand (the country was called Siam prior to 1939) is rich and varied with many unusual flavor combinations of spices, herbs, and fresh ingredients blending hot and sweet with sour and salty. The culinary influences are Chinese and Indian, a bit of Persian and Portuguese, and more recently, South American; the religious influences are Theravada, a form of Buddhism. All meals revolve around rice and fish and dishes that are stir-fried, steamed, or grilled over charcoal, or around curries, stewed dishes, and sweets. There is considerable use of *kapi,* a fermented shrimp paste particularly popular in a sauce called *nam prik.* Sauces are made not only with *kapi,* but also with garlic, onions, ginger, tamarind, lemongrass, and salt and pepper. Thai people like them and coconut milk, chili peppers, galangal, kaffir lime, coriander, mace, cumin, cloves, star anise, turmeric, salted beans and cabbages, fish sauces, palm sugar, and many herbs, including Thai basil in very piquant dishes, probably the most piquant in this region, if not in the entire world. Curry pastes are in constant demand and are made from the bird chili, considered the hottest anywhere. There are also milder curries, some of which use turmeric as their base. Thai people prefer grinding their spice mixtures daily, even though there are a plethora of ground varieties in the marketplace. Other popular spices are cinnamon, cloves, and nutmeg used in dishes that are not as piquant. The meat of four-legged animals plays a minor role in the cuisine, but chicken and shrimp are enjoyed and, along with fish, are used frequently with many vegetables, including a kind of celery in soups and stir-fried dishes called *ceun chai,* which is smaller and with thinner stems than the celery used in the United States. They also use a lot of *luk grawan,* which is cardamon ground from nongreen pods, and *krachai,* an unusual rhizome sometimes called lesser ginger.

14.2 EATING BEHAVIORS

Typical meals are served on mats on the floor or on low tables, either with bowls or on flat plates. Rice, known as *cow,* is the staple grain, and *kaengs* or curries, stewed, and other dishes accompany it, along with *nam pla,* an essential salt substitute that is poured onto foods. This liquid sauce is made from fish or shrimp fermented and then strained and bottled. Soups are popular at mealtimes as are sweets, the latter so popular that when guests are invited, two desserts are served, one sweet and the other dry, and both are made with egg yolks and sugar. The most popular fresh fruit are longans (many people grow their own), pomelos, mangoes, custard apples, guavas, pineapple, and durian. Fruits are served for dessert and are popular for snacks, too. Beverages vary, with much tea consumed, along with sweetened fruit drinks, beer, and *arak* (Thai whiskey). Foods are not consumed Chinese fashion with chopsticks; rather, forks push food onto large spoons that bring it to the mouth. The forks are held in the left hand, spoons in the right, and knives not used at the table. The sides of the spoons are used for cutting. Beverages are served in glasses. Some Chinese in Thailand use chopsticks, and Thai people can be seen using them for noodle dishes.

At breakfast, a soft soupy rice gruel, called *khao tom,* accompanied by many small well-seasoned or spicy small dishes, is served. Fish flakes and pickles might accompany these dishes. The old custom of eating many small meals, along with some larger ones during the day, persists for those not working and is popular on weekends and holidays. On workdays, lunch is almost always a noodle dish such as *kwaytiaow,* stir-fried with many spices, herbs, vegetables, meat, and/or crustaceans. *Pad thai* is popular at lunch, too, made with *tofu,* as is *mee krob,* the crispy fried noodles topped with a caramelized ground meat sauce. Dinner usually has three to five dishes and a soup seasoned with cilantro and lemongrass. The dishes and soup arrive together; another dish might come later, and dessert comes last. Sometimes at dinner, there might be a noodle dish similar to those at lunchtime, and almost always there will be a salad called *yam,* perhaps made with the young leaves of the mango tree or with edible flowers in season. *Yam neua,* made with a small amount of beef, is a favorite salad as it cools the palate, as does fresh fruit served at the end of the meal. Between-main-meal snack-meals might include *krathong,* the batter-fried molded cups that hold any number of sweet or savory snacks such as *martabak,* a ground beef omelette. It and various *satay* dishes of marinated skewered meat-on-a-stick are served with dipping sauces, many very piquant. Foods such as these and *pad thai* are made in homes and restaurants and used as small meal components. The foods to make them are found at *klongs,* the colorful popular floating markets on inland waterways in and outside of the main cities. They and supermarkets sell a plethora of fresh fruits, vegetables, spices, and herbs.

14.3 INTRACULTURAL DIFFERENCES

In the mountainous North, foods are heavier than in the South, and steamed glutinous rice might be rolled into balls used for dipping into stewed and curried dishes or used for dipping sauces with their raw vegetables and pork sausage. This sausage, called *nam,* is made from raw ground pork, ground pork rind, salt, garlic, and chilies. Regional foods in the North have Laotian and Burmese influences, such as *ook gai,* a curried chicken dish made with lots of lemongrass, and *nam prik noom,* a sauce for other foods. Northerners love a *kantoke* dinner served on bamboo mats on the floor or at low tables. At them, they eat small dishes, rice balls, and many small side dishes. In the Northeast, delicacies include items from the insect world, snails in curry, and foods made with exceptionally pungent fish sauces, many accompanied by a spicy green papaya salad and perhaps *larb,* a dip of minced raw chicken or fish and roasted rice powder eaten raw with accompaniments. *Larb* is also used in curried dishes, and one called *haw mok* is steamed in a banana leaf and then used to make a custard. In the South with its large Muslim population, seafood consumption is exceptionally high, and vegetarian foods abound. There is a strong Malay influence with foods, similar to those of Northern Malaysia's Indian-style such as the curries with cardamon. Fresh fruit is very popular as is a palace cuisine that was once only popular in Bangkok, which features *look choop,* or tiny imitation fruits made from bean paste and coconut milk, and *mee grob,* a crisp rice noodle dish served in a sweet and sour sauce.

14.4 FEASTS, HOLIDAYS, AND SPECIAL FOODS

Most occasions in Thai life have religious components, parades, and/or fireworks. All can be found at the most important holidays of *Songkran,* their New Year festivities in April; *Loi Krathong,* observed for the full moon in November; *Makha Puja* at the full-moon festival in February; and at the Flower Festival also in February. Early April has a *Pattaya,* or beach festival, and May a plowing festival and a Rocket festival to ensure plentiful rain. There is also *Visakha Puja,* the holiest of Buddhist holidays; *Charkri Day,* which celebrates the founding of the current dynasty; a Vegetarian Festival; an Elephant Round-up; *Charki Day* when people take flowers to the temple; and more. In addition to these, Thais celebrate the monarch's birthday with a royal banquet with foods made into floral shapes by a royal carver. Families, these days and special days such as when a son becomes a monk for some months and wedding days are occasions for prayer and feasting and serving delicate sweets called *look choob* made of a paste of soybeans, sugar, and coconut juice in the form of miniature fruits and vegetables. On ordinary days and less important holidays, fruits are peeled and pleasingly arranged on plates or platters, sometimes in the shape of flowers.

15. VIETNAMESE FOODS

15.1 FOOD SELECTION

The dietary of the people of Vietnam has been influenced by China, Japan, India, and Malaysia, and recently by France, by Buddhism, and by the Montagnards who live in remote mountainous regions. *Nuoc nam,* their soy sauce, is to the Vietnamese kitchen what salt is to the Western world. It serves as a basic seasoning for the country's numerous dishes. Second in importance is shrimp paste *(mam tom)* and then other pastes made from fish or coconut milk. Beef is the main meat, and *pho,* a soup with noodles that is ubiquitous throughout the country, is often made with it. Almost every meal has rice; a salad plate with fresh green leaves, or *rau;* dipping sauces; sweet and hot flavored foods; and *nuoc mam,* the fish sauce that characterizes this cuisine. What also characterizes Vietnamese cuisine are seasonings, most particularly chilies, pepper, garlic, lemongrass or lime, coriander, vinegar, sugar, and of course the pungent fish sauce above made from anchovies and *nuoc cham,* which is a sauce blend of fish mixed with the above-mentioned items, and *tuong,* another type of sauce somewhat related to soy sauce. Many foods, wrapped in a rice paper called *banh tranhg,* are flavored with lemongrass, basil, shallots, garlic, ginger, hot mint (not a true mint), coriander, tamarind, turmeric, and the cassia type of cinnamon. *Pandan* leaf or *la dua* is used to flavor or, more accurately, perfume desserts, and peanuts are favored in these and in many main dishes. Meats—chicken more so than pork—are preferred cooked or served with a *satay*-type sauce called *dau phong rang.* Foods such as eel fried with lemongrass and other fish dishes and/or grilled meat and poultry dishes are popular, as are meats and fish dishes with Indian curry flavors.

Rice is steamed and usually made with extra water to keep it soft. It is also popular ground and used in noodle or rice paper format, made into pancakes, and used as rice flour in a myriad of other ways. Spring rolls called *cha gio* are popular at meals and also a favorite snackfood, as are fruits of all varieties. Oranges, apples, coconuts, carambola or star fruit, bananas, mangoes, pineapples, pomelos, custard apples, and watermelons are but a few of them. Vegetables are used in varying amounts, and radishes, carrots, cucumber, zucchini, and eggplants are popular. Recently, there has been considerable use of asparagus and white potatoes. The Vietnamese eat a good many salads, with vegetables raw or minimally cooked, sometimes with a hint of meat or seafood for flavor. Because of the recent French influence, the peoples of this Asian country have developed a love of cafe au lait, yoghurt, ice cream, milk consumed as a few tablespoons condensed milk mixed with water, and a love of pâté, more frequently made with fish rather than meat.

15.2 EATING BEHAVIORS

Breakfast usually consists of a large bowl of noodle soup with condiments and herbs and, for some, a *chao,* or rice gruel, made with sticky rice. These can be served with meat or seafood in them or on the side. Lunch is almost always a *pho* of one sort or another with Asian basil (*rau que* or ocimum), bean sprouts, and other goodies to go with it. It might also be a *com dia* of barbecued meats and pickled vegetables on rice. Beef is the favorite meat in *pho* and in main dishes. Dinner, the main meal of the day, has individual appetizers and salads, much rice, and several communal dishes; perhaps one is a fish, another a meat, others vegetables and/or seafoods. The evening meal often ends with *caphe sua,* or French coffee with condensed milk, served in glasses, not cups. Between meals, people snack on *banh cuon,* a steamed ravioli-type food, or *banh mi gio,* which is French bread around a Vietnamese sausage. This and other sausages are made from finely ground pork, pork fat, shrimp, or fish with or without shrimp mixed with aromatic vegetables and Vietnam's pungent fish sauce. People might also have *che,* an exceptionally sweet Vietnamese pudding as a snack. Other snack and meal sweets are small cakes, candies, and cookie-type foods.

In Vietnam, foods are cooked and eaten with chopsticks, three meals a day are common, and snacks and sweet drinks intersperse them. Tea is a popular beverage, along with Asian beer, sweet soymilk, iced coffee, soda, and a local drink called *soda sua hot ga,* which is made with milk, eggs, and carbonated water. Rice, eaten with chopsticks, sometimes gets a little soup added to moisten it even more in the rice bowls and held near the mouth when eating. Mixing food from several dishes together with one's rice is considered socially unacceptable. Soup is consumed with a spoon, eaten at the end of a main meal, or sipped throughout the meal. At the table, when chopsticks are not in use, they are placed together across the rice bowl. For cooking, woks are shallower than those of the Chinese and common ones; woks, steamers, and woven baskets are used everywhere. Frying pans, clay pots, cleavers, coconut scrapers, molds to make bean cakes, and mortar and pestle join them as essential cooking and food preparation tools, along with a charcoal stove or barbecue.

15.3 INTRACULTURAL DIFFERENCES

The food of each region in Vietnam is a reflection of the country's history. Regional differences include fewer spices used in the North, very spicy foods popular in the central region, and sweeter foods popular in the South. There is also more French influence in the South than the North. Pork is more popular in the South and beef in the North, and seafood and poultry dishes are used more often in the South. The most popular dish in the north is *pho bac.* To make it, a boiling beef stock is poured over rice noodles and paper-thin slices of raw beef. Just before eating it, raw bean sprouts, green chili peppers, lemon juice, coriander, and mint leaves are added, as preferred by the individual diner. In the South, Saigon has a rival soup called *mee quang* that includes cucumbers, lettuce leaves, meat or shrimp, and broth; it is flavored with fish sauce and shrimp paste and is garnished with crushed peanuts. Pork sausages called *nems* can be purchased baked and packed in banana leaves. *Nems* and asparagus, potatoes, cauliflower, and artichokes are used in many dishes. At snacks and meals, everyone in the South enjoys *chao tom,* a paste of shrimp wrapped around a piece of sugar cane and then barbecued. There are localized dishes such as gecko lizards dipped in batter and pan-fried in one province or dog or snake dishes in another, some more popular than others. Another difference north to south can be seen in individual dishes. For example, the spring roll in the North often includes jicama, crab meat, and roe (if in season), while in the South it is filled with bean sprouts and shrimp. As to religious influences on this cuisine, Buddhism has had the most impact, adding a highly sophisticated vegetarian cuisine.

15.4 FEASTS, HOLIDAYS, AND SPECIAL FOODS

Tet, the Lunar New Year, is the most important holiday in Vietnam. It is a 3-day holiday in late January or early February where glutinous rice cakes are traditional and symbolic. There is also a 3-day festival called Water Buffalo Slaughter, National Day, Reunification Day, Teachers' Day, Children's Day, Christmas, religious holidays, and more. Special foods for holidays and special events include abalone soup called *sup bao ngu,* stuffed rice cakes filled with meat and beans that are boiled in banana leaves called *banh trung,* and *mui* or candied fruits. Foods are brought graveside to ancestors on the anniversary of their deaths. After the spirits enjoy the essence, the family eats the food there, sharing a meal with their loved one(s). At weddings, there are also offerings to the ancestors, and after them the couple shares a cup of tea and a betel nut to symbolize that they are one. On Children's Day, or *Trung Thu,* special cakes are served as are nuts and fruits. The cakes are made with a sticky rice exterior filled with raisins, peanuts, and tangerine peel. Sunday, or family gathering days, are popular, and many complex dishes are served for them, including a cook-it-yourself-type dish called *ta pli lu.* To eat it, everyone sits around a huge pot of bubbling stock and cooks raw chicken, beef, pork, shrimp, squid, and fresh vegetables; then they share the enriched soup stock adding bean sprouts, herbs, chilies, and the like. Jasmine or another scented tea is served after this Vietnamese hot pot, along with fresh fruit served whole or cut in pieces.

16. REFERENCES

Achaya, K. T. 1994. *Indian Food: A Historical Companion.* Oxford University Press, Oxford, England.

Anderson, E. N. 1988. *The Food of China.* Yale University Press, New Haven, CT.

Barer-Stein, T. 1979. *You Are What You Eat.* McClelland and Stewart, Toronto.

Bryant, C. A., Courtney, A., Markesbury, B. A. and DeWalt, K. M. 1985. *The Cultural Feast.* West Publishing Co., St. Paul MN.

Chang, K. C. 1977. *Food in Chinese Culture.* Yale University Press, New Haven, CT.

Chakravarty, I. 1972. *Saga of Indian Food.* Sterling Publishers Ltd., New Delhi, India.

Chakravarti, A. K. 1974. Regional preference for food: Some aspects of food habit patterns in India." *Can. Geogr.* 18(4):395–410.

Chu, G. 1985. Changing concept of self. In: *Culture and Self: Asian and Western Perspectives,* A. J. Marsella, G. DeVos, and F. L. K. Hsu, eds., Tavistock Publications, New York, pp. 252–277.

Chung, K. C., Haffner, P. G., and Kaplan, F. M. 1990. *The Korea Guidebook.* New York: Harper and Row.

Claudio, V. S. 1994. *Filipino American Food Practices, Customs, and Holidays.* The American Dietetic Association, Chicago, IL.

Dickenson, M. B. 1993. *National Geographic Picture Atlas of Our World.* National Geographic Society, Washington, D.C.

Far Eastern Economic Review Ltd., ed. n.d. *All-Asia Guide.* Far Eastern Review Ltd., Hong Kong.

Japan Travel Bureau, ed. 1991. *Eating in Japan,* 8th ed. Japan Travel Bureau, Tokyo, Japan.

Japan Travel Bureau, ed. 1989. *Festivals of Japan,* 4th ed. Japan Travel Bureau, Tokyo, Japan.

Kittler, P. G., and Sucher, K. P. 1998. *Food and Culture in America,* 2nd ed. West/Wadsworth, Belmont, CA.

Koo, L. C. L. 1982. *Nourishment of Life: Health in a Chinese Society.* Commercial Press Ltd., Hong Kong.

Korea National Tourism Corporation, ed. 1994a. *Travelers Korea.* Korea National Tourism Bureau, Seoul, Korea.

Korea National Tourism Corporation, ed. 1994b. *Korea: A Journey to Exotic Tastes.* Korea National Tourism Bureau, Seoul, Korea.

Korea National Tourism Corporation, ed. 1993. Bamboo salt. *Korea Trav. News,* 63:unpaginated.

Latsch, M. L. 1988. *Traditional Chinese Festivals.* Graham Brasch, Singapore.

Leads, M. 1991. *Ethnic New York.* Passport Books, Lincolnwoood, IL.

Lewis, G. 1994. Spam, South Korea, and the Pacific Rim. *Foodtalk* 17(3):1–3, 6.

Mariani, J. F. 1994. *The Dictionary of American Food and Drink.* Hearst Books, New York.

Millon, M., and Millon, K. 1991. *Flavours of Korea.* Andre Deutsch, London.

Morningstar, A., and Desai, U. 1990. *The Ayurvedic Cookbook.* Lotus Press, Wilmot, WI.

Nelson, S. M. 1993. *The Archeology of Korea.* Cambridge University Press, Cambridge England.

Newman, J. M. 1993. *Melting Pot: An Annotated Guide to Food and Nutrition Information for Ethnic Groups in America,* 2nd ed. Garland Publishing, Inc., New York.

Newman, J. M., Ludman, E. K. and Lynn, L. 1988. Chinese food and life-cycle events: A survey in several countries. *Chin.-Amer. Forum,* 4:16–18.

Kittler, P. G., and Sucher, K. P. 1998. *Food and Culture in America,* 2nd ed. Wadsworth Publishing Co., Belmont, CA.

Passmore, J. 1991. *The Encyclopedia of Asian Food and Cooking.* William Morrow and Company, New York.

Peterson, J., and Peterson, D. 1997. *Eat Smart in Indonesia.* Ginko Press, Madison, WI.

Seligman, L. 1994. The history of Japanese cuisine. *Japan Quart.* April–June: 165–176.

Shimoda, A. 1992. Dietary life in Japan. Paper presented at *Int. Fed. Home Econ. Mtg.,* Hannover, Germany, July.

Simoons, F. J. 1991. *Food in China: A Cultural and Historical Inquiry.* CRC Press, Boca Raton, FL.

Stepanchuk, C., and Wang, C. 1991. *Mooncakes and Hungry Ghosts.* China Books and Periodicals, San Francisco.

Storer, J. 1977. Hot and cold food beliefs in an Indian community and their significance. *J. Human Nutri.* 31:33–40.

Sun, H. 1987. More and better food for the future—Prospects to the year 2000. In: *Feeding a Billion,* S. Wittwer, Y. T. Yu, H. Sun, and L. Z. Wang, ed., Michigan State University Press, East Lansing, MI: pp. 425–436.

Tannahill, R. 1973. *Food in History.* Stein and Day, New York.

Toussaint-Samat, M. 1992. *Food in History.* Basil Blackwell, Cambridge, MA.

Trager, J. 1995. *The Food Chronology.* Henry Holt and Company, New York.

Von Welanetz, D., and Von Welanetz, P. 1982. *Guide to Ethnic Ingredients.* J. P. Tarcher, Boston, MA.

Wen, C. P. 1974. Food and nutrition in the People's Republic of China. In: *Chinese Medicine as We Saw It,* J. Quinn, ed., United States Department of Health, Education, and Welfare Publication No. NIH 75-684, Washington D.C.

WuDunn, S. L. 1997a. Korean lust after cures is a threat to species. *New York Times,* May 5th.

WuDunn, S. L. 1997b. Where dog can be a pet, or a dining experience. *New York Times,* January 2nd.

Zibart, E., Stevens, M., and Vermont, T. 1995. *The Unofficial Guide to Ethnic Cuisine and Dining in America.* MacMillan, New York.

18. ASIAN COOKBOOK RESOURCES

No more than 10 cookbooks per food section follow as do several multiethnic cookbooks. The ones selected have typical recipes, a glossary in many cases, additional historical or ethnic background information, and other types of valuable information that may be helpful. Consult them and the References and Recommended Readings listings above, your library and its interlibrary loan services, internet sources, book dealers, knowledgeable individuals, associations such as The Association for the Study of Food and Society, city culinary historian organizations such as the Culinary Historians of New York, Boston, or Ann Arbor for other materials.

18.1 MIXED ASIAN COOKBOOKS

Brennan, J. 1991. *One-Dish Meals of Asia.* Harper Perennial, New York.

Buck, P. S. 1972. *Pearl Buck's Oriental Cookbook.* Simon and Schuster, New York.

Granger, A., ed. 1987. *A Taste of the Orient.* Gallery Books, New York.

Howe, R. 1969. *Far Eastern Cookery.* The International Wine and Food Publishing Company, London.

Laas, W. 1967. *Cuisines of the Eastern World.* Golden Press, New York.

Moore, I., and Godfrey, J., ed. 1978. *The Complete Oriental Cookbook.* Marshall Cavendish Ltd, London.

Solomon, C. 1992. *The Complete Asian Cookbook.* Charles E. Tuttle Company, Rutland, VT.

18.2 CHINESE COOKBOOKS

Chang, W. W., Chang, I. B. Kutscher, H. W., and Kutscher, A. H. 1970. *An Encyclopedia of Chinese Food and Cooking*. Crown Publishers Inc., New York.

Chen, H. 1994. *Helen Chen's Chinese Home Cooking*. William Morrow and Company, New York.

Chen, P. K., Chen, T. C., and Tseng, R. L. Y. 1983. *Everything You Want to Know about Chinese Cooking*. Barrons Educational Series, Inc., Woodbury, NY.

Hom, K. 1990. *The Taste of China*. Simon and Schuster, New York.

Kuo, I. 1977. *The Key to Chinese Cooking*. Alfred A. Knopf, New York.

Leung, M. 1976. *The Classic Chinese Cookbook*. Harper and Row, New York.

Lo, E. Y. F. 1982. *The Dim Sum Cookbook*. Crown Publishers, Inc., New York.

Lo, K. H. C. 1979. *The Encyclopedia of Chinese Cooking*. A & W Publishers, New York.

18.3 INDIAN COOKBOOKS

Chowdhary, S. 1954. *Indian Cooking*. Andre Deutsch, London.

Dalal, T. A. 1985. *Indian Vegetarian Cookbook*. St. Martins Press, New York.

Dandakar, V. 1983. *Salads of India*. The Crossing Press, Trumansburg, NY.

Devi, Y. 1992. *Yamuna's Table*. Dutton, New York.

Jaffrey, M. 1980. *Madhur Jaffrey's Indian Cooking*. Barrons Educational Series, Inc., Woodbury, NY.

Morningstar, A. 1995. *Ayurvedic Cooking for Westerners*. Lotus Press, Twin Lakes, WI.

Ray, S. 1986. *Indian Regional Cooking*. Macdonald and Company, London.

Sahni, J. 1985. *Classic Indian Vegetarian and Grain Cooking*. William Morrow and Company, New York.

18.4 INDONESIAN COOKBOOKS

DeWit, A., and Borghese, A. 1973. *The Complete Book of Indonesian Cooking*. Bobbs-Merrill Company, Inc., New York.

Johns, Y. 1971. *Dishes from Indonesia*. Chilton Books, Philadelphia, PA.

Marks, C. 1989. *The Indonesian Kitchen*. M. Evans and Company, Inc., New York.

Von Holzen, H., and Arsana. L. 1995. *The Food of Indonesia*. Periplus Pte. Ltd., Singapore.

18.5 JAPANESE COOKBOOKS

Andoh, E. 1986. *At Home with Japanese Cooking*. Alfred A. Knopf, Inc., New York.

Downer, L., and Yoneda, M. 1986. *Step-by-Step Japanese Cooking*. Barrons Educational Series, Inc., Woodbury, NY.

Egami, T. 1971. *Typical Japanese Cooking*. Shibata Publishing Company, Tokyo, Japan.

Ortiz, E. L., and Endo, M. 1976. *The Complete Book of Japanese Cooking*. M. Evans and Company, Inc., New York.

Griffin, S. 1956. *Japanese Food and Cooking*. Charles E. Tuttle Company, Rutland, VT.

Kobayashi, S. K. and compiled by the Buddhist Bookstore. 1977. *Shojin Cooking: The Buddhist Vegetarian Cook Book.* The Buddhist Bookstore, San Francisco, CA.

Omae, K., and Tachibana, Y. 1981. *The Book of Sushi.* Kodansha International Ltd., Tokyo, Japan.

Slack, S. F. 1985. *Japanese Cooking.* HPBooks, Inc., Tucson, AZ.

Tsuji, S. 1980. *Japanese Cooking: A Simple Art.* Kodansha International Ltd., Tokyo, Japan.

18.6 KOREAN COOKBOOKS

Hyun, J. 1970. *The Korean Cookbook.* Follettt Publishing Company, Chicago, IL.

Marks, C., with Kim, M. 1993. *Classic recipes from the Land of the Morning Calm.* Chronicle Books, San Francisco CA.

Noh, C. 1985. *Traditional Korean Cooking.* Hollym International Corporation, Elizabeth, NJ.

18.7 LAOTIAN COOKBOOKS

Davidson, A. 1975. *Fish and Fish Dishes of Laos.* Charles E. Tuttle Company, Rutland, VT.

Sing, P. 1981. *Traditional Dishes of Laos.* Prospect Books, Inc., London.

18.8 MALAYSIAN COOKBOOKS

Fernandez, R. 1985. *Malaysian Cookery.* Penguin Books, London.

Hutton W., ed. 1995. *The Food of Malaysia.* Periplus Pte. Ltd., Singapore.

18.9 NEPALESE COOKBOOK

Association of Nepalis in America, ed. 1996. *The Nepal Cookbook.* Snow Lion Publications, Ithaca, NY.

18.10 PHILIPPINE COOKBOOKS

Alejandro, R. 1982. *The Philippine Cookbook.* Howard McCann, Inc., New York.

David-Perez, E. 1972. *Recipes of the Philippines.* National Bookstore, Manila.

18.11 THAI COOKBOOKS

Brennan, J. 1981. *The Original Thai Cookbook.* Richard Marek Publishers, New York.

Bhimichitr, V. 1988. *The Taste of Thailand.* Pavillion Books, London.

McDermott, N. 1992. *Real Thai.* Chronicle Books, San Francisco, CA.

Sananikone, K. 1986. *Keo's Thai Cuisine.* Ten Speed Press, Berkeley, CA.

Schmitz, P. C., and Worman, M. J. 1985. *Practical Thai Cooking.* Kodansha Publications Ltd., Tokyo, Japan.

Tang, T. 1991. *Tommy Tang's Modern Thai Cuisine.* Doubleday, New York.

18.12 VIETNAMESE COOKBOOKS

Ang, E. T. 1996. *Delightful Vietnamese Cooking.* Ambrosia Publications, Seattle, WA.

Duong, B., and Kiesel, M. 1991. *Simple Art of Vietnamese Cooking.* Prentice-Hall, New York.

Miller, G. B. 1966. *The Thousand Recipe Chinese Cookbook.* Atheneum, New York.

Miller, J. N H. 1968. *Vietnamese Cookery.* Charles E. Tuttle Company, Rutland, VT.

Ngo, B., and Zimmerman, G. 1979. *The Classic Cuisine of Vietnam.* Barrons Educational Series, Inc., Woodbury, NY.

Routhier, N. 1989. *Foods of Vietnam.* New York: Stewart, Tabori, and Chang.

Passmore, J., and Reid, D. 1982. *The Complete Chinese Cookbook.* Lansdowne Press, Dee Why West, NSW, Australia.

CHAPTER 16

Health Implication of Asian Diets and Supplements

CATHARINA Y. W. ANG

1. INTRODUCTION

FROM the early to mid-twentieth century, scientific interest and government concerns in many countries were aimed at increasing food production, reducing the cost of food, and ensuring a wholesome and safe food supply. During the latter part of this century, with the advancement in agricultural sciences, food technology, and economic development, foods and food products are becoming more adequate in many parts of the world. Abundant, essentially safe, and low-cost foods are available in developed countries, including some Asian locations. Health problems associated with malnutrition or food poisonings affect less of the population now than before. However, the advancement in science and technology and the development in economics has also changed lifestyles and dietary patterns of the general public. New food products and ingredients are being introduced. Diseases caused by nutrition deficiencies, infections, and general sanitation problems have been reduced, whereas chronic diseases possibly associated with dietary factors have increased (WCRF/AICR, 1997).

In most populations in the developed world, stomach cancer has been declining rapidly in recent decades, whereas rates of cancers of the colon, breast, and prostate have been rising. People in Western affluent countries may have higher risks in developing some types of cancers, heart diseases, obesity, and hypertension than people in Asian. Economically developing countries in Asia tend to have relatively high rates of cancers of the mouth and pharynx, esophagus, stomach, and liver in common (WCRF/AICR, 1997).

A number of developing countries in Asia have a rapidly developing economy. They tend to follow industrialized countries in Europe and North America in various aspects of technologies and lifestyles, including food and diets. For instance, soda beverages, hamburgers, dairy products, pizza, and fried chickens have become popular in recent years in Asian countries. How to improve the nutritional status of developing countries while avoiding problems associated with diets of industrialized countries is of the utmost importance to health science professionals and food technologists. Consumers in the late twentieth century are concerned about health aspects of foods. This concern is widespread, perhaps equal to or wider than the concerns related to food cost or food safety. Foods and diets that promote health and prevent diseases will be the utmost, universal interest in the next century.

This chapter will review the nutritional compositions of Asian foods, dietary intakes of major nutrients, dietary supplements, and their associations with health benefits and problems. Included in this chapter are also some comparisons between Western and Asian countries on food and health-related issues. The aim of this chapter is to provide updated information of food, nutrition, and health aspects of Asian diets; to guide health science professionals, as well as the public; to make appropriate dietary plans; and to aid food industry in producing healthful food products. The first section will address the nutritional quality and dietary pattern of Asia countries in general. Subsequent sections will address health effects of individual food groups. The last section will address dietary recommendations. Food composition tables of various Asian countries are listed in Appendix 1.

2. NUTRITIONAL QUALITY AND DIETARY INTAKE

2.1 FOOD COMPOSITION

One of the most important tasks for improving the nutrition-related public health status of the consumer is to establish accurate food composition information of various food items. Many Asian countries have published food composition tables of specific foods and food products in their regions. Examples of available references from different countries (e.g., China, India, Indonesia, Japan, Korea, Malaysia, the Phillippines, Singapore, and Taiwan) are given in Appendix 1. A representative publication is the *Food Composition Tables* compiled by the Nutrition and Food Hygiene Research Institute of the Chinese Academy of Preventive Medicine (1991). It includes 1358 food items and 26 nutrients.

In previous chapters, nutrient compositions of some food items have been listed under each product category. Because of the variations in production, climate, variety, species, and processing and preparation methods, the nutrient contents may vary greatly with time and place. In order to provide readers a general understanding about the nutritional quality of Asian foods, the compositions of some typical foods of plant and animal origins are presented in Tables 16.1 and 16.2, respectively. Only those nutrients that are of significant importance in each group are listed, such as fiber content of plant foods and cholesterol content of animal products. Further discussions concerning nutritional implications of various foods are included in Section 3.

Traditionally, the nutrient contents of food are grouped into two categories: the macronutrients, including energy, protein, carbohydrates, and fats, and micronutrients, including minerals, vitamins, and trace elements. Conventional food composition tables show the contents of these nutrients and may also include fibers, different types of fatty acids and several vitamins, minerals, amino acids, and cholesterol (selected compositions are shown in Tables 16.1 and 16.2). Besides these known essential nutrients for maintaining normal growth and sustaining lives, there are many other compounds that have biological functions. These microconstituents have been referred to as bioactive compounds. Many of these compounds, especially in foods of plant origin, may have some functions in preventing and reversing the development of chronic diseases. Dietary supplements are used by individuals for supplementing these micronutrients or bioactive compounds. At the present time, no food composition tables have included the quality and quantity of these compounds. Furthermore, beneficial effects of these bioactive compounds are derived from the combined functions of several compounds in a food rather than by any individual compounds. Many of these compounds have not been identified or quantified. Only a limited number of compounds in a few food materials has been reported. Much data are needed regarding bioactive microconstituents in foods and supplements.

2.2 DIETARY INTAKE

Details about the diets and food habits of many Asian countries have been described elsewhere in this book. Variations are observed between countries and between locations, cultures, and religions within a country. It is difficult to generalize the food consumption patterns in a country. Nevertheless, from available information, the Food and Agricultural Organization of the World Health Organization (FAO/WHO, 1991) was able to issue a report on dietary intake of major food groups in different parts of the world. Figure 16.1 shows the consumption of major food groups as percentage of total dietary energy in Asia countries, exclusive of Middle East Asia, which is not within the scope of this chapter.

Cereals or grains are the staple foods in Asia (Figure 16.1). Variations are related to per capita income. Generally, countries of low income consume the highest levels of cereals, with an average of 71% of the total dietary energy

TABLE 16.1. Composition of Selected Foods of Plant Origin (100 g Edible Portion).[a]

	Energy (kcal)	Water (g)	Protein (g)	Fat (g)	Carbohydrates (g)	Fiber (g)	Ash (g)
Cereals, Legumes, and Starchy Products							
Rice, raw	346	13.3	7.4	0.8	77.2	0.7	0.6
Rice, steamed	114	71.1	2.5	0.2	25.6	0.4	0.2
Wheat flour	349	12.0	10.4	1.1	74.3	1.6	0.6
Noodle, cooked	109	72.6	2.7	0.2	24.2	0.1	0.2
Instant noodle	131	67.9	4.0	0.2	27.5	0.2	0.2
Deep-fried dough strip	386	21.8	6.9	17.6	50.1	0.9	2.7
Tofu (soybean curd)	81	82.8	8.1	3.7	3.8	0.4	1.2
Dried *tofu*	140	65.2	16.2	3.6	10.7	0.8	3.5
Soymilk	30	94.0	2.5	1.5	1.8	1.3	0.3
Yam, cooked[b]	116	70.1	1.5	0.1	27.6	3.9	0.7
Vegetables and Fruits							
Bamboo shoots Cooked, drained[b]	12	95.0	1.5	0.2	1.9	0.6	0.6
Beans, mung sprouted Stir-fried[b]	50	84.0	4.3	0.2	10.6	0.7	0.5
Chinese cabbage Boiled, drained[b]	12	95.4	1.6	0.2	1.8	0.6	0.4
Snow peas Boiled, drained[b]	42	89.0	3.3	0.2	7.1	1.0	0.4
Water chestnut	59	83.6	1.2	0.2	13.1	1.1	0.8
Mushroom	19	91.7	2.2	0.3	3.3	1.9	0.6
Lychee	70	81.9	0.9	0.2	16.1	0.5	0.4

[a] *Source:* Chinese Academy of Preventive Medicine, 1991, except otherwise designated.
[b] *Source:* USDA, 1997.

TABLE 16.2. Composition of Selected Foods of Animal Origin (100 g Edible Portion).[a]

	Energy (kcal)	Water (g)	Protein (g)	Fat (g)	Carbohydrates (g)	Ash (g)	Cholesterol (mg)
Beef, flank, lean							
Short loin, braised[b]	244	68.3	28.0	13.8	0	0.8	71
Chicken, stewed							
Leg meat, no skin[b]	185	66.4	26.2	8.1	0	0.9	89
Breast meat, no skin[b]	151	68.3	29.0	3.0	0	0.9	77
Duck, Peking							
Roasted, meat with skin	436	38.2	16.6	38.4	6.0	0.8	89
Duck egg, preserved	171	68.4	14.2	10.7	4.5	2.2	608
Pork, roasted							
Loin, blade[b]	279	49.2	23.8	16.9	7.9	2.2	91
Pork, Chinese ham	318	48.7	16.4	28.0	0	6.8	98
Fish, carp, cooked[b]	162	69.6	22.9	7.2	0	1.9	84
Shrimp, steamed[b]	99	77.3	20.9	1.1	0	1.6	195
Milk *tofu*, Mongolia	305	31.9	46.2	7.8	12.5	1.6	36

[a]*Source:* Chinese Academy of Preventive Medicine, 1991, except otherwise designated.
[b]*Source:* USDA, 1997.

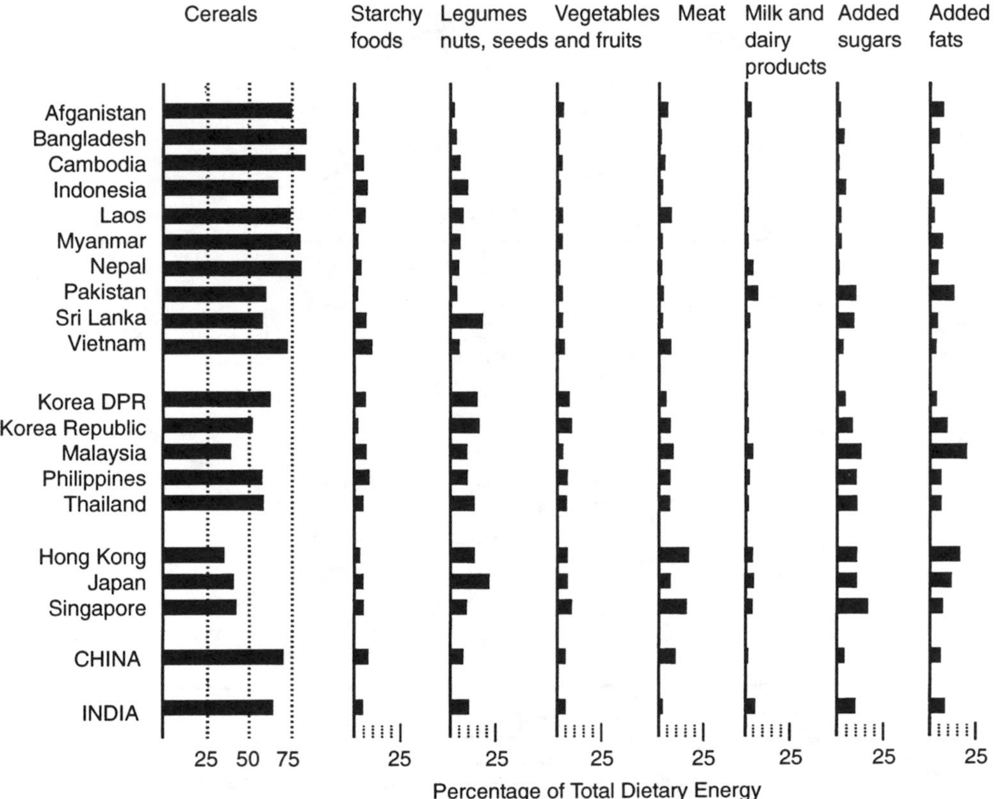

Figure 16.1 Consumption patterns of major food groups in East Asia as percentage of total dietary energy. Starchy foods include roots, tubers, and plantain; meat group includes red meats, poultry, fish, and eggs; added sugar includes syrups and honey; added fat includes vegetable oils and animal fats. [Adapted from WCRF/AICR, (1997) by kind

supply from cereals. Cereals are generally supplemented by small quantities of legumes. In the developing countries in Asia, meats and animal products (including poultry, fish, and eggs) supply less than 10% of total energy. Japan consumes about 13% total energy as meat and meat products (6% meat and 7% as fish). In China, meats represent about 8% total energy, and in India, meat provides less than 1% total energy. Low-income countries consume little meat. The amounts of milk and dairy products are generally low. Exceptions are the pastoral populations in inner and outer Mongolia, Tibet, and India. People in these regions usually consume much more meat and/or dairy products than other regions. The consumption of vegetables and fruits are generally related to income. In middle- and high-income countries, the consumption is about 5–6% total energy intake and is reduced to 2% of total energy in low-income regions. The amount of total dietary fat is mostly low, except for Hong Kong and Malaysia. The consumption of sugar and other sweeteners is moderately high, except in low-income countries, with several regions consuming more than 10% total energy.

From 1960 to 1990, there is a trend of dietary changes. Most countries show a decrease in cereal consumption. Most affluent countries showed an increase in meat, vegetable, and fruit consumption. The consumption of milk and dairy products has increased in a few countries. In comparison, cereals contribute less than 25% of total energy in Europe and North America, and meat and other animal products may account for 25% or more of the total energy intake in the United States, Canada, Australia, and New Zealand (FAO/WHO, 1991).

Just as there are differences in dietary patterns between developing and industrialized countries, there are also differences in the incidence and types of chronic diseases. The causes of chronic diseases include many factors, such as genetic, chewing and smoking of tobacco and chewing betel, air and water pollution, lifestyle, and diet. Some of these factors are difficult to control (e.g., genetic factor), but some other factors are controllable by individuals, such as smoking and dietary practices (Trichoponlos et al., 1996; Willett et al., 1996). Research in recent decades has shown that diet is an important risk factor in the development of some chronic diseases, such as cardiovascular disease and cancer. The following section will address each food group separately concerning the association of the dietary intake of these foods, health benefits, and adverse effects.

3. DIET AND HEALTH BY FOOD GROUPS

3.1 CEREALS

3.1.1 Consumption Pattern

The major types of cereals or grains in Asia are rice, wheat, millet, foxtail millet, pearl millet, sorghum, buckwheat, barley, pearl barley, oats, maize or corn, and rye. Traditionally, rice, other cereals, and starchy foods supply the majority of total energy intake. In recent years, along with economic development and urbanization, cereal consumption has decreased, particularly in high-income regions. Cereal-based diets tend to be bulky with a low-energy density. The consumption of cereals in Bangladesh and Cambodia accounts for 84% of the total energy supply (Figure 16.1). In China and India, cereal consumption provides 69% and 63% total energy, respectively. In higher income areas, Hong Kong, Singapore, and Japan, cereals provided about 40% total energy supply. Rice is by far the most important staple food in most Asian countries. In India and China, cereal foods also include appreciable amounts of wheat, millet, and other grains, with substantial regional variations within each country. As economic situations have improved in recent years, cereal consumption has generally declined in developing countries.

3.1.2 Nutrient Composition and Effect of Processing

Currently, polished white rice and refined white flour are the most common forms of cereal consumed worldwide. As shown in Table 16.1, rice and wheat flour contain about 74–77% carbohydrates (essentially starch) and 7–10% protein by weight, with lipid content of 1% or less. Cooked rice and noodles contain about 25–28% carbohydrates and 2–3% protein. There is no cholesterol and only a trace amount of saturated fatty acids. Cereals also contain non-starch polysaccharide, including cellulose, hemicellulose, pectin and gums, dietary fiber (lignans), B vitamins, vitamin E, iron and various trace elements, and bioactive compounds (Wu and Inglett, 1988).

Cereals are always processed (milled) and cooked for human consumption. The methods of processing and cooking can affect the nutritional values of cereal foods greatly. Processing of cereals increases the digestibility and palatability but also decreases some nutrients. The white rice of commerce is the scored and polished endosperm. Most of the nonstarch nutrients are concentrated in the hull and bran parts of the cereal grain, which are removed during processing (Wu and Inglett, 1988). Nutrients lost include small amounts of cellulose, hemicellulose (arabinose and xylose), proteins, fatty materials, several vitamins, and minerals. For instance, the conversion of brown rice to polished rice results in the following nutrient changes (dry weight basis): protein from 10.1% to 7.2–9.0%, fat from 2.4% to 0.3%, dietary fiber from 0.9% to 0.1%, and ash from 1.2% to 0.5% (Altschul and Planck, 1960). It is commonly recommended to store brown rice under refrigeration because it contains more lipids (susceptible to oxidation) than polished white rice. The rice bran and rice hull contain other micronutrients with antioxidant functions. One of these compounds that had been isolated and identified was the isovitexin, a *C*-glycosyl flavonoid. It has stronger antioxidant activity than α-tocopherol (Ramarathnam et al., 1988, 1989). Rice bran oil has been shown to have effects on lowering the total and low-density lipoprotein cholesterol in non-human primates (Nicolosi et al., 1991).

Whole wheat flour has essentially the same composition as the original wheat. During milling of wheat, wheat germ and bran are removed to various degrees. A small percentage of protein is lost because the germ is higher in protein than in the endosperm. Highly refined patent flours lost much of the original minor nutrients, such as oil, fiber, ash, B vitamins (in the germ and bran), and also vitamin E (Wu and Inglett, 1988). To restore some of the lost nutrients in cereals, an enrichment program was initiated by the U.S. Food and Drug Administration (USFDA) in 1957. In general, the methods of enrichment in the United States and Canada are also used in other parts of the world. The enriched rice and flour contain added thiamin, riboflavin, niacin, and iron. In 1996, the FDA announced the addition of folic acid in the enrichment program. Currently, many enriched cereal products in the United States have been fortified with folic acid. Information is not available regarding whether Asian countries have also added folic acid to their enriched cereal products.

3.1.3 Health Benefits and Adverse Effects

As discussed previously, cereals contain many carbohydrates, some dietary fibers, but very little total fat. The benefits of cereal consumption derive from, not only the nutrients in cereals, but also from the replacement of animal products in a diet (Ang and Hang, 1995; Popkin et al., 1993). One example of a large-scale project was the China Health Study, a collaborative effort of Cornell University (U.S.A.), the Chinese Academy of Preventive Medicine, and the University of Oxford (U.K.). It included almost 7000 Chinese rural families whose diets are mostly plant based (much lower in fat and animal protein and much higher in fiber than Western diets). Data showed that these people generally had lower blood serum cholesterol level and fewer incidences of cancers and heart diseases compared to the Western countries (Chen et al., 1990; Campbell and Chen, 1994).

Also as discussed earlier, some vitamins (thiamin, riboflavin, niacin, and folic acid) and iron are fortified in the enriched cereals. However, some other vitamins and minerals, dietary fiber, nonstarch polysaccharides, and bioactive compounds are also reduced during the milling process but not restored in the enriched products. Much of the nonstarch polysaccharides; dietary fiber; and the essential fats, vitamins, and minerals are removed in refining. Nutrient loss, such as vitamin loss from polished rice, prevails particularly in certain Asian rural areas where inexpensive mills are normally used. Such practices often lead to production of heavily polished rice, which can cause vitamin deficiency in certain populations, especially in youths. The beneficial effect of grain consumption may depend on the degree of refining. A number of studies have shown that diets high in whole grain products may possibly decrease the risk of stomach cancer. According to the review of Jacobs et al. (1995), six of seven control studies indicated a protective effect from consumption of whole grain cereals and cereal products.

The association between cereal consumption, such as maize (corn), wheat, and millet, and increased liver cancer risk has been studied. Research findings from epidemiological data in China and from several animal studies suggest that the association resulted from fungal contamination of stored grain (Hsieh, 1989; Yeh et al., 1989; Wilder et al., 1990). The most problematic cause involves various types of mycotoxins, which are metabolites of molds. Some mycotoxins (e.g., aflatoxin B_1) are highly carcinogenic in laboratory animals. Contamination by mycotoxins is more serious in regions where climates are hot and damp and there is a lack of proper storage facilities, such as some parts of Africa and Southeast Asia. Animal feeds may also be contaminated with aflatoxins and thus result in the secretion of a metabolite, aflatoxin M1, in animal's milk. Food and feed stored for long periods in hot, damp conditions are susceptible to pathogenic organism contamination, other than mycotoxins. Proper storage of cereals, as well as vegetables, fruits, legumes, and other foods under refrigeration, is essential in ensuring the safety of food supplies.

3.2 STARCHY FOODS

The starchy foods in Asia include roots, tubers, and plantains such as sweet potatoes, potatoes, and yams. This group of foods is high in carbohydrates. Starch content of sweet potatoes is about 24%, potatoes about 20%, plantains 31%, and yams about 28% by cooked weight (Table 16.1). The protein content is generally less than 2% and fat 0.5%. These foods and their products are generally good sources of nonstarch polysaccharides and dietary fibers, carotenoids, vitamin C, potassium, other vitamins and minerals, and other bioactive compounds. Starchy roots, tubers, and plantains provide about 5% total energy in some Asian regions in Indonesia, Vietnam, the Philippines, and China. Other countries in Asia consume less starchy foods (Figure 16.1). The consumption of starchy foods may be as high as 20% total energy in some regions of the Oceania Pacific islands (not a part of Asia), including Papua New Guinea, Samoa, and Tonga. Data at the present time are not sufficient to conclude any health relationship between specific starchy foods and cancer risks (WCRF/AICR, 1997).

3.3 LEGUMES

3.3.1 Nutrient Composition and Consumption Pattern

Legumes in Asia include soybeans, green beans, red beans, lentils, cowpeas, peas, mung beans, peanuts, common beans, and broad beans. They are considered a part of dietary staple in some regions of China and India, providing a significant portion of total protein intake. Legumes contain 6–11% protein by cooked weight. The nutritional quality of soybean protein is equal to animal proteins, with the digestibility corrected amino acid score of 1 (Messina, 1997). Examples in Table 16.1 show that soyfoods contain 6–8 % protein. In this regard, they are better than cooked cereals (e.g., rice or wheat products). Legume food products, except whole soybeans, are low in total fat (3–5%) and low in saturated fatty acids as compared to animal products. They are typically rich in nonstarch polysaccharides and dietary fiber, as well as a number of microconstituents and bioactive components such as isoflavones. The levels of unique soybean oligosaccharides (stachyose and raffinose) account for 5%, 0.5%, and 0.4% by weight in dry bean, soymilk, and *tofu,* respectively (Tomomatsu, 1994). Isoflavones are found particularly in soybeans, that is, 2–4 mg/g in raw beans and 1–3 mg/g in soyfood on a dry weight basis. One-half cup of *tofu* or 1 cup of soymilk contains 30–40 mg isoflavones (Wang and Murphy, 1994a, 1994b).

As described elsewhere in this book, legumes are prepared and eaten in a wide variety of ways, such as Japanese and Chinese *tofu,* Chinese mung bean sprouts, Indian *dahl,* and Indonesian cultured soybean cakes *(tempeh).* The consumption of legumes together with nuts and seeds has been estimated to provide about 5.6% total energy in economically developing countries and 10% in some areas of China, compared to 2.4% in developed countries (WHO, 1990).

3.3.2 Health Benefits and Adverse Effects

Legume foods are considered to contribute several beneficial health effects based on their protein content, relatively abundant nonstarch polysaccharides/dietary fiber, and low levels of total fat. Among the legume group, soyfood has been by far the most studied beans in relation to human health. In a case-control investigation in Singapore, a significantly protective effect was found from soy product intake on the risk of breast cancer in women. An inverse relationship was also found between the ingestion of soy products and other cancers (colon, lung, and stomach) in men and women (Lee et al., 1991). Other case studies in Asian countries have shown that one serving of *tofu* daily reduced the risk of lung, colon, rectal, breast, stomach, and prostate cancers (Messina, 1997). In clinical trials involving people with elevated levels of total and low-density lipoprotein (LDL) cholesterol and triglyceride levels, the consumption of soyfoods showed beneficial effects, that is, lowering the levels of these compounds (Anderson et al., 1995; Carroll, 1991).

Legumes are particularly high in nonstarch polysaccharides, which may protect against cancers of the stomach, pancreas, colon and rectum, breast, and perhaps other sites as well. The soybean oligosaccharides, including raffinose and stachyose, are not digestible by human digestive juices but can be utilized by bifidobacteria in the colon and suppress the activity of detrimental bacteria. The daily requirement of oligosaccharides (e.g., fructooligosaccharides, galactosaccharides, and soybean oligosaccharides) has been estimated as 3 g, and more health benefits have been found to be associated with the ingestion of oligosaccharides than other dietary fibers (Tomomatsu, 1994). Many nutritionists have been concerned about the possibility of a vitamin B_{12} deficiency in diets that include only plant foods. However, vitamin B_{12} can be produced by food fermentation. For instance, the vitamin content in fermented soyfoods, such as the Chinese soy paste *(jiang)* and fermented *tofu* cake, ranges from 0.051 to 0.715 mcg/100 g (Chinese Medical Academy, 1989). The body requirement of vitamin B_{12} has been estimated as 1 mcg per day (Herbert, 1988).

In a comprehensive review of the role of soyfood in disease prevention and treatment, Messina (1997) concluded that evidence ranges from fairly solid to highly speculative in support of a protective effect of soy against cancer, heart disease, kidney disease, and osteoporosis. He suggested that the most interest in soybeans and soyfoods was related to their isoflavone contents. Soy isoflavones (primarily genistein and daidzein) may have a function in modifying the estrogen levels in blood, and they can exhibit a synergistic effect in prevention of hormone-related cancers.

Raw soybeans contain some antinutrient factors such as saponins, phytate, lactins, protease inhibitors, oligosaccharides, and isoflavones. However, soybeans are always cooked or heat processed for human consumption. Varying amounts of these factors, such as protease inhibitors, are destroyed by heat. Several other factors, such as phytate, saponins, isoflavones, and phytosterols, may help lower the risk of certain cancers (Messina, 1997). Adverse effects from the ingestion of soyfood in adults are rare, if any. However, the isoflavone content in soy infant formula (Irvine et al., 1995) and dietary pills containing 20 mg isoflavones per pill could be a potential health concern.

3.3.3 Vegetarian Diets

Legume foods account for an important proportion in vegetarian diets. Depending on the degree of avoiding animal products, vegetarian diets can be classified into several types. According to the American Dietetic Association, these types include the lacto-ovo-vegetarian (diet with dairy foods and eggs), lacto-vegetarian (with only dairy foods, but no eggs), and the vegan (no animal foods of any type). None of these diets include animal flesh foods. As stated earlier, meat consumption in Asia is low, especially in India and some areas of China. A large population in India and certain groups in China are vegetarians or virtually vegetarians (lacto-ovo- or lacto-vegetarian) because of religious beliefs (Huang and Ang, 1992). The meat group provides only about 1% total energy in India. About 25–30% of the Indian population can be considered completely vegetarian. The percentage is higher in central and southern states, lower in the north, and lowest in the coastal areas because of the availability of fish (WCRF/AICR, 1997). Most vegetarians in India (90%) are lacto-vegetarians, with the exception of certain religious groups who are vegans. The average intake of meat, poultry, and fish in China accounts for about 8% total energy. However, in some low-income areas of China and for some Buddhist followers, no meat or only a little animal flesh foods are consumed.

A substantial number of studies have been conducted regarding the relationship of vegetarian diets and health. Vegetarian groups generally show lower overall mortality, lower risk of cardiovascular disease, lower rates of obesity, and longer life expectancy than general population comparison groups (Frenzel-Beyme and Chang-Claude, 1994; Key et al., 1996; Thorogood et al., 1994). Research findings have also identified plausible biological mechanisms for ways in which vegetarian diets may affect the risk of cancers of the colon, breast, and prostate (Bélanger et al., 1989; Dwyer, 1988; Pederden et al., 1991; Pusateri et al., 1990).

Regarding the association of vegetarian diets and cancer risk, an expert panel concluded that population groups following lacto-ovo, lacto-vegetarian, and vegan diets have a decreased incidence of cancers in general, as well as of cancers at several specific sites, such as colon, breast, and prostate (WCRF/AICR, 1997). The beneficial effects of vegetarian diets and semivegetarian diets containing small amounts of foods of animal origin are not only because of the exclusion of meat but also because of the inclusion of a larger number and wider range of plant foods containing a variety of potential cancer-preventive bioactive compounds. A fraction of vegetarian populations are at low income or poverty level; thus, their diets could be monotonous, imbalanced, and deficient in energy and some essential dietary nutrients. Both the poverty diet and the extremely limited vegetarian diet (e.g., fruitarians) are not likely to be beneficial to overall health.

3.4 MEAT, POULTRY, FISH, AND EGGS

3.4.1 Nutrient Composition and Consumption Pattern

Foods of animal flesh origin in Asia include pork, beef, lamb, chicken, duck, goose, and various types of fish. Generally, animal foods are good sources of dietary protein. As shown in Table 16.2, meat, poultry, and fish contain around 17–29% protein by weight, depending on the cuts and the amount of attached skin. Pork is the most popular animal meat in China. The fat content of meat, poultry, and fish ranges from a low of 3% in lean poultry breast to around 30–40% by cooked weight in fatty pork and roast duck. About 40–50% of the total fat in meat is saturated fat. Poultry contains a lower proportion of saturated fatty acids (35%) and a higher proportion of polyunsaturated fatty acids (15–30%). Fat from fish contains even less saturated fatty acids (20–25%). Meat and poultry also contain various B vitamins, such as thiamin in pork and B_6 in poultry (Chinese Academy of Preventive Medicine, 1991; USDA, 1997). The minerals in meat, including iron, zinc, and selenium, are readily absorbable. The amounts of B vitamins, iron, and zinc in fish are lower than in meat and poultry, but oily fish contains retinol and vitamin D. Eggs contain moderate amounts of protein and fat. Egg yolks are high in dietary cholesterol (USDA, 1997).

Meat and poultry intakes are relatively low in Asian countries, notably in India and other low-income Asian areas. The meat, poultry, fish, and egg group provides less than 10% total energy in most Asian regions, except Hong Kong and Singapore where the intake of these animal foods accounts for 15% total energy (Figure 16.1) and pastoral populations in Inner and Outer Mongolia where lamb meat is consumed more than other types of meat or cereals. The meat consumption has increased with economical development. Between 1960 and 1990, the meat intake in Japan, China, and South Korea has risen at least 2.5 times. During this period, the worldwide poultry consumption has risen an average of 50%, with an increase of 100–200% in Asia, even though it is still lower in most Asian countries than in the United States (WHO, 1990).

3.4.2 Health Benefits and Adverse Effects

The obvious benefits from the consumption of meat and poultry are derived from their high protein content, B vitamins, and readily absorbable minerals. The inclusion of meat in the diet may help eliminate some iron deficiency anemia, which is widespread in developing countries where iron losses are associated with intestinal parasitism. These foods also provide high dense energy. However, during the past 20 years, research findings have shown that high meat consumption is linked to adverse health effects. With meat intake between 15% and 25% as total energy, economically developed countries of Europe, North America, and Australia tend to have in common relatively high rates of cancers of the colon and rectum and of the hormone-related cancers of the breast, endometrium, and prostate. Cohort studies in Japan have also shown the increased risk of breast cancer with increased meat consumption (Hirayama, 1986). A case control study in China reported an increase in breast cancer risk with meat intake above 80 g/day (Qi et al., 1994). While undernutrition is still a problem in some segments of Chinese society, overnutrition has been observed among high income levels associated with a diet higher in fat and with problems of obesity (Popkin et al., 1993).

Numerous studies in different parts of the world have been conducted to assess the relationship between red meat intake and the risk of various cancers. A group of international experts has concluded that diets containing substantial amounts of red meat (essentially beef, lamb, and pork) probably increase the risk of colorectal cancer and possibly increase the risk of pancreatic, breast, prostate, and renal cancers (WCRF/AICR, 1997). However, evidence on cancer risks caused by the consumption of poultry, fish, and eggs is less substantial, except for a possible risk correlation of a high intake of eggs and colorectal cancer (WCRF/AICR, 1997). Higher intakes of red meat (containing saturated fat and cholesterol) have also been linked to cardiovascular diseases and other chronic problems in developed countries (Cannon, 1992).

A variety of processed meat and associated items are available in Asian, as well as world markets (Chapters 7 and 8). The advantages of processing meat, poultry, fish, and eggs are to provide a variety of products with improved quality, palatability, shelf life, and convenience. However, some processed products may inadvertently affect human health. For instance, cured meats and fish are generally the primary contributors of preformed *N*-nitroso compounds in the diet, which had been shown to be associated with the esophageal cancer mortality in China (Wu et al., 1993), and a high intake of cured meat possibly has increased the risk of colorectal cancer (Goldbohm et al., 1994; Willett et al., 1990; Wohlleb et al., 1990). Daily intake of salted and processed meat and fish has been shown to increase the risk of laryngeal cancer (Zheng et al., 1992) with data supported by a dose-response relationship. Most meats are cooked before eating. The cooking or heating process reduces or destroys harmful bacteria in raw, meaty materials. Undercooked meats, such as hamburgers, can cause illness or even fatality. Desirable meat flavors are also generated during cooking. However, meat, poultry, and fish cooked at very high temperatures (200°C or above), especially in flame, may produce polycyclic aromatic hydrocarbons (PAHs) and heterocyclic amines (HCAs). Frying of meats using added fat at a relatively high temperature can also generate PAHs. Many of these compounds and their derivatives are known to be carcinogenic in laboratory animals (Bogovski, 1983; Nagao et al., 1977; Skog, 1993). Diets high in meat and fish cooked at high temperatures (200°C) possibly increases the risk of stomach and colorectal cancers (WCRF/AICR, 1997). Asian foods such as the Chinese-style barbecued pork, Beijing roasted duck, Japanese-style beef steak grilled on a hot plate, and Malaysian barbecued meat and chicken are generally cooked at high temperatures and sometimes by open flame. These cooked meats may potentially generate PAHs and HCAs carcinogenic compounds, whereas stir-frying in a pan or wok normally heats foods at a high surface temperature only. Information is not available whether stir-fried meat, poultry, and fish items, such as by Chinese cooking techniques, have been associated with the risk of stomach and colorectal cancers. In view of this author, because of the short time involved in surface contact, foods cooked by stir-frying probably generate a lower amounts of PAHs and HCAs than grilling or broiling.

3.5 MILK AND DAIRY PRODUCTS

3.5.1 Nutrient Composition and Consumption Pattern

On an average, whole milk from cows contains about 3 g protein, 4 g fat, and 4.6 g lactose per 100 g. About 66% of the fatty acids in milk are saturated,

30% are monounsaturated and 4% polyunsaturated. Milk and dairy products, except butter and cream, are rich sources of calcium, riboflavin, and vitamin B_{12} (USDA, 1997). Traditionally, milk and dairy products are not important sources of dietary nutrients and energy in many parts of Asia. However, these products provide essential nutrients in regions where diets are otherwise marginal or deficient in protein, vitamin D, and/or calcium. This food group contributes less than 5% total energy in most Asian populations, with somewhat higher intakes reported in India and Pakistan (WHO, 1990). The lactose intolerance problem associated with many Asian adults is probably accountable in part for the low consumption of milk in Asia. However, in some traditionally pastoral regions of India, China, and Mongolia, the consumption of milk and dairy products may contribute around 10% total energy and typically may provide around 15–25% dietary protein and fat intake. During the past 30 years, the consumption of milk and dairy products has increased significantly, especially in Japan and Korea.

3.5.2 Health Benefits and Adverse Effects

In several case control studies and some cohort studies, milk, butter, cream, and/or cheese are associated with an increased risk of prostate cancer (Mettlin et al., 1989; Severson et al., 1989; Talamini et al., 1992). It has been suggested that fat in dairy products may be the constituent responsible for the possible increased risk in prostate cancer. Milk consumption also showed an association with the risk of kidney cancer (Kreiger et al., 1993; McLaughlin et al., 1992). In a Japanese prospective study, people drinking milk daily had about twice the risk of developing kidney cancer than nonusers (Hirayama, 1990). Diets high in milk fat and cheese have been shown to have a higher risk of cardiovascular disease in industrialized countries (Barnard, 1993). With increasing consumption of milk and dairy products in some Asian countries, such as Japan and Korea, it is important to understand the potential adverse health effects from the high consumption of this food group.

3.6 VEGETABLES AND FRUITS

3.6.1 Nutrient Composition and Consumption Pattern

Generally, fresh vegetables (excluding starchy roots and tubers and legumes) and fruits containing more than 90% water are low in energy, protein, and fat contents (Table 16.1). Foods in this food group are good sources of dietary fibers; nonstarch polysaccharides; various vitamins, especially vitamin C; folic acid; beta-carotene; minerals such as potassium (USDA, 1997); and other bioactive microconstituents. Dried fruits provide concentrated sources of energy, sugar, dietary fiber, and iron.

Vegetable and fruit consumption accounts for about 2% of total energy intake in lower income Asian countries. The average consumption in other countries supplies about 5% total energy, with Korea and Singapore consuming the highest amounts (Figure 16.1). Actual consumptions vary greatly with climate, season, location, and income levels within each country. In some parts of China, vegetables and fruits may provide over 10% total energy.

3.6.2 Health Benefits and Potential Adverse Effects

Because of the low fat content, diets high in vegetables and fruits reduce the chance of obesity and thus reduce the risk of cardiovascular diseases (Rimm et al., 1996). Numerous studies have shown the benefit of vegetables and fruits in reducing the risk of a number of diseases (WCRF/AICR, 1997). The antioxidant vitamins, carotenoid, vitamin C, tocopherols, and other microconstituents probably protect against cataracts and decrease the oxidation of cholesterol in the arteries and thus protect against cardiovascular disease. The nonstarch polysaccharides, oligosaccharides, and other dietary fibers may help control diabetes and high serum cholesterol levels and may protect against diverticular disease and other digestive disorders. Vitamin C helps prevent iron deficiency anemia. Some developing regions may have the problem of food shortages or monotonous diets. In such cases, the inclusion of even small amounts of vegetables and fruits containing beta-carotene and vitamin C in the diets may help prevent scurvy. The potassium content in vegetables and fruits may help prevent or control hypertension and thereby reduce the subsequent risk of stroke and heart disease.

Plant microconstituents with protective functions include not only a number of micronutrients, such as the so-called antioxidants (carotenoid, vitamins C and E, the mineral selenium), but also include an unknown number and combination of microconstituents, that is, bioactive compounds. These compounds include alliums, dithiolthiolthiones, isothiocyanates, terpenoids, isoflavones, protease inhibitors, phytic acid, polyphenols, glucosinolates and indoles, flavonoids, plant sterols, saponins, and coumarins. There are plausible research data showing the biological mechanisms of individual compounds and their roles in cancer prevention. However, more beneficial effects are associated with the ingestion of vegetables and fruits as food items as compared to individual bioactive compounds. The protective functions vary between the types of vegetables and fruits. For example, cruciferous vegetables probably contribute to the protection against the risk of colon, rectum, and thyroid cancers. There is convincing evidence showing that diets high in allium vegetables, tomatoes, or citrus fruits protect against stomach cancer. Carrots probably protect against lung, stomach, and bladder cancers (WCRF/AICR, 1997).

Hundreds of case control studies have shown diets high in vegetables and fruits are associated with the decrease in cancer risk of many sites. The National Academy of Science's (1989) report *Diet and Health* concluded that diets high in plant foods, including vegetables and fruits, are associated with lower occurrence of coronary heart diseases and cancers of the lung, colon, esophagus, and stomach. Five or more daily servings of a combination of vegetables and fruits, especially green and yellow vegetables and citrus fruits, are recommended. The U.S. Food and Drug Administration has authorized several health claims on food labels, such as that low-fat diets rich in fiber-containing grain products, fruits and vegetables may reduce the risk of some types of cancer, a disease associated with many factors (Giese and Katz, 1997).

Excellent literature reviews have been reported concerning the protective effects of vegetables and fruits against cancer risk. Examples of such reviews are those by Block et al. (1992), Potter (1996), Steinmetz and Potter (1991a, 1991b, 1996), and Ziegler (1991), and an example of a symposium proceeding is edited by Huang et al. (1994). An expert panel, consisting of international cancer scientists, has concluded that there is a strong and consistent pattern showing that diets high in vegetables and fruits decrease the risk of many cancers and perhaps cancers in general. The evidence is also convincing regarding the effect of such diets on decreasing the risk of mouth and pharyngeal, esophageal, lung, and stomach cancers. These diets probably protect against laryngeal, pancreatic, breast, and bladder cancers and possibly protect against ovarian, cervical, endometrial, and thyroid cancers. Diets high in vegetables convincingly protect against colorectal cancer and possibly protect against primary liver, prostate, and renal cancers. The evidence that green vegetables protect against lung and stomach cancers is convincing, and they probably protect against mouth and pharyngeal cancer.

Processing of fruits and vegetables changes the nutrient contents to some extent. Conventional studies have primarily focused on the evaluation of food processing methods (such as canning, freezing, drying, and chilling) on amino acids, vitamins, and minerals (Ang and Livingston, 1974; Ang et al., 1975; Karmas and Harris, 1988). Data are not as sufficient for other bioactive compounds as for those micronutrients. Nevertheless, micronutrients and bioactive compounds are not completely lost in processed products; their concentrations may be changed. In fact, as a result of food processing, perishable foods, either as raw or as further processed products, are made more available in various locations and/or seasons. The availability of these processed products should benefit populations in Asian countries, especially those regions where the consumption of fruits and vegetables is low.

There is a public health concern about chemical residues in fruits and vegetables in several Asian countries. While some farmers are reluctant to use chemical fertilizers and pesticides, some other farmers lack proper training for the usage of these chemicals. In countries where these chemicals are used, consumers are often worried about the residues and thus avoid eating vegetables and fruits. The application of chemical fertilizers helps increase agricultural production and reduces the food cost. Pesticides are indispensable in modern farming for reducing the crops lost to insect and pest infestation. In most developed countries, the applications of harmful pesticides, as well as their residual amounts in foods, are monitored by government regulation. Properly stored and cleaned vegetables and fruits have not been shown to present any significant adverse health effects. It is imperative that farming chemicals in developing countries are effectively regulated according to the international safety standards in order to provide consumers with safe and abundant supplies of vegetables and fruits.

3.7 FATS AND OILS

3.7.1 Nutrient Composition and Dietary Intakes

Fats from animal origins include butter from milk fat, lard from pig, and tallow or suet from cattle and sheep. Dietary oils from plant origins in Asia are derived from seeds or nuts, such as soybean, rapeseed, corn, peanut, sesame, palm, and coconut. Similar to Western countries, hydrogenated or partially hydrogenated vegetable oils are being used in industrialized regions of Asia. As described previously, animal fats contain cholesterol and a high proportion (41% in lard, 52% in tallow, and 66% in butter) of saturated fatty acids. Vegetable oils contain substantially more polyunsaturated and monounsaturated fatty acids than saturated counterparts, except palm oil, coconut oil, and hydrogenated oils. The saturated fatty acids contents in palm oil and coconut oil are 51% and 92%, respectively (USDA, 1997). Hydrogenated oils contain varying amounts of saturated fats, depending on the end uses.

The consumption pattern of added fats and oils in Asian countries is shown in Figure 16.1. Total dietary fats and oils include not only added fats and oils used in commercial processing and home preparation, but also the fats and oils originally contained in food materials. The total dietary lipids provide less than 15% total energy in most Asian populations. The exceptions are Hong Kong and Malaysia, where dietary fats and oils may supply approximately 30% total energy (WHO, 1990). In comparison, in developed countries of Europe and North America, the average consumption of dietary fats from various sources (meat, meat products, dairy products, margarine and butter, and baked goods) accounts for about 30%–40% total energy (WHO, 1990). Added fats and oils account for 15–20% total energy in these regions (Figure 16.1). Vegetable oils are the most common type of food oils used in Asia, except in some parts of China, where lard is the preferred cooking fat. There is a tendency that the consumption of meat and saturated fats increases with economic development in Asian countries. From 1961 to 1990, the contribution of total energy from dietary fats and oil increased about 50% in some parts of Asia. A higher increase in consumption of added fats was noted in Malaysia and Japan (WHO, 1990).

3.7.2 Health Benefits and Adverse Effects

Fats and oils provide high-density dietary energy and desirable flavor and texture to processed and home prepared foods. Dietary intake of linoleic acid at 1–2% total energy is required to prevent a chronic deficiency state with poor growth. However, excess caloric intake from fats and oils, either visible or invisible forms, as well as from other sources, is associated with obesity problems in some populations in China and Japan and possibly has increased the risk of obesity-related diseases, such as hypertension, cardiovascular diseases, and some cancers. Diets high in total fat, as well as saturated fats, possibly increased the risk of lung, colorectal, breast, and prostate cancers. Dietary cholesterol from animal fats is possibly related to the increased risk of lung and pancreas cancers (WCRF/AICR, 1997) and atherosclerosis (Barnard, 1993). Diets high in saturated fats, from either animal or plant origins, possibly increase the risk of elevated cholesterol level that is associated with increased risk of heart diseases (Anderson, 1987; Barnard, 1993). Coconut, palm kernel, and palm oils are products of tropical countries. In Southeast Asia, Malaysia and Indonesia are the major producers of coconut and palm oils. Since these oils are highly saturated, it may be advisable that people in tropical regions do not overuse these oils in their diets. There are still many unanswered questions about the roles of specific fatty acids in health and diseases. Perhaps the best strategy at present is to limit the total fat intake. Some food items are prepared by deep-frying in oil, such as the deep-fried dough strips in China (Table 16.1) and deep-fried plantains in Malaysia. Ingestion of these fried foods is likely to contribute to higher fat intake. Additionally, potential adverse effects could be derived from heated, oxidized fats. Studies have indicated that toxic effects of oxidized and heated fats may be induced by secondary lipid peroxides, and thermally stressed oils may have a role in the acceleration of atherosclerosis (Kubow, 1990).

3.8 SUGAR, SALT, AND MONOSODIUM GLUTAMATE

3.8.1 Sugar

The consumption of refined sugar in Asian countries is generally low. It has been estimated to be less than 10 g/day compared to 30 g/day in the western world. Sugar supplies energy but few nutrients. People with a high intake of sugar may tend to have a low intake of other foods (e.g., vegetables, fruits, and cereals), micronutrients (e.g., fibers, folate, other B vitamins, carotenoids), other antioxidants, and bioactive compounds that protect from colorectal cancer. Several epidemiological and experimental studies showed that the risk of colorectal cancer was positively proportional to the intake of refined sugar. There was an elevated risk for people consuming 30 g or more per day compared with those eating less than 10 g per day (Tuyans et al., 1988). A higher intake of sugar was also associated with an increase in body weight, hypertension, and dental caries. Caution should be exercised in both industrialized and developing Asian countries where the intake of refined sugar tends to increase because of the economical growth in recent years.

3.8.2 Salt and Salted Foods

Salt has been used in food preservation and food preparation for thousands of years. Many populations have acquired a taste for prepared foods with added salt. Salt is composed of sodium chloride, and sodium is one of the micronutrients required for normal biological functions. An estimated safe level of daily requirement of sodium is 500 mg for average adults and 115 mg for people who do not sweat much (NRC, 1989). The actual intake of sodium or salt varies greatly in different parts of the world. It has been estimated that salt intake is about 8–10 g per day in most industrialized countries and 10–12 g in Colombia, Japan, and Korea. The salt intake is as high as 14 g per day in Tianjin, China (WCRF/AICR, 1997). Dietary salt sources include table salt and salty and salted/cured meat, fish, vegetables, and processed foods. Salted foods are common in China, Korea, and Japan and other parts of Southeast Asia. Examples are salted fish in southern parts of China, pickled or salted vegetables in Japan, and pickled vegetables *(Kimchi)* in Korea. Table salt is not commonly used, but soy sauce is a popular condiment for cooking and for on the table. Another source of salt is *miso,* an indispensable ingredient for the preparation of Japanese *miso* soup.

Residents in some parts of the world have higher stomach cancer rates than other parts of the world. Regions with high rates include Japan, some parts of China, and Latin America, where traditionally prepared salty meats, fish, vegetables, and other foods are eaten regularly. Numerous studies have been conducted to determine the association of salt intake or salted foods and health effects. For instance, a prospective study suggested that a higher intake of salted fish was associated with an increased risk of stomach cancer (Kneller et al., 1991). Case control studies have shown that varying degrees of increased risk of stomach cancer with higher consumption of salted foods, such as salt-pickled vegetables, were associated with an increased risk of stomach cancer in Japanese living in Hawaii but not in Japan (Haenszel et al., 1972, 1976). However, another study showed that the consumption of dried, salty fish was positively related to the severity of intestinal metaplasia in men but not women (Nomura et al., 1982). Other case control studies examined the consumption of salted foods in general, salted vegetables, or specific salted foods, such as soybean paste and soy sauce, and found different degrees of increased risk of stomach cancer in different populations (Kono et al., 1988; Lee et al., 1995). A 50% increase in risk of stomach cancer was reported in Japan for high consumption of salted fish gut and cod roe in men, but not women, and no association for salted, dried fish in general for either sex (Kato et al., 1990). In China, studies have shown that high intakes of salt and soy sauce were related to the elevated risk of dysplasia (Kneller et al., 1992) and that daily consumption of salted, processed meat and fish, but not vegetables, increased the risk of laryngeal cancer (Zheng et al., 1992).

Cantonese-style salted fish, a common fish item in southern China, is usually prepared by softening under partial decomposition before or during salting. The salt content in the Cantonese-style salted fish could be as high as 5–10 g/100 g. Animal studies have shown an association of carcinomas of the nasal and paranasal regions in rats fed Cantonese-salted fish (Huang et al., 1978). Regular consumption of this type of fish has been found to be associated with an increased risk of nasopharyngeal cancer in nine of twelve case control studies. Positive correlations were found in China, Hong Kong, Malaysia, and Thailand. Risk is further increased if these foods were eaten frequently in early childhood (Ning et al., 1990; Sriamporn et al., 1992; Yu et al., 1985, 1986, 1988, 1989; Zheng et al., 1993). Several volatile *N*-nitrosamines and their precursors were identified in salted fish, and positive correlations were found between the levels of some of these nitrosamines and mortality from nasopharyngeal carcinoma (Huang et al., 1981; Zou et al., 1994). In view of the evidence from epidemiological and experimental data and plausible biological pathways, it has been concluded that diets high in salt or salted foods probably increase the risk of stomach cancer, and the evidence is convincing regarding diets high in Cantonese-style salted fish and their association with increased risk of nasopharyngeal cancer (WCRF/AICR, 1997). Additionally, sodium has also been identified as an important cause of hypertension and stroke. Diets high in salt, salted, and/or salty foods contribute to the high intake of dietary sodium.

3.8.3 Monosodium Glutamate

The use of monosodium glutamate (MSG) as a flavor enhancer in food is very popular in Japan and in Chinese communities in Southeast Asia. Generally, consumers in these regions do not suffer any severe adverse effects, if any, from the ingestion of MSG, and they appear to be adapted to the taste of soy sauce, which contains natural MSG. However, some other individuals, such as those in Western countries, who are not accustomed to the MSG, may have sensitive adverse reactions. Mild symptoms associated with the ingestion of MSG have been reported, such as headache; nausea; dizziness; facial tightness; chest pain; numbness in the face, temples, upper back, neck, and arms; drowsiness; and weakness for some individuals. The Federation of American Societies for Experimental Biology (FASEB, 1995) has concluded that the ingestion of MSG is safe for the vast majority of consumers. However, the

FASEB Panel also concluded that there is "sufficient evidence to support the existence of a subgroup of the general population of otherwise healthy individuals who may respond to large doses (≥3 g) under specific circumstances of use" and that "there may be a small subgroup of previously diagnosed unstable asthmatics who also may respond to large doses of MSG under specific conditions of use."

MSG is not only present in food as an added chemical, but also as a by-product of hydrolyzed vegetable proteins, which are used widely as seasonings and flavors in canned foods, dry mixes, sauces, and other manufactured products (Scopp, 1991). It has been recommended that processed foods be clearly labeled regarding their added MSG content either as MSG by itself or as part of hydrolyzed vegetable proteins. MSG contributes to dietary sodium intake. Individuals concerned with their sodium intake should also consider the amount of sodium supplied by MSG.

3.9 TEA, COFFEE, AND HERBAL SUPPLEMENTS

This group of dietary items is not considered foods; that is, consumption of these items is not for their energy or macronutrients. But these items play important roles in total diet by providing flavor, taste, or certain physiological effects derived from bioactive microconstituents, including terpenoids, isoflavones, caffeine, polyphenols, glucosinolates and indoles, flavonoids, plant sterols, saponins, coumarins, oligossacharides, and numerous other phytochemicals. The following sections briefly introduce health effects related to the ingestion of these items and/or their microconstituents.

3.9.1 Tea and Coffee

Tea plants of many varieties have been widely grown in Asia. Depending on the methods of preparation or processing, tea leaves are made into green, black, and oolong teas. Green tea is produced by the brief exposure of fresh tea leaves to a very high temperature, only long enough to deactivate enzymatic fermentation. To produce black tea, the leaves are first semidried at a warm temperature, then allowed to ferment, and finally roasted at a high temperature. Oolong tea is a semifermented tea. Because of the difference in fermentation process, oolong teas may vary greatly in color, flavor, taste, and chemical composition. Hot tea drinks are prepared by infusion of dried tea leaves (green, black, or oolong teas) with hot water. Hot tea by far is the most popular drink in many parts of Asia, except for certain populations in Malaysia and Indonesia, where coffee is preferred. In many Asian regions, such as most parts of China and Japan, tea is normally drunk without the addition of sugar, lemon, or cream. Exceptions are noted in Mongolia and Tibet where substantial amounts of milk and butter are added to hot teas. While green tea is especially popular in Japan, all teas (green, black, and oolong) are commonly drunk in China and Taiwan.

Green tea leaves retain significant amounts of the major leaf polyphenols, that is, catechins. During the processing of black tea leaves, polyphenols are oxidized to dark products, such as theaflavins. Being semifermented, oolong teas retain varying amounts of the original catechins. Most of the studies related to health effects were generally related to the green tea or black tea. Many animal studies have shown black, green, or oolong tea to inhibit experimentally induced tumors at various sites (Chen, 1992). Black and green teas can inhibit mutagenicity (Chen, 1992). Phytochemicals in teas and their inhibitory effects on carcinogenic processes have been reported in a number of studies (Ho et al., 1994). Nakane et al. (1994) found that tea polyphenols were strong inhibitors of human immunodeficiency virus (HIV)-reverse transcriptase. Four of five case control studies in Japan and China showed a decreased risk of stomach cancer with high intakes of green tea. Animal study data also support these findings. Green tea consumption possibly decreased the risk of stomach cancer. Data were somewhat conflicting regarding the relationship of black tea and cancer risk.

To produce coffee, green coffee beans are roasted by one of several methods and then ground. The coffee drink is then produced by boiling, brewing, or brewing under pressure (espresso) with water. Coffee is commonly flavored with added sugar, milk, or cream. Regular coffee drinking may be associated with raised blood low density lipoprotein (LDL) cholesterol and thus may result in an increased risk of heart disease (Cannon, 1992). A series of clinical trials have shown that the major hypercholesterolemic agents in coffee are the diterpenes cafestol and kahweol (Ratnayake et al., 1995). However, coffee brewed by different methods retains different amounts of these compounds, with "boiled" coffee containing the most (Urgert et al., 1995). Coffee is one of the most popular beverages in Malaysia. Young children sometimes start to drink coffee as early as 4 or 5 years old. The coffee-making method in Malaysia is traditionally by hot water infusion, that is, by adding hot water to coffee granules that are placed in a course cloth sack. After a few minutes, the sack is removed and the coffee is ready for drinking. Usually the coffee is very strong. In Malaysia, coffee may be mixed with condensed milk (from cans) and sugar to taste. It is not clear whether this preparation method would contain as much cafestol and kahweol as those prepared by some other methods. Nevertheless, the addition of condensed milk and sugar contributes extra energy and milk fat intakes.

3.9.2 Herbal Supplements

Tea and coffee are not the only hot beverages in Asia. There is a variety of herbal teas used widely and over a long time period. Some of these herbal materials are also used as dietary supplements in forms other than tea drinks. They can be used as ingredients in cooking, as additives for wine and liquor, or in capsule or tablet forms. These herbal products, along with conventional nutritional supplements (e.g., products of vitamins, minerals, and amino acids), have been referred to as "health foods" in China. These items may be referred to as "functional foods" in Japan and Singapore and as "nonnutritive health factors" in Korea.

Parallel with economic development, consumers are increasingly interested in supplementing their diets with extra "nutrients" to help them become healthier, stay young, and live longer. Many Asian people believe that these "health foods" can provide health benefits, such as to prevent and/or treat illness and to adjust body functions for health improvements. The physiological functions affected by these supplements may include cardiovascular, gastrointestinal, immune function, aging, and endocrine systems. In Asia, "health foods" are traditionally differentiated from other forms of medicinal herbal products that are mostly sold in traditional pharmacies instead of food stores. Medicinal herbal products are not taken regularly as vitamin pills because they may induce adverse health effects. However, in recent years, because of the rapid marketing expansion in the variety and quantity of herbal products, more consumers are taking supplements without the accurate information about these products. Consumers may encounter side effects and/or economic loss for products not suitable for their use.

Marketing of herbal products has also been expanding rapidly in Western countries. Many of the traditional Asian herbal products have gained popularity on other continents. During the last 15 years for instance, supplying the demand for botanical and herbals is a booming business in the United States. Until late 1997, according to the U.S. FDA's Dietary Supplements Health and Education Act of 1994 (DSHEA), all herbals in the United States have been categorized as dietary supplements as long as there are no claims made by the manufacturers on medicinal effects of these products. These products are sold in health food stores, supermarkets, and pharmacies. According to the Whole Foods surveys in 1995 and 1996, the top ten natural herbal products' sales in the United States included four products of the Asian origin or of Asian influence, namely, ginseng, ginkgo, *ma huang,* and Siberian ginseng (Landes, 1997). The market growth and promotion have been faster than the transmitting of accurate information on the function and uses of these Asian herbals. Misuse of herbal products can cause sickness and even death.

Hundreds of herbals have been used in Asian countries, and many have been documented in detail in literature. In this section, only a few Asian herbal supplements or herbal medicines are briefly described to serve as examples of how these items can be beneficial or detrimental to health if misused. Items included in this section are among the top-selling products in recent years (e.g., ginseng, ginkgo, Siberian ginseng, and *ma huang*).

3.9.2.1 Ginseng

Ginseng is probably one of the most famous "health foods" or dietary supplements in the Oriental countries. It has been used for thousands of years in China, Korea, and Japan. The plant part used for human consumption is the root of an herbal plant *Panax ginseng* C.A. May, also known as *P. schinseng,* Nees. The ginseng root with side branches resembles a human body with arms and legs and thus the name of this plant includes *gin,* which means people or human being in Chinese. Ginseng is also probably one of the most studied health-related food supplements in history. The earliest, detailed record describing ginseng's properties and biological effects was dated about 1900 years ago in China. From the mid-20th century, more and more scientific investigations have been conducted on ginseng, including its chemical constituents and clinic studies. More than 1000 papers on ginseng have been published.

In China, ginseng products are classified into two types, based on the nature of growth. Those ginsengs grown wild are named "mountain ginseng," and those grown by cultivation are called "round ginseng." It takes about 6–7 years of cultivation to develop the roots. Ginseng products are also divided into different types, based on the methods of processing or preparation. Generally, for the preparation of "raw dried ginseng" in China, the root is washed and dried under sunlight or by baking. The outer skin is light brown and yellow, whereas "red ginseng" is prepared by steaming the root followed by sun-drying or baking (Hsiu, 1992). The heating process inactivates enzymes to prevent the decomposition of ginsenosides. A specially "cured" root in Korea has been referred to as "Korean red ginseng," which is relatively expensive. There is also American ginseng (i.e., *P. quinquefolius* L), native to North America and some parts of Europe. The plant can be cultivated. The shape of American ginseng is similar to the China "raw dried ginseng." The composition and properties of ginsengs vary substantially between different types. For human consumption, the dried root can be slightly softened by heat, sliced, and then boiled or seeped in hot water. There are a variety of products, such as ginseng extract, instant ginseng, and ginseng capsules. Traditional methods of preparation involve simmering the roots with or without other ingredients, including other herbal medicines and alcoholic beverages.

The Chinese and Korean ginsengs are believed to provide "warming" or "heating" effects, while the American ginseng provides "cooling" effects. Depending on the variety and methods of processing and preparation, ginseng's property can vary substantially. Red ginseng is "warmer" than raw dried ginseng, and Korea red ginseng is even "warmer" or "hotter" and stronger than red ginseng. Oriental ginseng has been used as a remedy or medicine for various kinds of illnesses. It is an important ingredient in a number of traditional medicine formulas. Various biological functions of ginseng have been claimed, including the stimulation of the central nervous system, resistance to fatigue, increasing the synthesis of blood, lowering blood sugar levels and cholesterol, improving athletic performance, increasing the recovery from sickness, protecting liver and heart, adjusting blood pressure, and providing antitumor and anticancer effects (Hsiu, 1992). Useful pharmacologic effects in such conditions as anemia, atherosclerosis, depression, diabetes, edema, hypertension, and ulcers have been documented (Tyler, 1993). In Asia, the drug is held in esteem for the treatment of anemia, diabetes, gastritis, and many conditions arising from the onset of old age. It is used to replenish "vital energy" and to promote the secretion of body fluids (Lang et al., 1993). As a strong "heating drug," Oriental ginseng is only for weak people. It is not for people of regular health, especially not for youths and infants. Misuse of hot ginseng may cause sickness and side effects. In the West, it has become an extremely popular remedy in recent years, particularly for the improvement of stamina, concentration, and resistance to stress and diseases. Ginseng has been used as an ingredient in a variety of dietary products. Indiscriminate use of these products may not achieve beneficial effects. Sometimes, adverse side effects may occur, such as headaches, dizziness, nose dryness or bleeding, loss of appetite, and chest tightness, and children may develop overexcitedness, irritation, stress, and insomnia problems. Also, it is not recommended using ginseng with tea or turnips because they tend to counteract the ginseng activity (Hsiu, 1992). The American Association of Poison Control Centers has at least one report of a ginseng-related death. A 71-year-old man who drank undiluted ginseng extract for 2 weeks developed hepatitis and died from liver failure (Rand, 1994).

Based on the traditional medicine belief, the American ginseng is slightly bitter and "cool." Its power is lower than Oriental ginsengs. It can be used for quick recovery from sickness and as a remedy for minor throat and voice problems. American ginseng is often used to replenish the "vital essence" and to reduce the "internal heat" of the body (Lang et al., 1993). Depending on each individual's physiological status, people are advised by traditional doctors to take either the "warm" or "cool" ginsengs. One type of ginseng may benefit one type of individual but may cause adverse effects to others. Some people cannot use American ginseng just as some individuals cannot use Oriental ginsengs.

According to a review of Tanaka (1994), active chemical ingredients in ginsengs are a number of triterpenoid saponins, commonly named ginsenosides, and several acetylene alcohols. Biological functions of ginseng include anticarcinogenic effects, inhibitory effect on 12-*O*-tetradecanoylphorbol-13-acetate (TPA)-induced inflammation, and inhibition of pulmonary tumorigenesis; some data were derived from animal studies (Tanaka, 1994). Much research has confirmed that ginsenosides present in the ginsengs are the active constituents for sedative, antifatigue, antiinflammatory, antiplatelet aggregation, antioxidation, and antiaging effects. There are at least 13 types of ginsenosides identified. The ginsenoside Rf (6) was found only in the root and not the leaves of *P. ginseng.* It is also absent in the leaves or roots of American ginsengs *(P. quinquefolium).* Thus, different species of ginsengs can be differentiated by "fingerprints" obtained by high-performance liquid chromatographic techniques (Lang et al., 1993).

The variety and quality of commercial ginseng products often vary greatly between brands and even within a brand. These products include roots, tea bags, instant teas made from extracts, powders, capsules, tablets, and as an ingredient in other supplements. *Consumer Reports* (1995) analyzed 10 brands of ginseng capsules marketed in the United States and found various amounts of ginseng, as well as ginsenosides, in the capsules. The amount of ginseng per capsule ranged from 100 mg to 648 mg as labeled. The concentration of total ginsenoside (active ingredients) varied from 0.4 mg to 23.2 mg per capsule, which is equivalent to 0.2% to 7.5% ginsenosides in the capsules. Some brands have 10 times or 20 times as much as others, and one brand had very little ginsenoside. Furthermore, some products are not labeled with the specific types of ginsengs. It is difficult for an average consumer to make an intelligent selection among various products. Standardization of active components and proper labeling are necessary for the health benefit of consumers and for development of the herbal supplement industry.

3.9.2.2 Siberian Ginseng

Siberian ginseng is an entirely different plant. It does not even belong to the same genus as the true ginsengs; it is not a species of *Panax* although it is the same family of Araliaceae. The part used is the root of *Eleutherococcus senticosus* (Rupr. & Maxim.) Maxim., known as eleutherococcus in Russia, as Siberian ginseng in the United States and as *wujiaseng* or *ciwujia* in China. Thus, a more appropriate name for the Siberian ginseng is eleuthero. In some countries, Siberian ginseng is used to help cancer patients resist the side effects of radiation therapy (Tyler, 1993). It has also been claimed to have effects on fighting fatigue, boosting the immune system, improving athletic endurance, and helping the recovery from exercise-induced fatigue. Siberian ginseng has been used in Russia as an abundant and inexpensive substitute for ginseng. It is also cultivated in China for the roots, which are used as a tonic and sedative (Hsiu, 1992).

3.9.2.3 Ginkgo

Ginkgo (also ginko), like ginseng, is mentioned in the traditional Chinese pharmacopoeia. The pulp has ginkgo toxin, and it can cause a severe poison ivy-like dermatitis. The pulp is toxic if ingested, but the seed is edible. In

China, the edible inner seeds from ginkgo trees, *Ginkgo biloba* L., have been used for thousands of years as a traditional medicine for the treatment of asthma and bronchitis (Jian and Wu, 1993). The seed (nut) is slightly bitter and a little "warm" and may induce some minor side effects. It can be boiled and consumed. Ginkgo has antibacterial properties against several kinds of bacteria. Large doses of cooked ginkgo seeds can be toxic, especially for children ingesting 7–150 pieces and adults 40–300 pieces. Overdose can produce nausea, unconsciousness, slow reaction, shortness of breath, and diarrhea (Jian and Wu, 1993).

In Western countries, ginkgo leaves have been utilized in producing a concentrated, standardized *Ginkgo biloba* extract and is widely used in Europe for prescription drugs, as well as over-the-counter drugs. To produce *Gingko biloba* leaf extract, the multiliquid extract is dried to give 1 part extract from 50 parts raw leaves. The standardized extract contains 24% flavonoids (mostly flavonoid glycosides and quercetin) and 6% terpenoids (gingkolides and bilobalide diterpenes). The most important flavonoids are glycosides of kaempferol, quercetin, and isorhamnetin (Kleijnen and Knipschild, 1992).

Products of ginkgo leaf extract are available in both solid and liquid forms, and each tablet or capsule contains 40 mg of the extract. These products are marketed for their beneficial effects on the circulatory system and as an antiasthmatic. Many of the scientific and clinical studies concerning the effectiveness of leaf extracts were conducted in Germany by clinical pharmacologists. Their research findings showed the effectiveness of these products in treating ailments associated with decreased cerebral blood flow, particularly in geriatric patients with short-term memory loss, headache, tinnitus, and depression. A commercial extract of the gingko leaves has been shown to have pharmacologic activity, which is generally associated with vasodilation (blood flow). Gingko leaf extract may enhance blood flow to the brain and for the elderly, which supposedly can improve concentration and memory, absentmindedness, headaches, tinnitus, and ringing in the ears. It may also aid circulation to legs to relieve painful cramps (*Consumer Reports,* 1995).

A clinical trial study tested a *Ginkgo biloba* leaf extract, Egb 761, for treating mildly to severely demented outpatients with Alzheimer's disease and multiinfarct dementia at 120 mg per day for 6 months to 1 year. Results showed that Egb was safe and appeared to be capable of stabilizing the cognitive performance and, in a substantial number of cases, improving the social functioning of demented patients (Le Bars et al., 1997). The Egb extract has been used to alleviate symptoms associated with a range of cognitive disorders. Recently, ginkgo leaf extract was approved in Germany for the treatment of dementia. The product is regarded as a drug rather than as a dietary supplement.

The mechanism of biological action of the ginkgo extract has been reported by Kleijnen and Knipschild (1992). Ginkgolides include different types of diterpene lactones known as the ginkgolides A, B, C, and J. Very large doses may cause restlessness, diarrhea, nausea, vomiting, and other mild effects. The ging ko leaf extract has been shown to decrease platelet aggregation. Ginkgolide B is a potent inhibitor of platelet-activating factor (PAF). There has been a recent case of spontaneous hyphema in a 70-year-old man who was taking the leaf extract and aspirin (Rosenblatt and Mindel, 1997). In another case, long-term use of the leaf extract has been associated with bilateral subdural hematomas and increased bleeding time in a healthy 33-year-old Korean woman who developed bilateral subdural hematomas during chronic self-treatment with *Ginkgo biloba,* consisting of taking 60 mg twice daily for 2 years. It was then postulated that the subdural hematomas occurred in part because of the presence of abnormal platelet aggregability from the chronic ingestion of ginkgo extract (Rowin and Lewis, 1996). Chung et al. (1987) reported that a ginkgolide mixture given to human volunteers was found to cause significant inhibition of PAF-induced platelet aggregation after the ingestion of single 80-mg and 120-mg doses.

3.9.2.4 Ma Huang

While some herbal dietary supplements contain bioactive compounds that may be beneficial or detrimental to human health, depending on the dosages, some other herbal products may contain high concentrations of toxic constituents and can induce severe health problems. The latter group would be more appropriately classified as a prescription drug instead of a dietary supplement. An example in this category is *ma huang,* a dietary supplement in the United States, but it is commonly used in Chinese traditional medicine as a strong drug, not as a "health food."

Ma huang contains stimulant alkaloids (e.g., ephedrine and pseudoephedrine), which have been used in asthma and bronchodilator drugs and in decongestants. According to Zhang et al. (1989), high amounts of total alkaloids were identified in species of *Ephedra sincica* Stapf, *E. equisetina* Bunge, *E. monosperma* Gmel. ex Mey., and *E. intermedia* var. *tibetica* Stapf (over 2.0% in *E. equisetina* Bunge and *E. monosperma* Gmel. ex Mey.). Important differences of alkaloid amounts and alkaloid ratios among *ma huang* plants depend mainly on the species characters, rather than the collection locations. The major effective components in *ma huang* are three pairs of stereoisomer alkaloids such as 1-ephedrine, *d*-pseudoephedrine, 1-norephedrine, *d*-norpseudoephedrine, 1-methylephedrine, and *d*-methylpseudoephedrine. Although the structures of these six alkaloids are very similar, their pharmacological mechanisms are different and so are the intensity of the effects.

Based on the theory of Chinese traditional medicine, the application purposes and effects of *ma huang* vary in different prescriptions. Different *ma huangs* contain different types and quantities of alkaloids used for different purposes. For the purposes of sweating and the easing of asthma, one species of *ma huang* is used (such as *E. equisetina* Bunge, *E. monosperma* Gmel. ex May, and *E. sinica* Stapf) because of its high ephedrine content, while for the purposes of releasing moisture and enhancing digestion and for antiinfection and pain releasing, another type is prescribed (*E. intermdia* Schrenk ex May) because of its high levels of pseudoephedrine. Yet another species (*E. sinica* Stapf from Northeast of China and *E. intermedia* var. *tibetica* Stapf) is considered good for the medicinal prescription for antirheumatoid, asthma-releasing, and antiskin diseases, because the amount of norephedrine is higher in these species than other species.

In a traditional Asian community, it is probably common knowledge that *ma huang* is a strong drug, a poison, and it is not for general use. However, consumers in modern, industrialized societies in Asia and in Western countries apparently do not have the common knowledge about herbals, especially when those herbals are made into other forms. For instance, dietary supplements containing ephedrine alkaloids are widely sold in the United States under the names *ma huang,* ephedra, Chinese ephedra, and epitonin. The ingredient sources of the ephedrine alkaloids include raw botanicals and extracts from botanical sources, primarily from *Ephedra sinica* Stapf, *E. equistestina* Bung. Products containing *ma huang* have been marketed in a variety of forms, in capsules, tablets, powders, liquids, and energy-boosting formulas, sometimes with caffeine, which can augment the adverse effects. Most of them contain concentrated extracts. Many contain active ingredients of other botanicals. Some are promoted for weight loss, body building, increased energy, increased mental concentration, and sexual sensations. However, *ma huang* alkaloids are very toxic. They can raise blood pressure and cause palpitations, nerve damage, muscle injury and stroke, and more. Several deaths have resulted from the use of these drugs.

Since 1993, the U.S. FDA has received more than 800 reports of illnesses and injuries associated with the use of more than 100 different dietary supplement products that contained or are suspected to contain ephedrine alkaloids. Adverse health effects include cardiac-related symptoms, seizures, stroke, chest pain, myocardial infarction, insomnia, nausea, vomiting, fatigue, and dizziness. In 1997, the U.S. FDA issued a proposed rule (*Federal Register,* 1997) warning against the use of dietary supplements containing the herb *ma huang,* alkaloid ephedra (*Ephedra sinica* Stapf., Ephedraceae), and established dose limitations. The proposed regulations include the following provisions: The maximum allowable dosage of ephedrine alkaloids is 8 mg per unit dose, with a maximum dosage of 24 mg; maximum duration is 7 days; it cannot be combined with caffeine or caffeine herbs; it cannot make a claim for body building or weight loss; it should display multiple warning labels regarding heart attack, stroke, seizure, or death; ephedra is not allowed to be used in conventional foods; and the use of ephedra as a knock off street drug is banned.

The problem associated with supplements containing *ma huang* is one example that appropriate governmental regulations are needed for botanicals with medicinal functions. Currently, different countries have different rules and regulations for these products. Many national governments, such as Germany, France, and Canada, have already established special review systems for herbs and botanicals. Some countries in Asia regulate those products manufactured in capsule or tablet forms as over-the-counter or prescription drugs. Appropriate regulations and scientific and technological exchanges are needed for ensuring the effective and safe usage of herbal supplements in Asia and other parts of the world.

4. DIETARY RECOMMENDATIONS

Conventional health science professionals have been concerned with nutrient deficiency diseases. Since 1943, the U.S. National Research Council has published recommended dietary allowances for Americans, including the desirable intakes of various nutrients by age and sex groups. These recommendations specified the daily nutrient intakes for the maintenance of good health for the general population in the United States. Revisions of the recommended dietary allowance have been issued every several years. Many Asian countries have adapted the U.S. recommendations as a guide for their populations. For instance, in China, the recommended allowances include the total energy intakes of 60–70% from carbohydrates, 10–15% from protein, and 20–25% from fats and oils (Table 16.3).

Because of the economic development and agricultural advancement, food production has increased dramatically in this century, and nutrient deficiency problems have become less severe. However, a problem of "overnutrition" has been observed in industrialized and affluent countries. In the latter half of this century, it has been recognized that nutrient intake at higher levels may cause adverse effects. Recently, the Food and Nutrition Board of the National Academy of Science's Institute of Medicine has been developing tolerable upper limits (Miller, 1997). Research findings in recent decades have shown that diets and the incidence of some chronic diseases are closely related. Thus, problems associated with "overnutrition" could be more severe than "undernutrition" in Asian countries in the next century.

As research findings on nutrition and chronic diseases have been recognized, a new trend of dietary patterns and recommendations based on food groups, rather than individual nutrients, is formed in industrialized countries. These recommendations concern reducing the risk of cardiovascular disease and specify less fat from meat (Cannon, 1992). These patterns include an increasing consumption of vegetables and fruits and somewhat decreasing consumption of red meat, fat, full-fat milk, other dairy products, and refined sugar. In Japan, dietary recommendations include both individual nutrients and spe-

TABLE 16.3. Dietary Recommendations of China and Japan.

Food Group/Constituents	China[a] (1991)	Japan[b] (1984)
Carbohydrates	60–70% TE [c]	55–60% TE
Cereals and other plant foods	ns[d]	Choose whole grains, legumes, fungi, and seaweed
Sugar, refined	ns	Limit intake
Protein	10–15% TE	Take enough protein, 50% from plant sources
Meat, poultry, and fish	ns	Eat meat and fish equally (indirect limit meat and fat intake)
Fats and oils	20–25% TE ; <30% TE	Limit fat intake, 20–25% TE
Vegetables and fruits	ns	Choose red, green, or yellow vegetables once a day, and oranges
Vitamins and minerals	ns	Eat fruits and vegetables rich in carotene and vitamin C
Salt	ns	Limit intake, <10g/day
Other choices	ns	Eat a variety of foods. Take more than 30 food items a day. Eat soyfood or fish once a day.
Other limitations	ns	Avoid foods and drinks of very high temperature. Avoid burnt fish and meats.

[a]*Source:* Chinese Academy of Preventive Medicine, 1991.
[b]*Source:* Fujisawa, Y. and R. Itoh, 1984.
[c]TE = total energy intake.
[d]ns = no specific recommendation.

cific food groups (Table 16.3). They suggest that 55–60% and 20–25% total dietary energy to be provided by carbohydrates and fats, respectively, and diets to include whole grains, legumes, fungi, seaweed, equal amounts of meat and fish, and 30 food items a day. Contemporary dietary guidelines or recommendations are based on food groups and are used for the purpose of prevention of chronic diseases. A few examples are summarized below to serve as a reference for Asian countries where updated guidelines might not be available.

The World Health Organization's report, *Diet, Nutrition and the Prevention of Chronic Diseases* (WHO, 1990) recommended a diet provides between 50% and 70% of total energy from complex carbohydrates and provided 16–24 g/day of non-starch polysaccharides. This diet should include at least 400 g of vegetables and fruits daily, in addition to potatoes, including at least 30 g of legumes, nuts and seeds. This was focused on some types of cancer as well as coronary heart disease.

In the United States, *Nutrition and Your Health: Dietary Guidelines for the Americans* (USDA/USDHHS, 1990) and the *Food Guide Pyramid* (USDA, 1992) recommended that the daily diet include 6–11 servings of the cereal group (bread, rice cereals, and pasta); 3–5 servings of vegetables; 2–4 servings of fruits; 2–3 servings of milk, cheese, and yoghurt; 2–3 servings of the meat group (meat, poultry, fish, dry beans, eggs, and nuts); and sparing use of fats, oils and sugars. The Physicians Committee for Responsible Medicine in the United States proposed New Four Food Groups, recommending a healthy diet to be based on grains, legumes, vegetables, and fruits (Barnard, 1993). This plan suggested reducing the intake of total fat to about 10% of total energy while avoiding saturated fats from both animal and plant origin and cholesterol.

The expert panel of the World Cancer Research Fund and the American Institute for Cancer Research (WCRF/AICR, 1997) established recommendations applicable globally for the prevention of cancers, and they also are consistent with the prevention of other diseases. Essentially, the recommended diets are selected from predominantly plant-based foods rich in a variety of vegetables and fruits, legumes, and minimally processed starchy staple products. Specifically, daily diets should include a variety of starchy or protein-rich foods of plant origin, preferably minimally processed to provide 45–60% total energy (about 600–800 g of a variety of cereals or grains, legumes, roots, tubers, and plantains) and 7% or more total energy from vegetables and fruits (400–800 g) all year round. Limited consumptions are specified for refined sugar at less than 10% of total energy, total fats and oils at 15% to no more than 30%, and salt at 6 g for adults. It is recommended to use herbs and spices to season foods. If eaten at all, red meat intake is limited to less than 10% total energy (80 g daily). Fish, poultry, and meat from nondomesticated animals are preferred in place of red meat. Alcohol consumption is not recommended—limited to less than two drinks a day for men (<5% energy) and one for women (2.5%). Other recommendations include the appropriate storage of perishable foods to avoid fungal contamination and spoilage and a relatively low cooking temperature for meat and fish to avoid burning of meat juices and charring the food.

5. CONCLUSIONS

Most traditional Asian diets include a high proportion of grains and vegetables as compared to animal products. Contemporary scientific findings have shown that diets high in plant foods are associated with more health benefits, such as decreased risks of cardiovascular diseases, cancers of several sites, and risks of hypertension and obesity. A variety of cereals or grains, preferably minimally processed, are recommended to ensure the supply of micronutrients. The supply and consumption of vegetables and fruits year-round is encouraged.

As per capita income increases with industrial development, there is a tendency to increase the consumption of added sugar, fats, oils, and animal products. Consumers in Asian countries need to be cautioned to avoid increased consumption of these items. Certain regional consumers need to avoid the high intakes of coconut and palm oils, sodium, or salt in their diets. Appropriate labeling of products containing added MSG and/or hydrolyzed vegetables is encouraged. Cereals, including corns, are to be stored properly to avoid fungal contamination. Refrigeration is needed for the storage of perishable foods to retard the growth of microorganisms and as a means of reducing salt preservation of foods.

A variety of herbal items has been used in Asia for a long period. While some herbals may provide beneficial effects when used appropriately, some newer products in various forms containing one or several strong ingredients may induce adverse effects. Appropriate regulations and scientific research are needed to ensure the safety and effectiveness of herbal supplements as more consumers are interested in these products now than before.

6. APPENDIX: FOOD COMPOSITION TABLES IN ASIA

6.1 ASIA, EAST

Food Composition Table for Use in East Asia
WT Wu Leung, RR Butrum, FH Chang, MN Rao, W Polacchi FAO, Rome, Italy, and US Dept Health, Education, and Welfare, Washington, DC, USA 1972, 334 pp. (English)

6.2 BANGLADESH

Analysis of Some Food Stuffs of Bangladesh, Table I
KAhmad, MA Malek Pak *J Biol Agr Sc,* Vol 9, No 1, 1966 (English)

6.3 BURMA

The Nutritive Value of Burmese Foods
The Nutrition Project Directorate of Health Services, 36, Theinbyu St, Rangoon 1967, 41 pp. (English)

6.4 CHINA

Table of Food Composition
Department of Nutrition and Food Hygiene, The Chinese Academy of Preventive Medicine, Beijing, P.R. China, 1991, 264 pp. (Chinese)

Iodine Contents of Eighty-Five Kinds of Food in Shanghai District
Zhang Weihong, He Qianguong, Wu Qile *Acta Nutrimenta Sinica* 18,4:492–494. Shanghai Hygiene and Antiepidemic Centre, Shanghai 200335 1996 (Chinese and English)

6.5 INDIA

Balanced Diets and Nutritive Value of Common Recipes
M Swaminathan, K Joseph, M Narayana Rao, SV Chandiramani, L Subramanyam, K Indira. Central Food Technological Research Institute, Mysore. 1981, 117 pp. (English)

Nutritive Value of Indian Foods
C Gopalan, BV Rama Sastri, SC Balasubramanian. National Institute of Nutrition, Hyderabad; Indian Council of Medical Research, New Delhi. Reprinted 1982, 204 pp. (English)

6.6 INDONESIA

Nutrient Content of Indonesian Food
DS Slamet, I Tarwotjo. Pusat Penelitian dan Pengembangan Gizi 1980, 36 pp (Indonesian)

Food Composition Table
Direktorat Gizi Departemen Kesehatan Penerbit Bhratara Karya Aksara, Jakarta 1981, 57 pp. (Indonesian)

6.7 JAPAN

Standard Tables of Food Composition in Japan
Resources Council, Science and Technology Agency, Japan. 1982, Fourth revised edition, 707 pp., 2 copies. (Japanese with English)

Standard Tables of Food Composition in Japan: Fatty Acids, Cholesterol and Vitamin E (Tocopherols)
Resources Council, Science and Technology Agency, Japan. 1989, 208 pp. (Japanese with English)

6.8 KOREA

Food Composition Table
Rural Development Administration Rural Nutrition Institute, Seoul, 3rd ed 1986, 175 pp (Korean)

Food Composition Table
Office of Rural Development Rural Nutrition Institute, Seoul, 2nd ed 1981, 149 pp. (Korean with English)

6.9 MALAYSIA

Nutrient Composition of Malaysian Foods
Tee E Siong, MI Noor, MN Azudin, K Idris. ASEAN Food Habits Project, National Sub-Committee on Protein: Food Habits Research and Development, Malaysia. 1988, 150 pp. (English)

6.10 NEPAL

Nutrient Contents in Nepalese Foods
HMG, Ministry of Agriculture, Agriculture Development Department, Nutrition Programme Section. Babarmahal, Kathmandu. 1994 (Hindi and English)

6.11 PAKISTAN

Food Composition Table for Pakistan
T Hussain, Planning and Development Division, Ministry of Planning and Development, Department of Agricultural Chemistry and Human Nutrition, NWFP, Agricultural University, Peshawar 1985, Reprinted 1990, 70 pp. (English)

6.12 PHILIPPINES

Food Composition Tables: Recommended for Use in the Philippines: Handbook I.
IC Abdon, IF del Rosario, AR Aguinaldo, AV Lontoc, LG Alejo. Food and Nutrition Research Institute (FNRI), Dept. of Science and Technology, Manila. 1990, 6th Revision, 299 pp. (English)

The Philippine Food Composition Tables
Food and Nutrition Research Institute, Department of Science and Technology, Manila 1997, 163 pp. (English with Filipino alternative food names)

6.13 SINGAPORE

Eating Out—A Guide to Food Choice at Hawker Centres
Food & Nutrition Department, Ministry of Health, Singapore 1993, 59 pp. (English)

The Composition of 200 Foods Commonly Eaten in Singapore
Food & Nutrition Department, Ministry of Health, Singapore 1993, 54 pp. (English)

Food Consumption Study—1993
Food & Nutrition Department, Ministry of Health, Singapore 1994, 149 pp. (English)

6.14 SRI LANKA

Tables of Food Composition for Use in Sri Lanka
WDA Perera, PM Jayasekera, SZ Thaha World Health Foundation of Sri Lanka, Colombo 1979, 82 pp. (English)

6.15 TAIWAN

Table of Taiwan Food Composition
Food Industry Research and Development Institute, Hsinchu Revised 1974 (language unknown)

Nutritional Analysis of Main Fruits in Taiwan Area
CC Chou, KG Wei, SH Chen, YL Sheu. FIRDI, ISSN 0253-9039, 19 pp. (Chinese with English summary)

6.16 THAILAND

Amino Acid Content of Thai Foods
Division of Nutrition, Department of Health Ministry of Public Health Undated, 39 pp. (Thai with English)

Nutrient Composition Table of Thai Foods
Nutrition Division, Department of Health, Ministry of Public Health, Bangkok 1992, 97 pp. (Thai with English)

6.17 VIETNAM

Food Products in Vietnam, Composition and Nutritive Value
National Institute of Nutrition, Ministry of Health, Medicine Publisher, Hanoi, 1995, 555 pp. (Vietnamese with English)

7. ACKNOWLEDGMENT

The author thanks many colleagues for their comments and suggestions, which helped improve this article. The opinions expressed in this article are solely those of the author and not necessarily those of the U.S. Food and Drug Administration.

8. REFERENCES

Altschul, A. M. and Planck, R. W. 1960. Rice and rice products. In: *Nutritional Evaluation of Food Processing,* R. S. Harris and H. von Loesecke, eds., John Wiley & Sons, Inc., New York, pp. 204–214.

Anderson, J. W. 1987. Dietary fiber, lipids, and atherosclerosis. *Am. J. Cardiol.* 60:17G–20G.

Anderson, J. W., Johnstone, B. M., and Cook-Newell, M. L. 1995. Meta-analysis of the effects of soy protein intake on serum lipids. *N. Engl. J. Med.* 333:276–281.

Ang, C. Y. W., Chang, C. M., Frey, A. E., and Livingston, G. E. 1975. Effects of heating methods on vitamin retention in six fresh or frozen prepared food products. *J. Food Sci.* 40:997–1003.

Ang, C. Y. W. and Y. Hang. 1995. Potential health benefits of Asian vegetarian foods. Presented at a symposium entitled *Potential Health Benefits of International Foods and Beverages,* Institute of Food Technologists National Meeting, June 3–7, Anaheim, CA. Paper No. 71–3, p. 213, Book of Abstracts, 1996 IFT Annual Meeting.

Ang, C. Y. W. and Livingston, G. E. 1974. Nutritive losses in the home storage and preparation of raw fruits and vegetables. In: *Nutritional Qualities of Fresh Fruits and Vegetables.* P. L. White and N. Selvey, eds., Futura Publishing Co., Inc. New York, pp. 51–64.

Barnard, N. 1993. *Food for Life: How the New Four Food Groups Can Save Your Life.* Three Rivers Press, New York.

Bélanger, A., Locong, A., Noel, C., Cusan, L., Dupont, A., Prévost, J., Caron, S. and Sévigny, J. 1989. Influence of diet on plasma steroid and sex plasma binding globulin levels in adult men. *J. Steroid Biochem.* 32:829–833.

Block, G., Patterson, B., and Subar, A., 1992. Fruits, vegetables and cancer prevention: A review of epidemiological evidence. *Nutr. Cancer.* 18:1–29.

Bogovski, P., ed. 1983. Polycyclic aromatic compounds, Part l. *IARC Monogr. Eval. Carcinogenic Risks Hum.* International Agency for Research on Cancer, Lyon, France.

Campbell, T. C., and Chen, J. 1994. Diet and chronic degenerative diseases: A summary of results from an ecologic study in rural China. In: *Western Diseases: Their Dietary Prevention and Reversibility,* N. J. Temple and D. P. Burkktt, eds., Humana Press, Totowa, NJ, pp. 67–118.

Cannon, G. 1992. *Food and Health: The Experts Agree.* Consumers' Association, London.

Carroll, K. K. 1991. Review of clinical studies on cholesterol-lowering response to soy protein. *J. Am. Diet Assoc.,* 91:820–825.

Chen, J. Campbell, T. C., Li, J., and Peto, R. 1990. *Diet, Life Style and Mortality in China: A Study of the Characteristics of 65 Chinese Counties.* Oxford University Press, Oxford, UK; Cornell University Press; Ithaca, NY; People's Medical Publishing House, Beijing, PRC.

Chen, J. 1992. The antimutagenic and anticarcinogenic effects of tea, garlic, and other natural foods in China: A review. *Biomed. Environ. Sci.* 5:1–17.

Chinese Medical Academy. 1989. *Food Composition Tables,* 3rd ed. Institute of Hygien Research, Chinese Medical Academy, People's Hygien Publisher, Beijing, China (in Chinese).

Chinese Academy of Preventive Medicine. 1991. *Food Composition Tables (Average of National Values).* Nutrition and Food Hygiene Research Institute, Chinese Academy of Preventive Medicine, People's Hygiene Publisher, Beijing, China (in Chinese).

Chung, K. F., Dent, G., McCusker, M., Guinot Ph., Page, C. P. and Barnes, P. J. 1987. Effect of a ginkgolide mixture (BN 52063) in antagonising skin and platelet responses to platelet activating factor in man. *The Lancet* 1:248–251.

Consumer Reports. 1995. Herbal roulette. *Consumer Reports,* A publication of Consumer Union, November, 1995:698–705.

Dwyer, J. T. 1988. Health aspects of vegetarian diets. *Am. J. Clin. Nutr.* 48:712–738.

FAO/WHO. 1991. *Protein Quality Evaluation* (FAO food and nutrition paper 51) FAO, Rome.

FASEB. 1995. *Analysis of Adverse Reactions to Monosodium Glutamate (MSG).* D. J. Raiten, J. M. Talbot, and K. D. Fisher, eds., Life Science Research Office, Federation of American Societies for Experimental Biology, American Institute of Nutition, Bethesda, MD.

Federal Register. 1997. Dietary supplements containing ephedrine alkaloids; Proposed rules. Vol. 62, number 107, 87 pages proposed rules (21 CFR Part 111,), June 4, 1997, Washington, D.C.

Frenzel-Beyme, R. and Chang-Claude, J. 1994. Vegetarian diets and colon cancer: The German experience. *Am. J. Clin. Nutr.* 59:1416–1424.

Fujisawa, Y. and Itoh, R. 1984. *Dietary Guidelines and Nutrition Policies in Japan: Current Situation and Trends.* Japan Dietetic Association, Tokyo, Japan.

Giese, J., and Katz, F. 1997. Ethical marketing of functional foods. *Food Technol.* 51(12):58–61.

Goldbohm, R. A., van den Brandt, P. A, van't Veer, P., Brants, H. A. M., Dorant, E., Sturmans, F. and Hermus, R. J. 1994. A prospective cohort study on the relation between meat consumption and the risk of colon cancer. *Cancer Res.* 54:718–723.

Haenszel, W., Kurihara, M., Segi, M. and Lee, R. K. C. 1972. Stomach cancer among Japanese in Hawaii. *J. Natl. Cancer Inst.* 49:969–988.

Haenszel, W., Kurihara, M., Locke, F. B., Shimizu, K. and Segi, M. 1976. Stomach cancer in Japan. *J. Natl. Cancer Inst.* 56:265–278.

Herbert, V. 1988. Vitamin B_{12}: Plant sources, requirements and assay. *Am. J. Clin. Nutr.* 48:852–858.

Hirayama, T. 1986. A large-scale study on cancer risks by diet-with special reference to the risk reducing effects of green-yellow vegetable consumption. In: *Diet, Nutrition and Cancer,* Y. Hayashi, M. Magao, and T. Sugimura, eds. Japan Scentific Societies Press, Tokyo, Japan, pp. 43–53.

Hirayama, T. 1990. *Life Style and Mortality.* Karger, Basel.

Ho, C-T, Osawa, T., Huang, M-T. and Rosen, R. 1994. *Food Phytochemicals for Cancer Prevention. II. Tea, Spices and Herbs.* ACS Symposium Series 546. American Chemical Society, Washington, D.C.

Hsieh, D. P. H. 1989, Carcinogenic potential of mycotoxins in foods. In: *Food Toxicology,* S. L. Taylor and R. A. Scanlan, eds., Marcel Dekker, New York, pp. 11–30.

Hsiu, S. N., 1992. Ginseng. In: *The Complete Book of Medicinal or Food Supplements,* S. N. Hsiu, ed., Hobei Science and Technology Publishing House. Hobei, China, pp. 295–301 (in Chinese).

Huang, Y-W. and Ang, C. Y. W., 1992. Vegetarian foods for Chinese buddhists. *Food Technol.* 46(10):105–106, 108.

Huang, D. P., Ho, J. H. C., Saw, D., Teoh, T. B. 1978. Carcinoma of the nasal and paranasal regions in rats fed Cantonese salted marine fish. *IARC Sci Publ. 20,* Lyon: International Agency for Research on Cancer. (20):315–328.

Huang, D. P., Ho, J. H. C., Webb, K. S., Wood, B. J. and Gough, T. A. 1981. Volatile nitrosamines in salt processed fish before and after cooking. *Fd. Cosmet. Toxicol.* 19:167–171.

Huang, M-T. Osawa, T., Ho, C-T. and Rosen, R. 1994. *Food Phytochemicals for Cancer Prevention. I. Fruits and Vegetables,* ACS Symposium Series 546. American Chemical Society, Washington, D.C.

Irvine, C., Fitspatrich., M., Robertson, I. And Woodham, D. 1995. The potential adverse effects of soybean phytoestrogen in infant feeding. *N. Zealand Med. J.* 108:208–213.

Jacobs, D. R. Jr., Salvin, J., Marquart, L. 1995. Whole grain intake and cancer: A review of the literature. *Nutr. Cancer* 24:221–229.

Jian, Y. L. and Wu, H. S., eds., 1993. *Dictionary on the Resources of Chinese Natural Tonic Products.* Chung Zheng Publishing House, Beijing, China (in Chinese).

Karmas, E. and Harris, R. S. 1988. *Nutritional Evaluation of Food Processing.* Van Nostrand Reinhold, New York.

Kato, I., Tomimaga, S., Ito, Y., Kobayashi, S., Yoshii, Y., Matsuura, A., Kameya, A., and Kano, T., 1990. A comparative case-control analysis of stomach cancer and atrophic gastritis. *Cancer Res.* 50:6559–6564.

Key, T. J. A., Thorogood, M., Appleby, P. N., Burr, M. L. 1996. Dietary habits and mortality in 11,000 vegetarians and health conscious people: results of a 17 year follow up. *Br. Med. J.* 313:775–779.

Kleijnen, J. and Knipschild, P. 1992. Ginkgo biloba. *Lancet* 340:1136–1139.

Kneller, R. W., McLaughlin, J. K., Bjelke, E., Schman, L. M., Blot, W. J., Wacholder, S., Gridley, G., CoChine, H. T. and Fraumeni, J. F., Jr. 1991. A cohort study of stomach cancer in a high-risk American population. *Cancer* 68:672–678.

Kneller, R. W., Guo, W. D., Hising, A. W., Chen, J. S., Blot, W. J., Li, J. Y., Forman, D., Fraumeni, J. F., Jr. 1992. Risk factors for stomach cancer in sixty-five Chinese counties. *Cancer Epidemiol. Biomark & Prev.* 1:113–118.

Kono, S., Ikeda, M., Tokudome, S. and Kuratsune, M. 1988. A case-control study of gastric cancer and diet in northern Kyushu, Japan. *Jpn. J. Cancer Res.* 79:1067–1074.

Kreiger, N., Marrett L. D., Dodds, L., Hilditch, S, Darlington, G. A. 1993. Risk factors for renal cell carcinoma: Results of a population-based case-control study. *Cancer Causes Control* 4:101–110.

Kubow, S., 1990. Toxicity of dietary lipid peroxidation products. *Trends in Food Science and Technology* 1(3):67–71.

Landes, P. 1997. *Whole Foods Magazine's* 2nd annual herb market survey for U.S. Health Food Stores. *Herbal Gram.* 40:52–53.

Lang, W-S., Lou, Z-C., and But, P. P-H. 1993. High-performance liquid chromatographical analysis of ginsenosides in *Penax ginseng, P. quinquefolium* and *P. notoinseng. J. Chinese Pharmaceutical Sci.* 2(2):133–142.

Le Bars, P. L., Katz, M. M., Berman, N., Itil, T. M., Freedman, A. M., Schatzberg, A. F. 1997. A placebo-controlled, double-blind, randomized trial of an extract of *Ginkgo biloba* for dementia. *J. Amer. Med. Assoc.* 278:1327–1332.

Lee, H. P., Gourley, L., Duffy, S. W., Esteve, J., and Day, N. E. 1991. Dietary effects on breast cancer risk in Singapore. *Lancet* 337:1197–1201.

Lee, J. K., Park, B. J., Yoo, K. Y. and Ahn ,Y. O., 1995. Dietary factors and stomach cancer: A case-control study in Korea. *Int. J. Epidemiol.* 24:33–41.

Marles, R. J. and Farnswnth, N. R. 1995. Antidiabetic plants and their active constituents. *Phytomedicine* 2 (2):137–189.

McLaughlin, J. K., Gao, Y.-T., Gao, R.- N. Zhend, W., Ji, B.-T., Blot, W. J. and Fraumeni, J.F., Jr., 1992. Risk factors for renal-cell cancer in Shanghai, China. *Int. J. Cancer* 52:562–565.

Messina, M. J. 1997. Soyfoods: Their role in disease prevention and treatment. In: *Soybeans: Chemistry, Technology and Utilization,* K. S. Liu, ed., Chapman & Hall, now acquired by Aspen Publishers, Inc. Gaitherburg, Maryland, pp. 442–477.

Mettlin, C., Selenskas, S., Natarajan, N. and Huben, R. 1989. Beta-carotene and animal fats and their relationship to prostate cancer risk. *Cancer* 64:605–612.

Miller, S. A. 1997. Establishing tolerable upper limits for nutrient intake. *Food Technol.* 51(8):164.

Nagao, M., Honda M., Seino Y., Yahagi T., and Sugimura, T., 1977. Mutagenicities of smoke condensates and the charred surface of fish and meat. *Cancer Lett.* 2:221–226.

Nakane H., Hara, Y., and Ono, K. 1994. Tea polyphenols as a novel class of inhibitors for human immunodeficiency virus reverse transcriptase. In: *Food Phytochemicals for Cancer Prevention. II. Tea, Spices and Herbs,* C-T. Ho, T. Osawa, M-T. Huang, and R. Rosen, eds., ACS Symposium Series 546. American Chemical Society, Washington, D.C.

National Academy of Science. 1982. *Diet, Nutrition and Cancer.* National Academy Press, Washington, D.C.

National Academy of Science. 1989. *Diet and Health: Implications for Reducing Chronic Disease Risk.* Committee on Diet and Health, National Research Council (U.S.), National Academy Press, Washington, D.C.

Nicolosi, R. J., Ausman, L. M., and Hegsted, D. M. 1991. Rice bran oil lower serum total and low density lipoprotein choleserol and apo B levels in nonhuman primates. *Artherosclerosis* 88:133–142.

Ning, J. P., Yu M. C. Wang Q. S., and Henderson B. E. 1990 Consumption of salted fish and other risk factors for nasopharyngeal carcinoma (NPC) in Tianjin, a low risk region for NPC in the Peoples' Republic of China. *JNCI* 82:291–296.

Nomura, A., Yamakawa, H., Ishidate, T., Kamiyama, S., Masuda, H., Stemmermann, G. N. et al. 1982. Intestinal metaplasia in Japan: Association with diet. *J. Natl. Cancer Inst.* 68:401–405.

NRC, 1989. *Recommended Dietary Allowances,* 10th ed. National Research Council, National Academy Press, Washington, D.C.

Pan, W-H., Chin, C-J., Sheu, C-T., and Lee, M-H. 1993. Hemostatic factors and blood lipids in young Buddhist vegetarians and omnivores. *Am. J. Clin. Nutr.* 58:354–359.

Pederden, A. B., Bartholomew, M. J., Dolence, L. A., Aljadir, L. P., Netteburg, K. L. and Lloyd, T. 1991. Menstrual differences due to vegetarian and nonvegetarian diets. *Am. J. Clin. Nutr.* 53:879–885.

Popkin, B. M., Keyon, G., Zhai, F., Gno, X., Ma, H., Zohoori, N. 1993. The nutritional transition in China: A cross-sectional analysis. *Eur. J. Clin. Nutr.* 47:333–346.

Potter, J. D. 1996. Nutrition and colorectal cancer. *Cancer Causes and Control* 7(1):127–146.

Pusateri, D. J., Roth, W. T., Ross, J. K., and Shultz, T. D. 1990, Dietary and hormonal evaluation of men at difference risks for prostate cancer; plasma and fecal hormone-nutrient interrelationships. *Am J. Clin. Nutr.* 51:371–377.

Qi, X. Y., Zhang, A., Wu, G., Pang, W. 1994. The association between breast cancer and diet and other factors. *Asia Pac. J. Public Health* 7:98–104.

Ramarathnam, N., Osawa, T., Namiki, M., and Kawakishi, S. 1988. Chemical studies on novel rice hull antioxidants. 1. Isolation, fractionation and partial characterization. *J. Agric. Food Chem.* 36:732–737.

Ramarathnam, N., Osawa, T., Namiki, M., and Kawakishi, S. 1989. Chemical studies on novel rice hull antioxidants. 2. Identification of isovitexin, a *C*-glycosyl flavonoid. *J. Agric. Food Chem.* 37:316–319.

Rand, T. G. 1994. The healing power of herbs. *Redbook.* November: 102, 130, 132.

Ratnayake, W. M. N., Pelletier G., Hollywood R., Malcolm, S. and Stavric B. 1995. Investigation of the coffee lipids on serum cholesterol in hamsters. *Food and Chemical Toxicology* 33:195–201.

Rimm, E. B., Ascherio, A. Giovannucci, E., Spiegelman, D. Stampler, M. J. and Willet W. C. 1996. Vegetable, fruits and cereal fiber intake and risk of coronary heart disease among men. *J. Am. Med. Assoc.* 276:447–452.

Rosenblatt, M. and Mindel, J. 1997. Spontaneous hyphema associated with ingestion of *Ginkgo biloba* extract. *New Engl. J. Med.* April 10: 1108.

Rowin, J. and Lewis, S. L. 1996. Spontaneous bilateral subdural hematomas associated with chronic *Ginkgo biloba* ingestion. *Neurology* 46:1775–1776.

Scopp, A. L. 1991. MSG and hydrolyzed vegetable protein induced headache: Review and case study. *Headache* 31(2):107–10.

Severson, R. K., Nomura, A. M. Y., Grove, J. S. and Stemmermann, G. N. 1989. A prospective study of demographics, diet and prostate cancer among men of Japanese ancestry in Hawaii. *Cancer Res.* 49:1857–1860.

Skog, K. 1993. Cooking procedures and food mutagens: a literature review. *Food Chem. Toxicol.* 31:655–675.

Sriamporn, S., Vatanasapt, V., Pisani, P., Yongchaiyudha, S., and Rungpitarangsri, V. 1992. Environmental risk factors for nasopharyngeal carcinoma: A case control study in north eastern Thailand. *Cancer Epidemiol. Biomarkers and Prevention* 1:345–348.

Steinmetz, K. and Potter, J. D. 1991a. A review of vegetables, fruits and cancer. I: Epidemiology. *Cancer Causes Control* 2:325–357.

Steinmetz, K. and Potter, J. D. 1991b. Vegetables, fruits and cancer. II: Mechanisms. *Cancer Causes and Control* 2:427–442.

Steimetz K., and Potter J. D. 1996. Vegetables, fruits and cancer prevention: A review. *J. Am. Dietetic Assoc.* 96:1027–1037.

Talamini, R., Franceschi, S. and La Veccia, C., Serraino, D., Barra, S., and Negrin, E., 1992. Diet and prostate cancer: A case-control study in northern Italy. *Nutr. Cancer* 118:277–86.

Tanaka, O. 1994. Ginseng and its congeners. In: *Food Phytochemicals for Cancer Prevention. II. Tea, Spices and Herbs.* C-T. Ho, T. Osawa, M-T. Huang, and R. Rosen, eds., ACS Symposium Series 546. American Chemical Society, Washington, D.C., pp. 335–341.

Thorogood, M., Mann, J., Appleby, P., McPherson, K. 1994. Risk of death from cancer and ischemic heart disease in meat and non-meat eaters. *Br. Med. J.* 308:667–1670.

Tomomatsu, H. 1994. Health effects of oligosaccharides. *Food Technol.* 48:61–65; and personal communication.

Trichopoulos, D., Li, F., and Hunter, D. J. 1996. What causes cancer? *Scientific American,* September: 80–87.

Tuyans, A. J., Kaaks, R., Haelterman, M. 1988. Colorectal cancer and the consumption of foods: A case-control study in Belgium. *Nutr. Cancer* 11:189–204.

Tyler, V. E. 1993. *The Honest Herbal.* 3rd ed. Pharmaceutical Products Press, An Imprint of the Haworth Press, Inc., New York, London, Norwood (Australia).

Urgert, R., van der Weg, G., Kosmeijer-Schuil, T. G., van de Bovenkamp, P., Hovenier, R. and Katan, M. B. 1995. Levels of the cholesterol-elevating diterpenes cafestol and kahweol in various coffee brews. *J. Agric. Food Chem.* 43:2167–2172.

USDA. 1992. *Food Guide Pyramid.* Human Nutriton Information Service, U.S. Department of Agriculture, Home and Garden Bulletin, # 252. Washington, D.C.

USDA. 1997. *USDA Nutrient Data Base for Standard Reference,* Release 11–1, August 1997 (Web site address: http://www.nal.usda.gov/fnic). Nutrient Data Laboratory, Agricultural Research Service, U.S. Department of Agriculture, Riverdale, MD.

USDA/USHHS. 1990. *Nutrition and Your Health: Dietary Guidelines for Americans.* 3rd ed. U.S. Department of Agriculture and U.S. Department of Health and Human Services, Washington, D.C.

Wang, H-J. and Murphy, P. S. 1994a. Isoflavone composition of American and Japanese soybeans in Iowa: Effects of variety, crop year, and location. *J. Agric Food Chem.* 42:1666–73.

Wang, H-J. and Murphy, P. S. 1994b. Isoflavone content in commercial soyfoods. *J. Agric Food Chem.* 42:1674–78.

WCRF/AICR. 1997. *Food Nutrition and the Prevention of Cancer: A Global Perspective.* World Cancer Research Fund/American Institute for Cancer Research, Washington, D.C.

WHO. 1990. *Diet, Nutrition, and the Prevention of Chronic Diseases.* Technical Report 797, World Health Organization (WHO), Geneva.

Wilder, C. P., Jiang, Y. Z. Allen, S. J., Jansen, L., Hall, A. J., Montesano, R. 1990. Aflatoxin-albumin adducts in human sera from different regions of the world. *Carcinogenesis* 11:2271–2274.

Willett, W. C., Colditz, G. A., Mueller, N. E. 1996. Strategies for minimizing cancer risk. *Scientific American,* Sept.:88–95.

Willett, W. C., Stampfer, M. J., Colditz, G. A., Rosner, B. A. and Speizer, F. E. 1990. Relation of meat, fat and fiber intake to the risk of colon cancer in a prospective study among women. *New Engl. J. Med* 323:1664–1672.

Wohlleb, J. C., Hunter, C. F., Blass, B., Kadlubar, F. F., Chu, D. Z. J. and Lang, N. P. 1990. Aromatic amine acetyltransferase as a marker for colorectal cancer: Environmental and demographic associations. *Int. J. Cancer* 46:22–30.

Wu, Y. V. and Inglett, G. E. 1988. Effects of agricultural practices, handling, processing, and storage on cereals. In: *Nutritional Evaluation of Food Processing,* E. Karmas, and R. S. Harris, eds., Van Nostrand Reinhold, New York, pp. 101–118.

Wu, Y., Chen, J., Oshima, H., Pignatelli, B., Boreham, J., Li, J., Campbell, T. C., Peto, R., Bartsch, H. 1993. Geographic association between urinary execretion of *N*-nitroso-compounds and oesophageal cancer mortality in China. *Int. J. Cancer* 54:713–719.

Yeh, F. S., Yu, M. C., Mo, C. C., Luo, S., Tong, M. J. and Henderson, B. E. 1989. Hepatitis B virus, aflatoxin and hepatocellular carcinoma in Southern Guangxi, China. *Cancer Res.* 49:2506–2509.

Yu, M. C., Ho, J. H-C., Henderson, B. E. 1986. Cantonese style salted fish as a cause of nasopharyngeal carcinoma: Report of a case-control study in Hong Kong. *Cancer Res.* 46:956–961.

Yu, M. C, Ho, J. H-C., Henderson, B. E. and Armstrong, R. W. 1985. Epidemiology of nasopharyngeal carcinoma in Malaysia and Hong Kong. *Natl. Cancer Inst. Monogr.* 69:203–207.

Yu, M. C., Huang, T. B., Henderson, B. E. 1989. Diet and nasopharyngeal carcinoma a case control study in Guangzhou, China. *Int. J. Cancer* 43:1077–1082

Yu, M. C., Mo, C. C., Chong, W. X., Yeh, F. S. and Henderson, B. E., 1988. Preserved foods and nasopharyngeal carcinoma: A case-control study in Guangxi, China. *Cancer Res.* 48:1954–59.

Zhang, J. S., Tian, Z., and Lou, Z. C., 1989. Quality evaluation of twelve types of locally grown ma huang. *Acta Pharmaceutica Simica* 24(11):865–871 (in Chinese).

Zheng, T., Boyle, P., Willett, W. C., Hu, H., Dan, J., Evstifeeva, T. V., Niu, S., and MacMahon, B. 1993. A case-control study of oral cancer in Beijing, People's Republic of China; associations with nutrient intakes, foods and food groups. *Eur. J. Cancer* 29B:45–55.

Zheng, W., Blot, W. J., Shu, X. O., Diamond, E. L., Gao, Y. T., Ji, B. T., and Fraumeni, J. F., Jr. 1992. Risk factors for oral and pharyngeal cancer in Shanghai, with emphasis on diet. *Cancer Epidemiol. Biomarkers Prov.* 1:441–448.

Ziegler, R. G. 1991. Vegetables, fruits and carotenoids and the risk of cancer. *Am. J. Clin. Nutr.* 53:2525–2595.

Zou, X. N., Lu, S. H. and Liu, B. 1994. Volatile *N*-nitrosamines and their precursors in Chinese salted fish - a possible etiological factor for NPC in China. *Int. J. Cancer* 59:155–158.

Index

Biographies of Editors and Contributors

Catharina Y. W. Ang, Ph.D., is a Research Chemist, National Center for Toxicological Research, U.S. Food and Drug Administration, Jefferson, Arkansas, U.S.A. She received her Ph.D. degree from Michigan State University. Past experience includes food industry consultation and poultry food quality and safety research at USDA/ARS Russell Research Center. She is former Chair of the International Division, Institute of Food Technologists and President of the Chinese American Food Society.

Monisha Bhattacharya, Ph.D., is Postdoctor, Cereal Science Laboratory, Department of Botany, University of Hong Kong, Hong Kong. She received her Ph.D. degree in Food Science at the University of Hong Kong in 1997. Her research was on evaluation of wheat genetic resources for diversity in starch that might be useful in improving noodle quality.

Tien Chi Chen, Ph.D., is Professor Emeritus, Computer Science and Engineering, and formerly Head, United College, the Chinese University of Hong Kong. He is co-author of the book *Everything You Want to Know about Chinese Cooking* by Pearl Chen, T. C. Chen, and Rose Tseng, 1980, Barrons Educational Service, Hauptauge, NY.

Tsun-Chieh Chen, Ph.D., is Professor and Food Scientist, Poultry Science Department, Mississippi State University, Mississippi State, Mississippi, U.S.A. He received his Ph.D. in Food Science and Technology from the University of Massachusetts. His research is specialized in poultry products technology. Past accomplishments include serving as visiting professor at National Taiwan University, Associate Editor of *Poultry Science* (journal), and author of more than 260 scientific papers and two patents.

Guangseng Cheng is Professor, Institute of Microbiology, Chinese Academy of Sciences, Beijing, China.

Harold Corke, Ph.D., is Associate Professor, Cereal Science Laboratory, Department of Botany, University of Hong Kong, Hong Kong. He received his Ph.D. degree from the Weizmann Institute of Science, Israel, in 1988, and did postdoctoral work at the University of California, Davis, before joining the University of Hong Kong in 1992. His research is on starch quality for Asian food applications, and he serves as a director of American Association of Cereal Chemists.

Chung-yi Huang, Ph.D., is Associate Professor, Department of Food Science and Technology, National I-Lan Institute Technology, I-Lan, Taiwan. Her research experience is with Academia Sinica, National Taiwan University and Veteran General Hospital in Taiwan and the University of Georgia in the United States. Specialty areas include food microbiology, fermentation, and Chinese herbs.

Sidi Huang, is Senior Research Scientist, Asian Food Laboratory, Bread Research Institute of Australia Limited, North Ryde, NSW, Australia. He holds a master's degree in Food Chemistry from the University of Strathclyde, United Kingdom. His major interest has been in cereal chemistry. Before joining the Bread Research Institute of Australia, he worked in the Sichuan Grain Storage Research Institute.

Yao-wen Huang, Ph.D., is Associate Professor, Department of Food Science and Technology and Department of Marine Sciences, University of Georgia, Athens, Georgia, U.S.A. He received his Ph.D. from the University of Georgia in 1983. He is Consultant Professor at Shanghai Fisheries University (China) and provides consultant services internationally. He is past President of Chinese American Food Society, and is currently Chair-elect of the Product Development Division of the Institute of Food Technologists. He serves as editor of the *International Journal of Functional Foods.*

KeShun Liu, Ph.D., is Manager, Food Chemistry/Soyfoods, at Hartz Seed, a Unit of Monsanto Co., Stuttgart, Arkansas, U.S.A. Dr. Liu, a native of China, received his Ph.D. degree in Food Science from Michigan State University in 1989 and did postdoctoral work at the University of Georgia and the Coca-Cola® Company. His major interest has been food chemistry/soyfoods, with expertise in legume and oilseed storage, processing, and utilization. He is a frequent speaker at scientific conferences and the author of more than 30 publications, including the book *Soybeans: Chemistry, Technology and Utilization,* 1997, Chapman & Hall, now acquired by Aspen Publishers, Inc., Gaithersburg, Maryland.

Xingqiu Lou is a Ph.D. candidate at the Department of Animal Science, The University of Kentucky, Lexington, Kentucky, U.S.A. Mr. Lou is working towards his Ph.D. degree in Food Science, with a major interest in meat processing. He was a lecturer of Wuxi University of Light Industry, Wuxi, China.

Bor S. Luh, Ph.D., is Professor Emeritus, Department of Food Science and Technology, University of California, Davis, California, U.S.A. He received his Ph.D. degree from University of California at Berkeley. Research areas include fruit and vegetable processing and rice utilization. He is author of more than 12 books and book chapters and served as the first President of the Chinese American Food Society.

Seiichi Nagao, Ph.D., is Director, Wheat Flour Institute, Flour Millers Associations, Kabuto-cho, Nihonbashi, Chuo-Ku, Tokyo, Japan. He received his Ph.D. in agricultural chemistry from the Universities of Tsukuba. Prior to his retirement in 1997, he served as research director of Nisshin Flour Milling Co. since 1959. He continues serving Nisshin as an advisor. He was honored with Brabender, Geddes Memorial, and Fellow Awards from American Association of Cereal Chemists.

Jacqueline M. Newman, Ph.D., R.D., is Professor, the Department of Family, Nutrition and Exercise Sciences, Queens College-CUNY, Flushing, New York, U.S.A. She is author of 8 books, 6 chapters, and many articles, and her research focuses on Chinese foods, other Asian foods, and dietary habits. She serves as Editor-in-Chief of *Flavor & Fortune;* Board of Directors of American Institute of Wine and Foods and the Food Exhibition Museum in Suzhou, China; and president of the Association for Study of Food and Society.

John X. Shi, Ph.D., is Research Scientist, Food Research Program Southern Crop Protection and Food Research Center, Agriculture and Agri-Food Canada, Guelph, Ontario, Canada. He received his Ph.D. degree from the Department of Food Technology, Polytechnic University of Valercia, Spain, in 1994, and did his postdoctoral work at the Department of Food Science, University of Guelph and his research focus is on fruit processing technology.

Michael Tao, Ph.D., is Consultant at Tao, Tao and Chiu, Inc., Aptos, California, U.S.A. He received his Ph.D. degree from the University of Laval (Canada). He was formerly associated with Joseph E. Seagrams and Sons, Inc., Hunt-Wesson Foods, and Bristol-Meyer Squibb Company in Indiana. Specialty areas include food product development.

Peter J. Wan, Ph.D., is Research Chemical Engineer, Commodity Utilization Research Unit Southern Regional Research Center, USDA, Agricultural Research Service, New Orleans, Louisiana, U.S.A. He received his Ph.D. from Texas A&M University and is formerly associated with Best Foods, Anderson Clayton, and Kraft Company. His research focus is in lipid and protein chemistry, and he is editor of two books in fats and oil technology and solvents extraction of oil seeds. He is former President of Chinese American Food Society.

Samuel L. Wang, Ph.D., is Research Scientist, Horticultural Research Institute of Ontario, University of Guelph, Vineland Station, Ontario, Canada. He received his Ph.D. from Michigan State University. Industry experience includes association with several food companies in the United States, and his research specializes in fruits and vegetable processing. He is serving as current President of the Chinese American Food Society.

Youling L. Xiong, Ph.D., is Associate Professor, Department of Animal Science, University of Kentucky, Lexington, Kentucky, U.S.A. He received his Ph.D. in Food Science from Washington State University and did postdoctoral research at Cornell University. His research focus is on meat and poultry processing and food protein chemistry. He is former chairman of Muscle Foods Division of Institute of Food Technologists and received the Achievement Award from American Meat Science Association and Young Scientist Award from American Chemical Society.

Fang-Qi Yang is Associate Professor, Department of Food Science and Engineering, Wuxi University of Light Industry, Jiangsu, China. Mr. Yang is the director of Food Engineering Section in his department. His research interest is in meat processing.